FIFTY YEARS OF X-RAY DIFFRACTION

FIFTY YEARS OF X-RAY DIFFRACTION

DEDICATED TO THE
INTERNATIONAL UNION OF CRYSTALLOGRAPHY
ON THE OCCASION OF THE
COMMEMORATION MEETING IN MUNICH
JULY 1962

by

P. P. EWALD

EDITOR

AND NUMEROUS CRYSTALLOGRAPHERS

PUBLISHED FOR THE
INTERNATIONAL UNION OF CRYSTALLOGRAPHY
BY N.V. A. OOSTHOEK'S UITGEVERSMAATSCHAPPIJ
UTRECHT, THE NETHERLANDS

Published July 1962

The Editor expresses his sincere thanks to the many crystallographers whose contributions form the most valuable part of this commemorative volume. He apologizes to the authors for not having sent them proofs for lack of time, and to the readers for the misprints and shortcomings they will notice in the book, and of many of which he is aware. Manuscripts were expected to arrive in the summer months of 1961, and many did; others, however, were received as late as April and May 1962. Only thanks to the most obliging cooperation of the Publishers and their Printers was it possible to begin composition while the manuscript was still incomplete and to produce this 'Festschrift' just in time for the Munich meeting.

The Editor also gratefully acknowledges the help received by him on various occasions from Dr. E. A. Wood and Dr. D. W. Smits. Finally he expresses his great indebtedness to the John Simon Guggenheim Memorial Foundation for a grant which enabled him in the course of the last two years to devote most of his time to this publication.

Printed in the Netherlands

Contents

INTRODUCTION

Origin, Scope, and Plan of this Book

In July 1962 the fiftieth anniversary of Max von Laue's discovery of the Diffraction of X-rays by crystals is going to be celebrated in Munich by a large international group of crystallographers, physicists, chemists, spectroscopists, biologists, industrialists, and many others who are employing the methods based on Laue's discovery for their own research. The invitation for this celebration will be issued jointly by the Ludwig Maximilian University of Munich, where the discovery was made, by the Bavarian Academy of Sciences, where it was first made public, and by the International Union of Crystallography, which is the international organization of the National Committees of Crystallography formed in some 30 countries to represent and advance the interests of the 3500 research workers in this field.

The year 1912 also is the birth year of two branches of the physical sciences which developed promptly from Laue's discovery, namely X-ray Crystal Structure Analysis which is most closely linked to the names of W. H. (Sir William) Bragg and W. L. (Sir Lawrence) Bragg, and X-ray Spectroscopy which is associated with the names of W. H. Bragg, H. G. J. Moseley, M. de Broglie and Manne Siegbahn.

Crystal Structure Analysis began in November 1912 with the first papers of W. L. Bragg, then still a student in Cambridge, in which, by analysis of the Laue diagrams of zinc blende, he determined the correct lattice upon which the structure of this crystal is built. Soon afterwards he obtained the first complete structure determinations, namely of NaCl, KCl, KBr and KI, a series of alkali halides having similar structures. By this determination a scale for the measuring of atomic distances in crystals and, simultaneously, of X-ray wave-lengths was obtained. It led to the construction of the powerful instrument devised by W. H. Bragg, the X-ray Spectrometer, with which the majority of the early crystal structures were determined. It ushered in the access to a super-stereo chemistry, in three dimensions and with quantitative determinations of atomic distances and bond angles such as had been impossible up to then. This finally led to the close link which exists nowadays between X-ray structure analysis and the physical and chemical problems of the chemical bond, problems which dominate nearly every aspect of our scientific and industrial activities, as well as the study of metabolism in man, beast and plant.

X-ray Spectroscopy deals with the emission and absorption of X-rays by and in matter. In particular, Moseley's early work showed that the already vaguely recognized 'characteristic X-rays' were a direct outcome of atomic structure and could be used to identify an atomic species. At a time when modern atomic theory was in its earliest infancy—Niels Bohr's first paper appeared in the same year, 1913, as Moseley's—the spectroscopic data on X-rays, allowing a much simpler interpretation than those on visible light, gave invaluable support and guidance for the development of the principles of the quantum theory of the atom. From this the whole of physics and chemistry profited, and again today, the greatly refined methods of experimentation and discussion bring the research in X-ray spectroscopy very close to the problems of chemical bond and energy band structure of solids.

Among the later consequences of von Laue's discovery should be named the diffraction phenomena obtained in crystals by using beams of electrons and neutrons, instead of X-rays. These two applications of very nearly the same experimental procedures and theory as for X-ray diffraction, are rapidly developing along their own lines. Each of the means of obtaining diffraction—X-rays, electrons, neutrons, and even atoms—has it own peculiarities in interacting with matter. Therefore different information can be gained by using these methods judiciously; but they all spring from the source Laue opened up.

Let us then, after ten lustrums, look back on the discovery, and on the development of Crystal Structure Analysis, of Spectroscopy, and of Crystal Optics in the widest sense. This review forms the first and larger half of this volume.

In the second half, a more personal note will be struck. There is, nowadays, a general demand for more of the human touch in presenting science to the coming generation, for more detail about the men whose memory is handed down by the laws named after them (a few, like Röntgen, even achieve the status of becoming immortalized in a unit!) but whose personality is effaced as the circle of their students fades out. At which schools did they learn their art when they were young, with whom did they form friendships that lasted throughout their scientific life? What was their own evaluation of their work, what their hopes and their disappointments, their outlook on science and life?

Clearly the desire for a short autobiographical essay of his heroes is legitimate in one about to devote his own life to the continuation of the work they began. Would we older people not appreciate autobiographical essays from Hamilton, Kelvin, Rayleigh, Maxwell, Helmholtz; do we not appreciate those of Poincaré, Planck, Einstein, Hardy, and the popular writings of Boltzmann, Schuster and others?

Let us leave it to the historian of science at a later period to evaluate and weigh the merits of each of the pioneers of a new development with the distant objectiveness that becomes a disciple of Clio, for it would be unbecoming for us, nay, impossible, to be objective judges of our own times. But let us present him with a view, a personal and colourful view, that we have gained in our own experience and according to our own temperament. If the same facts, presented from different sides, appear contradictory, as well they may in independent autobiographical essays, let it be the historian's job to straighten this out. He will be grateful to us for attempting to offer him material and to disclose relations which he would find impossible to glean by combing the journals and books.

Part of this autobiographical collection concerns the main schools in which X-rays crystallography was developed and taught. Since these sections are written by prominent members of these schools, they may well pass as 'autobiographical' in a slightly wider sense. The autobiographical section should be considered only a first attempt of this kind of mosaic synthesis of the making of a science. Some of the main intended contributors died in the course of the preparation of this work—von Laue, Maurice de Broglie, W. T. Astbury, C. Hermann. Others could not be convinced of the usefulness of their contribution, and many who should have been asked to contribute were not approached because of the danger of exceeding limitations of space. It is the Editor's view that if this collection finds an approving response, the present commemoration volume should be enlarged and systematized in a second edition.

This book is not written for the first generation of X-ray crystallographers who grew up together with the subject, but rather for the second and later growth, and for other scientists. The Editor has tried to keep the presentation at a level so that it can be read by the non-crystallographer who is interested in learning what this 'New Crystallography' is about. Chapters 3–6, describing the setting of the discovery and the first few years of X-ray diffraction, should present no difficulties to a scientifically minded reader. After that, in describing the further development, some technical terms and some factual knowledge of the subject seemed unavoidable. Rather than to interrupt the historical account on every page in order to explain the terms and methods newly introduced, a short account of the whole subject has been given in Chapter 7. In a way this Chapter only aims at giving the reader a condensed course in the language used later; but since for the learning of a vocabulary a background of concepts is a necessity (except for a parrot), the Chapter gives, at the same time, a very condensed and incomplete, but, it is hoped, intelligible factual survey of the field. It should be read carefully and consulted repeatedly as the reader progresses to some of the later parts of the book.

PART II

The Beginnings

From W. C. Röntgen's Third Communication, *March 1897:*

'The experiments on the permeability (for X-rays) of plates of constant thickness cut from the same crystal in different orientations, which were mentioned in my first Communication, have been continued. Plates were cut from calcite, quarz, turmaline, beryl, aragonite, apatite and barytes. Again no influence of the orientation on the transparency could be found.'

'Ever since I began working on X-rays, I have repeatedly sought to obtain diffraction with these rays; several times, using narrow slits, I observed phenomena which looked very much like diffraction. But in each case a change of experimental conditions, undertaken for testing the correctness of the explanation, failed to confirm it, and in many cases I was able directly to show that the phenomena had arisen in an entirely different way than by diffraction. I have not succeeded to register a single experiment from which I could gain the conviction of the existence of diffraction of X-rays with a certainty which satisfies me.'

CHAPTER 2

X-rays

2.1. Physics at the Time of Röntgen's Discovery of X-rays

The first half of the nineteenth century was a period of tumultuous development of the exact sciences. The great mathematicians—Cauchy, Euler, Gauss, Hamilton, to name only a few—not only perfected the methods of analysis, but they also laid the foundations for a mathematical, quantitative, understanding of celestial and other Mechanics, of Hydrodynamics, Elasticity, Magnetism, and Optics. Following Lavoisier's introduction of the balance for checking reactions, Chemistry became a quantitative science. A series of brilliant experiments between 1820 and 1831 disclosed the relation of magnetism to galvanic electricity, and Faraday developed his notion of an electromagnetic field which was amplified and given mathematical expression by Maxwell in the 1860's. By 1848 the concept of Energy was clearly defined and the equivalence of energy and heat demonstrated. Clausius and Maxwell formulated the basic laws of Thermodynamics. The Kinetic Theory of Matter, long but vaguely foreshadowed in the works of Lucretius and of Boscovich, reached the first quantitative stage in the Theories of Gases of Maxwell and of Boltzmann. The discovery of the polarization of light (Malus, 1808) had proved that light was a transverse wave motion, and although hardly anything was known about the production of light, nearly all seemed to be known about its propagation. As a consequence, much improved telescopes, microscopes and other ingenious optical devices were being constructed and helped to open up vast new regions of the skies and of the animal and plant world. The application of the laws of physics to chemistry, engineering, and physiology made great strides and rational, quantitative and ever more precise relations replaced the former vague empiricism.

Considering the enormous advances in the mathematical description of nature, some scientists thought that science had reached such a stage of perfection that little more fundamental work remained to be done; working out new problems along the given lines was all that could be expected of future scientists.

Instead, in the last one or two decades of the century a hidden new world of physical entities and facts was discovered which stood quite apart from the classical system of physics. It turned out eventually to be the foreshore of the twentieth century physics. This discovery began in 1854 when, among other physicists, Julius Plücker in Bonn studied the spectra produced by the electric discharge in rarified gases. These brilliantly coloured and variable discharges in evacuated glass tubes, usually manufactured by the Bonn glassblower Geisler, were being very gradually classified and analysed in a descriptive way by their dark spaces, luminous band structure etc. A full understanding of the processes producing these effects came only in the 1930's when atomic theorie was well advanced. In 1859 Plücker observed that in highly evacuated tubes a bright luminescence occurred on the glass wall opposite to the cathode and that this was influenced in a peculiar way by the approach of a magnet. Johann Wilhelm Hittorf found in 1869 that with increasing evacuation of the discharge tube the dark space adjoining a disc-shaped negative pole (cathode) gains in length until it finally suppresses all the luminosity in the gas and reaches out to the glass wall opposite the cathode which then shines up in a bright green light called fluorescence. Hittorf in Münster, Crookes in London and other physicists investigating this form of discharge showed that the bright spot on the glass is produced by something that leaves the cathode surface at right angles and travels in straight lines, so that the shadow of an opaque metal cross is formed in the fluorescent spot. For this reason the name of cathode *rays* was given to the invisible something. If these rays fell on pieces of calcite or fluorite these minerals glow in beautiful colours, which differ according to the mineral species. Here then was a novel mode of producing light which attracted many investigators. Meanwhile two important developments took place regarding cathode rays: while Plücker had already indicated that the 'rays' were, perhaps, streams of electrically charged particles emitted by the cathode and deflected by a magnet, this view was shaken by experiments undertaken by Heinrich Hertz which showed no deflection of the rays by the electric field when they passed between the plates of a condenser. (Only much later the reason for this negative result was recognized in the electrical leakage between the condenser plates caused by too poor a vacuum.)

The second development came from Ph. Lenard, then a student of H. Hertz, who succeeded in letting the cathode rays pass out of the tube through a very thin aluminium foil or 'window'. The rays would traverse a few inches of air (the higher the voltage on the tube, the longer the path), while their intensity, as indicated by the brightness of a fluorescent screen, diminished exponentially as the traversed layer of air grew. The Lenard window permitted a much easier observation of fluorescence of minerals and other compounds, for no longer had a special tube to be constructed and evacuated for each observation.

It should be noted that the atomistic nature of the electric charge, which in our 'Electronics Age' is a familiar fact, was still unknown in the early 1890's. True, already in 1834 Faraday had shown that in the conduction of current through salt solutions, the charges were transported in a certain unit or a small multiple of this, and never in fractional or irregular quantities. But these electric charge units were carried by ponderable masses, say by the atoms of the silver deposited on the cathode of an electrolytic trough, and the appearance of a unit charge could be caused equally well by the carrying capacity of the atom as by some inherent property of charge itself.

In fact, the—apparent—absence of any deflection of cathode rays by electric fields, together with their power to penetrate through metal foils which are impervious to gas gave support to the view of Hertz and many other German physicists that cathode rays were a special form of electromagnetic field, perhaps longitudinal waves, rather than a stream of corpuscles. This view persisted until 1895 and 1896 when Jean Perrin in France and J. J. Thomson in Cambridge achieved electrostatic deflection of cathode rays, and the latter, soon afterwards, using a Faraday cage collected and measured the charge transported in the cathode ray. By deflection experiments, he also determined the ratio of the charge to the mass of the cathode ray particles, e/m; and found that, assuming the charge to be the same as that occurring in electrolysis, the mass of the particle would be only about 1/1800 of the smallest known atomic mass, that of the hydrogen atom. In 1891 finally, on the proposal of Johnstone Stoney, the name of electron was universally accepted for this unit of charge. Its absolute value was determined in 1910 by Robert Millikan in Chicago as $4.77 \cdot 10^{-10}$ el. static units and this value, one of the most fundamental ones in Nature, was revised in 1935 by E. Bäcklin as a consequence of Laue's discovery. The accepted value is today $4.803 \cdot 10^{-10}$ el. static units or $1.601 \cdot 10^{-19}$ Coulomb.

2.2. Röntgen's Discovery

Let us go back to the summer of 1895 and to the beautiful old Bavarian university town and former seat of an independent bishop, Würzburg. Here, six years earlier, Wilhelm Conrad Röntgen had been appointed Professor of Physics.

In the course of the summer of 1895 Röntgen had assembled equipment, such as a fairly large induction coil and suitable discharge tubes, for taking up work on the hotly contested subject of cathode rays. From Lenard's work it was known that these rays are absorbed in air, gases, and thin metal foils roughly according to the total mass of the matter traversed, and that the absorption decreases if a higher voltage is put across the discharge tube. It was also known that the intensity of the fluorescence excited in different crystals varies with the voltage used, fluorite being a good crystal for 'soft' cathode rays —those obtained with low voltage—and barium platino cyanide fluorescing strongly under bombardment by 'hard' cathode rays.

Röntgen never divulged what measurements he intended to make, nor what type of discharge tube he was using when he made his great discovery. The fact that the tube was fully enclosed in a light-tight cardboard box shows that he intended to observe a very faint luminescence. But the question of whether he was interested in the law of absorption of cathode rays or in the excitation of fluorescence in different media remains unanswered. The fact is that he noticed that a barium platino cyanide screen lying on the table at a considerable distance from the tube showed a flash of fluorescence every time a discharge of the induction coil went through the tube. This flash could not be due to cathode rays because these would have been fully absorbed either by the glass wall of the tube, or by the Lenard window and the air. Röntgen, in a breathless period of work between 8 November and the end of the year, convinced himself of the reality of his observations which at first he found hard to believe. He soon concluded that the fluorescence was caused by something, the unknown X, that travelled in a straight path from the spot where the cathode ray in the tube hit the glass wall; that the unknown agent was absorbed by metals and that these cast a shadow in the fluorescent area of the screen. He therefore spoke of X-*rays*; he showed that these rays were exponentially absorbed in matter with an exponent roughly proportional to the mass traversed, but very much smaller than the one found by Lenard for the corresponding cathode rays; he found the photographic action of X-rays and took the first pictures of a set of brass weights enclosed in a wooden box, and, soon after, the first photo of the bones in the living hand; he remarked that the output of X-rays can be increased by letting the cathode rays impinge on a heavy metal 'anticathode' (which may also be the anode of the tube) instead of on the glass wall and thereby started the development of the technical X-ray tube; he found that X-rays render air conductive and discharge an electrometer; he performed ingenious but entirely negative experiments for elucidating the nature of X-rays, in which he searched in vain for reflection or refraction or diffraction, the characteristic features of wave phenomena.

Röntgen was well aware of the fact that he had found something fundamentally new and that he had to make doubly sure of his facts. He hated nothing more than premature or incorrect publications. According to his habit he did the work single-handed and spoke not even to his assistants about it. Finally, in December 1895 he wrote his famous *First Communication* for the local Würzburg Scientific Society. In its 10 pages he set out the facts in a precise narrative, but he omitted—as also in all of his previous and his later work—all personal or historical indications, as transitory elements which he considered to detract from the finality of scientific publication. The paper was quickly set and Röntgen sent out proofs or reprints as New Year's Greetings to a number of his scientific friends.

After three months (March 1896) the *First Communication* was followed by a second one of seven printed pages. In it, Röntgen reported careful experiments on the discharge of charged insulated metals and dielectrics, by irradiation when in air, gases or vacuum; he finds that an anode of platinum emits more X-rays than one of aluminium and recommends for efficient production of X-rays the use of an aluminium cathode in form of a concave mirror and a platinum anode at its focus, inclined at 45° to the axis of the cathode. Finally he states that the target need not be simultaneously the anode of the tube.

A year later (March 1897) a third and final Communication appeared, slightly longer than the first two taken together and containing further observations and measurements. From it, the Motto on page 5 of this book is taken. Together these 31 pages of the three Communications testify to the classical conciseness of Röntgen's publications.

* * *

The response which this discovery prompted was unheard of at a time when, in general, Science was still a matter for the select few. In seeing on the fluorescent screen the bones of a living hand divested of the flesh around them, medical and lay public alike were overcome by an uncanny memento mori feeling which was vented in many serious and satirical contributions to the contemporary newspapers. The first medical applications were promptly made, and the demand for 'Röntgen Tubes' quickly initiated an industry that has been expanding ever since. Röntgen, a fundamentally shy and retiring character, was ordered by the young Emperor William II to demonstrate his discovery in the Berlin palace—an invitation Röntgen could not well refuse, as he did many other demands. The writer remembers the unveiling of the four seated figures on the buttresses of the remodelled Potsdamer Brücke in Berlin which on orders of the Emperor were placed there as representative of German Science and Industry: Carl Friedrich Gauss, Hermann von Helmholtz, Werner Siemens and Wilhelm Conrad Röntgen. This must have been in 1898 or '99 and there was much discussion in the family circle whether it was appropriate to put such a novel and poorly understood discovery on an equal footing with the well-established achievements of the three other figures.—The reader will find an entertaining account of the post-discovery period (and many interesting details besides) in O. Glasser's book *Wilhelm Conrad Röntgen and the History of X-rays.*

2.3. Progress in the Knowledge of X-rays up to 1912

In spite of the universal enthusiasm for X-rays and the great number of physicists and medical men working in the field, only very few fundamental facts were discovered in the next fifteen years. True, a constant technical development of the X-ray tubes and of high-tension generators took place in response to the increasing demands of the medical profession, especially when the therapeutic use of very hard X-rays began to be recognized at the end of this period. The commercial availability of fairly powerful X-ray equipment greatly facilitated Friedrich and Knipping's later experiments in 1912. But of experiments disclosing something of the nature of X-rays only four need be mentioned:

a. *Polarization of X-rays* (*Barkla 1905*). That X-rays are scattered, i.e. thrown out of their original direction, when passing through a body, was already noticed by Röntgen in his second communication.

Barkla used this property for an experiment similar to that by which Malus had detected the polarization of light. Malus (1808) had found that the rays of the setting sun, reflected on the windows of the Palais du Luxembourg acquired a new property by this reflection; for if they were once more reflected under a certain angle by a glass plate which could be rotated around the direction of the ray coming from the windows the intensity of the twice reflected ray would vary with the angle of rotation of the glass plate, being smallest when the twice reflected ray travels at right angles to its previous two directions, and strongest if it travels in their plane. This was a proof that light is a transverse wave motion, not, like sound, a longitudinal one, which has axial symmetry. Barkla repeated this experiment with X-rays, with the only difference that, there not being any specular reflection of X-rays, he had to substitute for the reflections the much weaker scattering under an angle of approximately 90°. He found the dependence he was looking for and concluded that *if* X-rays were a wave motion, they were, like light, transverse waves. This was fully confirmed by later experiments of the same type by Herweg (1909) and H. Haga (1907).

b. *Barkla's discovery of 'characteristic' X-rays (1909)*. X-rays could at that time only be characterized by their 'hardness', i.e. penetrating power. In general, the higher the voltage applied to the X-ray tube, the harder is the X-radiation emitted, that is, the smaller is its absorption coefficient in a given material, say aluminium or carbon. The absorption coefficient is, however, not a constant, because, since the soft components of the radiation leaving the tube are absorbed in the first layers of the absorber, the remaining radiation consists of harder X-rays. Thus the variability of the absorption coefficient with penetration depth is an indication of the inhomogeneous composition of the X-radiation. Barkla, studying tubes with anticathodes of different metals, found that under certain conditions of running the tube the emergent X-rays contained one strong homogeneous component, i.e. one whose absorption coefficient was constant. He found that the absorption coefficient decreased with increasing atomic weight of the anticathode material, and that this relation was shown graphically by two monotonic curves, one for the lighter elements and one for the heavier ones. He called these two types of radiation, characteristic for the elements from which the X-rays came, the K- and the L-Series. This discovery formed the first, if still vague, link between X-rays and matter beyond the effects determined by the mere presence of mass.

c. *Photoelectric Effect*. The photoelectric effect consists in the emission of electrons when light or X-rays fall on the atoms in a gas or a solid.

Its first observation goes back to Heinrich Hertz, 1887, who noticed that the maximum length of the spark of an induction coil was increased by illuminating the gap with ultraviolet light. In the following year W. Hallwachs showed that ultraviolet light dissipates the charge of a negatively charged insulated plate, but not that of a positively charged plate. This happens in air as well as in vacuum and in the latter case it was proved by magnetic deflection that the dissipation of the charge takes place by the emission of electrons. In 1902 Philipp Lenard found the remarkable fact that the intensity of the light falling on the metal plate influences the rate of emission of electrons, but not their velocity. Three years later Albert Einstein recognized the importance of this result as fundamental, and in one of his famous four papers of the year 1905 he applied Planck's concept of quantized energy to the phenomenon by equating the sum of kinetic and potential energy of the emitted electron to the energy quantum $h\nu$ provided by a monochromatic radiation of frequency ν:

$$\tfrac{1}{2}mv^2 + p = h\nu \quad (p = \text{potential energy})$$

At the time this was a very bold application and generalization of the concept of quantized energy which Planck had been proposing for deriving the laws of black body radiation, and whose physical significance was by no means assured. Einstein's equation was at first not at all well corroborated by the experimental results with ultraviolet light, because the unknown work term p in the equation is of the same order of magnitude as the two other terms. This is not so if the much larger energy $h\nu$ of an X-ray is used, and the fully convincing proof of Einstein's relation had therefore to wait until the wave-length and frequency of X-rays could be determined with accuracy by diffraction on crystals, and the equation could then in turn be used for a precision method of measuring the value of Planck's constant h.

Prior to this, in 1907, Willy Wien made a tentative determination of the X-ray wave-length (provided X-rays were a wave motion) by reversing the sequence of the photoelectric effect: he considered the energy of the electron impinging on the target as given by the voltage applied to the tube and, neglecting the small work term p, calculated the frequency and wave-length of the radiation released. Assuming a voltage of 20000 volt this leads to $\lambda = 0.6$Å.

W. H. Bragg interpreted the ionization of gases by X-rays (the amount of which served as a measure for X-ray intensity) as primarily a photoelectric effect on a gas molecule, with further ionizations produced by the swift ejected electrons. The fact that in this process a large amount of energy has to be transferred from the X-ray to the gas in a single act led him to consider this as a collision process and further to the concept that X-rays are a particle stream of neutral particles, or doublets of $\pm$ charge.

d. *Diffraction by a Slit.* Röntgen himself reports in his *First Communication* inconclusive attempts at producing diffraction effects by letting the X-rays pass through a fine slit. These attempts were repeated by the Dutch physicists Haga and Wind (1903). They claimed to have recorded faint diffraction fringes, but their results were challenged as possibly due to a photographic effect caused by the developing. In 1908 and 1909 B. Walter and R. Pohl in Hamburg repeated essentially the same experiment taking utmost care in the adjusting. The slit was a tapering one produced by placing the finely polished and gilded straight edges of two metal plates in contact at one end and separated by a thin flake of mica at the other. The X-rays fell normally on the slit and the photographic plate was placed behind the slit parallel to its plane. If diffraction took place, one would expect the narrowest part of the slit to produce the widest separation of fringes. On the other hand, they would be the least intense because of the narrowness of the slit. Since for complete absorption of the X-rays the plates forming the slit must have a thickness of the order of 1–2 mm and the slit width in the effective part is of the order of 1/50 mm, the slit is in reality a deep chasm through which the X-rays have to pass. Walter and Pohl's plates showed the otherwise wedge-shaped image of the slit to fan out at its narrow end into a brush-like fuzzy fringe system. Fortunately, in 1910, one of Röntgen's assistants, P. P. Koch, was engaged in constructing the first automatic microphotometer by using a pair of the recently improved photoelectric cells for the continuous registration of the blackening of a photographic plate. As soon as the instrument had been completed and tested, Koch traced several sections through the original plates of Walter and Pohl, and these showed variations which could be caused by diffraction.

So, once more, the probability rose that X-rays were a wave phenomenon. The order of magnitude of the wave-length could have been obtained roughly from the fringe separation and the width of the slit on any of the cross sections taken. But since the intensity profiles departed considerably from those obtained by diffracting light waves on a slit, Sommerfeld, the master-mathematician of diffraction problems, developed the theory of diffraction of light waves by a deep slit before discussing the Walter-Phol-Koch curves. Both papers, Koch's and Sommerfeld's, were published together in 1912. Sommerfeld's conclusion was that the fuzziness of the fringes was caused by a considerable spectral range of the X-rays, and that the centre of this range lay at a wave-length of about 4.10^{-9} cm. This possible but by no means unique explanation was known among the physicists in Munich several months before it appeared in *Annalen der Physik* in May 1912. The wave-length checked approximately with W. Wien's estimate.

e. *Waves or Corpuscles?* At the end of 1911 X-rays still remained one of the enigmas of physics. There was, on the one hand, the very strong argument in favour of their corpuscular nature presented by the photoelectric effect. The explanation of this concentrated and instantaneous transfer of relatively large amounts of energy from a radiation field into kinetic energy of an electron was utterly impossible according to classical physics.

On the other hand some phenomena fitted well with a field- or wave concept of X-rays. As early as 1896 a plausible explanation of X-ray generation had been given independently by three physicists: in Manchester by Stokes, in Paris by Liénard, and in Königsberg by Wiechert. They assumed the cathode rays to consist of a stream of charged particles, each surrounded by its electromagnetic field. On impact with the target (or 'anticathode') these particles are suddenly stopped and the field vanishes or changes to the static field surrounding a particle at rest. This sudden change of field spreads outward from the anticathode with the velocity of light and it constitutes the single X-ray pulse. In many ways X-rays seem then analogous to the acoustical report of shot hitting an armour plate. In order to work out this theory so as to check it on experiments, assumptions about the impact process in the target had to be made, the simplest being a constant deceleration over a few atomic distances in the target. The theory accounted readily for the non-periodicity or spectral inhomogeneity of X-rays as shown by their non-uniform absorption; also for the polarization as shown in the double scattering experiments. It was, however, desirable to obtain more information regarding the actual stopping process, and for this purpose measurements were made in 1909 by G. W. C. Kaye on the angular distribution of the intensity of X-rays generated in thin foils, where it seemed likely that only few decelerating impacts occurred. Sommerfeld, who was one of the protagonists of the impact or 'Bremsstrahl' theory, calculated the angular distribution and found as a general result that the higher the applied voltage and therefore the velocity of the electrons, the more the emission of the field was confined to the surface of a cone surrounding the direction of the velocity, the opening of which decreased with increasing voltage. This was well confirmed by the measurements for X-rays as well as for γ-rays, provided the conditions were such that no characteristic radiation was excited in the target. One has thus to distinguish between the general X-rays generated as 'Bremsstrahlen' or 'pulses' or 'white X-rays' i.e. through the decelerating impact, and those much more homogeneous ones with respect to absorption which are determined by the emitting material ('characteristic X-rays').

The problem arose whether polarization and directional emission could also be found for characteristic radiation. In order to study this experimentally, Sommerfeld, towards the end of 1911, appointed an assistant, Walter Friedrich, who had just finished his Doctor's Thesis in the adjoining Institute of Experimental Physics of which Röntgen was the head. The subject of his thesis had been the investigation of the directional distribution of the X-rays obtained from a platinum target; he was thus fully acquainted with the technique to be used in extending the investigation to a target, and a mode of operating the tube, which yielded strong characteristic rays, instead of Bremsstrahlung.

CHAPTER 3

Crystallography

3.1. Descriptive Crystallography

Quantitative crystallography began with Carangeot's invention of the contact goniometer (1780), an instrument with which the angles between the faces of a crystal could be crudely measured. The accuracy of this angular measurement was greatly increased, and extended to smaller crystals, by W. H. Wollaston's construction of an optical goniometer (1809). In this instrument light made parallel after passing through a slit in the focal plane of a collimating telescope is reflected by a crystal face, and focussed by a second telescope, so that the observer sees an image of the slit. The crystal is mounted in soft wax on an axis at right angles to the plane of the light path so that a 'zone axis,' i.e. the edge direction which is common to two or more crystal faces, coincides with the axis of rotation. If the crystal is rotated until a second face reflects into the telescope, then the angle of rotation, which can be read accurately on a graduated circle, is the angle formed by the normals of the two planes. With one setting of the crystal, only the normals of one 'zone,' which all lie in the plane at right angles to the zone axis, can be obtained. In order to measure the other angles the crystal has either to be re-set with a different zone axis coinciding with the axis of rotation, or a two-circle goniometer has to be used where this adjustment can be made without resetting.

The goniometer led to the discovery of the three fundamental laws of descriptive or morphological crystallography:

(i) Crystals grow naturally with plane faces.

(ii) Whereas the size of the crystal, the relative sizes of its faces, and thence the overall shape or 'habit' of a particular kind of crystal may vary widely according to the circumstances of its formation, the angles between the faces are characteristic for the chemical composition. It is true that the same angles occur in all crystals belonging to the so-called cubic system (see below), but apart from these crystals, angular measurements may be used to identify each kind of crystal. If the same chemical compound occurs in several crystalline forms (as in the case of SiO_2 which forms α– and β–Quartz, Cristobalite, Tridymite) these are distinct thermodynamical phases and the substance is called polymorphic.

(iii) Taking as axes any three edges of a crystal which do not lie in one plane, suitable unit lengths may be determined on these so that all observed faces of the crystal have rational positions. By this is meant that their intercepts on the three edges, measured each in the appropriate unit of length, stand in the ratio of three small integral numbers (Law of Rational Positions). The directions of the edges together with their unit lengths are called the crystal axes and are given the symbols a, b, c, or a_1, a_2, a_3. A change of scale common to the three unit lengths would clearly shift all planes parallel to themselves and leave the angles between them unaffected. It follows that in descriptive crystallography only the *ratio* of the axial lengths can be determined. It is only in X-ray crystallography that the absolute value of these lengths (in cm or in Å) makes sense. It also follows from the law of rationality that if axes are chosen along three other edges occurring on the same crystal then, by a suitable transformation of the unit lengths, the rational planes of the old axial system are again rational in the new system. The axial system is thus not uniquely determined; usually that axial system is adopted in which the principal observed faces can be described by the lowest integers—but there may be reasons for deviating from this.

Bravais (1850) introduced a dual expression of the Law of Rational Positions which can be shown to be mathematically equivalent to the previous one and which has the advantage of dealing directly with the directions of the face normals as obtained on the goniometer. The law is then stated as follows: Take any three non-coplanar directions of face normals; to each of these a unit length may be ascribed such that the direction of any other observed face normal is obtained by geometrical composition (vector addition) of small integer lengths along these directions. The axial system in the directions of the three chosen normals together with the unit length in each direction forms what Bravais called the Polar Axes; we denote them by a^*, b^*, c^* or by b_1, b_2, b_3. Again, since we are dealing only with directions, the absolute scale of the units on the polar axes remains arbitrary. Only in X-ray crystallography will the scale obtain a meaning, and, including this, the axes are usually called the Reciprocal Axes to those of the crystal.

3.2. Symmetry

Potters and architects used symmetry and periodicity for creating artistic values even in prehistoric times. All the great civilizations of the past offer examples of the intricate beauty and the refined complexity of their application. But essentially these examples are limited to two dimensions. Crystallographers extended the notion to three dimensions, and it took them the greater part of a century to formulate a correct and complete geometrical theory of symmetry in space.

By Symmetry of an object we understand the equivalence of directions within the object. To a chosen direction there exist one or more different ones which show the same geometrical relations to all other directions defined in relations to the geometry of the object. It is therefore not possible to define in general a direction uniquely within the geometry of the object. A vase produced on a potter's wheel has 'cylindrical' or 'axial' symmetry. Except for the axis of rotation itself, it is not possible to define a direction uniquely by the geometry of the vase; for to any direction making an angle α with the axis there exists an infinite number of equivalent directions, forming a cone of opening α around the rotation axis; these are indistinguishable from one another by their properties with respect to the object. The axis of rotation is called a symmetry axis of infinite order.

Take the centre line of the lead in a six-sided pencil. This is a symmetry (or rotation) axis of the sixth order because to every direction (except along the axis) there are five equivalent ones, which, neglecting the imprint on the pencil, are indistinguishable.

Or take a match-box (again neglecting the print). The normal directions to its faces, taken at the centres of the faces, are two-fold axes because after one half full rotation about these directions the object is in a 'covering position'. Besides, there are 'mirror planes' each of which passes through the mid-points of one of the three sets of four parallel edges of the box. Reflection on a mirror plane brings the box to a covering position, and reflecting twice across the same mirror plane restores the original position. For this reason a mirror plane is called a symmetry element of the second order.

Finally, as an example, consider a cube and an octahedron; the latter is the figure obtained by cutting away the eight corners of the cube until the former cube faces are reduced to their mid-points which form the corners of the octahedron. Both figures have the same symmetry, comprising the following symmetry elements:

3 fourfold axes (through opposite corners of the octahedron)
6 twofold axes (through the mid-points of opposite edges of the cube)
4 threefold axes (through opposite corners of the cube, or normal to the faces of the octahedron)
9 mirror planes (three midway between parallel faces of the cube and six through opposite parallel edges of the cube).

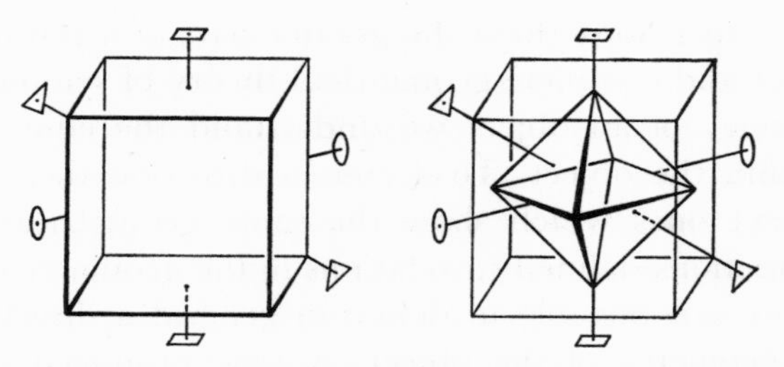

Fig. 3–2(1). Cube and octahedron with some of their symmetry elements.

Besides, we may distinguish a centre of inversion at the centre of the figure, that is, a symmetry element of order two which transforms any direction into the opposite one without change of length.

Not all symmetry elements are independent of one another. A fourfold axis always contains a twofold parallel one which results from the twofold application of the quarter-rotation. Two mirror planes intersecting at right angles always produce a twofold axis along their intersection. On the other hand symmetry elements may be incompatible with one another, such as a four-fold and a three-fold axis intersecting at right angles. For if this were the case, repeated application of the symmetry operations would show that *every* direction of space contains axes of both kinds—which is true for isotropy, but not compatible with crystallinity.

* * *

In the first half of the 19th Century the paramount symmetry problem was that of *Point Symmetry*: to enumerate all possible combinations of crystallographic symmetry elements which pass through a common point, the origin, and therefore leave this point single. The crystallographic symmetry elements were observed to be exclusively 2, 3, 4 and

6-fold axes, mirror planes, and centres of inversion. Fivefold axes, common in botany (flower petals) and zoology (starfish) do not occur in crystals; nor are there axes of 7th or higher order.

It was finally shown by Hessel in 1830 that geometrically there exist only 32 combinations of crystallographic symmetry elements, called the *symmetry classes*, and examples of substances were found for nearly all classes. For assigning a crystal to a class not only the geometrical symmetry of its faces is indicative but also the physical symmetry of the bulk crystal, namely of its dielectric and optical behaviour, its conductivity for electricity and heat, its elastic properties etc. The lowest common symmetry of the geometrical and physical behaviour is that proper to the crystal, namely that part of the symmetry of the physical observations which can not be accounted for by the inherent symmetry of the physical process itself. As an illustration take the optical property, the refractive index. Since the velocity of light is always the same in a direction and its opposite, the optical behavior always adds a centre of symmetry, whether the crystal possesses it or not. Therefore optical refraction alone does not allow distinction between those classes which differ only by the presence or absence of a centre of symmetry.

A cruder, and more easily effected assignment of a crystal than to one of the 32 symmetry classes is to one of the 7 *crystal systems* into which the classes can be divided according to the axial systems suitable for expressing their symmetry. There are, for instance, five classes which are all referred to a cubic (Cartesian) system of axes, i.e. three mutually orthogonal axes of equal length. Similarly, there are three classes referred to orthogonal axes of three different lengths, and these form the orthorhombic system. The seven systems are the cubic, tetragonal, orthorhombic, hexagonal, trigonal, monoclinic and triclinic. The face development and the optical properties are usually sufficient for assigning a crystal to one of the systems.

The establishment of the 32 classes provided an infallible framework by which the ever increasing data of observations on minerals and on other chemical compounds could be classified. Before the end of the nineteenth century, crystallography was mainly the domain of the mineralogists, but in the last two decades chemists took an increasing interest in the crystalline properties of the many new substances they isolated or synthesized, both organic and inorganic.

3.3. Theories of Crystal Structure

a. *General physical considerations.* Even before the laws of descriptive crystallography were fully explored, speculations were rife as to the peculiar nature of the crystalline state. They were prompted by the cleavage properties which were first noted about 1720 by metallurgists on the brittle fractures of metals, and then studied on rock salt and calcite (Iceland spar) by others. The father of crystallography, the Abbé René Just Haüy (1743–1826) concluded that a shape similar to the one that could be obtained by splitting must be preformed in the inner structure of the crystal. The cleavage planes, be they prominent in the usual development of the habit of the crystal or not, must exist in the crystal like the mortar joints in a brick wall. The crystal then is built up of ultimate crystal molecules or particles of the shape obtained by cleavage, and in cleaving, the surface common to such block-shaped elements of crystal structure is laid bare. If faces other than the cleavage faces occur on a naturally grown crystal, Haüy explained them as resulting from a stepwise growth on a sub-microscopic scale similar to the average inclination of stairs being formed by the off-setting of equal bricks. If this off-setting is by whole bricks only, in the ratio of step width to height of 1 : 1 or 2 : 1 or 3 : 2, etc., it accounts for the rational positions of the observable planes. Haüy did not hesitate to generalize this 'Theory of Decrescence' also to crystals which show no splitting because it offers such an easy explanation of the law of rational plane positions.

According to this view, then, the distinctive property of the crystalline state is its ultimate internal periodicity. Haüy could not decide what was the nature of the ultimate particles forming the repeat unit. This unit could clearly not be smaller than a molecule, but it could well consist of a whole cluster of molecules, that is, a multiple of the chemical formula. Haüy chose therefore for the repeat unit the name 'molécule intégrante'. This matter was settled by the first actual crystal structure determination, but that was not until 1913.

Meanwhile the assumption of a periodic internal structure of crystals came up in a long discussion between the founders of the Theory of Elasticity, notably Navier, Cauchy and Poisson, from 1821 onwards. In this year, Navier presented to the Paris Academy a paper (published in 1827) in which one of the pressing problems of the time received an answer, namely the establishment of equations governing the motion of an elastic deformation in a solid body, for instance the deformation of a struck plate or bell. In order to derive these equations,

Navier assumed the body to consist of randomly arranged molecules which exert central forces on one another. If the body is deformed, the distances and forces between the molecules are changed, and the change of the latter constitute the stress associated with the deformation. Objections were voiced against Navier's method of replacing sums of forces from individual molecules in random positions by integrals. Cauchy, who belonged to the referees of the paper, took up the matter in an independent and more formal way by replacing the use of a physical model of the solid by the mathematical assumption that deformations (strains) and forces (stresses) were proportional to one another—a generalization of Hooke's Law. For an isotropic medium Cauchy arrived at equations similar to Navier's, except for the fact that they contained two elastic constants (i.e. proportionality factors), whereas Navier's equations contained only one.

Was this discrepancy a consequence of the different physical assumptions, or was it introduced only by the approximations that had been made, in particular Navier's replacement of the sum by an integral? In order to decide this point, Cauchy adopted Navier's model of molecular force centres and made the model a fully determined one by assuming the molecules to lie at the nodal points of a lattice. This also enabled him to deal with the case of an anisotropic medium by assuming the lattice to be of low symmetry. The result of this second paper of Cauchy's was a system of equations of motion for the components of the deformation in which, for the general triclinic case, 15 elastic constants appear expressing the relations between the six components of strain and stress, respectively. The most general proportionality-assumption between two sets of six quantities requires 6×6 coefficients; in the case of central forces (or, more generally because of conservation of energy) this number is reduced from 36 to 21 in the triclinic, and to 2 in the isotropic medium. Cauchy obtained further 6 relations in the triclinic, and one in the isotropic case, by expressing the condition that the undeformed state of the medium is without stress. The equations for the isotropic medium are then the same as Navier's.

The assumption underlying the Cauchy relations seemed incontestible from the physical point of view,—but the reduction in the number of independent elastic coeffcients was not borne out by the measurements. The application of the experimental results of Wertheim (1848) was not immediate because of the two substances he used, brass is polycrystalline, and glass, with its strong elastic aftereffects is not a representative of a body fulfilling Hooke's Law. Not until 1887 was the non-validity of Cauchy's relations convincingly demonstrated by Woldemar Voigt's measurement of elastic deformations of anisotropic crystals. Yet, long before the convincing proof, it seemed probable that the Cauchy relations did not hold, and this opinion discredited the model from which the relations sprang, namely that in the natural state of a crystal its molecules are arrayed in a three-dimensional lattice.

Thus it came about that the concept of internal regularity and periodicity as a characteristic for crystalline matter, after having emerged in a very promising way, lay dormant for more than seventy years as a brilliant, but unfortunately not acceptable speculation which neither physicists nor crystallographers dared to use seriously.

Once Laue's discovery of 1912 had brought the irrefutable proof of the crystal's inner periodicity, it became the most urgent task to find where Cauchy's argument failed that had misled not only him, but all the other great mathematicians who had tried to escape his conclusion. It was Max Born's great achievement in 1913 to detect the flaw. Cauchy had assumed the molecules, that is, the force centres, to form a simple lattice. In this, each molecule is at a centre of symmetry of the entire (unlimited) system, and remains so in the case of a homogeneous deformation of the body. The forces exerted on a particular molecule by all others therefore balance, whatever the deformation may be. Born considered, instead, the more general case where each cell contains more than one molecule. A homogeneous deformation then consists of a change in the shape of the cell—which leads to the observable macroscopic strain—and a rearrangement of the molecules inside the cell, an 'inner displacement', which is not observable, except in some cases such as piezoelectric crystals where it leads to a change of the electric moment of each cell. The greater freedom gained by the crystal capable of inner displacements eliminates the interdependence of the elastic constants which is expressed in the Cauchy relations.—Only after having thus shown that Cauchy's results should not be applied to crystals of a sufficiently general structure was the way open for Born to develop his *Dynamik der Kristallgitter* (1914), the fundamental book on classical crystal dynamics.

b. *Space Group Theory*. The gist of Haüy's view was that a crystal is a periodic arrangement of equal particles, the 'molécules intégrantes', whatever these may be physically. The observation of symmetry puts certain restrictions on the arrangement and leads to the general, purely geometrical problem of finding all the types of symmetry that can be obtained by suitably arranging equal particles in space. In any such arrangement the particles have to remain equal, i.e. indistinguishable one from the other by any internal geometrical criterion. A medium of this kind has been called by P. Niggli (1919) a Homogeneous Discontinuum; it must, of course, be continued indefinitely filling all space, because otherwise the particles on or near to a boundary would be differently surrounded from those deep in the medium. The concept of a homogeneous discontinuum implies the periodicity of the internal structure, and therefore applies only to the crystalline state. If the demand of indistinguishability of the particles is restricted to average values, periodicity ceases to be required, and the resulting 'statistically homogeneous discontinuum' covers liquids and gases which are quite unordered, and fibres, high-polymers, and mesomorphic phases* which are partially ordered.

The problem of finding all symmetry types of homogeneous discontinua was solved in three steps, attached to the names of A. Bravais (1848), L. Sohncke (1867), and A. Schoenflies and E. von Fedorov (both 1891). Common to all of them are the following features:

(1) Symmetry axes of order 5,7 or higher are geometrically not compatible with a periodic structure; this is in accordance with the laws deduced from the morphology of crystals.

(2) Any periodic arrangement of particles is based on the repetition of a '*cell*', i.e. a parallelopipedon containing one or more particles; its shape and volume v_a is determined by three edges a_1, a_2, a_3 meeting in a corner point. The edges are called the *translations*, because by shifting the cell and its contents parallel to itself by integer multiples of the edges the entire structure is obtained from an original cell. The edges or translations may also be taken as the axial system of vectors $\mathbf{a}_1$, $\mathbf{a}_2$, $\mathbf{a}_3$ by which to describe the whole array of particles. If the cell contain n particles at positions given by vectors $\mathbf{x}^k$ ($k = 1 \ldots n$), these form the '*base*' within the cell. If then $\mathbf{x}_l = l_1\mathbf{a}_1 + l_2\mathbf{a}_2 + l_3\mathbf{a}_3$ (l_i integers) denotes the position of the origin of the lth cell (l standing for the three integers l_i), then the position of the kth particle of the lth cell is

$$\mathbf{x}_l{}^k = \mathbf{x}_l + \mathbf{x}^k,$$

i.e. the particle is reached from the origin by first finding the vector leading to the origin of the lth cell and then adding to it the base vector of the kth kind of particle.

(3) The description of a homogeneous discontinum by means of a

* These are also known under the name of liquid crystals.

particular cell, and its corresponding base and axes, is largely arbitrary and therefore void of physical consequence. For instance one of the axes, say a_1, might be doubled; then the cell volume and the number of base particles and base vectors would be doubled, but this is merely a different description of the same array of particles as before. Usually that cell is preferred which offers an easy visualization of the symmetry and is the smallest and therefore contains the least number of base particles.

(i) *Bravais*' fundamental contribution to structure theory is the proof that equal particles can be arranged in 14 types of 'lattices' differing by symmetry and geometry, such that each particle is translationally equivalent to any other. This last condition means that the system of particles can be brought to a covering position by a mere translation from one particle to any other particle. Or, to put it other-

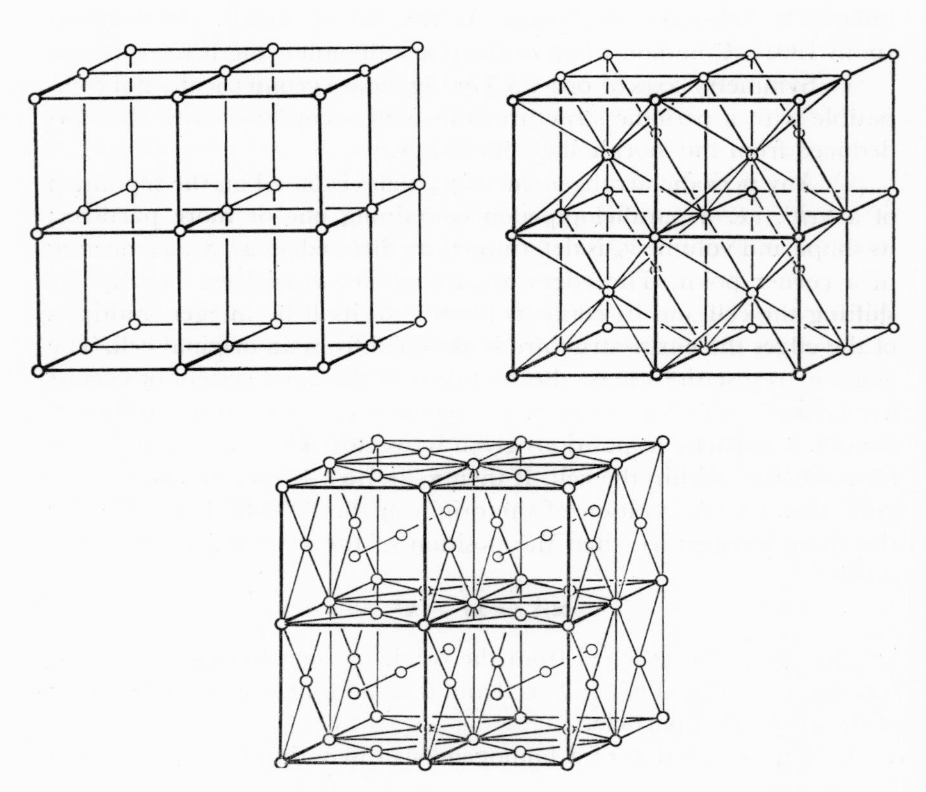

Fig. 3–3(1) a, b, c. The simple, body-centered and face-centered cubic Bravais lattices.

wise, if you were suddenly transported from one particle to any other, you would find the same view of the surroundings as before without turning your head. Fig. 3–3(1) illustrates a few cells of each of the three cubic Bravais lattices on which the correctness of the statement can be checked, while also showing that the arrangements are fundamentally different. (Compare the number of nearest neighbours in the three lattices, which are 6, 8, and 12, respectively.)

(ii) *Sohncke* saw the geometrical problem in greater generality. The condition of translational equivalence presents an unjustified restriction. It implies an external means of orientation, such as a compass or an external panorama in order to judge whether, while being transported from one particle to another, you have changed your direction of view. This extraneous orientation is alien to the definition of symmetry as given on pg. 19. Demanding then that the view of the system be the same from every particle, but not necessarily a parallel one, Sohncke found 65 different spatial arrangements. This answer of the problem included the introduction of novel symmetry elements in which a rotation about an axis, or the reflection on a mirror plane is coupled with a translation of the system along the axis, or in the plane of reflection. The first combination gives a '*screw axis*', the second a '*glide mirror plane*'. The translation contained in such glide symmetry elements is unnoticeable in an unbounded system of particles. As an example consider Fig. 3–3(2) where fourfold screw axes are arranged in a quadratic array. It is seen that a quarter full rotation of the system about one of the axes together with a translation along the axis of one quarter the pitch of the screw produces a covering motion of the un-

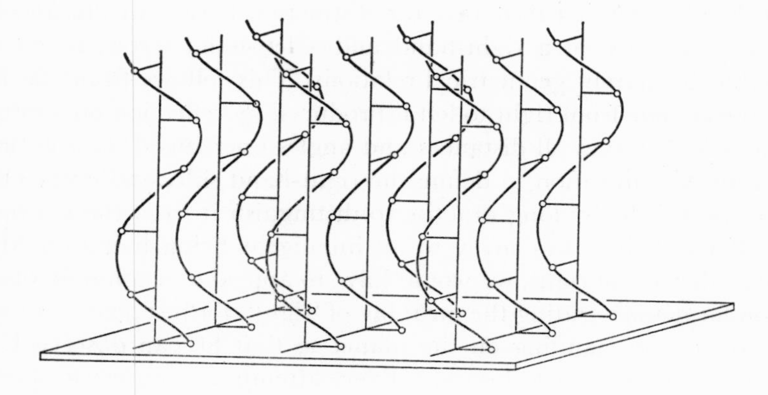

Fig. 3–3(2). A tetragonal Sohncke 'point system' with left-hand screw axes.

bounded system. It is also seen that no particle can be distinguished from any other by geometrical means, provided the aspects of the system may be compared after a suitable change of orientation. It is also easily seen that a similar arrangement might be shown using right-handed screws instead of the left-handed ones of the drawing. This second arrangement is the mirror image of the one shown, but it differs from it, since the two systems cannot be brought to coincidence —no more than a right-hand glove and a left-hand glove.

(iii) *Schoenflies*, then a lecturer in mathematics in Göttingen, worked out *230 Space Groups*, that is, different periodic arrangements of symmetry elements in space, and showed that they formed within the conditions given, a *complete* system. He considered the problem as one of geometrical group theory, a counterpart of the algebraic and the abstract Theories of Groups which began to be, in 1880–90, a very topical part of mathematics. Given the space pervading frame work of rotation and screw axes and of ordinary and glide mirror planes, a particle of any shape inserted anywhere will be reproduced by the symmetry elements, like the beads in a kaleidoscope, until the system of equivalent particles extends throughout space. Any one of the 230 arrangements, described either by the distribution of symmetry elements or by the coordinates of equivalent points, is called a Space Group. Sohncke's arrangements are Space Groups, but they formed only part of the complete system. The greater number obtained by Schoenflies is the result of abolishing yet another restriction, implicitly introduced in Sohncke's derivation, which contains an extraneous criterion not expressible by the geometry of the system itself.

It comes as quite a shock even to many scientifically trained minds to realize for the first time that the distinction between right and left and thence between a right-hand and a left-hand screw, is not expressible by purely geometrical relations. This follows from the fact that the change from right to left is produced by reflection on a mirror plane, which leaves all distances and angles unchanged. It is entirely a matter of convention to define the right-hand side, and every child has to be taught by long practise to distinguish it from the left-hand side. If we desired to convey to an intelligent being living on Mars which side we call right, we would have to appeal to a common observation for demonstrating the meaning of 'right', for instance by stating that for a man standing on the planet so that he can observe Ursa minor the sun will rise to his right. Every attempt at a purely geometrical definition would sooner or later beg the issue.

For this reason Schoenflies, in defining the equivalence of the particles, admitted that the view from one particle could change to its mirror image on transition to another particle. Of course the number of particles having a 'right' environment is equal to the number of those with a 'left' environment—otherwise there would be a distinctive feature between them.

Fedorov, the great Russian crystallographer, obtained independently and at the same time as Schoenflies the same system of 230 space groups.

* * *

In spite of the elaborately worked out theory the physical significance of these geometrical constructions remained obscure. What was the nature of the 'particle' inserted in, and multiplied by, the framework of symmetry elements? It could be any type of 'molécule intégrante' of Haüy's. Only conjectures were possible. Following the suggestion expressed by Sohncke in 1888, the eminent mineralogist and chemist P. Groth stressed the possibility of placing *atoms* in equivalent positions of a space group,—the various kinds of atoms of the crystal occupying not necessarily points of the same space group, but the interpenetrating space groups to be based on the same cell. Groth points out as a consequence of this that the molecular concept, while having a definite meaning in an amorphous body, loses this in the crystal since arbitrary atoms could be combined equally well to a 'molecule'.

With all this, the ideas about the size and contents of the cell remained rather vague. Also it was not possible to correlate an actual crystal with any space group, beyond choosing one of those which have the same point-symmetry (or class) as the crystal. Since there are 32 classes as against 230 space groups, this left many alternatives open.

The author's experience in 1911 may illustrate this statement. After having carried through in general terms a theory of double refraction caused by a simple orthorhombic arrangement of resonators he wanted to check it by a numerical calculation based on the axial ratio of some actual orthorhombic crystal. So he went to Professor Groth asking him what would be the most likely crystal to have its molecules arranged according to a simple orthorhombic Bravais lattice. Groth thought for a minute or two; then his face brightened and he said: 'there is only *one* crystal I can think of where this is nearly certain to be the case. This is Anhydrite. And I will give you the reason:

anhydrite shows excellent, good and fairly good cleavage on the three basic planes. I know of no other orthorhombic crystal having this property, and it means that it cannot be built according to any of the other three orthorhombic Bravais lattices.' The structure of anhydrite, determined in 1925, was found to be built according to a Bravais lattice in which one face is centered; there are four molecular units in the cell instead of one.

* * *

c. *Packing Theory of Crystal Structure.* A fundamentally different and much less systematic approach to crystal structure was made by the metallurgists, like Tammann in Göttingen and the chemists Barlow and Pope in Cambridge in the case of very simple compounds. They visualized atoms as spheres of a characteristic diameter which are closely packed so as to touch one another. The packing of equal spheres might be considered in the case of elements, while spheres of two sizes are required for the arrangement of the atoms in binary compounds such as the alkali halides NaF, NaCl, KCl etc. This theory, while offering by no means a complete geometrical-logical system of crystal structure, had certain features of physical reality which the other theory lacked. It was known to W. L. Bragg in 1912 when he discussed the photographs obtained with zincblende, and it gave him the clue for explaining why certain spots were missing which one would have expected to see on the diagrams. Pope, like others, had predicted a simple structure for alkali halides, consisting in the case of rock salt, NaCl, of an alternating arrangement of Na and Cl along the three cubis axes. On Pope's suggestion, W. L. Bragg took Laue photographs of these crystals and soon confirmed Pope's conjecture. This was the first full structure determination (published in *Proc. Roy. Soc.* June 1913). Since the wave-length of X-rays was at that time not yet known, the result could not have been obtained except by an inspired guess; by producing this, Pope's theory proved more fertile than the rival structure theories.

CHAPTER 4

Laue's Discovery of X-ray Diffraction by Crystals

4.1. Physics and Crystallography at the University of Munich in 1912

The University of Munich prided itself upon having the chairs occupied by eminent professors, well known beyond the confines of the city and of Germany. In 1912 some of the celebrities were H. Wölfflin for History of Art, L. Brentano for Economics, Amira for History, O. Hertwig for Zoology, P. Groth for Mineralogy and Crystallography, W. C. Röntgen for Experimental Physics and A. Sommerfeld for Theoretical Physics. The last three are of particular interest for our subject and may therefore be characterized in some detail. Each was head of an Institute with Assistants, Lecturers (Privatdozenten) or Assistant Professors (a.o. [= ausserordentlicher] Professor) and other staff attached to it.

a. *Röntgen's Institute* was by far the largest of the three and was situated in a separate three-story building in the main block of University buildings between the Ludwigstrasse and the Amalienstrasse. Besides the science students, the numerous medical students were supposed to go to Röntgen's lecture course and first-year laboratory and this demanded a large number of assistants and lecturers. P. P. Koch was already mentioned above, and E. Wagner will be mentioned later. E. von Angerer, who later became Professor at the Technical University in Munich and is well known as the author of several very useful books on the techniques of physical experimentation, was a third assistant. Röntgen had some 12–15 doctorands who were being looked after by the assistants and himself. As a 'doctor-father' Röntgen was, as in his own work, very exacting and 3–4 years full-time work on the thesis was not unusual. He demanded all possible precautions to be taken against errors and wrong interpretations and the maximum of accuracy to be obtained.

After his appointment to the chair of experimental physics of Munich University in 1900 Röntgen naturally maintained his interest in the clarification of the nature of his X-rays, and among the topics given out by him for thesis work there was usually an important one on X-rays:

E. v. Angerer (1905): Bolometric (absolute) energy measurement of X-rays.

E. Bassler (1907): Polarization of X-rays.

W. Friedrich (1911): Emission by a platinum target.

R. Glocker (1914): Study of interference.

Much of Röntgen's own work was spent on the electrical conductivity generated by X-ray irradiation in calcite (published 1907) and in other crystals. The greater part of this painstaking work was done together with A. Joffé, who had come to him after his graduation at the St. Petersburg Technological Institute in 1902, obtained his Ph. D. under Röntgen in 1905, and stayed with him for another year as assistant. It was, however, not before 1913 and 1921, respectively, that Röntgen felt satisfied with the checking of the measurements so as to release them for publication.

That a wide field of interest was covered by the work in Röntgen's institute is obvious to physicists from a list of a few others of the 25 doctorands graduating while Röntgen was director from 1900–22:

P. P. Koch (1901); J. Wallot (1902); A. Bestelmeyer (1902);
E. Wagner (1903); R. Ladenburg (1906); P. Pringsheim (1906);
P. Knipping (1913); J. Brentano (1914); R. Glocker (1914).

Because of the exceptionally high demands, graduation under Röntgen was attempted only by ighly devoted and serious students. They were expected to work independently, and even too much communication from door to door in the institute was not encouraged.

b. *Sommerfeld's* much smaller *Institute for Theoretical Physics* was an academic novelty. Sommerfeld had insisted on it before accepting the chair of theoretical physics in Munich which had been vacant for four years following L. Boltzmann's departure to Vienna. It had not been quite easy to overcome the obvious argument that theory demanded a library, and desks, but no experimental facilities. Sommerfeld, however, succeeded in convincing the faculty and the ministry of the necessity for a theoretician to keep in close touch with physical reality if his work was to obtain purpose and inspiration from physics. In contrast to an Institute of Experimental Physics which is equipped for experimenting in any field of physics, the Theoretical Physics

Institute would need only special equipment for supporting experimentally the lines of theoretical research.

At first, Sommerfeld's institute was situated in the 'Old Academy' or 'Augustinerstock' in the Neuhauserstrasse in Munich, where the Bavarian Academy of Arts and Science held its meetings and where also the zoological, geological and mineralogical Institutes of the University were housed. With the completion of the University extension along Amalienstrasse, the Institute moved in 1910 to part of the ground floor and basement there, in close proximity to Röntgen's Institute. It consisted of a small lecture theatre for about 60, a museum room for equipment (containing a.o. the models constructed by Sohncke from cigarboxes for demonstrating his 65 point systems), four offices, and, in the spacious basement, a workshop, a dark room, and four experimental and storage rooms. Apart from the occasional preparations of demonstrations for Sommerfeld's lecture course on Theoretical Physics, the main experimental work set up after the move to the new premises was an experimental investigation on the onset of turbulence in fluid motion in an open channel; Sommerfeld had been long interested in the problems of turbulence, and this particular work was performed by his doctorand Ludwig Hopf. In 1911, Sommerfeld appointed W. Friedrich as second assistant in order to make further experimental checks on the theory of X-rays, as mentioned above.

Up to then there had been only one assistant at the Institute, P. Debye, whom Sommerfeld had taken with him from Aachen to Munich, when he accepted the chair. Needless to say to those who know of his later development, Debye was, even then, an outstanding physicist, mathematician and helpful friend. He was, not less than Sommerfeld himself, a centre for the senior students and graduates frequenting the Institute and the Physics Colloquium. Of these about ten were actually working on theoretical subjects under Sommerfeld's guidance, while others, from Röntgen's and other institutes came in for occasional discussions of their problems. Even more efficiently and informally than at the Institute an exchange of views and seminar-like consultation on any subject connected with physics took place in the Café Lutz in the Hofgarten, when the weather permitted under the shade of the chestnut trees, and otherwise indoors. This was the general rallying point of physicists after lunch for a cup of coffee and the tempting cakes. Once these were consumed, the conversation which might until then have dealt with some problem in general terms, could at once be followed up with diagrams and calculations performed with pencil on the white smooth marble tops of the Café's tables—much to the dislike of the waitresses who had to scrub the tables clean afterwards. Sommerfeld and his friend R. Emden (Professor at the Technical University and well known for his pioneer work on stellar atmospheres) and also others like the mathematicians Herglotz, Carathéodory, Schoenflies when they happened to be in Munich, came to this unofficial centre of exchange of physical ideas and news. For the younger members of the group it was most exciting to watch research in the making, and to take sides in the first tentative formulation of experiments and theory. No need to say that Röntgen never came to this informal meeting—nor even to the regularly scheduled Physics Colloquium; he was dominated by a shyness that made him evade personal contacts wherever he could.

In the fall of 1909 Laue joined Sommerfeld's group. He was a pupil of Planck and had obtained his degree in Berlin. After two postdoctoral years in Göttingen he returned as assistant of Planck's to Berlin and became lecturer there for two years. He was Planck's favorite disciple, but for some personal or other reason he asked for being transferred to Munich University and this was arranged. Unmarried, and devoted to Physics as he was, he soon became a leading member in all the group's activities. His interests covered the whole of physics; he wrote the first monograph on the (special) Theory of Relativity, brought from his association with Planck a deep understanding of thermodynamics and the theory of radiation and had done some profound thinking on Optics. Sommerfeld was the editor of Volume 5 of the *Enzyklopaedie der mathematischen Wissenschaften* which dealt with Physics and contained many very important semi-original contributions such as those by H. A. Lorentz on the Theory of Electrons, by L. Boltzmann on Kinetic Theory of Gases, by Van der Waals on the Equation of State, etc. Laue, having finished his book on Relativity, agreed to write the chapter on Wave-optics, and set to work in 1911. This was also the year that he got married to a very charming and good-looking young girl coming from a Bavarian officer's family. They established themselves in the Bismarckstrasse and kept open house for the younger members of the Physics group.

c. As mentioned above, *Groth's Institute for Mineralogy and Mineral Sites* was in an old building near the centre of the city. This, originally an Augustine convent, had been taken over by the State during the period of secularization and had housed the Academy of Sciences and the Academy of Fine Arts. The latter obtained a handsome new building of its own at the end of Amalienstrasse, close to the main buildings of the University—see Gottfried Keller's description of its inauguration in *Der Grüne Heinrich*—whereupon the University, in great need of expanding, was given the vacated premises for some of its Institutes. On entering from the street one passed through a high hall and past covered stairways where the mail coaches formerly discharged their passengers, came to a large quadrangle and mounted on another broad flat stairways to Groth's institute. The balustrade, the stucco ornamentation of the walls, and the high double winged doors showed the typical 'Jesuit Style' of the first half of the eighteenth century. On passing through two very long and very high rooms where the practical classes of crystallography were held, one finally reached the door of the Geheimrat's room. After knocking and being asked inside the visitor would be confronted with a rather picturesque view. Except on the side where the tall windows admitted a flood of light and offered a fine view of roofs and parts of the old buildings near Munich's ancient cathedral, the Frauenkirche, all walls of the room were lined with Jesuit style, glass-fronted, high cases filled to the top with books, journals, manuscripts and occasional crystals. Two or three large tables stood in the room piled up with books, manuscripts, galley proofs and an odd goniometer, Bunsen burner and chemical glassware squeezed in among them. At the wall opposite to where the visitor entered he would finally detect the old-fashioned desk with a small worthy old gentleman facing the wall and turning his back to the visitor while eagerly entering the end of a sentence in a manuscript or a correction in a galley proof. This was the Geheimrat, P. von Groth, then in his early seventies. Once he was summoned from his work, Groth seemed eager to learn all the news his visitor could give him, both personal and scientific. But he was also willing to contribute to the conversation by reminiscing on his own experiences, or conversations he had had, or by offering his advice on problems about which he was consulted. His lively speech, with a strong Saxon intonation, belied his age and made the student lose the sense of distance—in strong contrast to what he felt in talking to Röntgen.

Groth's first great achievement was the classification of minerals according to chemical relationship and simultaneous occurrence in sites. The principles by which he re-arranged the mineralogical collection of the University of Strassburg, while he was professor there, was widely acclaimed. He insisted on including in crystallography not only the naturally occurring minerals, but artificially prepared chemical compounds as well, and in particular he fought for the wide introduction of crystallographic methods in organic chemistry. In order to facilitate this, he wrote a much used textbook *Physikalische Kristallographie*. In 1877 he founded the first *Journal of Crystallography and Mineralogy*, omitting from it Geology, Petrography and Palaeontology which at that time were often combined with the first two subjects. His personal relations to crystallographers and mineralogists all over the world were widespread, through correspondence, his own travels, and visits of foreign colleagues of shorter or longer duration to his laboratory. This personal contact as well as the large amount of active work he devoted to the *Zeitschrift für Kristallographie und Mineralogie* was the reason for his Journal's international success. Groth edited 55 volumes, from 1877 to 1920; only after the *Zeitschrift* had been firmly established by his work as the leading journal for Crystallography, and only after Crystallography itself had acquired a new depth through Laue's and the Braggs' work, was it possible to devote the *Zeitschrift* entirely to Crystallography, leaving the mineralogical part to be absorbed by a number of existing journals of mineralogy.

Groth's most stupendous work was the *Chemische Kristallographie*, five volumes which appeared between 1906 and 1919 with together 4208 pages and 3342 drawings and diagrams of crystals. The manuscript was written entirely by Groth in his tiny hand and corrected over and again by him until there was hardly a white spot left on the manuscript and again on the galley proofs. Oh for the admirable compositors in the Leipzig printing centres of the days before the general use of typewriters! The volumes contain a review of all crystallographic measurements, taking the substances in order of chemical complexity: Elements, Binary, Ternary and Higher Inorganic Compounds; Aliphatic, Aromatic and Mixed Organic Compounds. Each section is preceded by a survey of the crystal-chemical relations and includes many hints of gaps which should be filled in by further work. In many instances Groth doubted the correctness of work reported in literature, and, wherever possible, he got his pupils, assistants, or visiting colleagues to prepare the same substances again, and to crystallize and re-measure them. B. Gossner, H. Steinmetz and others carried out a great number of such assignments in the course of the years. Altogether measurements on between 9000 and 10 000 substances are critically discussed in *Chemische Kristallographie*, an astounding feat considering the small number of the team and the other work they had to do in routine training of students. The connection between Groth and his colleagues in Chemistry, Willstätter, Fajans and others was naturally stronger than with the physicists, but Groth's keen mind was always on the look-out for any method that could initiate a more direct approach to the problems of crystal chemistry.

4.2. *Ewald's Thesis*

Towards the end of the summer semester of 1910 the present author, Paul Ewald, had belonged to the group of students centering about Sommerfeld for about two years, and he felt that he could venture to ask his teacher to accept him as a doctorand. He came to Sommerfeld's light cherry-wood desk with this proposal, whereupon Sommerfeld took a sheet out of the drawer on which were listed some ten or twelve topics suitable for doctoral theses. They ranged from hydrodynamics to improved calculations of the frequency dependence of the self-induction of solenoids and included various problems on the propagation of the waves in wireless telegraphy—all of them problems providing a sound training in solving partial differential equations with boundary values. At the end of the list stood the problem: 'To find the optical properties of an anisotropic arrangement of isotropic resonators.' Sommerfeld presented this last topic with the excuse that he should perhaps not have added it to the others since he had no definite idea of how to tackle it, whereas the other problems were solved by standard methods of which he had great experience. In spite of the warning, Ewald was immediately struck by the last topic on the list, and even if he politely postponed the decision to the next appointment a few days later, he went home determined that it would be this topic or none. When this was agreed to, at the second interview, Sommerfeld gave Ewald a reprint of Planck's paper on the Theory of Dispersion (Berlin Academy 1902), and recommended him to study H. A. Lorentz's corresponding paper.

The problem just stated requires some explanation. The fact that a ray of light on entering a transparent body at an angle to the normal of the surface changes its direction is called *refraction*. The relation between the directions of the rays outside and inside the body is given by Snell's Law (1618), which introduces an optical property of the body, its *refractive index* n. This index n found a physical interpretation in the wave theory of light (in particular Augustine Fresnel, 1821) as being the ratio of the wave-velocity in free space to that in the body. Fresnel's wave theory not only predicted correctly the change of direction of the ray as a function of the angle of incidence, but also the ratios of the intensities of the incident ray to those of the two rays generated at the surface, namely the refracted ray inside the body and the reflected ray outside it (Fresnel's formulae).

The refractive index n varies within the optical spectrum. In most transparent bodies it increases from red to blue, i.e. with increasing frequency. In this case a fine pencil of white light passing through a prism will be spread into a coloured band, with blue suffering the greatest deflection. The change of refractive index with the frequency of the light is called *dispersion*, and the case considered is that of normal dispersion. The Danish physicist C. Christiansen discovered in 1870 that some intensely coloured transparent dyestuffs showed, in part of the spectrum, a decrease of n with increasing frequency of the light, and called this phenomenon *anomalous dispersion.* In such cases the curve showing the variation of n with frequency has many features in common with a '*resonance curve*' in Mechanics; that is a curve showing the response of a pendulum-like system to a periodic force of constant maximum magnitude as a function of the frequency of the force; in particular, a large increase of the amplitude of vibration occurs as the frequency of the impressed force approaches the natural frequency ('proper frequency') of the pendulum. The similarity led to the view that for optical theory a refractive body consists of pendulum-like 'resonators' which the light wave forces to oscillate with an amplitude depending on the frequency of the light. Each resonator acts as the source of a scattered optical field of the same frequency. This propagates from it in all directions and is therefore described as a '*spherical wave*'—or as we shall often call it because of its elementary nature, a '*wavelet*'. The nearer the frequency of the optical field lies to the proper-frequency of the resonators, the larger is their amplitude of vibration and that of the emitted wavelet; and the stronger is therefore the interaction between matter and light as measured by the value of the refractive index n or by the 'optical density' $n^2—1$ or by the 'Lorentz-Lorenz-Expression' $3(n^2—1)/(n^2+2)$. With regard to its refractive properties, a body is thus replaced by a system of as many resonators as there are molecules, these resonators floating in their positions surrounded by optically inert free space, held, as it were, like dew-drops on an invisible spider's web of forces which do not contribute to the optical properties. Having this model in mind, two features must be explained:

First, the existence of an index of refraction, and its dependence on frequency. This is the same as the question: how is it that the presence of the scattered wavelets changes the wave-velocity from its free space value c to a value q (which is smaller than c in the usual optical case of n greater than 1)?

And second: How does refraction and reflection arise at the surface of the body?

The first of these problems was the one treated in the theories of dispersion by H. A. Lorentz and M. Planck, though in a way whereby the wave-kinematical problem of the superposition of the wavelets to a slowly moving wave-front was eliminated. This study determined a possible mode of propagation in the interior of the body; the refraction and reflection problem was not considered and in as much as an 'incident wave' figured in these theories, its significance was obscure.

It should not be assumed that the division of the problem into that of dispersion and that of refraction was understood at the beginning of Ewald's investigation—it developed clearly only in the course of the work. What Sommerfeld had in mind was this: In Planck's and also in Lorentz' then known work, an amorphous medium had been assumed, characterized by a random distribution of the resonators in space. This led, naturally, to a single value of the refractive index, valid for all directions of the light ray travelling through the medium. If the same type of resonators were placed in a lattice array, with perfect regularity but different distances along the three coordinate axes—would the dispersive and refractive properties of this medium be those of a crystal? Would there result, for a general direction of propagation, two refractive indices whose magnitude depends on the direction and the polarization of the wave? In other words, would it be unnecessary to assume an inherent anisotropy of the resonators themselves for the explanation of crystal optics? These were the questions which preoccupied the author in the next two years. Heavy mathematics was involved in finding a general answer, and again in transforming this answer to a form where the magnitude of the effect could be calculated. All this mathematical technique was, much later, recognized as Fourier Transformation—a concept which had not yet been formed at the time—with the result that nowadays the mathematical derivations can be presented to a class of graduates in a two-hour session without undue strain. The model used for the theory was a simple orthorhombic lattice of isotropic resonators (or *dipoles* as they are also called); the positions of the resonators along the x, y, z Cartesian coordinate axes are $(X, Y, Z) = (la, mb, nc)$, where l, m, n are integers ranging independently from $-\infty$ to $+\infty$, and a, b, c are the axes or translations of the lattice.

Ewald showed that the model fulfilled the general laws of crystal optics. In order to check on the magnitude of the effect, he took, on the advice of Groth, the axial ratios of anhydrite ($CaSO_4$), $a : b : c =$ $= 0.8932 : 1 : 1.0008$. The result of the calculation was that in two directions the double refraction of the model was 3–4 times the observed one, and in the third direction it was six times smaller. Since no crystal structures were known at the time and it seemed unlikely that the resonators representing anhydrite should really have the simple arrangement assumed, an agreement between the observed and calculated values would have been most unexpected. The conclusion drawn from the calculation was, however, that the structural anisotropy was ample for producing double refraction of the observed magnitude, and that in any case its effect would have to be taken into account before ascribing an inherent anisotropy to the molecular resonators.

Ewald had finished his calculations and was writing out the thesis during the Christmas recess 1911 and in January 1912. In paragraph 3 of his presentation he stated the astonishing conclusion that his theory of dispersion, dealing with an unbounded crystal, had no use for an incident ray, even though this played a significant role in the existing theories of dispersion. The refractive index, like the proper frequency of a mechanical system, was determined by a free vibration of the whole system, without the need of any external excitation. Thence he concluded that in a bounded system, for instance a crystal lattice filling only the lower half of space, the incident wave must be shielded from the interior by action of the boundary, so as to allow the establishment of the self-supporting free vibration.

This conclusion was only later confirmed by direct calculation, in a sequel to the abbreviated re-publication of his thesis in *Annalen der Physik* 1916, vol. 49, pg. 1–38 and 117–143. At the time of writing the thesis, it seemed a rather radical departure from the traditional theory. For this reason Ewald meant to discuss it with Laue who had a strong leaning towards fundamental physical issues.

4.3. Laue's Intuition

Laue suggested that they meet the next day—it was probably late in January 1912—in the Institute and discuss before and after supper at his home. They met as arranged and took a detour through the Englische Garten, a park whose entrance was not far from the University. After having crossed the traffic on the Ludwigsstrasse, Ewald began telling Laue of the general problem he had been working on, because, to his astonishment, Laue had no knowledge of the problem. He explained how, in contrast to the usual theory of dispersion he assumed

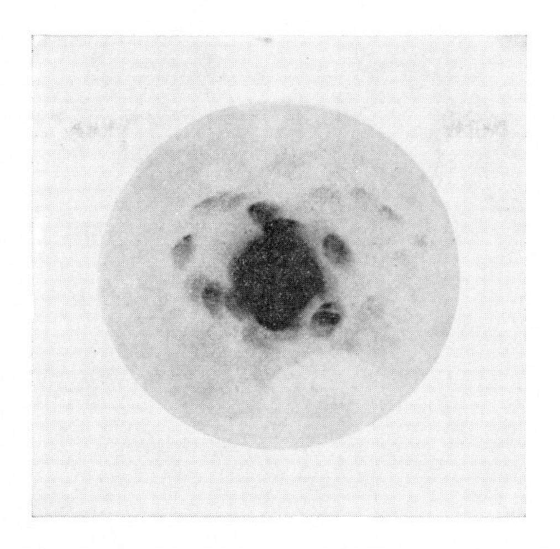

Fig. 4–4(1). Friedrich & Knipping's first successful diffraction photograph.

Fig. 4–4(2). Friedrich & Knipping's improved set-up.

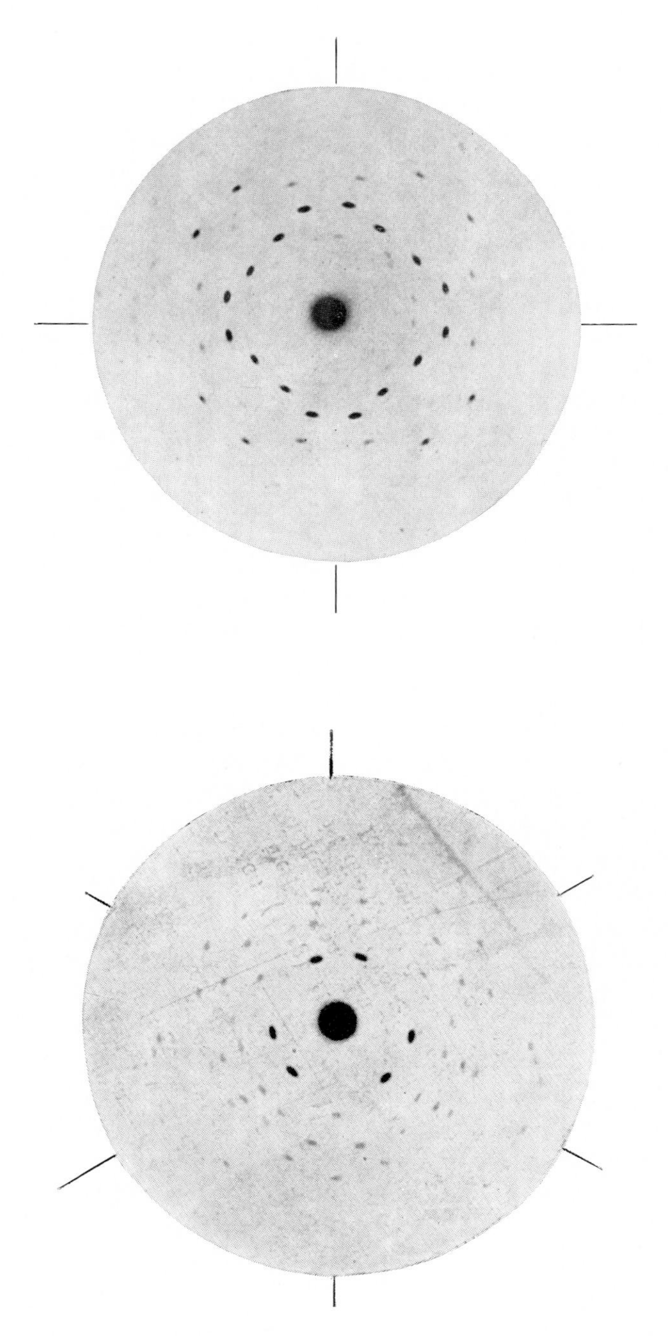

Fig. 4–4(3) and (4). Zincblende Laue photographs along four-fold and three-fold axes. (Laue, Friedrich & Knipping, *Sitz.ber. Bayer. Akademie d. Wiss.* 8. Juni 1912).

the resonators to be situated in a lattice array. Laue asked for the reason of this assumption. Ewald answered that crystals were thought to have such internal regularity. This seemed new to Laue. Meanwhile they were entering the park, when Laue asked: 'what is the distance between the resonators?' To this Ewald answered that it was very small compared to the wave-length of visible light, perhaps 1/500 or 1/1000 of the wave-length, but that an exact value could not be given because of the unknown nature of the 'molécules intégrantes' or 'particles' of the structure theory; that, however, the exact distance was immaterial for his problem because it was sufficient to know that it was only a minute fraction of the wave-length.

On the rest of the way, Ewald explained the technique of his treatment of the problem, leaving his main question over for the resumption of the conversation after supper. When this time came, he found Laue listening in a slightly distracted way. He again insisted on knowing the distances between the resonators, and when he received the same answer as before, he asked: 'what would happen if you assumed very much shorter waves to travel in the crystal?' Ewald turned to paragraph 6, Formula 7, of his thesis manuscript, saying: 'This formula shows the result of the superposition of all wavelets issuing from the resonators. It has been derived without any neglection or approximation and is therefore valid also for short wave-lengths. It only requires to be discussed for that case.—I, however, have to get my thesis delivered within the next few days and have then to do some reviewing for my oral examination—you are welcome to discuss the formula which I am copying out for you.'

Soon after this Ewald took his leave from Laue and Mrs. Laue since it was evident to him that his attempt at discussing his worries with Laue this time had failed. He submitted the thesis to the Philosophical Faculty (II. Division) on 16 February 1912 and stood the oral examination on 5 March, the date given on his Doctor Diploma. Over these events and the offers of two tempting jobs as assistant (either to Haber or to Hilbert) he forgot about Laue's interest in the passage of very short waves through a crystal. The next he heard of it was a report on Laue-Friedrich-Knipping's successful experiments which Sommerfeld gave to the Physical Society of Göttingen in June 1912. On coming home from it, Ewald at last looked at the formula recommended to Laue and found the same evening the obvious way of interpreting it geometrically for short waves by means of a lattice having translations proportional to 1/a, 1/b, 1/c, which he called the 'reciprocal lattice', and a sphere determined by the mode of incidence of the X-rays on the crystal, which in English is called 'sphere of reflection'. The paper containing this discussion appeared in *Physikalische Zeitschrift* 1913, vol. 14, pg. 465–472, and its equation (8) is the formula of the thesis recommended to Laue's attention but of which he never made use.

4.4. The Experimental Verification

Let us now return to Laue's further reactions. As he states in his Nobel Prize Lecture, 'On the Discovery of X-ray Interference', given in Stockholm on 3 June 1920, his question about the fate of short waves in a crystal was prompted by the expectation that if their wave-length is of a similar magnitude as the atomic distances the regular arrangement in a crystal must lead to some kind of diffraction effect. Through his work on the Encyclopedia article the theory not only of the simple diffraction grating but also that of a cross grating was fully present in Laue's mind. True, diffraction by a three-dimensional grating had never been considered, but, as he puts it: 'my optical intuition told me immediately that under such circumstances spectra must occur.'

There is no indication that Laue at that stage made any attempt at consolidating his 'optical feeling' by seeking to predict the kind of phenomenon that might be expected. Besides, the Easter vacations soon began and during that period a group of physicists traditionally met in the Alps for skiing. Here Laue discussed his idea with Sommerfeld, Wien and others with the result of encountering a strong disbelief in a significant outcome of any diffraction experiment based on the regularity of the internal structure of crystals. It was argued that the inevitable temperature motion of the atoms would impair the regularity of the grating to such an extent that no pronounced diffraction maxima could be expected. This objection may have been checked by a quantitative estimate of the magnitude of the thermal displacements—although this would have had to be based on a number of uncertain assumptions seeing that no crystal structure was as yet known. An evaluation of the thermal deformation of the crystal lattice could have been made by comparing the known average thermal energy of an oscillator at room temperature to that of an oscillator of amplitude A and frequency corresponding to a 'Rest-strahl' wave-length of, say, 50 microns as for rock salt or KCl. Assuming the mass of the oscillator to equal that of the chlorine atom, an amplitude A of about 0.75 Å is obtained. This is larger than the X-ray wave-length as given by Wien (0.6 Å) or Sommerfeld (0.4 Å), and thus the regular phase relations between the individual scattered wavelets, which are essential for the formation of a diffracted beam, would be destroyed. This or similar arguments seem to have weighed so heavily in Sommerfeld's mind that he was staunchly opposed to cede his newly appointed experimental assistant, Walter Friedrich, to Laue for the experiment. The situation was also discussed by Laue at the Café Lutz physics table, and here the opinion prevailed that experiment was safer than theory and that since the diffraction experiment required no elaborate set-up, it should at least be tried. Paul Knipping, who had just finished his thesis work in Röntgen's Institute, volunteered to assist, so as to reduce the time Friedrich would be taken off his work for Sommerfeld. The X-ray tube, the induction coil and the Wehnelt electrolytic interrupter had to be set up anyway for Friedrich's work, so that it was an easy matter to slip in a few unscheduled runs for Laue's experiment.

Once the three partners, Laue, Friedrich and Knipping had decided to go ahead, success came swiftly thanks to Friedrich's experience in X-ray experimentation. Led by the exposure times Herweg had required in his experiments on double scattering, Friedrich knew that exposures of several hours would be needed. This in turn meant careful screening of the crystal and photographic plate from the unwanted X-rays which come from the glass walls of the X-ray tube and from the mass of irradiated air. The tubes available at the time had a glass bulb of 10 cm radius and the glass wall acquired a high charge and potential while the tube was running. Any grounded lead diaphragm had to be at least 17 cm from the target in order to avoid a breakdown of the tube. The minimum distance target-crystal thus came to be about 25 cm, and this meant that only a very small fraction of the total output of the tube was used.

Friedrich constructed a lead box containing the crystal and the photographic plate. It consisted of a tray of lead sheet about 12 $\times$ 7 cm with a turned-up rim, and a cover in the form of an open box about 6 cm high which could be placed with the open side on the tray, and whose side facing the tube had a hole of 3 mm diameter for admitting the X-rays. There may have been a second hole on the opposite side through which the strong primary beam passed out of the box without generating secondary X-rays by hitting on lead.

For crystal, a piece of copper sulfate was used as it was found in the laboratory. In fixing the crystal on its holder by means of wax no particular orientation was aimed at.

The photographic plate was placed between the X-ray tube and the crystal on the assumption that the crystal would act like a reflexion grating.

The first exposure gave no effect. Thinking this negative result over, Friedrich and Knipping came to the conclusion that better success might be achieved by placing the plate behind the crystal, as for a transmission grating. Knipping insisted on placing plates all around the crystal.

The result of the second attempt was positive. On the plate behind the crystal, surrounding the imprint of the direct or primary ray, rings of fuzzy spots appeared, each spot of elliptical shape with the minor axis pointing to the overexposed and therefore solarized centre of the black area produced by the primary ray. No similar spots were produced on the other plates. Crude as the picture was, it contained an unmistakable proof that some property of X-rays had been found which had escaped all previous investigators. It also gave strong support to the correctness of Laue's idea of diffraction of X-rays by crystals. Laue learned of this result in Café Lutz; he hurried to the Institute and convinced himself of the correctness of his 'optical feeling'. Going home in deep thoughts he suddenly perceived the theory of the diffraction effect—so suddenly that in his autobiography he mentions the street and house in passing which his illumination occurred. He writes (*Autobiographie*, see pg. 294 of this book): 'Only shortly before this, when writing an article for *Enzyklopaedie der mathematischen Wissenschaften*, I had given the old theory of diffraction by an optical grating, which went back to Schwerd (1835), a new formulation in order that by applying the equation of the theory twice over, the theory of diffraction by a cross-grating could be obtained. I had only to write out this equation three times, corresponding to the three periodicities of a space lattice, so as to obtain the interpretation of the new discovery. In particular the observed rings of rays could thus be related to the cones of rays demanded separately by each of the three conditions of constructive interference. When, a few weeks later, I was able to give this theory a quantitative testing on a more suitable photograph and found it confirmed—then this became to me the decisive day.'

Röntgen was among the first who came to look at the experimental set-up and at the results. He was deeply impressed by the photographs but held back on their interpretation as diffraction. As true experimentalists Friedrich and Knipping had made sure that the crystalline nature of the sample was involved in forming the pattern.

They had pulverized the crystal and kept the powder in place in a little paper box, while exposing it; the photograph showed the central spot formed by the primary beam and the rings of large spots were absent. Only very small freckles, of about the size of the powder particles, were to be seen in the region surrounding the primary spot. They also had convinced themselves that only the primary spot was formed if the sample was removed altogether. In view of these findings Röntgen had to concede that the spots on the successful photograph were caused by the presence of the crystal, but he held in abeyance his verdict as to diffraction.

It must be made clear that at this time the three investigators were themselves under a misapprehension which, retrospectively, is hard to understand. They were convinced that the diffracted rays would consist of characteristic radiation emitted by the crystal under the influence of the primary, incident, ray. 'Thus', Friedrich and Knipping write in their paper, 'a crystal had to be chosen containing a metal of considerable atomic weight, in order to obtain intense and homogeneous secondary rays, as these seemed the most suitable ones for the experiment. According to Barkla, metals of atomic weight between 50 and 100 were to be taken into consideration. Since initially we had no good crystal containing such metals, we used for the preliminary trials a fairly well-developed copper sulfate crystal.' Copper has atomic weight 63.5; zinc, its neighbour in the periodic system of the elements, has weight 65.4. It is therefore very likely that the same considerations prompted the ordering of a zincblende plate from the well-known firm of Steeg and Reuter in Bad Homburg as soon as the first results were obtained. Before this plate arrived and during the period of construction of a more elaborate camera, diagrams were taken with cleavage plates of zincblende (ZnS), rocksalt (NaCl), and galena (PbS). They confirmed the experiences gained on copper sulfate.

The preliminary tests had included shifting the crystal parallel to itself showing that all parts of the crystal gave the same pattern; fixing a second photographic plate behind the crystal at double the distance of the first one, on which a picture of double the size was obtained, proving that there were really secondary rays fanning out from the crystal; and finally, changing the orientation of the crystal by a few degrees and finding that the position of the spots is very sensitive to the orientation of the crystal with respect to the incident X-rays.

These last observations showed the desirability of constructing an improved apparatus so that not only the direction and delimitation of the incident ray, but also the orientation of the crystal were sharply defined. Friedrich achieved this in an unsurpassed way by setting up the crystal on an accurate goniometer and using a collimating system consisting of a first, wider hole of about 3 mm diameter in a 10 mm thick lead plate followed at a distance of 70 mm by a much finer hole of 0.75 mm diameter also drilled in 10 mm thick lead. Since the shape of the fine hole actually is that of a long cylinder, it was essential to mount the second lead plate so that its plane could be adjusted by three screws; this also offered the advantage of enabling wider holes drilled in the same plate to be rotated in position for the rough adjustment.

The adjustment made use of the telescope of a cathetometer which stood fixed, once for all, at a distance of about 3 m from the tube. Its optical axis was directed to the focus of the tube and formed the line on which first the coarser and then the fine hole were centered. Then, by means of a small piece of fluorescent screen and with the tube running, the shape of the primary beam was checked; the plane of the second lead piece had to be adjusted by means of the three screws until a circular picture of uniform intensity was achieved. Next came the setting of the crystal plate with its main face perpendicular to the optical axis. For this purpose a little plane mirror, mounted (on a metal strip held with wax) at the center of the front lens of the telescope, reflected light from a sideways source along the axis of the instrument. If the crystal plate reflected the light back into the telescope then it was perfectly oriented. The necessary tilt of the crystal could easily be achieved with the motions provided by the head and circle of the goniometer. Finally, the photographic plate, wrapped in black paper, was aligned in the same way as the crystal by means of a mirror temporarily clamped to its back side. The collimator, goniometer, and plate holder all stood on a lead tray, and when the adjustments were completed, a heavy leaden hood was lowered from above so as to eliminate secondary radiation. The legs of the tall wooden tripod and part of the hood suspended by it appear on the photograph of the apparatus.

The working of the old-fashioned X-ray tube depends on the remnants of gas left in it, from which ions are being formed whose impact in turn liberates the electrons forming the cathode ray. The less gas there remains in the tube, the higher is the voltage that must be applied in order to pass a certain current, and the greater the energy of the electrons and the 'hardness' or penetrating power of the X-rays. Gas is emitted from the target and other metal parts of the tube as they heat up, and on the other hand gas is removed and occluded on the walls of the tube by the sputtering of cathode metal. If the tube is run at too high power, heating and with it loss of resistance of the tube prevails and if a constant voltage is maintained, the increase of current through the tube, the heating-up, and the release of gas continue until the target or some other metal part melts—and this is the end of the tube. If, on the other hand, the tube is run with too little power, scattering of metal and depletion of gas prevail; the resistance increases, and so does the peak voltage accepted from the induction coil or transformer, until the insulation suffers a breakdown, either in the coil or transformer, or by sparking through the glass of the tube —and that again is the end of the experiment. Because of this inherent instability the running of the old X-ray tubes was like walking on a mountain ridge with a precipice on either side. On the tube shown in Fig. 4–4(2) the little side tube to the utmost left is an automatic 'regeneration device.' It contains a rolled-up mica sheet, and if the main tube begins to offer too much resistance, a spark will jump from the lower pole of the tube, the cathode, to the wire brought near to it from the lower end of the regenerating device, and pass to its upper end which is permanently connected to the anode and anticathode, thereby liberating some gas occluded on the mica. In spite of such devices, the old X-ray tubes needed constant vigilance and the successful end of a long exposure for a diffraction photograph drew a sigh of relief from the operator.

It was a great advance when W. D. Coolidge invented a new type of X-ray tube which has a much higher vacuum and provides the electrons for the cathode rays by a well regulated thermal emission from an incandescent wire. One of the first tubes of its kind was donated to Sommerfeld's Institute by Dr. Coolidge and was used for taking some photographs. But this work was soon interrupted by the war in 1914. The entire X-ray plant was set up in an emergency hospital and used for medical purposes.

4.5. The Publication of the Work

Laue, Friedrich and Knipping's research was communicated to the Bavarian Academy of Sciences at the meetings of 8 June and 6 July 1912 by A. Sommerfeld as Fellow of the Academy. Röntgen seconded the acceptance and stressed the importance of the work. Two papers were published in the *Proceedings* (*Sitzungsberichte*) of the Academy, namely pg. 303–322: 'Interferenz-Erscheinungen bei Röntgenstrahlen' by W. Friedrich, P. Knipping and M. Laue; and pg. 363–373: 'Eine quantitative Prüfung der Theorie für die Interferenzerscheinungen bei Röntgenstrahlen' by M. Laue. In reprint form both papers were sent out with a common paper cover.

Simultaneously with Sommerfeld's presentation in Munich, Laue himself reported on his discovery to his old group of Berlin physicists at the meeting of the Berlin Physical Society of 8 June 1912. While he still resented the initial cautious-critical attitude of some of his colleagues and elders in Munich, he was warmed up by the unrestricted acceptance of his theory by the group in Berlin, and especially by the enthusiasm of the highly gifted astronomer-physicist Karl Schwarzschild.—On his return journey, Laue stopped over in Würzburg and gave a report there to Willy Wien's physics group. Erwin Madelung from Göttingen happened to be present and obtained from Laue the loan of his slides in order to show them to the colleagues in Göttingen.

The first of the papers published in the Bavarian Academy contains on $8\frac{1}{2}$ pages under Laue's name an introductory paragraph, and the theory of diffraction by a three-dimensional lattice. The remaining 11 pages are signed by Friedrich and Knipping and describe the preliminary and the final experiments. Eleven Laue diagrams are beautifully reproduced by heliogravure on five plates.

In the second paper, Laue applies the formulae of the general theory to a discussion of the ZnS diagram with X-rays incident along the axis of fourfold symmetry. It is, evidently, this paper and the quantitative confirmation it seemed to contain, which gave Laue the final certainty of having to do with the expected diffraction effect. The gist of the experimental part has been given in the preceding section, but Laue's theoretical contribution has yet to be reviewed.

* * *

We begin with the first part of the first paper. Using Cartesian coordinates and the (x, y, z)-components of the triclinic axial vectors or translations $\mathbf{a}_i$ (i = 1, 2, 3), the coordinates of an atom (m, n, p) (m, n, p integers) are first written out. The assumption, usual in optics, is made that the wavelet emitted by an atom is of a definite frequency; otherwise a monochromatic Fourier component of the emission would have to be considered. It is further assumed that the excitation is in form of a plane wave progressing in the crystal and that this produces a phase factor of the wavelet according to the position of the atom. The only assumption here made, as Laue points out, is that all atoms react in the same way to the excitation. Whereas in visible optics the atom is small compared to the wave-length, and the emission is uniform to all sides, 'one has here to consider the possibility—and the experimental results seem to confirm it—that the emission depends on the direction because the distances within the atom are comparable to the wave-length'. Laue therefore introduces an amplitude factor ψ of the wavelet, which depends on the direction and wave-length. This ψ corresponds to what is nowadays called the atomic factor f.

The optical field generated by the crystal can now be expressed as a sum over all wavelets. By assuming the observer to be at very great distance from the crystal, the spherical wavelets can be replaced, near the observer, by fragments of plane waves whose phases are determined by the emitting atom, and the summation over all atoms can be carried out. To simplify the summation (which actually is that of a simple geometrical series), Laue assumes the (finite) crystal to form a block of $(2M+1)$, $(2N+1)$ and $(2P+1)$ cells, respectively, in the directions of the crystal axes. This results in the famous expression for the intensity I observed at great distance R from the crystal in an arbitrary direction given by the three cosines (α, β, γ) of its angles with the (x, y, z) axes when the direction of incidence is fixed similarly by $(\alpha_0, \beta_0, \gamma_0)$:

$$I = \frac{|\psi(\alpha,\beta)|^2}{R^2} \frac{\sin^2 MA}{\sin^2 \frac{1}{2}A} \frac{\sin^2 NB}{\sin^2 \frac{1}{2}B} \frac{\sin^2 PC}{\sin^2 \frac{1}{2}C}, \tag{1}$$

where

$$\begin{aligned} A &= \frac{2\pi}{\lambda} \{a_{1x}(\alpha - \alpha_0) + a_{1y}(\beta - \beta_0) + a_{1z}(\gamma - \gamma_0)\} \\ B &= \frac{2\pi}{\lambda} \{a_{2x}(\alpha - \alpha_0) + a_{2y}(\beta - \beta_0) + a_{2z}(\gamma - \gamma_0)\} \\ C &= \frac{2\pi}{\lambda} \{a_{3x}(\alpha - \alpha_0) + a_{3y}(\beta - \beta_0) + a_{3z}(\gamma - \gamma_0)\} \end{aligned} \tag{1'}$$

The expression for I corresponds exactly to what Laue visualized on his way home after having seen the first successful photograph, namely the thrice applied summation of wavelets issuing from a linear lattice; for each of the $\sin^2$-quotients refers to only one of the axial directions of the crystal and the atoms alined on it.

The denominators of the $\sin^2$-quotients are much slower changing functions of the direction of observation than the numerators whose arguments contain the large numbers M, N, P. Maxima of I will occur when all three denominators are zero, i.e. the arguments $\frac{1}{2}$ (A, B, C) are integer multiples of π, say $(h_1, h_2, h_3)\pi$, respectively. Given the incidence, this condition determines the direction cosines (α, β, γ) of the rays generated by diffraction. The condition is

$$\begin{aligned}
\tfrac{1}{2}A &= \frac{\pi}{\lambda}\{a_{1x}(\alpha-\alpha_0) + a_{1y}(\beta-\beta_0) + a_{1z}(\gamma-\gamma_0)\} = h_1\pi \\
\tfrac{1}{2}B &= \frac{\pi}{\lambda}\{a_{2x}(\alpha-\alpha_0) + a_{2y}(\beta-\beta_0) + a_{2z}(\gamma-\gamma_0)\} = h_2\pi \quad (2) \\
\tfrac{1}{2}C &= \frac{\pi}{\lambda}\{a_{3x}(\alpha-\alpha_0) + a_{3y}(\beta-\beta_0) + a_{3z}(\gamma-\gamma_0)\} = h_3\pi
\end{aligned}$$

Laue separates the known from the unknown quantities (the latter being α, β, γ) by writing these equations

$$\begin{aligned}
a_{1x}\alpha + a_{1y}\beta + a_{1z}\gamma &= h_1\lambda + a_{1x}\alpha_0 + a_{1y}\beta_0 + a_{1z}\gamma_0 \\
a_{2x}\alpha + a_{2y}\beta + a_{2z}\gamma &= h_2\lambda + a_{2x}\alpha_0 + a_{2y}\beta_0 + a_{2z}\gamma_0 \quad (3) \\
a_{3x}\alpha + a_{3y}\beta + a_{3z}\gamma &= h_3\lambda + a_{3x}\alpha_0 + a_{3y}\beta_0 + a_{3z}\gamma_0
\end{aligned}$$

As the direction of observation is changed, the right hand sides remain constant. Now the left-hand sides can be interpreted geometrically as the projections of a unit vector in the direction of observation (α, β, γ) on to the axes a_1, a_2, a_3, respectively. Each projection equals the right-hand side if the vector (α, β, γ) lies on a cone of a certain opening having one of the axes a_i as central line. As the integer h is changed, the opening will change; each axis is therefore surrounded by a system of co-axial cones, and these are the geometrical loci for the directions under which a diffracted ray can be observed. Now the two sets of cones around two of the axes always have a number of lines of intersection, provided their opening is wide enough. The two conditions fulfilled on such common lines are those for diffraction maxima formed by a cross-grating of atoms in the plane of the corresponding axes. Such cross-grating rays therefore will *always* exist; but the effects of the parallel cross-gratings, generated by the third axis or translation would be destructive, unless the third equation is also fulfilled. It will be a rare event that the direction of intersection of the first two cones also belongs to a cone of the third set. Only when this happens will there be a diffracted ray coming out of the three-dimensional lattice. Laue indicates that an approximate fulfilment of the third condition may be sufficient to produce a diffracted ray, without, however, discussing the necessary degree of approximation.

Narrowing the discusssion down to photograph number 5 of the paper, which was obtained with the cubic ZnS crystal plate by incidence along the fourfold symmetry axis, Laue points out that a photographic plate normal to this axis intersects two of the systems of cones in hyperbolae, and the third system in circles. These circles which are caused by the periodicity of the lattice in the direction of the incident beam are compared to Quetelet's rings in light optics: these are obtained if light falls on the surface of a dusty glass mirror and are caused by the interference of light scattered directly with that having been reflected (before or after scattering) by the silvered back side of the mirror. The two rays interfering in the optical case give rise to rings of the same radius but much greater width than those caused by the many cross-gratings of the lattice.

Laue further points out that a given lattice can be described by translations in a great variety of ways and that each of these leads to a different system of conic sections on the photographic plate as locus for the spots registered. This is confirmed by the diagrams obtained with different orientations. On the whole, so Laue concludes, the diagrams seem to be explainable on the assumption that they are caused by a number of discrete wave-lengths in the range of 0.038 *a* to 0.15 *a*, where *a* is the axial length of the zincblende lattice.

There remains, however, a difficulty in understanding how it is that the thermal motion does not wash out the interference phenomena, since it displaces the molecules over considerable fractions of the lattice constant *a* and therefore in some cases over several wave-lengths. It should certainly affect the sharpness of the intensity maxima. That the observed maxima form only acute angles with the incident ray (except for diamond) is most likely due to the atomic factor ψ, possibly in conjunction with the thermal agitation.

The paper ends with some general conclusions concerning the bearing of the observations on the nature of X-rays, in particular whether they are composed of waves or corpuscles.

Both the sharpness of their intensity maxima and their great penetrating power are strong arguments for the wave nature of the diffracted rays; it would be hard to understand these properties on the assumption of corpuscular rays. One could, however, doubt the wave nature of the primary X-rays. But if these are assumed to be corpuscular (or photons), then, in the case of incidence along the four-fold symmetry axis of ZnS, only those atoms could scatter coherently which lie on a lattice row parallel to the symmetry axis, whereas the scattering of neighbouring rows would occur without any phase relation; that is, the scattering would be as of independent linear gratings. Thus there would be only one condition for maxima, expressed by the circles on the photographic plate. We would then expect the circles to show a uniform blackening instead of being reduced to a few localized maxima. Also the primary and the diffracted rays are so similar that the proven wave nature of the latter is most likely to hold also for the former. 'One difference, however, remains: the radiation which leaves the crystal certainly has a considerable spectral homogeneity, i.e. a certain periodicity. The primary radiation on the other hand, as much as it is 'Bremsstrahlung', has to be considered, following Sommerfeld's views, as consisting of quite unperiodic pulses; the experiments at least are compatible with this supposition. Let it remain undecided at present whether the periodic radiation is only formed in the crystal by fluorescence or whether it already exists in the primary ray besides the pulses and is only separated by the crystal. It is to be expected, however, that further experiments will soon elucidate this point.'

* * *

In writing the first paper, Laue clearly had before him the results of the second paper which will now be abstracted. It contains the first *'indexing'* of a diagram, i.e. the assignment of three integers (h_1 h_2 h_3) to each diffracted spot. They are defined by the conditions of obtaining maximum intensity, as given above, and the set of the three numbers is called the *order of the diffracted ray* (often written (hkl)), and may be interpreted as the Miller indices of a reflecting net-plane of the crystal (see under 5). The discussion which Laue uses for finding the order of each spot is basically correct and has served as example in many other cases, although easier methods were soon developed. But there is one wrong assumption made which confuses the results. In calculating the cell size of the cubic lattice, Laue assumes that each elementary cube of edge length a contains one molecule ZnS. If N is Avogadro's number, i.e. the known number of molecules in a mole ($0.6025 \cdot 10^{24}$), then the volume of a mole of the crystal is Na^3; on the other hand, if m is the molecular weight of ZnS ($=65.4+$ $+32 = 97.4$) then m gram is the mass of one mole, and, dividing this by the density δ of the crystal (4.06 g cm^{-3}) the mole volume is also m/δ. Comparing the two expressions, one finds for the lattice constant

$$a = (m/N\delta)^{1/3}.$$

For zincblende Laue obtains $a = 3.38 \cdot 10^{-8}$ cm, and a cubic cell of this size is used for the indexing, and for the conversion of the ratios λ/a, as determined from the photographs, into values of the wave-lengths λ.

Actually zincblende is not based on a simple cubic lattice with one molecule per elementary cube but on a face centered cubic lattice which has four molecules per cubic cell. Laue's value of a is therefore too small by a factor of $\sqrt[3]{4}$. This change of scale of the lattice constant can be countered by a change in λ with respect to all geometrical properties of the diagrams (which depend only on λ/a), and it does therefore not affect the assessment of orders to the spots; but all absolute values of wave-lengths are too small by the same factor $\sqrt[3]{4} = 1.5874$.

For the assignment of orders, Laue discusses the zincblende diagram of fourfold symmetry. The direction from which the incident ray comes, which is also that of the fourfold symmetry axis normal to the crystal plate, is taken as the z-direction of coordinates, and the x- and y-directions and crystal axes a_1 and a_2 are at right angles to one another and in the plane of the crystal plate. Thus the direction cosines of the incident ray (α_0, β_0, γ_0) are $(0, 0, -1)$, and the components of the axial vectors of the crystal are

$$a_1 = (a, 0, 0) \quad a_2 = (0, a, 0) \quad a_3 = (0, 0, a).$$

Equations (3) reduce to

$$\begin{aligned} a\alpha &= h_1\lambda & \qquad \alpha &= h_1\lambda/a \\ a\beta &= h_2\lambda & \text{or} \qquad \beta &= h_2\lambda/a \\ a\gamma &= h_3\lambda - a & \qquad \gamma &= h_3\lambda/a - 1 \end{aligned} \tag{4}$$

Since γ must be larger than -1, h_3 must be positive. Squaring and adding the last set and seeing that the sum of the squares of three direction cosines of a direction is 1. one obtains after cancellations

$$\lambda/a = \frac{2\,h_3}{h_1^2 + h_2^2 + h_3^2}; \tag{5}$$

that is, once the integers (h_1, h_2, h_3) have been attributed to a spot, the λ/a ratio contained in the spot is fixed.

Using this result in the previous equations, these give the direction cosines expressed by the indices of the spots:

$$\alpha = \frac{2h_1h_3}{\Sigma h^2}; \quad \beta = \frac{2h_2h_3}{\Sigma h^2}; \quad \gamma + 1 = \frac{2h_3^2}{\Sigma h^2}. \tag{6}$$

Now the ratio of the x- and y-coordinate of a spot on the plate is the same as $\alpha = \beta$ and therefore as $h_1 : h_2$; this can be taken from the plate. The distance r of a spot from the centre of the primary spot is given by

$$r = Z \tan \varphi,$$

where Z is the distance of the photographic plate from the crystal (3.56 cm) and φ the angle whose cosine is γ. Therefore γ can be obtained by measuring r/Z and thence from the last of the three equations above the value of Σh^2 is found, assuming a value for h_3. Inserting numerical values for the innermost of the strong spots of the diagram and trying $h_3 = +1$, Laue finds $\Sigma h^2 = 36$. This makes $h_1^2 + h_2^2 = 35$ and on squared paper it is easy to find two integers having very nearly this sum, namely $h_1 = 3$, $h_2 = 5$ or $h_1 = 5$, $h_2 = 3$. The assignment can be checked by noting that the ratios 3/5 and 5/3 correspond to the actual angular positions of the two symmetrically equivalent spots shown in one quadrant of the photograph. Assuming then the indices (3, 5, 1), the value $\lambda/a = 0.0564$ follows from (5). Could the ring on which this spot is situated also be interpreted as a second order ring? In that case the indices would be (6, 10, 2), and $\Sigma h^2 = 140$. There would be a ring very close to this, namely with $\Sigma h^2 = 141$ on which the spot (4, 11, 2) would be expected to be formed. Of this, and its companion (11, 4, 2), however, no trace can be found.

Laue next asks whether a second order ring of the observed wavelength can be seen. For it $\Sigma h^2 = 70$ from (5), and $h_1^2 + h_2^2 = 66$. The nearest to this is $h_1 = 2$, $h_2 = 8$ giving $\Sigma h^2 = 72$. This ring, of radius 1.84 cm on the plate, actually contains two spots in the quadrant under the expected azimuths. The spots (1, 8, 2) and (4, 7, 2), with $\Sigma h^2 = 69$ which is even closer to the above value of 70, are not to be seen.

A third order ring of radius 2.26 cm, formed by the same wavelength, should give $\Sigma h^2 = 105$. Actually a spot (7, 7, 3) giving $\Sigma h^2 = 107$ can be seen, whereas no trace of (4, 9, 3) with $\Sigma h^2 = 106$ was found. Again a fourth order ring shows (8, 8, 4), whereas the spots (11, 3, 4) and (7, 9, 4), which could be expected, are absent.

In this way the discussion continues, assessing indices and wavelengths to the strongest and most of the weaker spots, and pointing out that some further spots should be expected to appear because they would lie close to the observed rings, and their λ/a-values fall within the range of the observed ones, namely $\lambda/a = 0.0377$ to 0.143.

The fact that all of the observed spots can be accounted for by low order interferences ($h_1\, h_2\, h_3$), where none of the h exceeds 10, is considered by Laue a convincing argument for the correctness of the interpretation. He is aware of the fact that the ratios of the h_i are based on more direct arguments than their absolute values. But multiplication with common factors, which he tried out, seemed to give no improvement.

Laue is rather indefinite about the degree of accuracy with which the three conditions for maximum intensity have to be fulfilled. For this reason he is not too much perturbed by the absences of expected spots. In the 12 observed independent spots he finds 5 different wave-lengths, and these, he remarks, stand approximately in the ratio of 4 : 6 : 7 : 11 : 15. He turns down the possibility of making all wave-lengths integral multiples of a wave-length one quarter of the shortest one found, because then all indices would have to be multiplied by 4 and one would have to expect many more than the observed spots. Also this common wave-length would be $3.20 \cdot 10^{-10}$ cm and the rays should be much less absorbed than they actually are. The range of wave-lengths, $1.3 \cdot 10^{-9}$ to $5.2 \cdot 10^{-9}$ cm, is what would be expected from Sommerfeld's determination.

But how is it to be explained that these five sharply defined wave-lengths emerge from the crystal? In answering this question Laue as well as his co-workers missed the point. If they had only applied the assumption underlying Laue's theory, namely that each atom scatters the incident wave as it is reached by it, then they could not have failed to see that the sharply defined wave-length is identical with periodicity in the diffracted ray produced by the periodically repeated scattering of the incident wave on successive atoms. This periodicity-producing action of a grating had been pointed out in the optical case by Lord Rayleigh and was discussed at length in Sir Arthur Schuster's *Textbook on Optics*. Laue, who had spent much thought on 'coherence' of light most probably had come across this discussion.

But in the first papers, and for some time afterwards, Laue and Friedrich were so strongly impressed with the analogy between the monochromatic interference rays and the only other known 'homogeneous' and therefore very likely monochromatic X-rays, namely Barkla's characteristic rays, that they considered the wave-lengths obtained to be those of characteristic or fluorescence X-rays of the crystal. It will be remembered that the choice of crystal was already made with a view to production of characteristic radiation. The statement from the end of Laue's theoretical introduction of the first paper, which was quoted earlier, also clearly expresses the situation. The persistence of this misapprehension at the end of a period of the most strenuous and successful work is like a sign of exhaustion. It led, for several precious weeks, to speculations along a wrong line, and distracted from the further vigorous exploration of the points left unexplained by the remarkable quantitative discussion in the second paper.

It is of interest to note in this connection that when Laue's paper was republished in *Annalen der Physik* (1913, *41*, 998–1002) he appended three notes dated March 1913. In the first of these he develops the theory of diffraction in a crystal containing several atoms per cell. This leads to the first formulation of the Structure Factor (without introducing this name) and stresses its importance for the explanation of absent spots on the diagrams, especially in patterns produced by hemihedral crystals. The discussion is not carried through in detail, and in particular the systematic absences due to the centering of faces of the cell seem not to have been recognized.

In the second note Laue explicitly refutes the idea that the monochromatic nature of the diffracted rays could be explained by a selective effect of the crystal. He argues that *any* direction of diffraction can be approximated by three integers h_1, h_2, h_3, and, provided the incident X-ray has a sufficiently large spectral range, one would have to expect the photographic plate to be blackened everywhere. His conclusion is that the production of discrete wave-lengths in the diffracted rays is a property of the atoms and included in his ψ-factor, and not a property of the lattice.

In the last note, indices and wave-lengths are assigned to three further weak spots on the ZnS diagram of fourfold symmetry, and it is shown that the wave-lengths, like the previously determined ones, are relatively low multiples (up to 38) of a fundamental $\lambda = 0.032$ Å. It is announced that the discussion of the diagram with threefold symmetry leads to the same fundamental λ.

Finally Laue replaces the lattice constant of ZnS, $a = 3.38$ Å, which had been calculated on the assumption of one molecule per cell, by $\sqrt[3]{16}\, a = 8.53$ Å. According to the article by A. E. C. Tutton in *Nature* 1912, *90*, November issue, 'The Crystal Space Lattice Revealed by Röntgen Rays', structure theory makes it very likely that zincblende crystallizes with 16 molecules per cubic cell. This illustrates once again the confusing statements to which physicists were exposed in listening to the acknowledged authorities on crystals. It also serves to make clear what relief was brought about by W. L. Bragg's intrepid and direct approach.

CHAPTER 5

The Immediate Sequels to Laue's Discovery

5.1. W. H. Bragg and his Studies of Ionization by Gamma and X-rays

In 1912 William Henry Bragg was Cavendish Professor of Physics at the University of Leeds. Born in 1862 in Wigton (Cumberland) he was Laue's senior by seventeen years. His career was unusual in that he began research only after his fortieth year, although his great mental ability appeared from his early childhood on. After studying mathematics in Cambridge, and finishing there in 1884 with high honours, he was appointed, at the age of 22, as Professor of Mathematics and Physics at the young University of Adelaide, then in its tenth year. Here, Bragg's activity and interest was directed to physics, and to the perfection of his teaching and lecturing, in which he became one of the great artists. It was here also that he set up, soon after Röntgen's discovery became known, the first X-ray tube in Adelaide. Seventeen years passed after Bragg had become the head of the physics laboratory before the spark of original research reached him,—but from there on a mighty and steady flow of scientific results emanated from him until shortly before his death. The occasion which brought this change about was the need of reviewing, for a presidential address to the Australian Association for the Advancement of Science, the recent advances in radioactivity. He was struck with the possibility that a decision between the hypotheses of J. J. Thomson and of Ph. Lenard on the constitution of the atom might be obtainable from measurements of the absorption of α-rays in matter. In his paper of 1904 with R. Kleeman 'On the Ionization Curves of Radium' he showed that the exponential law which had been tacitly assumed for the decrease in intensity of an α-ray passing through matter was far from correct, and the characteristics of the range of α-particles were established.

The twelve papers Bragg published in the next four years led to the offer of the Cavendish Chair in Leeds which he accepted in 1908.

Here he extended the study of ionization in gases from α-rays to X-rays, using in the latter case the characteristic radiations, which, because of their homogeneous absorption, were considered to lead to more fundamental observations. He strongly favoured a corpuscular interpretation of X-rays for the reason explained in Chapter 2, and summarized his results on α- and X-rays in a book *Studies in Radioactivity* which appeared in 1912.

It is characteristic for W. H. Bragg's unbiased way of thinking, as well as for the impact of Laue's experiment, that it took Bragg only a very short transition period for accepting the pure wave theory of X-rays in explaining the diffraction experiments. Only in his first letter to *Nature*, dated 18 October 1912, does he make an implicit attempt to save the corpuscular idea by proposing an alternate explanation to Laue's for the zincblende diagram of fourfold symmetry, namely that 'all the directions of the secondary pencils in this position of the crystal are "avenues" between the crystal atoms' ('assumed to be arranged in a rectangular fashion') (cf. the discussion of this letter in the next section.). The same idea was expressed at the same time by another famous physicist whose previous work had also stressed the corpuscular aspect of radiation, Johannes Stark. His paper in *Physikalische Zeitschrift* 1912 (*13*, 973) assumes propagation of the radiation along 'Kristallschächte'—tunnels or pit shafts formed by the regularity of the atomic arrangement.

It is not quite certain how the news of Laue's discovery reached England, and, in particular, W. H. Bragg. Laue himself thought that it was through the off-prints of the Academy papers which he sent out very soon to all those whom he considered to be immediately interested. But these reprints may not have been immediately available. The daily press played no part. The *London Times*, in those years, carried a section 'Science, Arts, Music and Drama', but the first of these items hardly ever was considered a suitable topic for the *Times*' readers; the space of this column was filled with reviews of concerts, plays, exhibitions and auctions of art objects and silverware. No mention of Laue's discovery is to be found.

In July of 1912 the Royal Society celebrated in London its 250th anniversary. Among the representatives from Germany was Woldemar Voigt from Göttingen.* He must have known of Laue's work through Madelung, and through the talk Sommerfeld gave to the Physical

* P. v. Groth took part in the celebrations as representative of the Bavarian Academy, and it seems most unlikely that he should have abstained from publicising the important papers presented at the recent meetings of his academy.

Society of Göttingen not long after 8 June. It would only be natural that Voigt spoke of it to his British colleagues, especially J. J. Thomson. W. H. Bragg, who had been elected Fellow of the Royal Society in 1907, might also have been present at the celebrations. Full and detailed information on the work seems, however, to have come to Bragg through a lecture which J. J. Thomson gave to the physics group in Leeds (or Manchester) somewhat later. By that time reproductions of Laue's diagrams were available, and W. H. Bragg interested his son in them and together they undertook an independent discussion.

5.2. W. L. Bragg and the Origin of Crystal Structure Analysis; X-ray Spectroscopy

W. H. Bragg's son, William Lawrence Bragg, was following in his father's footsteps by taking physics as his main subject. Born in 1890 in Adelaide, he went to school there in 1900–05. He studied Mathematics (major) and Physics (minor) at the University of Adelaide and obtained his Bachelor's degree in Mathematics at the age of 18. On the return of the family to England, he entered Trinity College in Cambridge and received his final training in Physics at the Cavendish Laboratory under the 'grand old man' of physics, Sir J. J. Thomson, and the famous members of his staff, including C. T. R. Wilson, F. W. Aston and others. In 1911 he obtained his first appointment, to a lectureship in Trinity College.

But let us hear in W. L. Bragg's own words what the exciting sequence of events was after Laue's paper had reached W. H. Bragg in form of an offprint. He tells the story in an address given in 1942 in Cambridge at the First Conference on X-ray Analysis in Industry (held under the auspices of the Institute of Physics), which was published in 1943 in the series *Science in Britain*.

'At that time father held the view that X-rays had the properties of material particles rather than those of electromagnetic waves like light. He was led to this view by his experiments on the knocking of electrons out of atoms by X-rays (ionization), which he had shown to be a hit or miss affair affecting only a very small proportion of the atoms, not a general effect on all atoms as one would expect if X-rays were waves. I was a young student at Cambridge at the time, and of course an ardent supporter of my father's views. During the summer of 1912 we had discussions on the possibility of explaining Laue's patterns by some other assumption than that of diffraction of waves, and I actually made some unsuccessful experiments to see if I could get evidence of 'X-ray corpuscles' shooting down the avenues between the rows of atoms in the crystal. On returning to Cambridge to ponder over Laue's paper, however, I became convinced of the correctness of his deduction that the effect was one of wave-diffraction—but also convinced that his analysis of the way it took place was not correct. It is small clues that often lead to a solution, and perhaps I may be forgiven for repeating a figure (Fig. 5–1(1)) from my paper in the *Proceedings of the Cambridge Philosophical Society* (November 1912) which shows the clue I followed.

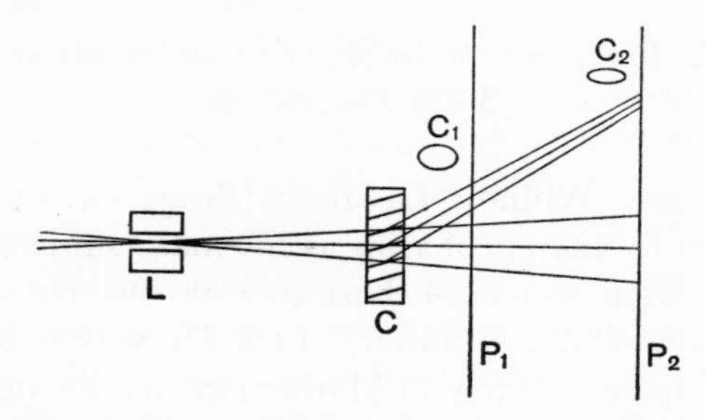

Fig. 5–1(1). Origin of the shape of the spots on a Laue-diagram.

'When the plate was placed at P_1 near the crystal the spots were almost circular like C_1, but when placed farther back at P_2 they became very elliptical (C_2). Now Laue had ascribed his pattern to the diffraction of certain specific wave-lengths in the X-ray beam by the regular pattern of the crystal. Given a fixed wave-length, optical theory tells us that the diffraction must take place at a definite angle, and this means that the diffracted rays drawn in the picture should all have been parallel. I had heard J. J. Thomson lecture about Stokes' theory of the X-rays as very short *pulses* of electromagnetic radiation. I worked out that such pulses of no definite wave-length should not be diffracted only in certain directions, but should be *reflected* at any angles of incidence by the sheets of atoms in the crystal as if these sheets were mirrors. A glance at the geometry of Fig. 5–1(1), in which the rays are drawn as if reflected, shows that they close together again vertically while continuing to spread horizontally, thus explaining why the spots get more elliptical as the plate is placed farther away. It remained to explain why certain of these atomic mirrors in the zinc-blende crystal reflected more powerfully than others, a difficulty which had led Laue to postulate a group of definite wave-lengths. Pope and Barlow had a theory that the atoms in simple cubic compounds like

ZnS were packed together, not like balls at the corners of a stack of cubes, but in what is called cubic close-packing, where the balls are also at the centre of the cube faces. I tried whether this would explain the anomaly—and it did! It was clear that the arrangement of atoms in zincblende was of the face-centered type. I was careful to call my paper on the structure of zincblende 'The Diffraction of Short Electromagnetic Waves by a Crystal', because I was still unwilling to relinquish my father's view that the X-rays were particles; I thought they might possibly be particles accompanied by waves.

'Pope, who was Professor of Chemistry at Cambridge, was very pleased at this support of his theories, and at his suggestion I tried crystals of NaCl, KCl, KBr and KI. The Laue pictures which they gave were simpler than those of zincblende, and led to a complete solution of their structure. These were the first crystals to be analysed by X-rays (*Royal Society Proceedings*, June 1913).

'At about this time C. T. R. Wilson suggested to me that I might try the direct experiment of reflecting X-rays from a cleavage face, because such a face must be parallel to dense sheets of atoms in the crystal. I tried the experiment with mica, and I well remember J. J.'s excitement when I showed him the still wet photographic plate with a mirror reflection of X-rays on it (*Nature*, December 1912). My father thereupon examined a reflected beam, measuring its ionization and absorption, and proved conclusively that the diffracted waves had in fact all the properties of X-rays. As he put it, 'The problem then becomes, it seems to me, not to decide between two theories of X-rays, but to find... one theory which possesses the capabilities of both', a point of view with which quantum theory has now made us familiar, but which seemed very paradoxical at the time.

'In order to examine the reflected X-ray beam more thoroughly, my father built the X-ray spectrometer. In this instrument, a crystal face can be set so as to reflect the X-rays at any angle (it is actually the sheets of atoms parallel to the face which reflect), and the strength of the reflected beam is measured by an ionization chamber. With this instrument he made the next great discovery. In addition to the 'white' X-radiation of all wave-lengths which I had called the X-ray pulses, he found that each metal used in the X-ray tube as source of radiation gave a characteristic X-ray spectrum of definite wave-lengths, just as elements give spectra in the optical region (*Royal Society Proceedings*, April 1913).

'The X-ray spectrometer opened up a new world. It proved to be a far more powerful method of analysing crystal structure than the Laue photographs which I had used. One could examine the various faces of a crystal in succession, and by noting the angles at which and the intensity with which they reflected the X-rays, one could deduce the way in which the atoms were arranged in sheets parallel to these faces. The intersections of these sheets pinned down the positions of the atoms in space. On the other hand, a suitable crystal face could be used to determine the wave-lengths of the characteristic X-rays coming from different elements as sources. A 'pure' beam of monochromatic X-rays could be selected by reflection from a crystal and its absorption in various substances measured. It was like discovering an alluvial gold field with nuggets lying around waiting to be picked up. At this stage my father and I joined forces and we worked furiously all through the summer of 1913, using the X-ray spectrometer. Although the description of this instrument was published in our joint names, I had no share in its design. The capital I brought to the family firm was my conception of reflection and the application in general of the optical principles of diffraction, and my success in analysing the first crystals by the Laue method. It was a glorious time, when we worked far into every night with new worlds unfolding before us in the silent laboratory. My father was at first far more interested in X-rays than in crystals, and left the determination of crystal structure to me, with the exception of a paper on diamond which showed the power of the instrument he had devised. He measured the wave-lengths of the X-ray spectra given by the elements platinum, osmium, iridium, palladium, rhodium, copper and nickel. He identified them with Barkla's K and L radiations. He calculated their energy quanta according to Planck's relationship, and showed that this agreed with the energy of the cathode rays required to excite them. He showed that the shortest wave-lengths ($K\alpha$ and $K\beta$) from various elements were similar, and that they were approximately inversely proportional to the square of the atomic weight. This was in fact a first hint of the subsequent brilliant generalization of this principle by Mosely, who used it to determine the atomic numbers. He measured *absorption edges*, the critical wave-lengths at which a sharp step in the absorption of X-rays by an element takes place. In fact, he laid the firm foundations of X-ray spectroscopy, afterwards to be so brilliantly developed by Mosely and Siegbahn. I remained true to my first love, the determination of crystal structure. By using measurements on crystals made with the X-ray spectrometer, many of them due to my father, I was able to solve the structures of fluor spar, cuprite, zincblende, iron pyrites, sodium nitrate and the calcite group of minerals. I had already solved

KCl and NaCl, and my father had analysed diamond. Between them, these crystals illustrated most of the fundamental principles of the X-ray analysis of atomic patterns. These results were produced in a year of concentrated work, for the war in 1914 put an end to research. I have gone into these early experiments in some detail because it is a story which I alone can tell, and which I wish to put on record.'

* * *

Let us add some detail to this account by going over the letters to *Nature*. The first, in *Nature* of 24 October, is from W. H. Bragg and dated 18 October. In it an explanation of the zincblende photograph of fourfold symmetry is attempted according to the following rule: 'The atoms are assumed to be arranged in a rectangular (= simple cubic?) fashion; any direction which joins an atom to a neighbour at a distance na from it (where a is the distance from the atom to the nearest neighbour and n a whole number) is a direction which a deflected (or secondary) pencil will take, and it will, in doing so, form one of the spots. In other words, we have to seek for all the cases in which the sum of three squares is also a square, and we then recover the positions of all the spots on the diagram. For example, secondary spots take the directions (2,3,6), (4,1,8) and so on.'

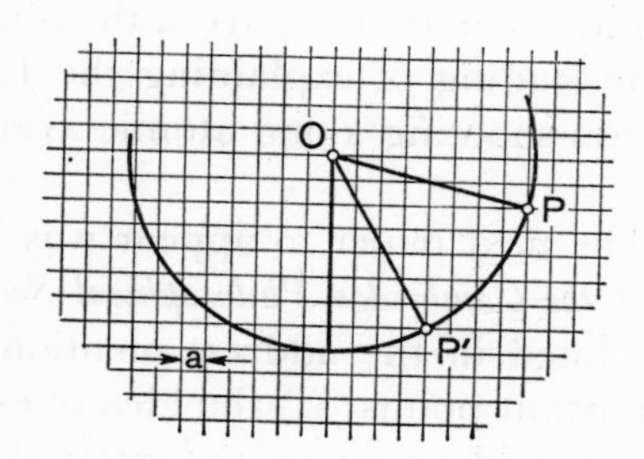

Fig. 5–2(2). Case of diffracted rays travelling along atomic avenues.

It is then pointed out that (5,7,11) is present, although the sum of the squares, 195, is one short of a perfect square, and that (2,5,14) which should be there, is absent.

The letter concludes: 'The rule suggested itself to me as a consequence of an attempt to combine Dr. Laue's theory with a fact which my son pointed out to me, viz. that all the directions of the secondary pencils in this position of the crystal are 'avenues' between the crystal atoms.'

The construction proposed by W. H. Bragg is the analogue of the two-dimensional construction (Fig. 5–2(2)). Let a be the distance of nearest neighbours and na the radius of the circle drawn. If this circle passes through an atom P with integer components (n_1a, n_2a), then n^2a^2 =

$= n_1^2a^2 + n_2^2a^2$, or $n^2 = n_1^2 + n_2^2$; adding the third dimension, the sphere of radius na passes through an atom P with components $(n_1, n_2, n_3)a$ whenever the equation $n^2 = n_1^2 + n_2^2 + n_3^2$ can be solved with integers. The direction OP thus determined is, according to the proposed rule, that of a possible secondary ray. It can be called an avenue between the atoms in as much as it is a rational direction in the lattice and the nearest parallel lines passing through atoms will be at a finite distance. (In the drawing it would be the parallel lines passing through atoms (4,1) or (4,7).) This minimum distance is characteristic for rational directions; irrational directions through one lattice point may never again meet a lattice point (otherwise they would be rational, even if expressed by large integers) and they lie densely side by side.

Now it is true that in the case of X-rays incident on a cubic crystal along the cube edge, the directions of all diffracted rays are rational— but the construction proposed does not prove it. It was purely empirical and covered only part of the diffraction effect. Evidently the hope was misleading of explaining the Laue diagram in terms of corpuscular rays shooting along atomic avenues without collisions.

The next paper to appear was W. L. Bragg's paper in the *Proceedings of the Cambridge Philosophical Society*, November 1912, from which he quoted in the address mentioned above. It contains three very important points: (i) the idea of explaining the Laue spots as *reflections* of the incident ray on the internal atomic net-planes; (ii) the assumption of a *continuous spectrum* of the incident ray and the *selective action* of the sets of reflecting planes in reinforcing only those wave-lengths which fit into their distances of repeat; and (iii) the proof that the lattice for zincblende is not the simple cubic one of Bravais, but his *face-centered cubic lattice*.

The concept of internal reflection does not contradict Laue's concept of diffraction, of which Bragg says he became convinced; rather it is only a different form of expressing the same results, and, as the successful application proved, a form simpler to visualize. It is in this paper that the Bragg formula $n\lambda = 2d \sin \theta$ occurs for the first time, though in the now less familiar form $n\lambda = 2d \cos \Theta$, where $\Theta = 90° - \theta$ is the angle of incidence against the normal to the plane. It is stated that Θ has the same value for λ, $\lambda/2$, $\lambda/3$ etc.

Points (ii) and (iii) are closely connected. (ii) did away with the search for explaining the sharpness of the diffracted spots by assuming the generation of monochromatic radiation by crystal fluorescence. Instead, the crystal acts as a pass filter of great selectivity for the incident radiation. This name was not in use in 1912, but the selective action of regularly spaced reflecting elements was well known in optics from such devices as Lummer-Gehrcke plates or echelon interferometers.—By trying the assumption of a face-centered instead of the simple cubic lattice, Bragg was able to account for the absences of spots which both his father and Laue had noted without being unduly worried. The face-centering atoms form their own system of planes which need not coincide with the parallel net planes supported by the corner atoms, and, if not coincident, they halve the distances of repeat. This may lead to destructive interference and the absence of a spot. Bragg gave the proof of the assumption of face-centering by showing that all the spots, and no more, were present which the crystal could pick out of a continuous spectrum, given certain angular and other restrictions.

Although this early paper does not yet contain a full structure determination, it comes very close to one, in the case of such a simple compound as ZnS.

W. L. Bragg presented his paper at the meeting of the Cambridge Philosophical Society on 11 November 1912. An abstract is given as a report on the meeting in *Nature* of 5 December.

In the discussion following the meeting C. T. R. Wilson suggested that if the internal planes reflected, external crystal planes might reflect as well, provided their roughness was small enough. This condition seemed likely to be fulfilled for crystals with good cleavage planes and W. L. Bragg reports successful experiments in a letter to *Nature* of 12 December 1912 (dated 8 December). He mounted a strip of mica of about 1 mm thickness on thin aluminium foil and exposed it to an X-ray beam at a grazing incidence of about 10° to the surface. After exposure of only a few minutes the trace of the reflected beam showed up on the photographic plate. That it was a specular reflection was shown not only by varying the angle of incidence, but also by bending the crystal to an arc and obtaining a fine focus.

In this experiment, the crystal was used as a reflexion grating—as had been tried without success in Friedrich and Knipping's first attempt. The reason why they failed was that they used practically normal incidence. Bragg notes that the reflection increases in strength as the glancing angle is diminished.

Bragg also investigated whether this was a surface or a body effect. A mica slip 1/10 mm thick proved as good a reflector as a thicker one, yet, he writes, the effect is almost certainly not a surface effect, only the critical thickness has not yet been reached.

The idea of reflection of X-rays on the atomic net-planes of the crystal appealed to all English physicists by its simplicity, and the explanation given by W. L. Bragg of the occurrence or non-occurrence of diffracted spots on the Laue diagrams according to the selective action of the crystal was considered more reasonable than Laue's ad hoc assumption of characteristic wave-lengths of the crystal. Finally the introduction of the face-centered cubic lattice as underlying the structure of zincblende was the first step beyond the merely formal application of the lattice idea in Laue's theory, and therefore the first step in crystal structure analysis. The importance of this paper by W. L. Bragg for the further development of the field to which Laue had found the access can hardly be over-estimated.

W. L. Bragg's observation of specular reflection on the surface of the crystal opened up a period of closest collaboration between father and son which is perhaps unique in the history of Science, both for its lasting intensity and the importance of the resulting discoveries. It was a partnership of two outstanding scientists belonging to different generations, with an unrestricted give and take of ideas on both sides and a fundamental respect for the other's achievements.

* * *

The first to be stimulated was the father, W. H. Bragg, who made certain that the reflected rays could be detected in an ionization chamber as well as photographically (letter to *Nature 91*, 23 January 1913). This experiment must have been the conclusive preparation for the construction of the X-ray spectrometer, the instrument which for decades to come was the main tool for crystal structure analysis throughout the British school and in many other laboratories. This was described in the joint paper 'The Reflection of X-rays by Crystals, I', *Proc. Roy. Soc.* A 1913, *88*, 428–438. In this instrument the crystal is carefully adjusted on a goniometer head which can turn about a vertical axis; the horizontal X-ray beam falls through a slit system on to a vertical crystal face and is there reflected into an ionization chamber which can be swung round the same axis about which the crystal turns. Two types of measurement can be made: in the first the crystal face is set at a definite angle θ to the incident beam, and the angular position of the entrance slit of the ionization chamber is varied about the position 2θ; if it is moved to angle $2\theta + \varphi$, the ionization current is plotted as a function of φ, thus producing a 'reflectivity' curve. This curve is strongly influenced by the geometry of the slit system, but in any case it reaches large values only if θ is one of the 'Bragg angles' which are determined by the relation now written

$$n\lambda = 2d \sin \theta$$

(n = integer called the order of reflection; λ wave-length; d the identity period of the set of reflecting net-planes of atoms and θ the 'glancing angle').

The other type of measurement is performed by opening wide the slit of the ionization chamber, so that at each position θ of the crystal the integral value of the reflectivity curve is measured by the ionization current, provided the chamber is always set at the angle 2θ. This gives a much quicker way of obtaining values for the overall strength of reflection. An important focussing property of the Bragg reflection was clearly recognized which was later much used in the construction of instruments; it has been fundamental ever since as a means for increasing the intensity and improving the resolution.

The spectrometer revealed for the first time the existence of very nearly monochromatic components in the X-ray emission from platinum targets, if the tube was run in a very 'soft' condition. By measuring the absorption coefficients of these monochromatic radiations Bragg found them to have values close to those known from other measurements for the L-series of platinum. While this gave the correct identification of these spectral peaks, the value of the wave-length, obtained by applying the Bragg Law, came out wrong because of a mis-interpretation of the meaning of a face-centered cubic lattice: in each of the four equivalent sites of the cubic cell of NaCl an atom of the average atomic weight $\frac{1}{2}(35.5 + 23)$ was placed instead of the whole molecular weight.

W. H. Bragg soon continued this investigation in a paper signed only by himself (*Proc. Roy. Soc.* A 1913, *89*, 246), in which the correct structure of rocksalt is being used, that had meanwhile been determined by W. L. Bragg. Neglecting the splitting of the peak into two very close lines, Bragg now found the wave-length of the Pt L-radiation to be 1.10 Å, and those of nickel and tungsten to be 1.66 and 1.25 Å, respectively. These values received confirmation by the following tests: the energy $h\nu$ associated with wave-length 1.10 Å is $1.78 \cdot 10^{-8}$erg, and this should be the minimum energy of the cathode-ray electrons capable of producing these X-rays. Now measurement of the absorption coefficient of the rays forming the peak gave 23.7cm^{-1} in aluminium, and according to Barkla this absorption is found for the characteristic K-radiation emitted by a target of atomic weight 74, or the L-radiation from an atom of weight 198. The atomic weight of platinum, 195, checks with the latter interpretation. Again, the energy required to produce radiation of this absorption quality is, according to results of Whiddington for the K-series, $2.14 \cdot 10^{-8}$erg. This also stands in fair agreement with the minimum energy as obtained above from the wave-length.

It is interesting to note what a round-about procedure was necessary at that time for obtaining an independent checking of the first wave-length determination. With its success, the determination of the first crystal structure, that of NaCl by W. L. Bragg, also received an independent confirmation.

W. H. Bragg continued the exploration of X-ray spectra in several more papers, studying the L-spectra of the elements from which good targets could be made, like Os, Ir, Pd, Rh, Ni, and Cu; he investigated how the intensity distribution among the three peaks of the L-reflections depend on the composition of filters, or of the reflecting crystals themselves. In this paper (*Proc. Roy. Soc.* A 1914, *89*, 430) he also discusses the relation between the scattering and absorbing powers of an atom. Whereas absorption changes abruptly at certain wave-lengths, scattering does not. Simple measurements show that scattering coefficients of different atoms are roughly proportional to the atomic weights. This is illustrated by comparing the 222 reflections in zincblende and diamond, and, in a most elegant way, by the vanishing of the 222 reflection in fluorite, CaF_2, where the combined weight of the two fluorine atoms (2×19.0) just counteracts that of calcium (40.1).—These structures had meanwhile been determined by W. L. Bragg.

* * *

Independently of the Braggs, two young physicists in Rutherford's Manchester laboratory had meanwhile begun work on the lines of W. L. Bragg's idea of reflection, H. G. J. Moseley and C. G. Darwin. Their main interest lay in the direction of making X-ray spectra useful for the infant atomic theory that was then being weaned in Manchester by Niels Bohr. The most suggestive, if not convincing, argument for the ideas which Bohr was applying to Rutherford's model of the atom was his success in interpreting the simplest and most fundamental of all optical spectra, that of Hydrogen. Moseley and Darwin sensed correctly that further support for Bohr's novel and rather shocking application of principles derived from Planck's concept of energy quanta might be gained if the X-ray spectra were known. This soon proved to be true to an unforeseen extent, so that their work became one of the experimental pillars around which the quantum theory of the atom grew. An account of their work is given in C. G. Darwin's Personal Reminiscences in Part VII of this book.

* * *

The great break-through to actual crystal structure determination and to the absolute measurement of X-ray wave-lengths occurred in W. L. Bragg's paper 'The Structure of some Crystals as indicated by their Diffraction of X-rays' in *Proc. Roy. Soc.* A 1913, *89*, 248. The argument in the first part of the paper is based entirely on Laue diagrams taken by the author on KCl, NaCl, KBr, ZnS (blende), CaF_2(fluorite), and $CaCO_3$(calcite). First it is shown that any secondary ray, which in Laue's theory is described by three path differences $(h_1, h_2, h_3)\lambda$ between the wavelets from neighbouring scatterers, can also be regarded as split off from the primary beam by specular reflection on a set of atomic planes of Miller indices (h_1, h_2, h_3). Next, a convenient graphical method is developed for the attribution of indices to the spots on symmetrical Laue diagrams by making use of the evident arrangement of such spots on nearly circular ellipses. The general theory of the Laue diagram is further discussed on the assumption that the primary ray is a 'pulse', that is, lacks any particular periodicity. This approach differs from Laue's in that the latter considered the diffraction of only one Fourier component of the pulse, so that actually the crystal is acted upon by a monochromatic incident wave rather than a pulse. At a time when the reasons for the occurrence of only certain definite wave-lengths in the diffracted rays were not yet fully clarified, it was essential to carry through the theory without the assumption of monochromasy of the incident X-ray.

The analysis of the KCl diagram of fourfold symmetry is taken first. It is shown to be a 'complete' diagram, that is, that all the spots are present which can be expected to occur by diffraction in a simple cubic lattice by variation of the two first indices in (h, k, 1) within certain limits. This completeness proves that the scattering centres in KCl are arranged according to a simple cubic lattice.

The same completeness is not found for the diagrams obtained with KBr, KJ, CaF_2, and ZnS. If, however, for ZnS only odd values of h and k are admitted, the 'index field' showing all combinations of these odd values is completely filled within a certain region. This is correctly explained by assuming the scattering centres to be arranged according to a face-centered cubic lattice, and this result is corroborated by a discussion of the trigonal Laue diagram of zincblende in the Laue-Friedrich-Knipping paper. To that effect a rhombohedral axial system is introduced, consisting of the half face-diagonals of the cube, which include 60° angles with one another. It is remarked that a diagram of KCl, taken along the cube body diagonal can be referred to the same pattern of intersecting circles which yields the ZnS diagram, except for a change of scale of the filled-in region of the index field. This conclusion is found to be correct. If, on the other hand, a face-centered lattice is taken for the discussion of the KCl picture, and a simple cubic lattice for that of ZnS, no 'complete' regions in the index fields can be found; this disproves the suitability of this lattice choice.

The Laue patterns obtained with NaCl are neither of the KCl type nor of the type common to KBr, KJ, or ZnS. This is interpreted by assuming the scattering power of an atom to be proportional to its weight. Then K and Cl, of weights 39 and 35.5, are so similar that the crystal practically contains only one kind of scatterer; for NaCl (23 and 35.5) the difference can not be neglected, so that the crystal contains two kinds of scattering centres; whereas KBr and KJ again only have one kind of scatterer, because compared to the weight of the halogen (80 and 127, respectively) that of K (39) can be neglected.

This argument then leads to the correct structures of the four alkali halides. But Bragg declares that in spite of the apparent rigour of the argument the question remains unsettled whether the individual scattering centres are actually atoms or more complex units. The last third of the paper is devoted to the discussion of this point. It begins with a comparison of the trigonal Laue diagrams of fluorite, calcite and rock salt. The first two show great similarity between them and with the zincblende diagram, the last one differs. This leads to the conclusion that the scattering centres in ZnS, CaF_2 and $CaCO_3$ are arranged according to the same lattice, except that in calcite the angle between the axes is somewhat greater. Now it is characteristic of a lattice that every lattice point is surrounded by others in the same way. This could not be the case for the scatterers in the three crystals, except if scattering is concentrated in the only heavy atom which each of the crystals possesses. The conclusion is therefore that the active centres of scattering are the heavy atoms themselves. Scattering by NaCl is due to both atoms, and this makes its trigonal diagram different from those of the other crystals.

A check on this conclusion can be obtained by spectrometer measurements of the angles under which the various crystals reflect the same X-radiation. This gives the ration d/λ, where d is the spacing of the reflecting net planes, and λ the unknown wave-length. Geometry then gives V/λ^3, where V is the volume of the cell. Using the known density ρ and molecular weight M, the expression $V\rho/(\lambda^3 M)$ is obtained which is proportional to the number of molecules in the cell. This is found to be very nearly the same for NaCl, ZnS, CaF_2, $CaCO_3$, and only half this value for KCl. Thus the conclusion is reached that KCl carries two equivalent centres of scattering, in contrast to the other molecules which contribute only one, the heavy atom.

Even so, Bragg continues, the proof of single atom scattering is not complete. It is not excluded that groups of four molecules each are associated with each lattice point or scattering centre. This would even be in keeping with the views expressed by Barlow and Pope on the dense packing of binary compounds, but it would be very difficult to visualize such arrangement in the cases of CaF_2 and $CaCO_3$. Thus, while no strict proof can be given of the correctness of the simple structures and of atomic scattering centres, the odds are for it.

A further argument can be obtained from the spectrometer measurements by comparing the total intensities reflected in various orders by the different crystals. It is seen that the strength of the reflection is influenced by the presence of the lighter atoms in a way depending on whether these fall on the same atomic net planes as the heavy atoms, or midway between them ('halving of distances'). This explains the differences of the spectrograms of KCl and NaCl; and in CaF_2, where the F-planes sandwich the Ca-planes at $\pm \frac{1}{4}$ the spacing, it suppresses some of the reflections.

Finally, W. L. Bragg derives from the knowledge of the NaCl structure the absolute value of the wave-length as $\lambda = 1.10 \cdot 10^{-8}$ cm; with the observed d/λ ratio this corresponds to a distance Na-Cl of $a = 2.783$ Å (instead of the better value 2.814 Å).

In the series of fundamental papers published by both Braggs in 1913 and 1914 this paper by W. L. Bragg unquestionably brings the greatest single advance. With its well documented, if not rigorously established answer to the challenge of the first structure determinations, it made all future structure determinations very much easier by providing an absolute wave-length scale. Henceforth it was possible to find the number of molecules per cell directly from the Bragg angles measured on the spectrometer. Together with the observed subdivisions of the repeat distances of certain sets of reflecting planes, an idea also introduced in this paper, the number of atoms or molecules per cell fixed the positions of the scattering centres in the simple crystals that were first investigated. It would, however, be an invidious undertaking to single out any one of the early papers as the most important one, so closely were they all interlinked and so rapid was the progress at the time of their writing which formed a background for their formulation.

* * *

W. L. Bragg's just analyzed paper was directly followed in *Proc. Roy. Soc.* A 1913 (*89*, 277) by a joint paper with his father 'The Structure of the Diamond'. According to a remark in Sir Lawrence Bragg's Personal Reminiscences in Part VII of this book, this paper was mainly his father's work. But it employed all the arguments developed in the preceding paper, and, if only for this reason, the joint authorship seems justified. The reflection by the octahedral planes of diamond shows the second order reflection to be absent, while first, third and fifth orders are observed. This absence is interpreted as meaning that between the simple series of 111-planes of a single lattice a second set of equally dense planes is interleaved with a shift of one quarter of the repeat distance. It is next shown from the values of the diffraction angles that the cubic cell contains eight carbon atoms, distributed over two interpenetrating face-centered lattices. The only way to comply with these demands is to let the two lattices have a displacement of one quarter the body diagonal of the cube between them. This gives each atom a position in which it is surrounded by four nearest neighbours lying at the corners of a regular tetrahedron. A Laue diagram of threefold symmetry confirms this structure by showing the absence of other reflections from planes with spacings similar to those of the 111-planes.

Diamond was the first example of a structure in which the effective scattering centres did not coincide with the points of a simple (Bravais type) lattice. The determination of this structure was acclaimed as a great triumph of the new methods. Whereas in the structures of rocksalt, zincblende and fluorite the absence of molecules in the accepted sense created an element of bewilderment, the beautiful confirmation of the tetravalency of carbon on purely optical principles made this structure and the method by which it was obtained immediately acceptable to physicists and chemists alike.

In further papers published before the work was interrupted by the outbreak of war, August 1914, the laws of absorption and the influence of absorption on the intensities of the reflections were investigated by W. H. Bragg, and the structures of CaF_2 (fluorite), FeS_2 (pyrites) and $CaCO_3$(calcite) determined by W. L. Bragg. This latter paper with the title 'The Analysis of Crystals with the X-ray Spectrometer' (*Proc. Roy. Soc.* A 1914, *89*, 468) shows remarkable progress in a number of ways.

(i) It is clearly recognized that for a complete structure analysis the intensities of the reflections have to be known and evaluated.

(ii) In view of the great difficulty of obtaining a meaningful theory of intensities, a practical 'normal' intensity ratio of 100 : 20 : 7 : 3 : 1 for the successive orders of reflection on a simple set of atomic planes is abstracted from experience. Deviations from these ratios show that the set of reflecting planes is a composite of several interleaving simple sets, possibly formed by different kinds of atoms.

(iii) The positions of the S atoms in FeS_2, or of the O atoms in $CaCO_3$, are not fixed at certain points of the cell by the number of molecules contained in the cell. They remain displaceable on certain lines of intersection of symmetry elements. In FeS_2, for instance, the S atoms have to lie on the body diagonal of a cube formed by Fe atoms, and symmetrically to the cube centre; but their distance apart remains undetermined by symmetry and has to be found from a discussion of the intensities of reflection. The ratio of the S-S distance to the length of the cube diagonal is a 'parameter' of the structure. Pyrites and calcite were the first structures containing parameters, and only one each. For many years structure determinations remained restricted to cases where not more than two or three parameters occurred—tackling other structures was a hopeless endeavour.

(iv) Pyrites was for another reason a markstone in the development of crystal analysis. It was the first cubic crystal in which the axes of threefold symmetry in the four directions of the body diagonals of the cube, which are a characteristic feature of all cubic crystals, do not all intersect at the centre of each cubic cell; they are offset so as to pass one another skew without ever intersecting. The writer remembers receiving in Germany, long after the outbreak of war in 1914, a postcard from England from W. L. Bragg—the censor had seen to it that it could not contain any recent information—saying: 'I have nearly finished finding the correct structure of pyrites, but it is *terribly complicated*.'

5.3. C. G. Darwin's Work

While the series of discoveries contained in the early papers by W. H. and W. L. Bragg was by far the most important contribution to the field opened up by Laue's experiment, there were other physicists, of course, who took up the matter independently. The spectroscopic work of Moseley and Darwin has already been mentioned. Darwin alone published in 1914 two very fundamental papers 'X-ray Reflection' (*Philos. Mag.* 1914, *27*, 315 and 675) which are entirely based on Bragg's reflection idea. Here, for the first time, is the successive reflection of an incident ray by a set of equidistant atomic planes treated with due consideration of the back- and forth reflections between the planes. This is, as will be discussed in Ch. 15, the beginning of a 'dynamical' theory of X-ray diffraction because the repeated scattering is taken into account, and not only the scattering of the incident beam, as in Laue's and Bragg's theories. These latter theories are correct only if a crystal is so minute that the building-up of large scattered amplitudes cannot occur. For a large crystal of undisturbed growth, called by Darwin a '*perfect crystal*', the reflection curve is found quite different from the one for a small crystal. Darwin calculated the reflection coefficient of a single atomic plane along the lines of a famous calculation by J. J. Thomson for the scattering of a single electron, and expressed the result for the crystal in terms of this coefficient. The measured reflectivity of the crystal could thus be compared to a theoretical value—and it was found to be too big by a factor of 10. It is much to the credit of the author that he was so convinced of the correctnes of his calculated coefficient of reflection for the single plane, that he considered the factor of 10 to indicate that something was wrong in one of his fundamental assumptions, namely the model of the perfect crystal. He remarked that if the outwardly uniform crystal were supposed to be broken up into small domains which are not fully alined with each other, then a *greater* reflected intensity is to be expected than for a perfect crystal. The reason for this is that even if the incident ray does not form the Bragg angle with a domain at the surface, it may still find a suitably oriented domain at greater depth which reflects it. Very small domains and de-orientations suffice for producing agreement between the theoretical and observed reflectivities. A crystal of such broken-up structure was later called a '*mosaic crystal*'. Darwin's formula for the reflected intensity became fundamental for crystal structure analysis, and a crystal obeying it was termed an '*ideal mosaic crystal*.' While the ordinary grown crystal lies somewhere between the perfect and the ideal mosaic type, and its reflected intensity is therefore, within limits, unpredictable, it was shown later by W. L. Bragg that by a suitable surface treatment of grinding most crystals can be turned, superficially, into the ideal mosaic type, so that measured intensities may then be used in crystal structure determinations. Without such standardization of the measuring technique, crystal analysis would have been much delayed. —The formula for the reflection by perfect crystals became of importance only much later, see Chapter 15.

5.4. Early Work in Other Countries

Only a short synopsis of the development of X-ray diffraction work in some other countries will be given here; for more detail the reader is referred to Part VI.

In France, the earliest work on X-ray diffraction was done by Maurice de Broglie, who published no less than twelve notes in the *Comptes Rendus* of the French Academy between 31 March 1913 and 27 July 1914. He started out with the Laue-Friedrich-Knipping arrangement and showed, for instance, that the diffraction pattern of magnetite is not changed by applying a strong magnetic field to the crystal (which, incidentally, is mentioned as a proof that the diffracted rays are not formed of electrons). Other points in the first four papers are the use of two superimposed photographic plates in order to evaluate the absorption coefficient of the rays forming the spots, and the obtainment of diffraction by metals and organic compounds. De Broglie also tried in 1913, but with no success, to obtain diffraction of γ-rays; this was achieved some months later by Rutherford and Andrade in Manchester in a very ingenious and efficient way (see in Part VII Andrade's Reminiscences).

The series of de Broglie's brilliant papers on X-ray spectroscopy begins with a note of November 1913 describing a primitive photographic spectrometer—the fist rotation diagram camera—with which spectra were registered, and with a second note of December 1913 containing the description of the Pt and W spectra obtained on an improved spectrometer with exposure times of the order of 15 minutes and a variety of analysing crystals. A great improvement in the technique of obtaining spectra is contained in a note of 25 May 1914 under the joint authorship of M. de Broglie and F. A. Lindemann (the later Lord Cherwell). This note introduces the method of secondary excitation of the characteristic spectra by irradiating a sample placed outside the X-ray tube with sufficiently hard X-rays. In this way substances which are not suitable as targets can be investigated. It is shown that brass emits the characteristic lines of copper and zinc. The method is also convenient for obtaining absorption spectra because of the ease with which different wave-lengths can be obtained. The systematic study of X-ray spectra which Moseley had started is then extended to spectra which he was unable to obtain with the method of primary excitation. Finally, as an example of chemical analysis, a sample of some 40–50 mg of rare earth oxides, obtained from Urbain, is shown to be a mixture of gallium and germanium oxides. (For further details of de Broglie's work see Part VI, French Schools, and Part VII, Trillat.)

* * *

Another centre of early activity in X-ray diffraction development was Japan. (See also Part VI, Japan, and Part V, Nishikawa.) Here it was T. Terada in Tokyo who read a paper to the Tokyo Physico-mathematical Society at the meeting of 3 May 1913 in which he described the production of Laue diagrams of a large variety of crystals, among them alum, borax, sugar, turmaline, epidote etc. By using only a single large diaphragm with a hole of 5–10 mm diameter in front of the tube (at about 17 cm from the target), the intensity of the diffracted spots was such that they could be observed directly on the fluorescent screen. Visual observation made it easy to follow the changing pattern of ellipses of spots as the crystal was rotated. The idea that the spots were obtained from the primary beam by reflection on densely populated atomic net-planes was gained from these observations without knowledge of W. L. Bragg's papers which had not yet reached Japan at the date of the presentation of the paper. Why the densely populated planes should be preferred, 'regardless of the fact that the volume of different primitive parallelopipeda is invariant', is not obvious. Either it could be assumed 'that each of the crystal molecules placed at one point of the lattice consists, in its turn, of a group of chemical molecules arranged in the form of a similar lattice with a finite boundary', or else one could consider the radiation 'to consist of an assemblage of detached entities,—though it seems rather difficult to proceed in this way'.—These quotations illustrate well how many alternatives there were as long as one was still groping about for a satisfactory explanation of the fundamental processes.

Terada's junior colleague in Nagaoka's laboratory, S. Nishikawa, took up the study of X-ray diffraction effects in only partially ordered matter (paper with S. Ono on 'Transmission of X-rays through Fibrous, Lamellar and Granular Substances', read 20 Sept. 1913). Diagrams of asbestos and fibrous gypsum showing star-like streaks radiating out of the central spot are reproduced and explained on the assumption that the fibres contain elementary crystal arranged in all possible azimuths round the common axis of the fibre. Organic fibres, such as silk, wood, bamboo, or hemp gave very similar patterns, whereas flaky minerals like talcum and mica gave patterns of a different type which, by tilting the sample to the X-ray beam, could however be transformed into the first type. Very finely powdered samples ('granular substances') gave a ring or halo surrounding the primary spot similar to those already observed by Friedrich (*Phys. Zs.* 1913, *14*, 317) with bees wax. Finally, in extension of observations by E. Hupka (*Phys. Zs.* 1913, *14*, 623) of star-shaped figures obtained by letting the primary beam pass through platinum foil, a systematic search establishes the dependence of the diffraction pattern of poly-crystalline metals on the mechanical treatment and the strains set up in the sample. 'Rolled zinc, for example, gives rise to six patches at a distance from the centre where a ring is observed in the case of zinc dust.'

Considering the fact that all this work had to be done with equipment borrowed from institutions all around, one can but admire the independent approach and the valuable ideas contained in these and other early Japanese papers. Terada's (English written) papers bear testimony of the author's thorough acquaintance with the German crystallographic literature; expressions like Zonenaxe, Punktreihe are added, in brackets, for clarifying the English terms. It seems likely that Terada knew Schoenflies' book on structure theory, and he recommended its study to Nishikawa* who later drew Wyckoff's

* Professor S. Miyake kindly writes to the author that he remembers Prof. Nishikawa as saying: 'It was suggested to me by Professor Terada to study Schoenflies' book. Prof. Terada seemed to anticipate that this theory would play a role in the structure problems of the future. So I went to the library of the Mathematics Department and found that the book had been bought for the library. There was no indication of its having been read by anyone before. I began studying it. At first, it was somewhat tedious to work through the theory, but I soon came to realize its implications for structure analysis. I wonder how Terada had acquired his knowledge of space group theory, or at least of its existence, at that time.

'When I went to England in 1919 and met Sir William Bragg, I handed to him a paper I had written, including a table of space groups, and asked him to publish it somewhere if he considered it to be meaningful. But it seems that my paper remained in his drawer. I suppose he did not well understand my idea at the time.'

attention to it. This was an important transmission of information much in advance of the time when the British crystallographers recognized the value of the systematic structure theory.

* * *

Of other sequels to the first Laue and Bragg papers the following should be mentioned here because they do not appear in the later Chapters on the development in the various countries after 1920.

The identity of all geometrical results in Laue's diffraction and Bragg's reflection theories was first formally demonstrated by G. Wulff (*Phys. Zs.* 1913, *14*, 217), though both Laue and Bragg seem to have been fully aware of this result.—At the second Solvay Conference in Brussels, 1913, Sommerfeld gave a discussion of the original zincblende photograms with the twofold objective of showing that Laue's assumption of characteristic crystal radiation was unnecessary, and that the spectral distribution of the insensity in the Bremsstrahl radiation, i.e. the 'white' X-rays, could be reconstructed in a qualitative way from the estimated intensities of the diffraction spots. To that end the spots had to be split into groups according to the order of reflection, the absorption, and geometrical factors. Ewald worked this out more fully in a paper in *Ann. d. Physik* 1914, *44*, 257. Looking back, the main result of this work was the full confirmation of W. L. Bragg's structure determination and of his physical picture of the pass-filter action of the crystal.

A very important contribution to the understanding of intensities was made by H. A. Lorentz when he lectured to his students in Leiden on the recent discovery and Laue's theory. He showed that the diffracted intensity of order h (= h, k, l) is proportional to h^{-2}, that is, that it diminishes with higher order. This 'Lorentz Factor' in the intensity helps to account for the limitation of observed spots to those of low orders. Lorentz' calculation, which was published in one of

Professor Miyake adds: 'As a guess, it might be that Prof. Terada had heard of space group theory from his mathematical colleagues in the Faculty of Science, unfortunately all dead now. Prof. T. Takagi, a famous algebraist, had been in Göttingen and belonged to Hilbert's school.' (See also Part V, *In Memoriam* for S. Nishikawa; Schoenflies was in Göttingen from 1884 to 1899.)

'By the way, Prof. Terada, who died in 1935, is very famous in Japan not only as a scientist, but even more so as a man of letters. He ranks among the top essayists since the Meji era in this country, and his collected literary works, amounting to about tenthousand pages including letters and diaries, have had repeated editions, even up to the present, and so have many collections of his selected essays. He disseminated plenty of original ideas, scientific and non-scientific, fruitful and not fruitful.'

Debye's papers mentioned below, also made it clear why the intensity of a diffracted ray should be proportional to the scattering volume, i.e. to the total number of atoms in the crystal, N, rather than to its square as would be the case if the ray were formed by a superposition of wavelets scattered under the very best conditions of re-inforcement which are expressed by the Laue-Bragg conditions. In fact, integration over directions for which the re-inforcement is only partial has to be considered in order to obtain the observable intensity. This was well known in the theory of the optical grating, but its extension to the three-dimensional case showed H. A. Lorentz' great mastery of mathematical physics.

Meanwhile Debye, who had been called away from Sommerfeld's institute in 1911 to Zürich University (where he was succeeded in 1912 by Laue) had just settled as Professor of Theoretical Physics at the University of Utrecht in Holland. Here he tackled a problem which appeared to others hopelessly complicated at the time: the influence of the temperature motion of the atoms on the diffraction of X-rays. It will be remembered (but it was not known to Debye) that this influence had been expected by Laue's seniors in Munich to blot out any clear signs of diffraction. Debye was particularly well prepared to deal with this problem because he had already repeatedly combined statistical and classical methods in physics—then a relatively novel combination,—notably in his theory of the specific heat of solids. In four papers in 1913/14 he answered the problem of diffraction in a temperature-disturbed crystal. This resulted in the Debye 'Temperature Factor' for the diffracted intensity which has the form exp $(-Mh^2)$, where h, as before, is the order of diffraction, and M a constant which can be expressed by means of the elastic properties of the crystal or their combination, the 'Debye Temperature', which occurs in the theory of the specific heat. As I. Waller in Uppsala showed in 1923, Debye's expression for M was wrong by a factor of two which became important for the quantitative relation of X-ray scattering and specific heat; but apart from this slip, Debye's achievement was a most impressive one and of great consequence not only in view of the future quantitative evaluation of diffracted intensities for crystal structure analysis, but also by paving the way for the first immediate experimental proof of the existence of zero-point energy, and therewith of the quantum statistical foundation of Planck's theory of black-body radiation. (Cf. Part VII, James.)

A brilliant star like Debye moved quickly from one university to the other in those days, in spite of the war. Thus, in 1917, we find

Debye occupying the chair of Experimental Physics in Göttingen. Here, with P. Scherrer, he developed the method of X-ray diffraction by powders. This was, simultaneously, but under the conditions of war quite independently, found in U.S.A. by A. W. Hull of the General Electric Co. It is interesting to compare the motivations leading to the successes in both cases, and the reader is referred to Part VII, the Reminiscences of Hull and of Scherrer. Neither Friedrich, when studying the wax halos, nor Hupka or Nishikawa when obtaining diffraction from polycrystalline metals, hit upon the powder method; the main reason for this is that none of them had tubes suitable for producing strong monochromatic X-rays, so that all they obtained were blurred halos.

It should not be forgotten that one of the greatest contributions from which all fields of X-ray diffraction profited, was that of William D. Coolidge of the General Electric Research Laboratories in Schenectady, N.Y. His new type of X-ray tube eliminated the necessity of juggling between a deficient and an excessive gas content by providing an independent electron source through emission from a white-hot filament. It thus made possible to run a tube in a stable régime with independent adjustments of power (milliampères) and voltage (kilovolts), which had been impossible with the old-time gas-containing tubes. It is obvious how important this was for the generation of characteristic radiation and for measurements requiring constant conditions over hours. It is less obvious, today, what a wealth of novel technological ideas and developments had to precede the construction of the first marketable tube. This was the time when high-vacuum began to emerge from the laboratories and enter into manufacturing processes, owing to the invention of powerful pumping systems (Gaede in Karlsruhe), getters (electric lamp industry), metals that could be outgassed (tungsten made malleable by Langmuir), and glass that could be sealed to metal (cf. Part VII, Hull)—in short it was the period when modern electronics was born.

PART III

The Tools

CHAPTER 6

The Principles of X-ray Diffraction

6.1. X-ray Reflection according to W. L. Bragg

Consider a set of N+1 equidistant atomic planes of spacing d, and a monochromatic plane X-wave falling on it at a glancing angle θ (Fig. 6–1(1)). It is assumed that each atomic plane reflects a very small fraction of the incident amplitude, small enough so that the weakening effect of this reflection on the incident amplitude may be neglected throughout the crystal. Under most angles of incidence, θ, the waves reflected from neighbouring planes will show a phase difference, and where all the reflected waves come together at great distance from the crystal, the superposition of these waves of systematically increasing phases will lead to a cancellation of amplitudes and to optical field zero. There exists, then, only the transmitted wave. If, however, the phases of all the reflected waves arrive within less than one half wave-length phase difference, then all reflected amplitudes will build

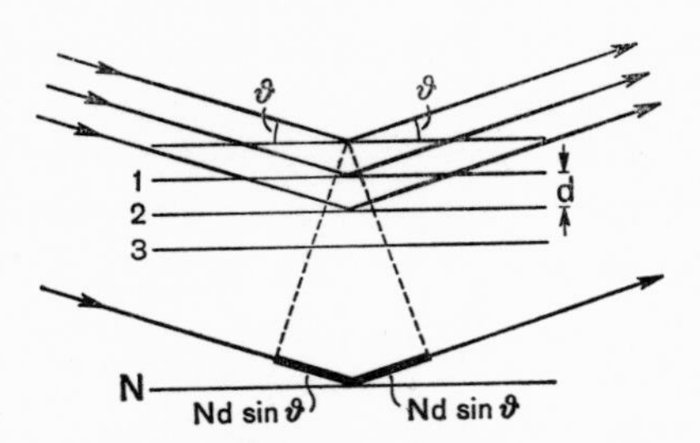

Fig. 6–1(1). Bragg reflection on a set of N atomic planes.

up together to an optical field in the direction of reflection, without any actual cancellations of contributions. Should all waves arrive in the *same* phase, then full re-inforcement of the waves takes place to an amplitude of N+1 times that of the single reflected wave.

Now the difference of optical path for the top and bottom wave is shown by the heavy-drawn path lying between two parts of the wave-fronts of the incident and reflected waves. Its length is $2Nd \sin \theta$. The path difference between reflections on neighbouring planes is $2d \sin \theta$. If this equals a whole number, n, of wave-lengths λ, then the phase difference is zero throughout the crystal. Therefore maximum amplitude of the reflected wave is obtained for angles θ_n such that

$$2d \sin \theta_n = n\lambda. \tag{1}$$

This 'Bragg Equation' determines the angles θ_n under which the first, second, third,... order reflections occur, for $n = 1, 2, 3, \ldots$ The greater the wave-length, the larger the glancing angle for reflection on the same plane; the greater the spacing, the smaller is the glancing angle for a given wave-length. If λ is known and θ_n measured, then the value of n/d follows, and if the order can be found, the value of the spacing of the set of reflecting planes is determined. By putting together the information on various sets of reflecting planes, obtained in this way with the X-ray spectrometer, the first crystal structure determinations were made.

As the angle of incidence is slightly varied from the Bragg angle, a phase difference develops between reflections from neighbouring planes and re-inforcement of the reflected waves becomes less perfect. Their effect cancels out to optical field zero if the maximum phase difference throughout the crystal corresponds to an entire wave-length path difference, or, indeed, to any multiple of it, say $s\lambda$, where s is an integer. For then there will be for any reflected elementary wave one of opposite phase superimposed. The condition for the angles under which these 'secondary zeros' of the reflection curve occur is

$$2Nd \sin \theta_{n,s} = (Nn + s)\lambda, \tag{2}$$

and the angular distance of the sth zero from the Bragg angle θ_n is given by

$$2d \sin \theta_{n,s} - 2d \sin \theta_n = (s/N)\lambda,$$

or, since the difference of angles is usually very small, this can be approximated by

$$\theta_{n,s} - \theta_n = \Delta\theta_{n,s} = \tan \theta_n \cdot (s/N). \tag{3}$$

If we plot the reflected amplitudes as a function of the difference of the angles of incidence and reflection from the Bragg angle, $\varphi = \theta - \theta_n$, each Bragg angle is seen surrounded by zeros of the reflected amplitude and corresponding zeros of intensity obtained by squaring the curve. The principal or zero order maximum lies between the zeros for $s = \pm 1$ and has therefore a width of $(2 \tan \theta_n)/N$.

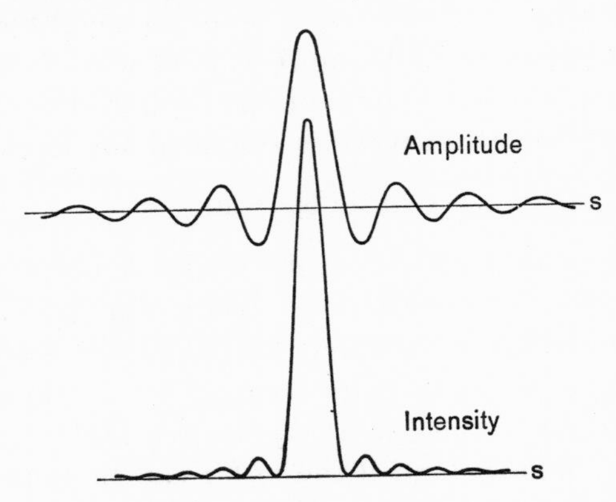

Fig. 6–1(2). Amplitude and intensity curve of the wave reflected by a set of N atomic planes in the neighbourhood of the Bragg angle.

The finite angular width of the Bragg reflection is a consequence of the limitation of thickness of the crystal to Nd. The thinner the crystal, the wider is the angular range of the principal maximum, and the poorer is the resolving power of the crystal as an optical instrument for distinguishing between neighbouring wave-lengths in spectroscopy. The total reflected intensity in any order of reflection is proportional to the area under the principal maximum of the intensity curve. Since the amplitude maximum is proportional to $N+1$, or essentially to N if this is large, the maximum intensity is proportional to N^2. With the width of the principal maximum given above, the total reflected intensity is then proportional to $2N \tan \theta_n$. This shows that the reflected intensity is proportional to the thickness or volume of the crystal,—a result that is true only as long as the crystal is so thin that the reflection leaves the incident, and transmitted, ray at practically its full strength.

The consideration of the secondary maxima and of the finite resolving power of the crystal is essentially what H. A. Lorentz introduced into the theory, although using Laue's diffraction, not Bragg's reflection language. The measurement of the angular width of the principal maximum of reflection forms the basis of the determination of particle size by means of X-rays—in the particular case considered the thickness of a crystal flake could be determined from such measurements.

Bragg's formula can be applied to reflections on atomic planes which are not parallel to the crystal surface, because the condition of re-inforcement does not contain the orientation of these planes to the surface. If a small general refraction of X-rays in a crystalline medium is taken into account, λ and θ_n in the formula have to be interpreted as the internal values which are related to the values outside the crystal by the ordinary laws of refraction. Since the refractive index for X-rays differs from 1 only by about one part in hundredthousand, this refraction need be considered only in measurements of very high precision; it was first found by W. Stenström of M. Siegbahn's school in 1919, and leads to a correction of wave-lengths or lattice constants when the usual Bragg formula is applied within the above range of accuracy.

* * *

The great simplicity of Bragg's theory is achieved by the introduction and use of the spacing d of the reflecting planes. Starting from the axial system, or the cell of the crystal, the determination of d for different kinds of planes requires some geometry. On the surface Laue's theory appears more complicated, but it contains this internal geometry of the crystal as it were built-in. There were, besides, some objections to Bragg's idea of reflection which, in the early stages, made its acceptance not obvious. While densely populated planes of atoms are well defined, with wide spacings between them, the sparsely populated planes have little physical reality. Their spacing must be very small, in order to produce the fixed number of atoms per unit volume. Therefore it seems rather artificial to consider with Bragg that the few atoms on each plane reflect like a mirror, independently of the much closer atoms which belong to neighbouring parallel planes. Furthermore the regular arrangement of the atoms in any one of these planes is known, from the diffraction point of view, to produce many diffracted rays (cross-grating spectra), of which the specular reflection is only one. What becomes of the others?

6.2. *X-ray Diffraction according to Laue*

a. *The linear grating.* We first consider a row of equally spaced atoms, each of which becomes the source of a scattered spherical wavelet under the stimulation by an 'incident' monochromatic plane wave. The mode of propagation of a plane wave is represented by its 'wave-vector' $\mathbf{k}$; this has length $1/\lambda$ (λ the given wave-length), and its direction is that of the normal to the planes of equal phase, or, as we may call these, the wave front. Let the three axial vectors by which the crystal is described be $\mathbf{a}_i (i = 1, 2, 3)$, then any point in space can be described by the 'coordinate vector' $\mathbf{x} = x_1\mathbf{a}_1 + x_2\mathbf{a}_2 + x_3\mathbf{a}_3$, where the x_i are called the 'coordinate numbers'. It will be convenient to use the notion of the 'scalar product' of two vectors, for instance $\mathbf{k} \cdot \mathbf{x}$. This is defined as meaning the product of the lengths of the two vectors (indicated by $|\mathbf{k}|$ and $|\mathbf{x}|$ respectively) times the cosine of the angle between their positive directions:

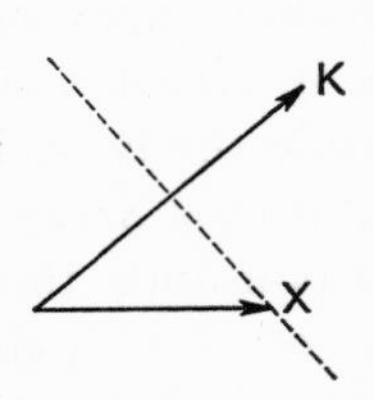

Fig. 6–2(1). Scalar product of wave vector and coordinate vector.

$$\mathbf{k} \cdot \mathbf{x} = |\mathbf{k}| \cdot |\mathbf{x}| \cdot \cos (\mathbf{k}, \mathbf{x}).$$

Now $|\mathbf{x}| \cos (\mathbf{k}, \mathbf{x})$ is the length of the projection of $\mathbf{x}$ on the direction of $\mathbf{k}$; this projection is often called the resolved part of $\mathbf{x}$ along $\mathbf{k}$. Evidently the resolved part is the same for all points $\mathbf{x}$ lying on the dotted line of Fig. 6–2(1), and, in three dimensions, for all points of the plane normal to $\mathbf{k}$ which contains the dotted line. Therefore the expression $\mathbf{k} \cdot \mathbf{x} =$ constant can be used to describe the planes of equal phase of a wave. The value of the constant is the length of the optical path, expressed in wave-lengths, from the wave front passing through the origin to that passing through point $\mathbf{x}$. The argument $(-\nu t + \mathbf{k} \cdot \mathbf{x})$ is characteristic of a wave of frequency ν travelling in the direction of $\mathbf{k}$; $\nu/|\mathbf{k}| = q$ is the velocity with which the phase travels.

Consider now the wavelets scattered by the equally spaced atoms under the stimulus of the incident wave of vector $\mathbf{k}_1$. Can a direction be found in which they all arrive in the same phase at a very distant point? We enumerate the atoms by an index l and call the vector from an atom l to its neighbour $l + 1$ the 'translation' $\mathbf{a}$. For full cooperation of *all* wavelets it is sufficient to find the condition for wavelets l and $l + 1$ to arrive at the point of observation without any difference of phase. In Fig. 6–2(2) the wave fronts through atoms l and $l + 1$ respectively, are shown dotted. The optical path, measured in wave-lengths, through $l + 1$ is shorter than that through l by the resolved part of $\mathbf{a}$ along $\mathbf{k}$, and longer by the resolved part of $\mathbf{a}$ along $\mathbf{k}_1$. Therefore it is shorter by

$$\mathbf{k} \cdot \mathbf{a} - \mathbf{k}_1 \cdot \mathbf{a} = (\mathbf{k} - \mathbf{k}_1) \cdot \mathbf{a}$$

For best re-inforcement this must be an integral number of wave-lengths, say h wave-lengths. Thus a diffracted maximum occurs if **k** is such that

$$(\mathbf{k} - \mathbf{k}_1) \cdot \mathbf{a} = \mathrm{h} \tag{2}$$

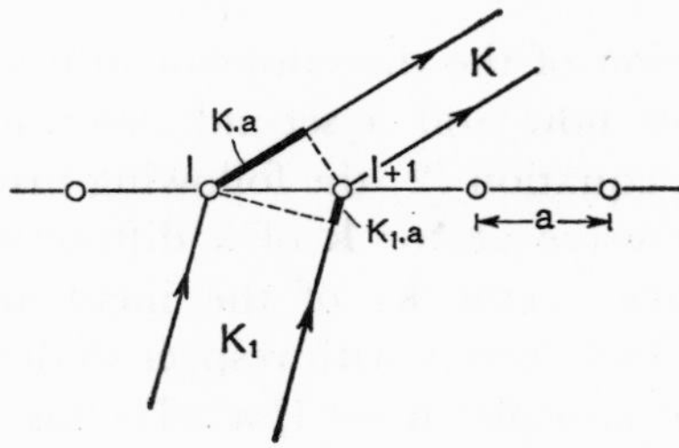

Fig. 6–2(2). Diffraction by linear grating (row of atoms); difference of optical paths through neighbouring atoms.

By putting h = 0, a solution of this is seen to be $\mathbf{k} = \mathbf{k}_1$; that is, there always occurs optimum re-inforcement of the wavelets in the direction of incidence. To the right of this direction (in the above figure) lie the directions with positive integers or '*orders*' h, to the left those for negative h. The values of h are limited by the condition that the diffracted waves have to move away from the row of atoms. The fact that the individual scattered wavelets are spherical will make the result of their superposition the same in all planes containing the row of atoms. The direction of **k** is therefore only the representative of a cone of directions surrounding the row, and maximum amplitude is achieved in all directions of observation along this cone. If we imagine the rays issuing from the row of atoms to be made visible by

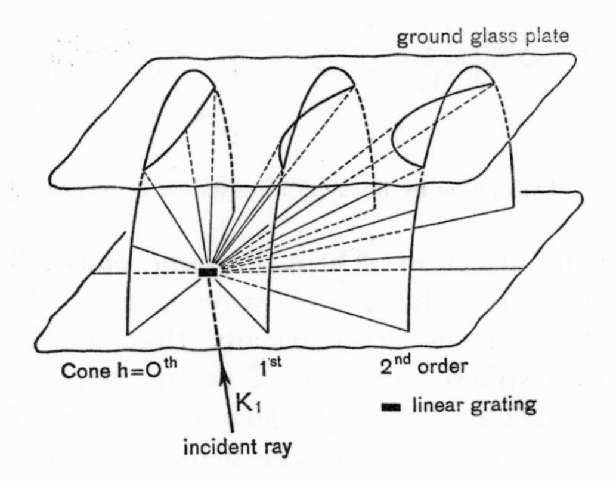

Fig. 6–2(3). Diffraction by linear grating shown on ground glass plate in physical space.

their intersection with a ground glass plate which is parallel to the row, the result will be a series of bright hyperbolae as in Fig. 6–2(3); the same pattern would be obtained on a glass plate placed beneath the row— or, in fact, on *any* plate held parallel to the row.

To this first description of the directions of diffraction in relation to the row of atoms we now add a second description in a '*reciprocal space*'. We take from equation (2) the following instruction for finding the direction of the wave vector $\mathbf{k}$ of a diffracted wave, given the direction of the wave vector $\mathbf{k}_1$ of the incident wave: make the resolved parts of the two vectors with respect to the translation $\mathbf{a}$ such that they differ by a multiple (h) of $1/|\mathbf{a}|$. To this end we construct a series of equidistant planes normal to the row of atoms and with spacing $1/|\mathbf{a}|$ between them. If the given wave vector $\mathbf{k}_1$ is laid down so that it ends in the origin 0, its starting point T is determined. If now from T we draw *any* vector $\mathbf{k}$ ending on one of the planes, say that labelled h, then condition (2) is fulfilled and *all* wavelets arrive at a

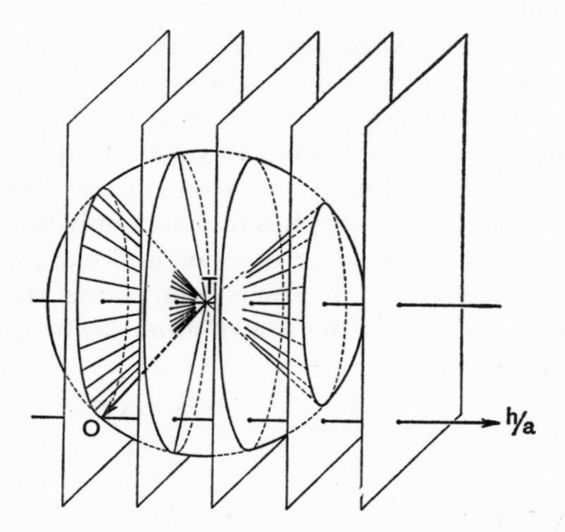

Fig. 6–2(4). Diffraction by linear grating shown in reciprocal space.

distant point in the direction of $\mathbf{k}$ in the same phase. But in classical scattering there is no change of wave-length; therefore $|\mathbf{k}| = |\mathbf{k}_1| = 1/\lambda$. This means geometrically that the end-point of $\mathbf{k}$ must also lie on a sphere about T of radius $1/\lambda$, i.e. which passes through the origin. This '*sphere of reflection*' is intersected by the set of planes in circles, and these are the geometrical loci for the ends of the diffracted wave-vectors. Connecting T, which will be called the '*tie-point*', to all points of the circle in the plane numbered h, one obtains the cone of diffracted directions of order h, as in Fig. 6–2(4).

In this construction the distances at which the planes are drawn represent $1/|\mathbf{a}|$, and the radius of the sphere of reflection is $1/\lambda$; thus we are using a space in which the distance represents the inverse of a length in physical space and is measured in cm^{-1} or $1/Å$ ($= 10^8 cm^{-1}$). There is nothing more unusual in this than if we represent, for instance, a velocity [cm/sec], force [$g \cdot cm/sec^2$] or an electric field strength [volt/cm] by the length of an arrow in a suitably labelled space. The space we are operating in above is called 'reciprocal space' because the product of a length in this space and a length taken in physical space is a dimensionless quantity, i.e. a pure number.

We shall now extend the same two representations of the directions of diffraction, in physical space and in its reciprocal, to two- and three-dimensional lattices of atoms.

* * *

b. *The cross-grating*. This is another name for an array of scattering centres filling a plane in a periodic manner. All atoms—if we take these to be the centres—are obtained from an original one by applying two translations, given vectorially by $\mathbf{a}_1$ and $\mathbf{a}_2$. If l_1 and l_2 are independent integral numbers, ranging from $-\infty$ to $+\infty$, the positions of the atoms are

$$\mathbf{x}_{l_1, l_2} = l_1\mathbf{a}_1 + l_2\mathbf{a}_2. \tag{3}$$

The condition for maximum re-inforcement of *all* wavelets issuing from these points is that there be no phase difference between the wavelets coming from an atom and its two neighbours by the $\mathbf{a}_1$ and $\mathbf{a}_2$ translations, respectively. But this is exactly the condition (2) applied twice over, once using the translation $\mathbf{a}_1$ and an integer h_1, and again with $\mathbf{a}_2$ and an independent integer h_2. Thus the condition for the wave-vector $\mathbf{k}$ of a diffracted wave is

$$\begin{aligned}(\mathbf{k} - \mathbf{k}_1)\cdot\mathbf{a}_1 &= h_1\\ (\mathbf{k} - \mathbf{k}_1)\cdot\mathbf{a}_2 &= h_2.\end{aligned} \tag{4}$$

We visualize this condition with the help of the ground glass plate of Fig. 6–2(5) which we imagine to be parallel to the plane of the cross, grating. Each of the above equations is fulfilled along the hyperbolcf intersections of the plate with the cones surrounding the directions oi $\mathbf{a}_1$ and $\mathbf{a}_2$. The directions for which both conditions are fulfilled are those leading to the points of intersection of the hyperbolae. Each diffracted ray is named as a two-digit 'order', (h_1, h_2), the integers of which tell the whole number of wave-lengths path difference of the wavelets scattered by an atom and its neighbours along $\mathbf{a}_1$ and $\mathbf{a}_2$. Cross-grating spectra always exist, for any wave-length or angle of incidence. Should $\mathbf{a}_2$ be considerably larger than $\mathbf{a}_1$, then the second set of hyperbolae have a much narrower spacing than the first, and the points of intersection mark out the first set very clearly. Starting out from the central spot which gives the direction of the incident ray and has order (0, 0), one can easily '*index*' the diagram by assigning each point its (h_1, h_2) values.

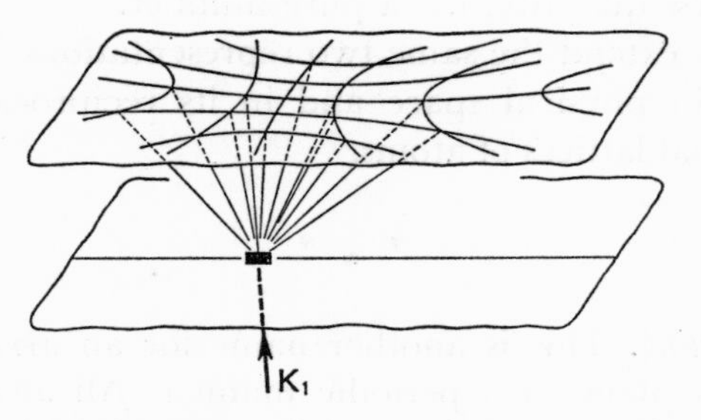

Fig. 6–2(5). Cross-grating diffraction shown on ground glass plate in physical space.

The spherical wavelets combine equally on both sides of the cross grating; a second ground glass plate beneath the grating would show the same design. To the transmitted beam on the upper plate corresponds a reflected beam on the lower plate, and this has again the order (0, 0). These two beams (0, 0) are *always* present, whatever the atomic distances, or even their regularity may be.

Let us now look at the corresponding construction in reciprocal space. The first of the two conditions (4) leads to the set of planes normal to $\mathbf{a}_1$ with spacings $1/|\mathbf{a}_1|$ between them which we know from Fig. 6–2(4). The similar set of planes representing the second condition (4) is normal to $\mathbf{a}_2$, with spacing $1/|\mathbf{a}_2|$. Both conditions are simultaneously fulfilled on the intersections of these sets, that is, on a periodic array of straight lines normal to the plane of $\mathbf{a}_1$ and $\mathbf{a}_2$. Additional to this, the condition of scattering without change of wave-length requires that the vector $\mathbf{k}$ of a diffracted wave end on the sphere of reflection of radius $|\mathbf{k}_1|$ and centre at T. The wave-vectors $\mathbf{k}_{h_1, h_2}$ of the diffracted waves therefore start out from T and end in the points of intersection of the sphere of reflection with the straight lines of orders

h_1, h_2. Provided the radius of the sphere of reflection is not too small, intersections will always exist on the upper and lower hemisphere, corresponding to the cross-grating spectra of the same orders emitted to both sides of the cross-grating, as explained above.

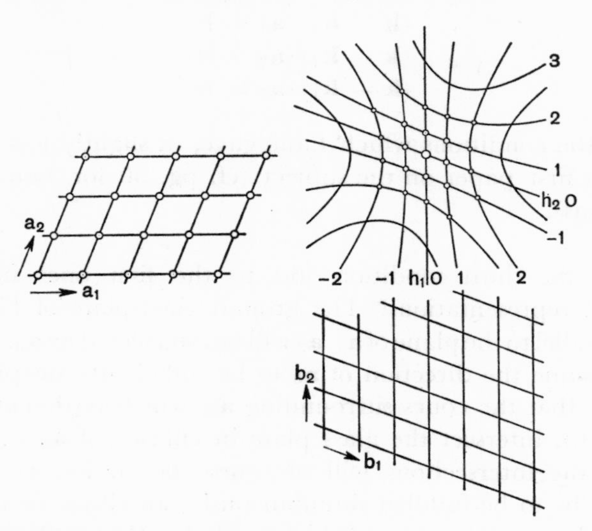

Fig. 6–2(6). Left: cross-grating in physical space; upper right: pattern of cross-grating spectra in physical space; lower right: trace of the planes normal to $\mathbf{a}_1$ and $\mathbf{a}_2$ for construction in reciprocal space.

Fig. 6–2(6) illustrates the cross-grating with translations $\mathbf{a}_1$ and $\mathbf{a}_2$, and its relation to the pattern of crossed hyperbolae in physical space; it further shows in reciprocal space the ground plan of the sets of planes normal to $\mathbf{a}_1$ and $\mathbf{a}_2$, and the distribution of their lines of intersection. The area of the parallelograms between these lines can be transformed into a rectangle whose one side is $1/|\mathbf{a}_1|$, while the other (vertical) side is $1/(|\mathbf{a}_2| \sin \alpha)$, where α is the angle formed by $\mathbf{a}_1$ and $\mathbf{a}_2$. Thus the area is $1/(|\mathbf{a}_1|\, |\mathbf{a}_2| \sin \alpha)$, and this is the inverse of the cell (or parallelogram) of the cross-grating.

* * *

(c). *Lattice in three dimensions.* If cross-gratings, formed by two translations $\mathbf{a}_1$ and $\mathbf{a}_2$, are stacked on top of one another by a third translation $\mathbf{a}_3$, not lying in the plane of $\mathbf{a}_1$ and $\mathbf{a}_2$, then a lattice is obtained with atoms lying at

$$\mathbf{x}_l = l_1\mathbf{a}_1 + l_2\mathbf{a}_2 + l_3\mathbf{a}_3 \ (l_i \text{ integers}). \tag{5}$$

The condition for the diffracted rays from all cross-gratings to be superimposed with the same phase in the direction of a wave vector $\mathbf{k}$ is once again of the form (2). Therefore we have now to fulfill three conditions, containing three arbitrary integers h_i, namely

$$\begin{aligned}(\mathbf{k} - \mathbf{k}_1) \cdot \mathbf{a}_1 &= h_1 \\ (\mathbf{k} - \mathbf{k}_1) \cdot \mathbf{a}_2 &= h_2 \\ (\mathbf{k} - \mathbf{k}_1) \cdot \mathbf{a}_3 &= h_3\end{aligned} \tag{6}$$

These are the conditions which Laue gave, in slightly less condensed form, in his first paper on the subject (cf. pg. 50 for Laue's form of the equations).

What does the third condition add to the first ones in the two geometrical representations? The ground glass plate of Fig. 6–2(5), which is parallel to the plane of $\mathbf{a}_1, \mathbf{a}_2$ will be intersected by $\mathbf{a}_3$. For clarity we may assume the direction of $\mathbf{a}_3$ to be sufficiently steeply inclined to $\mathbf{a}_1, \mathbf{a}_2$ so that the cones surrounding $\mathbf{a}_3$, which represent the third condition (6), intersect the glass plate in ellipses; if $\mathbf{a}_3$ is normal to the plate, the intersections will of course be circles. For all three conditions (6) to be fulfilled simultaneously, an ellipse or circle must pass through an intersection of the hyperbolae (Fig. 6–2(7)). This will in general not occur, except for the direction of incidence, and then the incident wave will travel through the crystal without being diffracted. However, as we decrease the wave-length, the cross-grating pattern contracts towards its centre, the direction of the incident ray, and the ellipses or circles contract about the point where the axis $\mathbf{a}_3$ intersects the plate. These contractions take place at different rates, and this means that for certain wave-lengths three curves intersect simul-

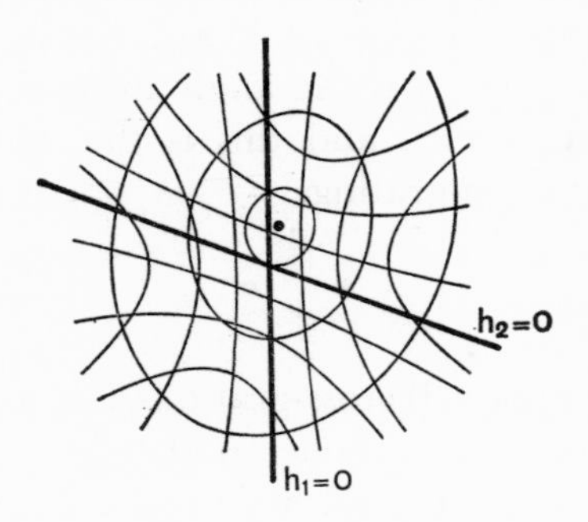

Fig. 6–2(7). Diffraction by three-dimensional lattice shown in physical space on ground glass plate. The ellipses represent the third condition of re-inforcement of the wavelets.

taneously and a diffracted ray of order (h_1, h_2, h_3) flashes up. This is the pass-filter action of the crystal for white X-radiation, since, given the direction of incidence, only a certain wave-length can appear in a diffracted ray, other wave-lengths not being admitted at this particular angle of diffraction. This statement has to be modified, however, because by multiplying the lengths of $\mathbf{k}$ and $\mathbf{k}_1$, as well as the order numbers h_i by the same integral factor n, the equations (6) remain fulfilled without change of direction of $\mathbf{k}$ or $\mathbf{k}_1$. In other words, if a wave-length λ is diverted by diffraction into a direction $\mathbf{k}$, then $\lambda/2$, $\lambda/3, \ldots$ can be contained in the same direction by diffraction at correspondingly higher orders, provided, of course, these shorter wave-lengths are contained in the spectrum of the incident white radiation. This multiplicity of the wave-lengths in the spots of Laue diagrams makes these less suitable for crystal structure determination than the diagrams obtained with monochromatic radiation.

In reciprocal space, the third of the conditions (6) adds a set of equidistant planes normal to $\mathbf{a}_3$, and of spacing $1/|\mathbf{a}_3|$, to the array of straight lines representing the first two conditions. The intersections of these lines or rods with the planes yields a three-dimensional lattice of points called the '*reciprocal lattice*'. The condition of diffraction is now that the sphere of reflection intersect a point of the reciprocal lattice. Again, with the direction of incidence and the wave-length given,—in short, given the tie-point—,the chances are that the sphere of reflection will not pass through *anv* point of the reciprocal lattice, except the one at the origin through which it passes by definition. If this is so, only the 'primary' beam (000), the continuation of the incident beam, is formed and there is no diffraction in the usual sense. If, however, we vary the position of the tie-point in any way, the sphere of reflection will sweep over some of the lattice points, and every time such intersection occurs, a diffracted ray will be flashing up. In the Laue method the direction of incidence is fixed, but the length of the vector $\mathbf{k}_1$ is variable, inversely as the wave-length; therefore the tie-point moves on a straight line passing through the point (000), and the radius of the sphere varies accordingly. In the methods working with monochromatic X-rays, T is free to move on a sphere of radius $1/\lambda$, the 'wave number sphere'. If there is incidence in all directions with respect to the crystal lattice, as in powder diagrams, T moves over the entire surface of this sphere. In wide-angle diagrams, T moves only within that part of the sphere that is cut out by the solid angle of incidence. In a rotation or oscillation diagram, finally, T is restricted to a great circle, or part of it, which is normal to the axis of rotation or oscillation. In each case, the process of 'indexing' a diffraction diagram is essentially the reconstruction of this geometry from the diagrams, and it thereby leads to the determination of the shape of the reciprocal lattice. This process is made easier by any additional information obtained about the spots on the diagram. It is hardest in the case of powder diagrams, where nothing is known; the higher the crystal symmetry is known to be, the easier it becomes. Oscillation diagrams are easier to index than rotation diagrams on which more overlapping of spots may occur. It is of further great help to know the position of the crystal at the instant of reflecting each spot of a rotation or oscillation diagram. This is achieved in the so-called goniometer methods by coupling a displacement of the film to the rotation of the crystal and limiting the recording of spots to certain types of reflections by 'layer-line diaphragms', as in the Weissenberg and Schiebold-Sauter goniometers. In the De Jong-Bouman and the Buerger precession camera film and crystal are moved in such a way that the spots appear in a pattern which is a section through the reciprocal lattice itself, so that there is no problem of indexing. In all cases, however, indexing is a routine procedure, once it has been done a few times.

Let us add some detail to the geometry of the reciprocal lattice. We go back to the third part of Fig. 6–2(6) which indicates the cross-section of the array of parallel rods normal to the plane of $\mathbf{a}_1$, $\mathbf{a}_2$. The same is shown in perspective in Fig. 6–2(8) together with the third axis, $\mathbf{a}_3$. Two successive planes normal to $\mathbf{a}_3$ are shown (which are at distance $1/|\mathbf{a}_3|$ from one another), and the points marked in which they intersect the rods. The eight intersections suspend a parallelopipedon which is the repeat unit or cell of the reciprocal lattice. Let us determine its volume. We know already that the ground plan of the cell had an area $1/A$, where A was the area defined by the axes $\mathbf{a}_1$ and $\mathbf{a}_2$. The volume of the cell is obtained by multiplying this area with the height of the cell, OQ. But we know that the resolved part of OQ in the direction of $\mathbf{a}_3$, that is, OP has length $1/|\mathbf{a}_3|$. Therefore OQ = $= (1/|\mathbf{a}_3|) : \cos\gamma$, where γ is the angle POQ. Thus OQ is the inverse of the thickness of the crystal cell, namely $1/(|\mathbf{a}_3| \cos\gamma)$, and this makes the volume of the cell of the reciprocal lattice the inverse to the volume of the crystal cell.

The faces of the reciprocal cell are formed of planes normal to $\mathbf{a}_1$, $\mathbf{a}_2$, and $\mathbf{a}_3$, respectively. The edges of the cell are the translations of the reciprocal lattice, and we call them $\mathbf{b}_1$, $\mathbf{b}_2$, $\mathbf{b}_3$; each of them is therefore normal to two of the **a**-axes. This fact can be expressed by stating that the scalar products vanish,

$$\mathbf{b}_i \cdot \mathbf{a}_k = 0 \text{ for } i \neq k \tag{7}$$

(see the definition of the scalar product on pg. 86).

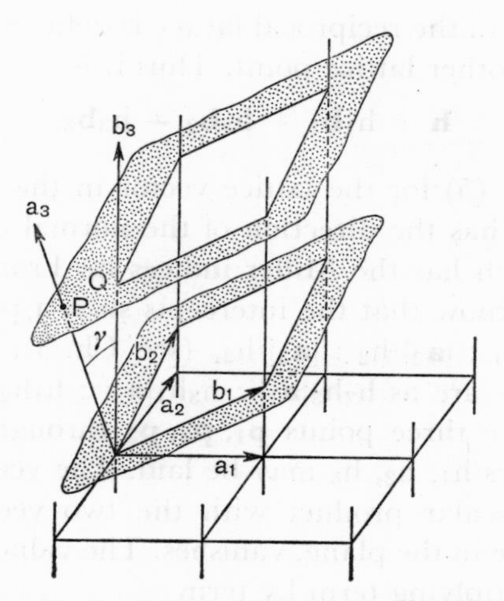

Fig. 6–2(8). The reciprocal lattice in three dimensions. The origin O is the corner from which the vectors $\mathbf{a}_i$ and $\mathbf{b}_i$ are drawn. $\mathbf{a}_1$ and $\mathbf{a}_2$ are actually shown representing length $1/|\mathbf{a}_1|$ and $1/|\mathbf{a}_2|$, and OP $= 1/|\mathbf{a}_3|$. Then $\mathbf{b}_1$ and $\mathbf{b}_2$ are as shown and $\mathbf{b}_3$ = OQ. These are the reciprocal vectors to the $\mathbf{a}_i$, fulfilling Eqs. (9), and they support the reciprocal lattice.

This property of the **b**-axes is shared by the 'polar axes' which Bravais introduced in the 1840's, but Bravais added an unsuitable definition of the length of the axes. For the reciprocal axes, which were first introduced by Willard Gibbs in his lectures at Yale University, the normalization can easily be seen from Fig. 6–2(8) for $\mathbf{b}_3$ = OQ, the length of which was discussed above. The result is that the resolved part of OQ in the direction of OP is the inverse of the length of OP; in other words, the scalar product of $\mathbf{a}_3$ and $\mathbf{b}_3$ has value 1. Since the decomposition of the crystal lattice into cross-gratings might have been performed with the choice of any two **a**-axes, analogous equations must hold for all axes, namely

$$\mathbf{b}_i \cdot \mathbf{a}_i = 1 \ (i = 1, 2, 3) \tag{8}$$

The conditions (7) and (8) can be condensed to the elegant statement of the relation between the **a**-axes and their reciprocals, the **b**-axes

$$\mathbf{b}_i \cdot \mathbf{a}_k = \delta_{ik}, \tag{9}$$

where δ_{ik}, called the Kronecker symbol, signifies 1 if $i = k$ and 0 if $i \neq k$. In this form the equivalence of the two sets of axes appears very forcibly. Just as we obtain the **b**-set of axes from the **a**-set, so we construct the **a**-set from a given **b**-set. In crystallography, many of the optical or X-ray measurements lead in the first instance to the **b**-axes.

A lattice vector in the reciprocal lattice is defined as the vector from the origin to any other lattice point. Thus it is

$$\mathbf{h} = h_1\mathbf{b}_1 + h_2\mathbf{b}_2 + h_3\mathbf{b}_3, \tag{10}$$

in analogy to Eq. (5) for the lattice vector in the crystal lattice. We now show that **h** has the direction of the normal of the net plane in crystal space which has the Miller indices h_i. From the definition of these indices we know that the intercepts such a plane makes on the axes are as $|\mathbf{a}_i|/h_1 : |\mathbf{a}_2|/h_2 : |\mathbf{a}_3|/h_3$, (see Ch. 3), and, multiplying with $h_1h_2h_3$, they are as $h_2h_3|\mathbf{a}_1| : h_3h_1|\mathbf{a}_2| : h_1h_2|\mathbf{a}_3|$. Thus $h_2h_3\mathbf{a}_1$, $h_3h_1\mathbf{a}_2$, $h_1h_2\mathbf{a}_3$ are three points $\mathbf{p}_1, \mathbf{p}_2, \mathbf{p}_3$ through which the plane with Miller indices h_1, h_2, h_3 may be laid. The vector **h** is normal to this plane if its scalar product with the two vectors $\mathbf{p}_2 - \mathbf{p}_1$ and $\mathbf{p}_3 - \mathbf{p}_1$, which lie in the plane, vanishes. The values of these products are found by multiplying term by term

$$\mathbf{h}\cdot(\mathbf{p}_2 - \mathbf{p}_1) = (h_1\mathbf{b}_1 + h_2\mathbf{b}_2 + h_3\mathbf{b}_3)\cdot(h_3h_1\mathbf{a}_2 - h_2h_3\mathbf{a}_1)$$
$$\mathbf{h}\cdot(\mathbf{p}_3 - \mathbf{p}_1) = (h_1\mathbf{b}_1 + h_2\mathbf{b}_2 + h_3\mathbf{b}_3)\cdot(h_1h_2\mathbf{a}_3 - h_2h_3\mathbf{a}_1).$$

Because of the relations (9) both products vanish, and this proves the statement.

If in (10) the component numbers h_i have no common factor, we denote them by h_i^* and the lattice vector by $\mathbf{h}^*$. This vector ends in the first point of a row of equidistant points whose positions are obtained by letting a common factor n of the component numbers h_i assume all integral values, positive and negative. This is indicated by writing

$$\mathbf{h} = n\mathbf{h}^*. \tag{11}$$

The linear lattice of spacing $|\mathbf{h}^*|$ is, in reciprocal space, the image of the set of planes in physical space which are normal to **h**, and it is easy to show that their spacing is inverse to the spacing along the row of points along **h**:

$$d_h = 1/|\mathbf{h}^*|. \tag{12}$$

This spacing d_h is the one entering in the Bragg formula $n\lambda = 2d \sin\theta$.

To the vector $\mathbf{h} = n\mathbf{h}^*$ there corresponds the spacing d_h/n, or, in the Bragg formula, the nth order reflection on the set of planes.

The relation between the reflection and diffraction terminologies becomes evident in Fig. 6–2(9) which shows a plane of the reciprocal lattice, the tie-point T and two diffracted rays. For one of these, the wave-vector $\mathbf{k}_h$ is entered. The fact that the sphere of reflection passes through the lattice point $\mathbf{h}$ can be stated in the equation

$$\mathbf{k}_h - \mathbf{k}_1 = \mathbf{h} \text{ (Laue Equation)} \tag{13}$$

This one-vectorial equation expresses the same facts that are contained in the three scalar equations (6), or in the original equations of Laue quoted on page 50.

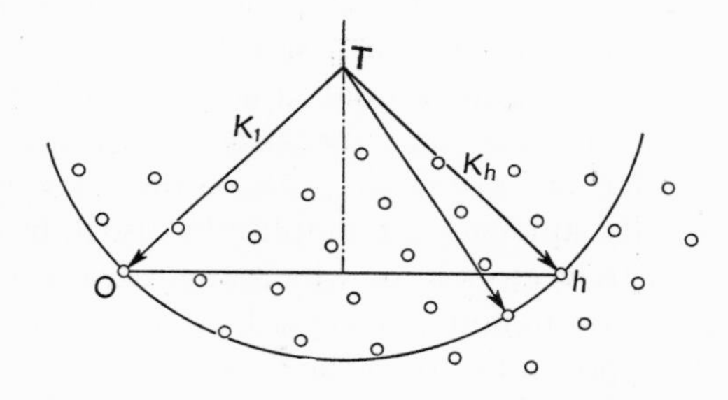

Fig. 6–2(9). Relation between 'diffracted' and 'reflected' rays.

Because $\mathbf{k}_1$ and $\mathbf{k}_h$ have the same length, the plane at right angles to the vector $\mathbf{h}$ and passing through its mid-point contains the tie-point. $\mathbf{k}_h$ can thus be said to be the reflected image of $\mathbf{k}_1$ by this plane—and this plane belongs to the set of atomic net planes with Miller indices (h_1, h_2, h_3). This shows that every one of Laue's 'diffracted' rays is simultaneously one of Bragg's 'reflected' rays.

6.3. Fourier Space

So far, the reciprocal lattice has been introduced as a convenient means for visualizing the directions under which alone the perfectly periodic crystal gives diffraction. Since in all these directions the scattered wavelets combine without phase difference, the amplitude generated by a finite crystal in these directions is proportional to N, the number of atoms of the crystal, and the energy, which is always proportional to the square of the amplitude, would be proportional to N^2, i.e. to the square of the volume of the crystal if the latter is 'bathed' in X-rays. We have already discussed in 6.1 that this is not the intensity which would normally be observed, because the larger the (perfect) crystal, the greater is its resolving power and the sharper its filter action, so that, with increasing N, the angular width of the principal maximum and the spectral range of admitted wave-lengths decreases. If we consider a finite, and therefore in the mathematical sense non-periodic crystal, there is a certain latitude in the fulfilling of the Laue-Bragg conditions, because for a very slight infringement the wavelets coming from the crystal will continue to re-inforce one another. This means that the vector **k** of the diffracted wave need not end on the lattice point of the reciprocal lattice, as long as it ends nearby, and that in this case the amplitude will be less than the maximum one. We can therefore plot the amplitude distribution in the space between the lattice points, similar to the amplitude distribution shown for the one-dimensional case in Fig. 6–1(2). Again, as in 6.1, the observed intensity may be taken as proportional to the integral value under the principal peak of the energy distribution, which is obtained by squaring the amplitude distribution. The limits of this peak are set by vectors **k** for which one extra wave-length path difference develops throughout the crystal, beyond that existing when **k** ends at the lattice point. For then the crystal can be divided into two halves, so that to each reflecting plane in the first half there exists one in the second half whose reflections are by $\lambda/2$ out of step with those of the first. If the areas of these corresponding planes were equal, full cancellation would occur; if they are unequal, only a surface effect remains over, instead of a volume effect.

Let us assume that the crystal is a parallelopipedon oriented according to the crystal axes, i.e. containing N_1 cells along $\mathbf{a}_1$, N_2 along $\mathbf{a}_2$, and N_3 along $\mathbf{a}_3$. Considering the direction $\mathbf{a}_1$, the path difference between neighbouring wavelets is $h_1\lambda$ for the diffraction of order (h_1, h_2, h_3). Therefore between the two extreme wavelets coming from the crystal the path difference is $N_1h_1\lambda$ if **k** ends at the lattice point (h_1, h_2, h_3). If we increase this path by $s\lambda$, where s is an integer, we get near-zero field. But this means that between any two neighbouring planes h_1 and h_1+1 of the reciprocal lattice, which have a distance $1/|\mathbf{a}_1|$ between them, there are interleaved N_1 parallel planes which are the geometrical loci for those wave-vectors **k** for which the resultant amplitude is zero, or nearly so. The same can be done in the two other directions with corresponding spacings $1/(N_i|\mathbf{a}_i|)$. We thus obtain a division of the cell of the reciprocal lattice into $N_1N_2N_3$ subcells whose shape is reciprocal to the shape of the whole crystal. On the walls of these subcells end the vectors **k** for which the diffracted optical field nearly vanishes; in the interior of each subcell the diffracted amplitude will have a (positive or negative) maximum value, and the intensity distribution a positive maximum, but the height of these maxima falls off rapidly with increasing distance from the lattice point, because an ever smaller fraction of the crystal produces those wavelets which are not cancelled out by others in opposite phase. All this is but the extension of Fig. 6–1(2) into three dimensions.

We have now filled reciprocal space with an amplitude distribution which is the same about each point of the reciprocal lattice. Whereas these points themselves are indicative of the crystal lattice, the subcell walls and the amplitude distribution contained between them have nothing to do with the inner periodicity of the crystal; they are, instead, determined by the external shape of the crystal. In fact, they would remain unchanged about the origin point (000) if the external shape were filled with an amorphous distribution of scattering matter —liquid or glassy—,while in this case the reciprocal lattice points outside the origin loose their significance, and with them the surrounding amplitude distribution. The construction of the zero amplitude subcell walls can be obtained easily by what is known in optics as Fresnel Zones.

Fig. 6–3(1) shows in a schematic way how the lattice points in reciprocal space are surrounded by the amplitude distribution (the 'shape transform' of the diffracting crystal). The division in the drawing is in

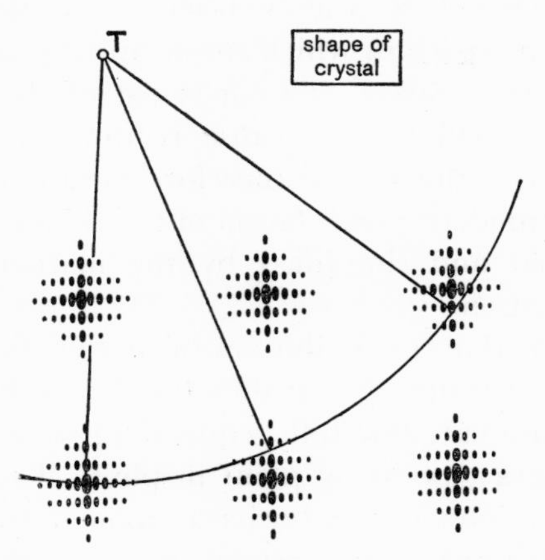

Fig. 6–3(1). Amplitude or intensity distribution surrounding the reciprocal lattice points in the case of diffraction by a finite rectangular crystal (Crystal Shape Factor).

twentieths and in tenths of the axes $\mathbf{b}_1$ and $\mathbf{b}_2$ and the shape of the crystal is indicated in the drawing. This corresponds to a crystal of 20×10 cells. The sphere of reflection intersects the distribution about the origin, and this would give rise to small angle diffraction. Besides, the sphere passes close by two other lattice points and the diffracted field in these orders will be given by the amplitude (or intensity) distribution on the surface of the sphere. In the upper intersection we see that the spot will be split into two parts, where the sphere intersects the two central rows passing through the point (h_1, h_2, h_3). These rows are principal peaks for either the h_1 or the h_2 directions, and as such have double width, as we know from the one-dimensional case of 6–1. Such 'intensity spikes' always occur when the shape of the crystal contains relatively large flat faces, and their occurrence is easily understood by applying a Fresnel zone argument. Laue was the first to point this out and to explain the splitting of diffracted spots in electron diffraction pictures of very small and regular, electrolytically deposited metal crystals of octahedral shape. If the crystals are larger, the whole amplitude distribution contracts around the lattice points and becomes unobservable; the integrated value of the intensity remains measurable, and the integration produces the Lorentz factor.

* * *

The above may be taken as an example of the way in which continuous amplitude and intensity distributions can be plotted in the space surrounding the lattice points of the reciprocal lattice. Because this is very closely connected to the mathematical theory of Fourier Transformation, the space in which the reciprocal lattice is imbedded is best called Fourier Space. This gets rid of the term 'reciprocal space', which is a bad term because reciprocity is a symmetrical relation between two things and therefore unsuitable for designating one of them. The modern presentation of the (kinematical) diffraction theory is governed and simplified by the mathematical notion of Fourier transformation.

It can be shown that a distribution of matter, or electron density, or any other property throughout the crystal, can be described in two fundamentally different, but fully equivalent ways. One is, to give this function, say $\rho(\mathbf{x})$, point by point in physical space, that is, as a function of the position vector $\mathbf{x}$ (or its component numbers x_1, x_2, x_3). The alternative one is to give, instead, a complete description of *all* the diffraction effects obtainable from this distribution of mass with all possible wave-lengths, and this means to give the value of the diffracted amplitude F throughout Fourier space, that is for all values of the position vector $\boldsymbol{\eta} = \eta_1\mathbf{b}_1 + \eta_2\mathbf{b}_1 + \eta_3\mathbf{b}_3$ in this space. The transition from one description to the other is performed by an integration and is therefore a straightforward mathematical operation to write down, though in many cases difficult to carry out. In the case that $\rho(\mathbf{x})$ is strictly periodic (which includes its extending everywhere to infinity) $F(\boldsymbol{\eta})$ is zero except at the points of the reciprocal lattice where $\boldsymbol{\eta} = \mathbf{h}$; this may be expressed by saying that Fourier space is reduced to an 'index space', namely the points of the reciprocal lattice corresponding to the periodicity of $\rho(\mathbf{x})$. In that case the integral over Fourier space degenerates to a sum, or the Fourier integral to a Fourier series.

Since we cannot observe $F(\boldsymbol{\eta})$ but only $|F(\boldsymbol{\eta})|$ from the intensity $|F(\boldsymbol{\eta})|^2$, we cannot perform the Fourier transformation which would lead us back directly from Fourier space to crystal space; instead, the 'Phase Problem' looms large between the two spaces (see also the next chapter).

CHAPTER 7

*Methods and Problems of Crystal Structure Analysis**

7.1. Various Forms of the Problem

a. *Description of the structure.* In the preceding chapter we have paid little attention to the actual configuration of a crystal. We have, in fact, considered the atoms to be point centres of scattering of the incident field which are situated at the lattice points $\mathbf{x}_l = \Sigma l_i\, \mathbf{a}_i$ ($i = 1, 2, 3$; l_i integers varying from $-\infty$ to $+\infty$). There is thus only one atom per cell. In a crystallized compound there must be at least as many atoms in the cell as the chemical formula indicates, or a multiple of this number, corresponding to, say, Z molecules. These atoms form the '*base*' associated with the cell. We distinguish them by a superscript s and their positions relative to the lowest-indexed corner of each cell are given by the 'base-vectors'

$$\mathbf{x}^s = x^s{}_1\mathbf{a}_1 + x^s{}_2\mathbf{a}_2 + x^s{}_3\mathbf{a}_3, \qquad (1)$$

where the $x^s{}_i$ are fractional coordinate numbers. The position of the atom of sort s in cell l as seen from the origin of the crystal is then

$$\mathbf{x}^s{}_l = \Sigma_i(l_i + x^s{}_i)\mathbf{a}_i. \qquad (2)$$

b. *Atomic factor.* The scattering power for X-rays is not the same for atoms of different sorts. Besides, since the size of the atoms is comparable to the X-ray wave-lengths, the angular distrubution of scattered amplitude is not the same for the atom as for the point scatterer which was considered so far. The wavelets issuing from various parts of the electron cloud of the atoms arrive with phase differences in the direction of observation and this makes the total amplitude received a function of the angle of scattering and of the size and distribution of the electron density in the atom. This function differs from the simple amplitude which, according to a classical calculation by J. J. Thomson, a single electron would give if it were

* See also parts of Ch. 10.

substituted for the atom; the ratio of the actual amplitude to this fictitious one is called the '*atomic factor*' f. Like all quantities that depend on phase differences caused by differences of optical path, f depends only on the order of diffraction, that is, on the vector **h** if we have to do with a crystal, or on the continuously variable position vector in Fourier space, $\boldsymbol{\eta}$, if we consider a single atom instead of a periodic array of such. The definitions are

$$\begin{aligned} \boldsymbol{\eta} &= \eta_1\mathbf{b}_1 + \eta_2\mathbf{b}_2 + \eta_3\mathbf{b}_3 \\ \boldsymbol{\eta}_{\mathrm{h}} \equiv \mathbf{h} &= \mathrm{h}_1\mathbf{b}_1 + \mathrm{h}_2\mathbf{b}_2 + \mathrm{h}_3\mathbf{b}_3. \end{aligned} \tag{3}$$

The experimental determination of atomic factors from X-ray observations was, and still is, an important aim, even though from 1926 onwards, with the advent of wave mechanics, these factors could be calculated with a fair, but not always sufficient accuracy (Hartree's method of self-consistent fields). By introducing the f-factor explicitly the remaining function of the atom is that of a point-scatterer.

c. *Structure factor*. Since the order of diffraction (h_1, h_2, h_3) indicates the differences of optical path for wavelets scattered by an atom and its neighbours along the directions of the axes $\mathbf{a}_i$, an atom in the cell which is removed only by fractions of $\mathbf{a}_i$ sends out a wavelet whose path length is compounded of the corresponding fractions of h_1, h_2, and h_3, respectively. That is, for the atom of sort s in (1) the path difference, measured in wave-lengths, will be

$$\mathrm{h}_1\mathrm{x}_1^s + \mathrm{h}_2\mathrm{x}_2^s + \mathrm{h}_3\mathrm{x}_3^s = \varphi^s(\mathbf{h}) \text{ (as in } \varphi^s(\mathrm{h}^s)\text{).} \tag{4}$$

For reasons which need not be explained here the path difference of φ wave-lengths against some standard wave is described mathematically by the exponential function exp with an imaginary argument, namely by exp $(-\mathrm{j}\varphi)$, where $\mathrm{j} \equiv 2\pi\sqrt{-1}$. Using this symbolism we can now write down the factor by which the amplitude diffracted into an order $\mathbf{h}$ ($= \mathrm{h}_1, \mathrm{h}_2, \mathrm{h}_3$) will be modified through the superposition of the fields generated by each sort of atom separately. Owing to the definition of the atomic factors f^{s}, this factor F, which is called the '*structure amplitude*', compares the amplitude of the diffracted wave received from the crystal in any direction to that amplitude which would be obtained if the atoms were all replaced by single electron scatterers. The formal expression of F is

$$\mathrm{F}(\mathbf{h}) = \Sigma_{\mathrm{s}}\, \mathrm{f}^{\mathrm{s}}(\mathbf{h}) \exp\left(-\mathrm{j}\varphi^{\mathrm{s}}(\mathbf{h})\right), \tag{5}$$

the summation extended over all atoms in the cell (the base), be they chemically alike or not.

It should be noted that in the precise direction of the diffraction (h_1, h_2, h_3) all the atoms in the crystal of any one sort s cooperate without any phase differences, and if the direction of observation is changed very slightly, secondary maxima and zero values will be generated by each kind of atom under the same angles. For this reason the factor F multiplies not only the amplitude obtained at the reciprocal lattice point $\mathbf{h}$ but also the entire distribution of amplitude in the neighbourhood of $\mathbf{h}$. The measured intensity of diffraction is therefore proportional to the square of the absolute value of the complex quantity $F(\mathbf{h})$, which is indicated as $|F(\mathbf{h})|^2$. It is from this quantity, and *only* from it, that we gain information about the contents of the cell, that is about the atomic structure of the crystal. Experimentally, between 100 and 20 000 observed intensities may be available, the higher number for the most complex crystals with large cells; each intensity has to be scaled up by division with angle-dependent factors like the Lorentz factor or the temperature factor (see Chapter 6) so as to yield a value proportional to $|F|^2$. If these values are entered as 'weights' at the points of the reciprocal lattice, this weighted lattice shows the data available for the structure determination.

If $F(\mathbf{h})$ itself (i.e. including the sign or the complex phase) could be used as weight, then a simple Fourier synthesis, consisting of the summation of a series of sin and cos functions with the $F(\mathbf{h})$ as coefficients, would lead back to the mass distribution in the crystal. Actually it is the scattering power or electron density which produces the diffraction effect and which we determine in a structure analysis; but we use the term 'mass' in crystal space as opposed to 'weight' in Fourier space. Exploration of the entire Fourier space by experiment would require the use of very long wave vectors, i.e. very short wave-lengths, and this is not feasible, even in principle, because for the shorter wave-lengths, such as gamma rays, the angles of diffraction become very small and, furthermore, classical scattering without change of wave-length gives way to quantum effects involving changes of the states of the atoms, the emission of photoelectrons, and so forth. This means that only a limited region of Fourier space, surrounding the origin, is explorable by X-ray diffraction. Factors like the Lorentz and temperature factors also limit the observable intensities to those of lower orders. Thus the 'aperture' through which we look at reciprocal space is limited and the picture in it incomplete. This means that if this information is transformed back into physical space it will give a somewhat blurred mass distribution. This blurring is often called 'break-off effect' or 'termination of series effect'. It is one of the minor troubles of actual structure analysis. From the large number of intensities observed in most cases it is seen that the available region in Fourier space usually contains many more data than the number of atomic coordinates that have to be determined.

It is this excess of data over the number of unknowns which makes up for the lack of information on the phases of the $F(\mathbf{h})$, but a systematic way of exploiting this overdetermination is still in its trial stage (Karle & Hauptmann's method of 'joint probability', see below).

d. *Algebraic and analytical structure determination.* In practically all of the early work—up to about 1930—the aim of structure analysis was to find the positions of the centres of the atoms in the cell, that is, the base vectors for its Z atoms $s = 1 \ldots Z$. This requires the determination of 3Z coordinates, or rather of $3(Z-1)$ coordinates, since only relative positions count. Besides, the distribution of masses over these sites must be found. In order to speak of point-sites, the atomic factors must be used in reducing the intensities. The finite number of unknowns is hidden away in the reduced intensity values via the structure factor. Their extraction from there is essentially a problem of algebra, but one for which no general solution has been found. Even in the very much simplified case when the same atomic factor may be used for all atoms, as in many organic substances because of the similarity of the atomic factors for C, N, and O, the simultaneous equations to be solved are of a high degree and quite forbidding. One has therefore to proceed by 'trial and error' methods, i.e. by systematic guessing and approximation.

On the other hand, the crystal may be considered as a periodic continuous distribution of scattering power or mass density, $\rho(\mathbf{x})$, and one can see the aim of structure analysis in the determination of this *function*. This is, of course, equivalent to an infinity of unknowns. The structure amplitude produced by such a continuous distribution differs from (5) only in that the summation over individual atoms is replaced by an integration over the contents of the cell, whereby the continuously variable $\varphi(h, x) = \mathbf{h} \cdot \mathbf{x}$ takes the place of $\varphi^s(\mathbf{h})$ from (4); also the lumped atomic factor $f^s(\mathbf{h})$ is replaced by the electron density itself which causes the scattering power in each element dv of the cell volume. Thus the structure amplitude in this case is

$$F(\mathbf{h}) = \int \rho(x) \exp(-j\varphi(h, x))\, dv. \qquad (6)$$

Again, the intensity is proportional to $|F(\mathbf{h})|^2$, and in this 'structure factor', which is derivable from the experiments, is hidden away the unknown function $\rho(\mathbf{x})$.

Clearly, the determination of this continuous function is an even worse proposition than that of a finite number of coordinate values in the algebraic version of the problem. And yet for many purposes the electron density $\rho(\mathbf{x})$ is what we are really aiming at, for instance in studies of the chemical bond. True, bond lengths and angles can be obtained from the algebraic locating of atomic centres, but only the analytical treatment reveals the actual electron densities on and near the directions of the bonds, for instance an excess density near the midpoints of the C—C bonds in diamond, or the small but definite differences of electron density surrounding the carbon atoms at the end and in the middle of the anthracene molecule (Fig. 10(3c)). $\rho(\mathbf{x})$ also contains information about the spreading of the atoms by the temperature motion.

Since the analytical approach includes the determination of the shape of the electron cloud of each atom, the measured intensity may not be reduced to that of a point scatterer by reduction of the intensities with the atomic factors. On the other hand, it is necessary to attribute the correct sign or phase to each of the $|F|$ before taking them as Fourier coefficients. In many cases this knowledge is obtained from the preceding location of the atomic centres by means of an algebraic structure determination. The two methods are therefore both employed in most cases.

7.2. The Algebraic Structure Determination

The indexing of the geometrical data obtained by any of the methods using monochromatic X-rays leads to the acceptance of a definite shape and size of the cell. Knowledge of the chemical formula, i.e. of the actual weight of the molecule, and of the specific weight of the crystal then tells the number Z of the molecules which are in the cell. This determination is not unique; we could, for instance, double one of the axial lengths, say a_3, and obtain a cell of double the volume and a content of 2Z molecules. With this cell all the h_3-values would have to be doubled and the bigger cell opens up the possibility that between these even h_3-values odd ones could be observed. Only if these are not found, even on close inspection of overexposed photographs, can we be satisfied that the smaller cell is the true one. In this sense the determination of the cell contains an element of intensity discussion besides the use of the geometrical data. In alloys, silicates, long-chain compounds and other structures that are close to possessing an internal periodicity of the base the search for the 'true' cell by means of 'inter-layer lines' or other forms of weak reflections is a very essential step in the crystal analysis.

The next step, which may be determining the placement of some of the atoms, is the determination of the space group. This again does not require a quantitative discussion of intensities, but only the observation of certain zero intensities occurring systematically, the '*absences*'. These are found whenever the structure contains glide symmetry elements because these lead to the interleaving of reflecting atomic planes with similarly populated ones at $\frac{1}{2}$, $\frac{1}{3}$, or $\frac{1}{4}$ of the spacing required by the cell. Not all space groups show absences, and even if such are observed, it does not fully determine the space group, but it restricts the choice.

The early structure determinations dealt with cases where there were only few atoms in the cell, and then their positions were often fully fixed by the knowledge of the space group or groups, especially in cases of high symmetry. Each symmetry element crossing the cell will reproduce an atom placed in the cell, except it is placed on the symmetry element itself; therefore, if the number of atoms in the cell is small, they have to lie on the symmetry elements and often on their points of intersection. In suitable cases, this fully determines the possible positions of the atoms, and since these points of intersection may be the same in related space groups, this may be a unique determination in spite of some indeterminacy of the space group. In other cases, the atomic positions may be restricted to lie on symmetry elements without being fixed otherwise. Any undetermined atomic coordinate is called a '*parameter*', and the value of a parameter can only be assigned from a discussion of the intensities. All modern crystal analyses deal with structures containing a large number of parameters—meaning anything beyond four or five parameters, so that their direct determination from a discussion of the intensities is not possible.

It is here that the '*trial and error*' method sets in, together with methods of '*structure refinement*'. The gist of this procedure is that for any assumed positions of the atoms it is a straightforward, if sometimes lengthy, matter to calculate the structure amplitudes F_h and thence the theoretical intensities $|F_h|^2$. These can then be compared to the observed ones. If the positions chosen were the correct ones the $|F|_{calc}$ and the $|F|_{obs}$ values should show a definite close correlation; coincidence is not to be expected because of the many factors influencing $|F|_{obs}$ which are not too well known (atomic factor, temperature factor, etc.). However, if the assumed atomic positions are in part correct or approximate, some parallelism between the calculated and observed F-values will be noticeable which indicates in which way, by shifting the atoms, a better agreement may be produced. As a measure of the overall agreement a '*residue*' or '*reliability number*' R is formed, usually of the form

$$R = \frac{\Sigma||F_h|_{obs} - |F_h|_{calc}|}{\Sigma|F_h|_{obs}}, \tag{7}$$

the sums being extended over all observed orders of diffraction. It will be noticed that R is really the inverse of a figure for the correctness of the structure proposal, since the correlation is the better the smaller R is.

Special methods have been developed for the refinement of a model structure once this seems to be not too far off the truth. The influence which certain shifts in the atomic positions have on the R-value are studied and improved positions calculated therefrom. This process is simple only in rare cases when the shifts of different atoms can be considered separately; usually all shifts are coupled together and the minimum R-value sought for is a true multidimensional minimum problem in the space of all atomic coordinates. This is a typical computer problem and it has been programmed for a variety of electronic computers:

trial structure – F^2_{calc} — R_1 – proposed shifts – new F^2_{calc} — R_2 – second shifts – third F^2_{calc} — R_3 – etc.

The machine can be programmed to do all this without help, and also to watch that the R-values decrease, and to stop when this is no longer the case or when a certain value of R (often 0.1, i.e. 10%) is reached. Beyond this, it is considered useless to go with algebraic determinations.

7.3. The Analytical Structure Determination

If the positions of the atomic centres have been determined in an algebraic determination, the signs or phases which are needed for converting $|F_h|$ into F_h itself are known from the last cycle of approximation. This information may now be applied to the $|F_{obs}|$ values, uncorrected for atomic factors and, possibly, temperature factor, in order to obtain a set of F-values which can serve as coefficients of a Fourier series representing the actual distribution of the electron density in the crystal. The summation, the *Fourier synthesis*, is required at the points of a conveniently narrow grid in the crystal cell. If the electron density curves are drawn on clear sheets of plastic for the consecutive layers of the grid, and the sheets stacked in the proper positions, the atoms appear floating in the cell as balls of high electron concentration.

The numerical work for a three-dimensional synthesis is so large that it has become a practical proposition only with the advent of the electronic computers. Before that, syntheses were restricted to two dimensions, and these were taken up soon after the first one-dimensional syntheses by Duane and Havighurst showed the usefulness of the method in 1925. The two-dimensional series require observation of intensities only in one zone of crystal planes, that is, in the reciprocal lattice, the knowledge of weights only on one plane passing through the origin. The summed series then represents the mass distribution obtained by projecting the spatial electron density on to a plane in the crystal which is normal to the zone axis. In order to obtain a significant projection, the direction of projection must be a very simple rational one which makes successive layers of cells match in projection. Of course the pictures of atoms may lie very close to one another in projection, even if the atoms are widely separated in the direction of projection. It is therefore not so easy to recognize atoms in the contour lines of the projection as it would be in a three-dimensional synthesis. The early diagram of anthracene (Ch. 10, Fig. 10(3b)) shows in a favourable case of two-dimensional synthesis what kind of indication of atoms can be expected; it may be contrasted with the next diagram showing a section through the same molecule based on a three-dimensional synthesis.

The Fourier summation in two dimensions was first carried out by W. L. Bragg (*Proc. Roy. Soc.* A 1929, *123*, 537) for the monoclinic crystal diopside, $Mg(SiO_3)_2$, using 30–40 reflections and summing at a grid of 24×12 points covering one eighth of the (a, c) face of the cell; the remaining parts of the cell follow from symmetry. The signs of the F-values were taken from a previous algebraic determination of the structure by Bragg & Warren. Even in this first try-out the number of numerical operations was quite high. It was inevitable that the repeated application of Fourier synthesis led to the invention of methods for abbreviating this procedure. Beevers and Lipson developed their 'Fourier strip' method which did more for the general introduction of Fourier methods than any other improvement, and remains, for its simplicity and educational value, one of the chief means for evaluating two-dimensional Fourier series of limited complexity. An account of the origin of the method is given by Lipson in Part VII.

The density calculation is always performed for the points of a grid in the cell, which must be fine enough to show individual atoms by their high electron density. Once the densities, measured in electrons per square ångström for the projection, are inscribed in the grid, the curves connecting points of equal density are easily drawn with a convenient height of the contour step.

A different kind of two-dimensional map of electron density is used for building up the mass distribution in three dimensions. For this purpose, *sections* through the crystal cell have to be constructed, each showing the contour lines of the spatial density (in electrons per cube ångström), so that by piling up such sections drawn on plates of clear plastic a vivid impression of the atoms is obtained as they float in the interior of the cell. Models of this type were first shown by Dorothy Hodgkin for the penicillin structure (cf. the remarks in Ch. 8 and 10). One of the many relations of duality between physical and Fourier space is that sectioning in one space corresponds to projecting in the other. We saw above how the projected electron density was obtained from the intensities lying in a cross section through Fourier space; we now obtain the sections through the crystal cell from Fourier series having as coefficients certain sums of the F-values normal to the plane of sectioning. These sums can be formed only if the phases or signs are correctly attached to the observed $|F|$-values.

7.4. *Methods of Phase Determination*

We have, so far, assumed that the phase of F_h is known from a preceding algebraic placement of the atoms. This, however, is not always necessary.

If the crystal structure has a centre of symmetry, this can be taken as the origin in crystal space, and, as is easily seen from the expression (5) or (6) of the structure amplitude, F_h will be a *real* quantity, so that only a + or − sign remains undetermined after observing the intensity $|F_h|^2$. The general phase problem is thus greatly reduced, and more centro-symmetric structures have been determined than corresponds to their natural occurrence. One could imagine that it might be possible with the help of computers to synthesize the Fourier series with all possible combinations of signs, and pick out thereafter the most attractive result. But even if only 30 Fourier terms are used, the different combinations of sign will number about 2^{30}, which is of the order of 10^9. Thus, if only one second were allowed for the inspection of each synthesis, it would take about 35 years to get through with it!

A more effective way is the '*heavy atom technique*'. The atomic factors of heavy atoms reach out to higher orders than those of light atoms; this is due to the high concentration of scattering power in the small inner electrons shells of the heavy atoms. By using only high-order reflections it is therefore often possible to locate the heavy atoms alone, of which there are usually few in the cell. From the known positions the phase of the heavy atom contribution to each order of diffraction can be calculated, and with it that part of $|F_h|$ which is due to them. If the observed $|F_h|$ is greater than this value, the likelihood is that the light atom contributions are additive, if it is smaller, they must be subtractive, and that fixes the phase. It is often possible to compare crystals which have essentially the same structure (isomorphous crystals) but contain heavy atoms of different weights. The difference of F-value produced by such changes then show clearly whether the phase to be attributed to a particular $|F_h|_{obs}$ is that produced by the heavy atoms alone, or not.

The classical example for the application of this kind of argument is J. M. Robertson's determination of the phthallocyanine structures in 1935/36. This case is particularly favourable because it is possible to compare the diffracted intensities of the purely organic molecule $C_{32}N_8H_{18}$ to those obtained after insertion of a nickel or platinum atom at the centre of the organic group. The heavy atom contribution outweighs that of the other part in determining the phases and the Fourier synthesis can proceed without any previous model. Even the assumption of the existence of atoms need not be made, since no atomic factor is used; the well rounded-off balls of high electron density appear in the course of the synthesis, and the existence of atoms is thus shown on purely optical grounds.

A similar statement can be made for the most complex structure so far analysed, namely myoglobin whose molecule contains some 2500 light atoms; here various heavy atom groups can be attached to the surface of this enormous globular molecule without appreciably disturbing either its internal constitution or the crystal structure it forms by interaction with adjacent molecules. It is clear that in a case of such complication any attempt at an 'algebraic' structure determination would be hopeless.

While the method of heavy atom substitution has been the most efficient means of supplying the signs or phases required for a Fourier synthesis, other methods have been helpful and are being improved continually. The scattering power of an atom undergoes a change when the wave-length of the incident radiation is shifted across an absorption edge of an atom. This is not a very large change, but Bijvoet showed that with accurate intensity measurements it can be traced and put to the same use as the substitution of a heavy or light atom. A different kind of phase relation near the absorption edge of an atom has been used by Peerdeman, working with Bijvoet, to distinguish between the diffraction by a right-handed and a left-handed structure. In this case measurements are made at only one wavelength, for which one of the atoms has a strong absorption. This is reflected in an imaginary part of the atomic factor of this atom which is always positive. If a right- and a left-handed structure are compared, the geometrical structure factor will change the sign of its phase, since the transition from one structure to the other is made by an inversion, i.e. replacing $+ (x, y, z)$ by $- (x, y, z)$. The addition of the always positive absorption part of the sensitive atomic factor will increase the $|F_{\mathbf{h}}|^2$ in one case and decrease it in the other. A comparison of the intensities from a d- and an l-crystal thus allows to tell which of these is built with right-handed screw axes. This important and rather unexpected method was applied to strychnine by Peerdeman. Instead of using two crystals, the author compared the intensities of several (h, k, l) reflections to those on the back faces, $(\bar{h}, \bar{k}, \bar{l})$; this serves the same purpose since only the product $\mathbf{h} \cdot \mathbf{x}$ appears in the geometrical structure factor.

7.5. *The Patterson Method*

What happens if the measured $|F_{\mathbf{h}}|^2$ themselves are used as coefficients of a Fourier series? The answer was given in 1934 by A. L. Patterson: the summed series represents the '*convolution*' or '*fold*' of the electron density distribution in the crystal.

To explain this term let us assume that the base of the crystal contains point atoms of masses m^s at positions $\mathbf{x}^s$. We can then draw all the vectors connecting one atom with any other one, i.e. $\mathbf{x}^{ss'} = \mathbf{x}^s - \mathbf{x}^{s'}$, from one origin, letting each end in a weight which is the product mass $m^{ss'} = m^s \cdot m^{s'}$. If the same distance occurs between several atoms, the total weight at the end of this vector will be the sum of the individual product masses. It is easily seen that all vectors will lie within a cell equal to that of the eight crystal cells which have the origin in common. We call the space in which we perform this construction Patterson space, and the distribution of product masses in it the self-convolution or fold of the mass distribution in the crystal cell.

If instead of condensed masses we have a density function $\rho(\mathbf{x})$ in crystal space, the corresponding fold is a continuous function $P(\mathbf{X})$ which is called the Patterson function. Its significance is that it indicates for every vectorial distance $\mathbf{X}$ in Patterson space what total product density will be picked up in crystal space if we let a vector $\mathbf{X}$ roam over all points of the crystal cell, that is,

$$P(\mathbf{X}) = \int \rho(\mathbf{x})\rho(\mathbf{x} + \mathbf{X})dv_{\mathbf{x}}. \tag{8}$$

The origin of Patterson space carries the product mass

$$\Sigma(m^{s})^{2} \quad \text{or} \quad \int(\rho(\mathbf{x}))^{2}dv_{\mathbf{x}},$$

as the case may be.

If the same construction is continued beyond the confines of one cell for the entire crystal, then the product-mass distribution in Patterson space will be periodic with the same cell as the crystal. For to every pair of masses lying at x^{s} and $x^{s'}$ in the same cell there exists a pair lying in different cells, so that their relative distance is greater than that of the other pair by a lattice vector x_1 of the crystal lattice. This periodic product-mass distribution, appropriately normalized so as not to become infinite, is the self-convolution of the crystal mass distribution; as such it contains the desired information folded away in a multiply superimposed fashion.

The great virtue of the Patterson function is that it can be constructed by Fourier summation with the real, positive, observed values $|F_{\mathrm{h}}|^{2}$ as coefficients. Therefore also projections and sections can easily be calculated in Patterson space, in contrast to crystal space. Patterson maps, i.e. two-dimensional representations of the projected product mass distribution by means of contour lines, are thus really an alternative representation of the observational data. As such they do not help solve the phase problem, but they change it to a geometrical and often more appealing form, namely to the problem of unfolding the folded distribution.

Much thought has been spent on the problem in this changed form and great progress has been made in its clear perception and in practical methods of unfolding, especially by M. Buerger (see next section).

7.6. *The Mathematical and Instrumental Approach to Structure Determination*

a. *Discussion of the problem.* Imagine the structure to be known from some source, so that the correct density $\rho(\mathbf{x})$ is known. By means of a

Fourier analysis we can deduce therefrom the coefficients F_h, including signs or phases. The $|F_h|^2$ check with the values obtained from the diffracted intensities. Now imagine the signs or phases to be altered in an entirely arbitrary way. This does not change the agreement with the observed values, but it destroys all the relations between the terms of the Fourier series which, with the correct signs, led to the building-up of very high values of the electron density in the central parts of the atoms. In short, instead of obtaining maps with layer lines showing well rounded and separated atoms, like a dish of fried eggs, we obtain from the changed series a disorganized picture corresponding to scrambled eggs. And yet, the reliability index is unchanged, and would remain so, even if more diffraction data were added.

This discussion shows that there is more to a structure determination than the mere conversion of ever so many intensity data from Fourier space to crystal space. It shows that in the actual procedure of structure determination we add essential information of a non-optical kind, often without being fully aware of its importance. Such information is the *existence* of atoms; their individual size, number and distribution of electrons as expressed by the atomic factor; their kind and number according to the chemical formula. A structure analysis would be rightly rejected if it did not show well rounded atomic peaks, correct electron density distributions in the light and heavy atoms, and acceptable values of bond lengths and angles. In fact, therefore, the criterion for accepting a structure as correct is not so much that the observed intensities of diffraction are well accounted for, but rather that this is the case within the limitations of a whole set of conditions drawn from experience.

Any attempt at making structure analysis a straightforward mathematical procedure which could finally be left entirely to computers has to incorporate at least some of the above restrictions of the non-optical type. The most usual one is the assumption that the observed intensities can be rendered through an algebraic structure determination based on the use of known atomic factors.

b. *Bond distances and bond angles*. The notion that atoms in crystals could be thought of as spheres of certain diameters which are packed so as to touch one another was elaborated by Barlow and Pope before the days of X-ray diffraction. Lothar Meyer's significant curve of the distribution of atomic volumes over the periodic system made such an assumption plausible. With the rapidly accumulating precise data on atomic distances in crystals attempts to assign fixed radii to atoms could be followed up in greater detail, and this was done in various places. W. L. Bragg was the first to use these radii for the construction of trial structures with which to start the trial and error course of determination. V. M. Goldschmidt showed that the 'radius' will vary with the state of ionization and the number of nearest neighbours, the coordination number, of the atom. Finally, some years after the advent of wave-mechanics in 1926, Linus Pauling deduced radii from the approximation of wave-mechanical calculations.

With the wealth of data on atomic distances and bond directions between atoms now established, the structure analyst of today has a fair idea of what values to expect in a particular case, and he generally becomes suspicious of a structure proposal if it leads to deviations of more than a few tenths of an Ångström from the expected distances, or to unusual bond angles or coordination configurations.

It would be an enormous help if this knowledge could be incorporated in some form as a non-optical condition at the beginning, instead of after the completion of a Fourier synthesis, but so far all attempts to achieve this have failed. The only form in which this condition can be made useful from the beginning of the analysis is in the construction of a model of what the molecule might eventually look like, and seeing how this can be fitted into the available space in the cell. This gives an initial structure for the trial and error process which has a chance of being not too far off the truth.

c. *Positiveness.* The electron density is an essentially positive function throughout the cell, and in the case of centrosymmetric structures this leads to some restrictions on the signs of the Fourier coefficients. The first to find how to use the positiveness were D. Harker and J. S. Kasper in 1948. They showed that this property led to relations such as

$$F_H^2 \leqq \tfrac{1}{2} F_O(F_O+F_{2H}) \tag{8a}$$

$$(F_H+F_{H'})^2 \leqq (F_O+F_{H+H'})(F_O+F_{H-H'}) \tag{8b}$$

$$(F_H-F_{H'})^2 \leqq (F_O-F_{H-H'})(F_O-F_{H-H'}) \tag{8c}$$

(H stands for h, k, l; H+H′ for h+h′, k+k′, l+l′, and 0 for 0, 0, 0.) Now F_O is known from the chemical formula and the density, since it is the total number N of electrons in the cell. (In the forward direction (000) all electrons scatter in phase, without any destructive interference.) The first of the inequalities establishes a restrictive connection between two Fourier coefficients, namely any F_H and F_{2H}.

This may serve to determine the sign of F_{2H}. Let for instance $|F_H| = 0.8$ N, $|F_{2H}| = 0.45$ N. Then allowing either sign for F_{2H} one gets from (8a)

$$0.64\ N^2 \leqq \tfrac{1}{2}N\ (N \pm 0.45\ N)$$

or

$$0.64 \leqq \tfrac{1}{2}\ (1 \pm 0.45)$$

The two values which the right hand side can take are 0.725 and 0.275, of which only the first, which corresponds to the + sign of F_{2H}, is possible for the inequality to hold.

The two lower inequalities (b) and (c) involve four Fourier coefficients besides F_o, and in order to be used the signs of at least two coefficients should be known. It will be noticed that inequality (a) follows from (b) when H is put equal to H′. Many special inequalities can be formed with regard to the symmetry of various space groups and they have proved extremely useful in opening up a way to structures to which other approaches had been unsuccessful, like Harker & Kasper's original object, dekaborane $B_{10}H_{14}$. Their applicability is restricted to the large F-values; for the smaller F they are indecisive.

Mathematically the condition that a Fourier series represent a function which is everywhere positive can be expressed as the non-negativity of a series of determinants of increasing order formed from the Fourier coefficients. These more general inequalities include as the simplest cases the ones discussed above, which seem to be the most powerful ones. Much interesting work has been done on the general lines by Karle & Hauptman, McGillavry, Goedkoop, Bertaut, von Eller and others, but it can not be said to have increased the applicability of inequalities for practical purposes.

* * *

There are three further developments which have proved their value for structure analysis. One is the computational aid obtainable nowadays by means of electronic digital computers. A great many lengthy calculations, which it would take years to perform by hand machines, have been programmed for several types of machines, and this service is increasingly used. From the administrative point of view it requires getting used to the idea that the 'desk work' involves expense like the experimental work for which nobody would doubt it. If the man-hour work performed by the machine is converted into salaries, the savings would become apparent, but of course the fact remains that with machine work projects of a complexity can be undertaken which would never be tackled without it.

Of the analogue computers Ray Pepinsky's XRAC (for X-ray analogue computer) is by far the most important. In this ingenious electronic device the magnitudes of a large number of $|F|$-values are set on dials and they can enter the Fourier synthesis with a positive or negative sign according to the flipping of a switch. The machine then performs instantaneously not only the summation of the two-dimensional Fourier series with these coefficients, but also displays the result in the accustomed form of contour lines on a television screen from where they can be photographed if desired. Besides, the magnification can be varied instantaneously, so that either several cells of the projection are displayed, or one only, or a part of it. With this machine it is possible to try out on the spot whether the change of sign of a Fourier coefficient makes the electron projection more acceptable or less. As a criterion for this, a low background between well separated atoms is usually taken.

A counterpart of this machine is one for Fourier analysis, called SFAC (for Structure factor analogue computer) in which the Fourier coefficients or structure amplitudes are instantaneously calculated after atomic coordinates and atomic factors have been set on dials.

A second important advance in the technique of structure determination is to be seen in Martin Buerger's profound studies of unfolding or deconvoluting of the Patterson function. This function was obtained by searching the cell for vectors leading from one atom to another and then shifting these vectors to the origin of Patterson space and labelling them with the product mass. Therefore, if we imagine one of the atoms to be shifted to the origin of Patterson space, the relative vectors from it to the other atoms of the base will be among the vectors in Patterson space. With Z atoms forming the base, there are $Z-1$ such relative vectors, whereas the Patterson function contains $Z(Z-1)$ peaks because any one of the Z atoms has to be shifted to the origin. (Many of these vectors may coincide, and this makes the unfolding harder.) Buerger succeeded in constructing several 'image seeking functions' with which to find those Patterson peaks which belong together and form an image of the base. His theory is a great step in the direction of systematic unfolding and has been successfully applied in a large number of structure determinations; it presents no automatic solution because it requires attention and ingenuity in its application, especially to the unfolding of projections to which it has been mainly applied.

Finally a few words should be said about the statistical approach to the problem of structure analysis. Since the amount of data for an algebraic determination often exceeds the number of unknowns considerably, one can afford to use the data is a statistical way rather than as single and independent pieces of evidence. A. J. C. Wilson made important contributions by showing how symmetry elements of the structure, including the directly not observable inversion centres, could be deduced from suitable averages of $|F|^2$-values. In a series of papers beginning in 1953 J. Karle and H. Hauptman have developed a method of determining probable sign relations between Fourier coefficients from 'joint probabilities' for the simultaneous appearance of certain sums of $|F|^2$-values. This method has been successfully applied to a number of structure determinations and comes closest to the ideal of a fully automatic derivation of a structure.

PART IV

The Growing Field

CHAPTER 8

*The Growing Power of X-ray Analysis**

by W. L. Bragg

It is now nearly fifty years ago that von Laue, whose recent sad death has caused us so much grief when we were all looking forward to having him with us here, made his discovery of X-ray diffraction by crystals. I propose in this talk to survey the stages by which the science of X-ray analysis has developed since that time. It is interesting to trace the influences which determine the rate of development of a science. It was a favourite saying of my father's that, after a year's research, one realizes that it could have been done in a week. This saying must of course not be interpreted literally. There are many cases in research where data have to be laboriously collected before one can arrive at an answer. He was, I think, referring to the fumblings of thought, the exploration of alleys which turned out to be blind, the lack of knowledge of some vital bit of information which was there for the asking if one had known where to ask. All of us must have said to ourselves at one time or another, when at last a solution was reached, 'How blind I have been'.

X-ray analysis is particularly a subject where a review of this kind is fascinating, because it is so well integrated. It is a branch of research with a discipline of its own, and one with features rather unlike those of any other science. Those who pursue it are a very closely united band, who talk the same language and keep closely in touch with each other. At the same time it is a typical border-line science, which has had a great influence on many other sciences, and has used their body of knowledge as a basis for its discoveries.

It is convenient to divide the story into certain periods and I propose to take each in turn, discuss the important advances during the period, and sum up the stage which the subject had reached at its end. Our first period, starting with von Laue's discovery, may be taken to end in

* Congress Lecture given at the Fifth International Congress of the International Union of Crystallography in Cambridge, 1960 (adapted).

1920. It includes the First World War, a time when little research was possible, but even during the war there were some outstanding contributions, and at the end of it a foundation had been laid on which all subsequent work has been based. The next decade, 1920–1930, witnessed X-ray analysis taking shape as a quantitative science, with the result that it could be extended to more complex structures. The next decade, 1930–1940, saw the Fourier series developing into maturity as a standard method of attack; its principles had been realized much earlier but only in that decade did it begin to play an essential role.

The Second World War interrupted research far more completely than the first. Science played a far larger part in it, and scientists had to abandon all thought of research, so in a general survey of this kind we may consider it as a blank period and take our next decade to be 1945–1955. To me this decade appears as the flowering of the trial and error methods of X-ray analysis, which except in a few favourable cases had been characteristic of X-ray work from the start. The phase difficulty was accepted as a necessary concomitant of X-ray analysis, and the outstanding researches were characterized by brilliant *tours de force* in which success was attained in spite of it. Finally, the decade in which we are now living, from 1955 onwards, is one in which a great stride forward has been made. The structures which are tackled are so complex that no trial and error methods could possibly succeed. X-ray analysis has assumed a new aspect in which the phase difficulty must be and has been defeated and since the new methods are direct methods, machines can be our slaves to handle vast masses of data because they can be given clear and precise orders. We may perhaps liken these stages to geological eras, and say the primary extended until 1920, the early and late secondary from 1920 to 1940, the tertiary up to 1955, and say that we are now living in a quaternary age of X-ray analysis. Any such divisions are of course arbitrary, and only justified as a skeleton plan on which to base our review. I intend to use as illustrations outstanding contributions which are landmarks, but it will be realized that it is impossible to refer to all such events and if the work of a particular school is reviewed in some detail, it is because it affords a measure of the advance of the subject, not because it was the only work of its kind at the time.

Let us now review the first period during which the foundation of X-ray analysis was laid, and try to trace the interplay of the various lines of thought and experience which gave it birth. The story of its start in 1911 and 1912 has been told at length in Chapter 4 and need not be repeated here. I start the account therefore from the moment my father began his work.

My father was intensely interested in von Laue's results because at the time he upheld the theory that X-rays were a type of corpuscular radiation, the neutral pair theory. This theory was not so strange as might appear. He had been led to it by deducing, what we now know to be true, that when X-rays ionize a gas the energy is handed over to individual electrons here and there in the gas, as if they had been struck by a projectile. It must be remembered that this was before the days of the Wilson Cloud Chamber, and before the dual wave and particle nature of radiation had become a familiar idea. We had many discussions about the results in the summer holiday of 1912, and when I went back to Cambridge as a research student in the autumn I pored over von Laue's pictures. Here again chance played a part. It must be difficult nowadays to imagine how utterly unaware physicists were at that time of the geometry of three-dimensional patterns. I happened to know a little about it, because Pope and Barlow had proposed a valency-volume theory of crystal structure which although incorrect was highly suggestive, and a member of a small scientific society to which I belonged had given a paper on this theory. It brought home to me the fact that atoms in space lattices were arranged in planes! J. J. Thomson had been talking about the pulse theory of X-rays, and to cut a long story short I explained von Laue's results as due to the reflection of a band of 'white' radiation by the planes of the crystal. I published a note in *Nature* showing the reflection of X-rays by a mica sheet at a series of angles. Further, I showed that the spots von Laue had obtained with zincblende were characteristic of a face-centered cubic crystal, having heard about such a thing by studying Pope and Barlow's papers. Pope was intensely interested and encouraged me to try Laue photographs of NaCl and KCl, in which he also believed the atoms to be in cubic close packing. These results established the structure of the sodium-chloride group of crystals (NaCl, KCl, KBr, KI).

This tentative start of crystal analysis, however, was soon completely superseded by my father's development of the X-ray spectrometer at Leeds, and I think one can trace the reason why the lead in crystal analysis so rapidly passed to this country although it had started in Germany. My father was supreme at handling X-ray tubes and ionization chambers. You must find it hard to realize in these days what brutes X-ray tubes then were. One could not pass more than a milliampere through them for any length of time or the anticathodes got too hot. The discharge drove the gas into the walls; one then held a match under a little palladium tube which allowed some gas to diffuse through and so softened the tube. The measurement of ionization with a Wilson goldleaf electroscope was quite an art too, and my father had thoroughly mastered all the techniques in his researches. The great strategical importance of the ionization spectrometer was that it enabled reflections to be studied one by one. The discovery of three characteristic lines A, B, and C, of Platinum was announced in April 1913 and my establishment of the rocksalt structure enabled wave-lengths to be assigned to these lines. It is not generally realized, I think, to what an extent my father founded the science of X-ray spectroscopy in 1913. He had tubes made with anticathodes of Platinum, Osmium, and Iridium, of Palladium and Rhodium, and of Copper and Nickel. He found the wave lengths of their characteristic lines, and by measuring absorption coefficients he identified them as Barkla's L radiation in the Platinum group of elements, and K radiation in the others. Whiddington had established the energies of the cathode rays required to excite Barkla's K and L radiations in atoms of different atomic weight, so my father was able to link wave-length with energy and show that the relation was in agreement with Planck's quantum law. He found the K line pattern and the L line pattern to be similar for successive elements, and showed that the frequencies increased roughly as the square of the atomic weight (June 1913). He identified the absorption edges, as they are now termed. The link with Moseley's work is interesting. Moseley and Darwin had surveyed the spectrum emitted by a platinum anticathode, but missed the characteristic lines owing to setting their slits too fine. My father's success encouraged Moseley to review the lines for a continuous series of elements from aluminium to gold. Bohr was then at Manchester, and Moseley's triumph lay in explaining the succession of frequencies by Bohr's theory of spectra and so identifying the atomic number as the number of positive electronic charges in the nucleus (April 1914), as had been suggested by Van den Broek in the previous year.

I had of course the heaven-sent opportunity at this stage of joining in the work with the ionization spectrometer. My father was more interested in X-rays than in crystal structure, and I was able to use the spectrometer for investigating the latter. We published the diamond structure together, and I established the structures of ZnS, FeS_2, CaF_2, $NaNO_3$, and $CaCO_3$. The structure of pyrites, FeS_2, provided the greatest thrill. It seemed impossible to explain its queer succession of spectra until I discovered, going through Barlow's geometrical assemblages, that it was possible for a cubic crystal to have non-intersecting trigonal axes. The moment of realization that this explained the iron pyrites results was an occasion I well remember. I tried to explain it to an aunt who happened to be in the room, with indifferent success.

I will select three further papers which appeared during this period as standing out from the rest. The first was a paper by my father, 'The Intensity of Reflection of X-rays by Crystals'. He showed that one could get a quantitative measure of intensity reflection, which was not affected by variations in orientation of the blocks of an imperfect crystal, by sweeping the crystal at a uniform rate through the reflecting position. Comparing the reflections from rocksalt with h, k, l even, he found that the intensities fell on a single curve when plotted against sin θ. This sweep procedure was the start of accurate quantitative measurement, and is still the standard method. My father also measured the Debye temperature effect on rocksalt. The precision of the measurements would be considered quite respectable today, and is remarkable when one considers the X-ray equipment available at that time.

The second contribution was made by Darwin's two theoretical papers in 1914 on the intensity of X-ray reflection. Darwin calculated that a perfect crystal should give complete reflection over a range of a few seconds of arc. He realized that the actual reflection was too strong to be accounted for in this way, correctly explained the discrepancy as being due to the crystal consisting of slightly disordered blocks, and obtained the formula for the mosaic crystals. They were remarkable papers. It was not till after the war that X-ray analysis reached a point where Darwin's results could be used; in fact Ewald then quite independently calculated the same formulae without realizing that Darwin had established the theory at the very start of X-ray analysis.

The third contribution appeared in my father's Bakerian Lecture in 1915; a quotation from it will show its significance: 'Let us imagine then that the periodic variation of density (in the crystal) has been analysed into a series of periodic terms. The coefficient of any term will be proportional to the intensity of the reflection to which it corresponds.' It was the start of Fourier analysis, and my father used it to get some idea of the distribution of scattering matter in the atoms.

To sum up, then, the achievements of this first period from 1912 to 1920 when active work was resumed after the war:

(a) The wave-length of X-rays had been established.

(b) A number of simple crystals had been analysed, including several with one parameter, and it had been shown that this parametef could be fixed with a high accuracy by comparing the orders or spectra.

(c) A method for the accurate measurement of intensity had been found.

(d) The Debye effect had been measured.

(e) We had Darwin's formulas for reflection by perfect and mosaic crystals.

(f) It had been realized that each crystal diffraction corresponds to a Fourier component of the density of the crystal.

(g) Finally, a whole new range of crystalline substances had become available through the powder method, developed in 1916 by Debye and Scherrer and independently a year later by Hull. In order to get a sufficient intensity, we used enormous crystals with the ionization spectrometer. We asked our colleagues to provide specimens an inch or two across if possible, though we had to be less ambitious in the case of diamond, which fortunately had a low absorption coefficient and gave very strong reflections. The powder method opened the way to examining microcrystalline material.

The decade 1920–1930 witnessed a great advance in the application of X-ray analysis to more complex crystals, particularly inorganic crystals. I shall abandon the strictly chronological order and summarize the inorganic field first, leaving until later to recount advances in other fields. The advance was made possible in the first place by the development of measurements of absolute intensities, in the second place by the earlier results leading to an understanding of the principles which govern the structure of inorganic crystals, and this in its turn to analysing ever more complex structures.

The absolute quantitative methods were based on Darwin's theory of diffraction by the mosaic crystal, which related the integrated reflection when a crystal was swept at a given angular rate through the reflecting position to the intensity of the X-ray beam falling per second on the crystal. The crucial quantity is $E\omega/I$, when E is the integrated reflection, ω is the rate of turning, I the intensity of the incident X-ray beam. A. H. Compton in 1917 had tested Darwin's formula with rocksalt and found that the absolute intensity of reflection was of the right order of magnitude. Bragg, James, and Bosanquet in 1921 and 1922 made a thorough examination of rocksalt, in which they estimated the effects of primary and secondary extinction, and determined the scattering curves of sodium and chlorine. Rocksalt was chosen for the experiment because a face can be reduced to an almost ideal mosaic state by light grinding. The importance of the variation with angle of the scattering by an atom (f-curves) became apparent, and Hartree in 1925 calculated f-curves based on Bohr orbits. Later in 1928, as the wave functions replaced the Bohr orbits, he recalculated their curves with his well-known 'self-consistent fields'. In the case of heavier atoms, their curves are very similar to f-curves calculated for the 'Fermi atom' which treats the electrons as an atmosphere around the nucleus. The peak of this line of work is represented by the paper by James, Waller, and Hartree on the Zero Point Energy in the Rocksalt Lattice. By accurate measurements of intensity of reflection at different temperatures, and by using Hartree's values for the f-curves based on self-consistent fields, they found that 'the f-curves so calculated agree very closely with the experimental curves on the assumption that the crystal possesses zero-point energy of half a quantum per degree of freedom, as proposed by Planck'.

This understanding of the law governing absolute intensity was concurrently applied to problems of crystal structure. The integrated reflection for rocksalt (400), 1.09×10^{-4}, for molybdenum $K\alpha$ radiation, formed a convenient standard for establishing absolute intensities for other crystals. The first structure, I believe, which was based on the quantitative approach was that of barytes, $BaSO_4$, analysed by James and Wood in 1925. It is an orthorhombic structure, with eleven parameters. A simplification was introduced by assuming the SO_4 group to be tetrahedral, but the assigned structure was rigorously confirmed by comparing calculated and observed absolute values. One only has to recall that only a few years before a crystal with one or two parameters had been thought of as at the limit of determination to realize the power of the new methods. The tables in papers began to contain numerical comparisons of amplitude, instead of the mystic letters v.s., s, m, w, v.w. for very strong to very weak.

In 1928 West and I wrote 'A technique for the examination of crystal structures with many parameters' in which we summed up our experience—a short guide to the art which might interest the reader nowadays who wishes to assess the state of knowledge at that time.

We went on at Manchester to apply the new technique to a series of silicates, not because we were interested in silicates, but because their structures were of the right order of complexity and we could get ground sections perpendicular to the main zones and conveniently measure their integrated reflections. The first analyses were laborious frontal attacks. There were, however, three aids which increasingly enabled short cuts to be made. Niggli, Wyckoff, and Astbury and Yardley, had drawn up their tables for determining the space group of symmetry. In the first analyses the need for such a systematic survey was not felt; the structures were simple and we worked out the space-group in each special case *ad hoc*. Now we realize the usefulness of a directly determined frame-work of symmetry elements to which the atoms must conform. Next, ionic radii had been established. I put forward evidence in 1920 that they were additive, and Wasastjerna in 1923 corrected my radii to a more realistic set. The exhaustive survey of simple structures by V. M. Goldschmidt and his school established the radii securely. A classic case of the determination of a structure by space group and atomic radii alone was that of beryl. When West had found the space-group, I well remember that we worked out the structure in less than an hour, which surely must be a record for a quite complex crystal. The scheme of the silicate structure in general was beginning to emerge when Pauling in 1928 published his great paper on 'The Coordination Theory of the Structure of Ionic Crystals'. This paper was a real landmark, it revealed the fundamental principles underlying all inorganic crystals. It had its most striking illustration in the silicates, and I have always felt that the order and simplicity in their structures revealed by X-ray analysis is one that gave the deepest aesthetic satisfaction. I think the turning point in understanding silicates was the solution of diopside $CaMgSiO_3$, by Warren and myself in 1928, because it showed the tetrahedral SiO_4 groups to be linked by corners to form a chain of composition SiO_3. Before that, chemists and mineralogists had postulated the existence of a number of acid radicles with different Si/O ratios. Taylor's analysis of the felspar structure in 1933 completed our knowledge of the main silicate structures.

Two other lines of research started in this period, the study of organic crystals and the effective use of Fourier representation. The first investigations of organic crystals by my father were not really straight determinations of structure by X-ray analysis, a problem at that time far too complex for direct solution. Rather, it was shown that molecules of the shape predicted by organic chemistry, with bonds like those in diamond, could be fitted into the measured cells. My father examined anthracene and naphthalene in 1921, and long-chain compounds were measured by Muller and Shearer, Piper and Grindly, and de Broglie and Friedel in 1923. Later, in 1929, Mrs. Lonsdale determined hexamethylene benzene. These investigations were important as a first breaking of new ground.

Towards the end of the period, in 1925, a new technique for crystal analysis took form, the two-dimensional Fourier analysis. I confess that the events at that time have always caused me a feeling of remorse. One-dimensional series were first used by Duane and Havighurst in 1925 to determine the electron distribution in the sodium and chlorine atoms, and they had been applied in a number of cases to establish electron distribution in sheets parallel to a crystal plane, based on the measurement of successive orders. Now about this time West and I had measured all the reflections, within a maximum value of sin θ, for the three principal zones of diopside. I had tried without success to see (what seems so obvious now) how such measurements could be used to get a two dimensional projection as a sort of tartan plaid. My father wrote to me to suggest the solution ought to come from a criss-crossing of Fourier elements in all directions, and a trial with our diopside data showed peaks representing the metal, silicon, and oxygen atoms. I wished to publish the paper with my father, but he insisted I should publish it alone. The first two-dimensional Fouriers appeared therefore as the three principal plane projections of diopside, though credit for its start is due entirely to my father.

To sum up, I think the outstanding feature of the 1920–30 period was the development of accurate and absolute intensity measurements, which led in particular to a wide knowledge of complex inorganic compounds and so to a much deeper understanding of the principles which govern their structure.

In reviewing the next period, 1930–40, I see it as the great era of the two-dimensional Fourier series, in particular as adding geometrical precision to the molecules of organic chemistry. Concurrently the powder method came to maturity in a field where it was particularly applicable, the structure of alloys.

Two-dimensional Fourier projections demand a large number of measurements, since all the reflections around a zone are needed up to a maximum sin θ value which decides the resolution. It so happened that in the case of diopside these reflections had been laboriously accumulated with the spectrometer, but in general this instrument had achieved its successes by the accurate measurement of a limited number of reflections. Now more rapid methods of data-accumulation became essential. The rotation photograph was first proposed by Niggli in 1919, and the Weissenberg camera supplied the need, combined with accurate photometric measurements of the spots. A paper by Bernal in 1926 had systematised the interpretation of these photographs, and a definite technique was established. The organic crystals examined were centro-symmetrical and one only needed signs in phase determination. Trial and error, based on stereochemistry, gave a rough solution which yielded the principal signs, and with them a Fourier projection was calculated and refined by deducing further signs. As an example of this body of work I may cite the series of papers by Robertson and his school. In 1932 he refined anthracene, then analysed durene, dibenzyl, benzoquinone, resorcinol, hexamethylene benzene, and other molecules. These structures had a number of parameters in the 20–30 range. A landmark was the direct determination of phthalocyanine with 60 parameters, made possible by the substitution of a heavy atom at the centre of the group. The molecules in each case confirmed the structure predicted by the organic chemists. The new contribution was the accurate determination of bond-lengths and bond-angles, and their interpretation in terms of the nature of the chemical bond. The Fourier series thus became the major method of attack, using measured amplitudes and postulated signs, as opposed to the older methods of moving atoms about till the diffraction effects were explained.

Patterson's famous paper 'A Direct Method for the Determination of the Components of Interatomic Distances in Crystals' appeared in 1935. The 'Patterson' was the first of a series of discoveries about the ways Fourier series can be manipulated for the purposes of X-ray analysis, a series which seems to come from an inexhaustible mine. New applications are continually being found. We think now in terms of Fourier series and reciprocal lattices, and it is in the present era of our review that this new trend in crystal analysis became established.

Westgren in Sweden started his study of alloys in 1921, and was the pioneer in this field. He was followed by Bradley who worked with Westgren and published the well-known structure of γ-brass and α-manganese in 1926 and 1927. But it was in this 1930–1940 period that Bradley and his school made their great contributions to the studies of alloys and alloy phase-diagrams in a brilliant series of papers. The powder method was employed with a virtuosity which has perhaps never been excelled since, if indeed it has been equalled. Highly complex structures were analysed. The knowledge so gained enabled Hume-Rothery's electron-atom ratio postulate to receive an explanation by Jones in terms of Brillouin zones. Phase boundaries in binary and tertiary systems were accurately outlined. The order-disorder transformation was explored by Bradley and by Sykes, and led to theoretical treatments by Dehlinger and by Williams and myself, and even proved to be so fascinating to the theorists that Bethe and Peierls were led into giving it their consideration. It is no exaggeration to say that the principles of metal chemistry for the first time began to emerge.

This brings the story up to the beginning of the Second World War, and it is convenient to resume it at the end of the war and take as our next decade the period 1945–1955. *Acta Crystallographica* started publication in 1948, providing the X-ray crystallographers with a journal of their own. The excellent way in which the journal has been edited by Ewald and his co-editors has had the most beneficial effect on the advance of our subject. In particular we in this country owe a great debt to R. C. Evans who gave so much of his time and energy to *Acta*. Instead of having to look for new work in many journals, we could be sure of keeping abreast of new developments by finding all the most important work in this journal. A glance through the numbers of the journal shows the general trend during the period under review. The majority of papers represent the exploitation of the methods which had been developed or at least foreshadowed before the war, applied to organic compounds, inorganic compounds, and alloys. Many improvements to the techniques were made, resulting in the successful solution of more complex compounds, in increasing accuracy of the determination of the atomic arrangement, the detection of the position of hydrogen atoms, and the measurement of individual thermal vibrations. The volume of work is so great that it is hard to single out individual achievements without a feeling that injustice is being done by not mentioning many others. I shall have to give up any attempt to summarize what may be described as the vast extension of crystal chemistry by the use of the established methods of analysis, and confine myself to what I see to be the outstanding achievements which broke new ground.

In the organic field, most determinations of structure had hitherto confirmed the structural models of the organic chemist. Now a very difficult kind of 'sound barrier' began to be passed, that of telling the organic chemist something he did not already know.

I may perhaps select the determination of the structure of cholesteryl iodide by Carlisle and Crowfoot (Mrs. Hodgkin) in 1945 as an early example of such an achievement. Bernal's studies of the sterols in 1932 had given some indications of their structure, and in the interval the chemical nature of the sterol structure had been largely established by the organic chemists. Carlisle and Crowfoot, starting with clues to the phases indicated by the heavy iodine atoms, were able to fix the positions of the thirty-three carbon atoms in the asymmetric molecule, and so establish the sterol skeleton with precision.—Then there was the neck-and-neck race with the organic chemists for the solution of the structure of strychnine. Bijvoet in 1947 obtained a first rough Fourier synthesis of the sulphate on the 010 plane, getting signs by comparing sulphate and silicate. In 1948 he attained a precise projection, with which only one model was compatible, unaware that just previously Sir Robert Robinson had succeeded in making the right choice between alternative models on chemical grounds. These papers which established the structure of strychnine are models of brevity, some two pages in length, which might well be studied by our more diffuse writers! The results were confirmed by Robertson and Beevers.

'The X-ray Crystallographic Investigation of the Structure of Penicillin' by Crowfoot, Bunn, and colleagues, is famous. The investigation lasted over four years. Success was the result of team work in which the chemists and crystallographers made essential contributions. To quote from the paper: 'Throughout the whole of the X-ray crystallographic investigation of Penicillin we have been working in a state of much greater ignorance of the chemical nature of the compounds we have had to study than is common in X-ray analysis.' A large variety of X-ray techniques was employed to establish the structure both of the degradation products and of the complete molecule. The description of the analysis is in itself a small text-book on X-ray analysis.

The 'sound barrier', to which I have likened the surpassing of the bounds of the organic chemists' knowledge, was broken through with a loud report in the case of Vitamin B_{12}. Crowfoot and colleagues published their completed analysis in 1957 after eight years study. The asymmetric molecule has the composition $C_{63}H_{88}N_{14}O_{14}PCo$, and a vital role in the analysis was played by the heavy cobalt atom. In some respects the chemical evidence was definitely misleading, and the establishment of the structure was mainly due to X-ray analysis. We all recognize this analysis of B_{12} as an achievement which stands in a class of its own.

A noteworthy event in this period was the establishment of methods for determining all the characteristics of the space-group symmetry of a crystal. After we had been pointing out for many years in our lectures and papers that it was impossible to detect whether a crystal had a centre of symmetry by X-ray methods alone, Wilson showed us how to do it by a statistical survey of the distribution of diffracted intensities.

Like all brilliantly original ideas, it seems so obvious when pointed out. The last element of uncertainty was removed by Bijvoet and his colleagues who used the phase-change in scattering to distinguish between a right and left handed configuration of atoms. They established the absolute configuration of the tartrate ion in sodium rubidium tartrate. It was of course an even chance whether the actual configuration corresponded with the 'd' and 'l' convention of the organic chemist; actually it turned out to be in accord with the convention.

Another very interesting body of work in this period was the series of papers by Karle and Hauptmann, Harker and Kasper, Goedkoop, Sayre and others which appeared from 1950 onwards. They constituted an attempt to solve a structure by direct methods. The phases of the structure factors must be such that, combined with the measured amplitudes, they give a 'reasonable' answer. They must, for instance, produce a Fourier in which the densities are everywhere positive, and outline nearly spherical atoms with an appropriate electron distribution. This consideration is of course the basis of all trial and error methods which start by assuming the presence in the structure of such atoms, and seek to find their positions. These authors attacked the problem from the other end, seeking to get general relationships between the structure factors which must hold because of the physical nature of the diffracting body, and which would lead directly to establishing a number of phases sufficient to determine the structure. It is perhaps fair to say that this fine body of work has not, in actual fact, had the influence on structure determination it deserved. It had the misfortune to come too late. The relationships can only be based on F-values whose strength is in reasonably high proportion to the maximum possible value. As crystals increase in complexity, such high values become increasingly rare, and by the time their methods were established the power of X-ray analysis had so grown that it had outflanked and taken the positions assailed by the direct attack.

To sum up the 1945–55 era, I see it as the extensive development of a large body of techniques based on the earlier trial-and-error methods, but now becoming so powerful that they were breaking new ground in all fields of crystal chemistry, and reaching a stage where they could tackle problems of chemical constitution which had proved impossible to solve by other techniques.

This brings me to the present era, and here the thrilling feature has been the successful application of X-ray analysis to biological structures. The model of the spiral structure of the polypeptide chain, the α-helix of Pauling, Branson and Corey, was first deduced on theoretical grounds supported by a study of the form of amino-acids and simple polypeptides. It soon received confirmation from the treatment of X-ray diffraction by helical structures in general by Cochran, Crick and Vand, and its application to the diffraction pictures yielded by artificial polypeptides. This analysis was used by Crick and Watson to explain the excellent pictures of nucleic acid (DNA) which had been obtained by Wilkins, and in 1953 they published their structural scheme of the atomic arrangement in DNA which has aroused such intense interest and has led to further researches in a number of laboratories which have confirmed its essential correctness. The transform of the helical structures, expressible in Bessel functions, also led Watson to deduce the helical nature of rod-like viruses. The diffraction pictures of the globular viruses led Caspar to suggest that they were a cluster of similar protein molecules in a symmetrically arranged point group. Further developments by Crick and Watson and by Franklin and Klug have given us a much deeper understanding of the general architecture of virus particles. X-ray analysis has also been applied to the structure of collagen and of muscle. These cases present the novel feature that one is dealing with the transform of asymmetrical arrangement which is not part of a space lattice. Advance at the present time is so rapid that I must be content with a very brief reference to these very important researches.

The analysis of the proteins myoglobin and haemoglobin by Kendrew and Perutz is more in line with former work in that the goal is the complete determination of the atomic positions in molecules of definite structure arranged according to space-group symmetry. The new feature is the high complexity of the structures which have been successfully analysed. The molecule of vitamin B_{12} contains 181 atoms, and its analysis represented a landmark in the solution of a complex molecule by what might be termed classical methods. Myoglobin of molecular weight 18 000 contains some 2500 atoms, and haemoglobin is four times as large. It would have seemed almost impossible to tackle such structures only a few years ago. The model of the myoglobin molecule is something quite new in science, and poses fascinating problems of the interplay of forces in this vast array of atoms and the way in which they determine the chemical behaviour of the compound.

One feature is the complete absence of trial and error methods, and the directness of the approach with no assumptions about the kind of answer to be expected. Nature has, as it were, given us an unexpected break. On the one hand there is the formidable complexity of a molecule containing many thousands of atoms. On the other hand, the very size of the molecule makes it possible to attach heavy atoms, often associated with quite large complexes, in various places without a disturbance of the arrangement of the molecules in the crystal. As it were, the molecule takes no more notice of such an insignificant attachment than a maharaja's elephant would of the gold star painted on its forehead. The attachment of such atoms in a sufficient number of places makes possible a completely direct determination of phases and so a solution of the structure. Taking advantage of the presence of a heavy atom, or of the possibility of isomorphously replacing one heavy atom by another, is of course a classical device, but this addition of heavy atoms in various places is new.

Next, because the approach is direct, clear instructions can be given to our slaves, the electronic computers, to work out the vast number of calculations required. The collection of data has also become automatic, and the main application of the investigator's judgement is to the interpretation of the data turned out by the machines. I am not minimizing the ingenuity which has led to the discovery of how to attach the heavy atoms chemically, to the design and accuracy of the recording apparatus, and to the programming of the data fed into the machines. The new feature is that the element of guesswork has gone, and been replaced by the handling of vast masses of measurements and calculations.

The nearly complete solution of the myoglobin structure by Kendrew, and the first outline of haemoglobin which will no doubt soon be also extended to a higher resolution, represent the result of some twenty years' study of these two molecules. One cannot but admire the tenacity and faith which led to the final success after so many years with few encouraging results.

The other outstanding feature, to my mind, is the bridge which has been established between two ways of studying the structure of living matter. The electron microscope has been improved until now it is able to resolve bodies which are of the order of ten atoms across, so containing thousands of atoms. X-ray analysis has now reached a point at which it can place the atoms in molecules containing thousands of atoms. For the first time we shall be able to study the structure of matter right down to the atomic scale.

As I have worked for so long in this field, it will easily be understood how deep a gratification it is to me to witness the growing power of

X-ray analysis and see how far it has progressed from the early days I remember so well.

I end this lecture with a graph, in which the years are plotted horizontally, and the number of parameters in typical structure determinations are plotted vertically on a logarithmic scale. The curve starts with one parameter in 1913. It shows a sharp break at the exciting present time, and in tracing its course I have ventured with some confidence to assume that the haemoglobin structure will be completed next year. If we prolong the graph we conclude that we shall reach the million mark in 1965. This is not so improbable as might seem; is there not hope that by that time we shall know the atomic architecture of a simple virus? How the million parameters will be listed in *Acta* is a problem I leave to the editors at the time. I have high hopes that I shall see the great day.

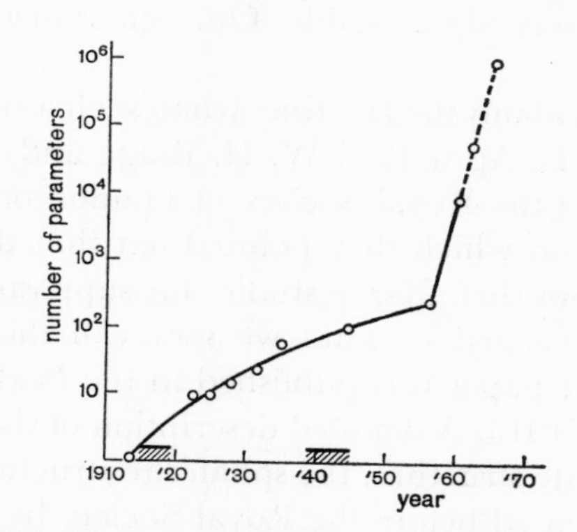

Fig. 8–1. The growing power of X-ray crystal analysis.

CHAPTER 9

Problems of Inorganic Structures *

by Linus Pauling

In June 1913 a leading American physical chemist, H. C. Jones, in his book *A New Era in Chemistry*, pointed out how little was known about the solid state. He said 'We do not know what is the formula of rock salt, or of ice; and we have no reliable means at present of finding out these simplest matters about solids. Our ignorance of solids is very nearly complete.'

June of 1913 was about the last time when such a statement could be made. Already on 17 April 1913 W. H. Bragg and W. L. Bragg had read a paper before the Royal Society of London on the reflection of X-rays by crystals, in which they pointed out that they had obtained strong evidence from their X-ray studies for supposing that the atoms in rock salt are arranged in what we now call the sodium chloride arrangement. Their paper was published in the *Proceedings of the Royal Society* in the fall of 1913. A detailed description of the sodium chloride structure, the fluorite structure, the sphalerite structure, and the pyrite structure was presented before the Royal Society by W. L. Bragg on 27 November 1913, and published early in 1914. A powerful and reliable method for determining the structure of crystals had thus become available, and during the past 50 years it has been principally responsible for the development of a powerful and extensive theory of structural inorganic chemistry, and has also, of course, contributed greatly to other fields of chemistry.

9.1. Simple Inorganic Structures

The English mathematician and astronomer Thomas Harriot, who was tutor to Sir Walter Raleigh and who travelled to Virginia in 1585,

* Contribution No. 2574 of the Gates and Crellin Laboratories of Chemistry.

apparently knew about the structure representing closest packing of spheres; he said that the bodies with greatest density are those in which every atom touches twelve surrounding atoms, whereas those with least density have only six contacts. The crystallographers such as Haüy did not make as much use of crystallographic information as they might have, because instead of attempting to account for the forms of crystals in terms of aggregates of spherical atoms, they in general represented the fundamental unit as a polyhedron.

The structure of diamond was determined by W. H. and W. L. Bragg in 1914, and recognized at once as corresponding to the theory of the tetrahedral carbon atom that had been proposed by Van 't Hoff and LeBel just forty years before and that had become one of the important principles underlying the structure theory of organic chemistry. It seems to me astounding that in this period of forty years, between 1874 and 1914, no scientist had suggested that the diamond crystal has the diamond structure.* In this period a number of investigators (W. Barlow, W. J. Pope, W. J. Sollas, F. M. Jaeger) had attempted to discover reasonable arrangements of atoms in space and to assign them to various crystals on the basis of the available information about the properties of the crystals. Barlow was the most successful of these investigators. In 1883 he suggested five 'very symmetrical' structures, the sodium chloride, cesium chloride, and nickel arsenide arrangements, cubic closest packing, and hexagonal closest packing. He was strongly criticized by L. Sohncke, who pointed out that Barlow seemed not to know what the adjective symmetrical meant. This criticism apparently stimulated Barlow to make a study of symmetry, and led to his independent development of the 230 space groups during the following decade. Barlow and Pope assigned copper, silver, and gold to the cubic closest packed arrangement, and pointed out that magnesium and beryllium had the correct symmetry and axial ratio to correspond to hexagonal closest packing. However, their discussion of diamond was incorrect: they assigned diamond to the cubic closest-packed arrangement.

After the discovery of X-ray diffraction rapid progress was made in the determination of simple structures. By January 1915, when W. H. and W. L. Bragg sent the manuscript of their book *X-Rays and*

* This statement may not be formally correct because the true diamond structure was reportedly predicted by A. Nold in 1891 on the ground of tetravalency (I have, however, been unable to trace the paper). That this prediction found no acceptance is excusable because the remainder of about a dozen structures of elements, proposed on similar principles, makes no sense whatever (cf. P. P. Ewald, *Kristalle und Röntgenstrahlen*, pg. 96 and 316, J. Springer, Berlin, 1923).

Crystal Structure to the publisher, they were able to report complete structure determinations representing nine structure types: sodium chloride (NaCl, KCl, KBr, KI, PbS), diamond, sphalerite (ZnS), wurtzite (ZnO, CdS), cesium chloride (NH_4Cl), cubic closest packing (Cu), fluorite (CaF_2), pyrite (FeS_2, MnS_2, CoAsS), and calcite ($CaCO_3$, $MgCO_3$, $MnCO_3$, $FeCO_3$, $ZnCO_3$, $NaNO_3$). They had also made good progress toward the determination of the cuprite structure, the hematite structure, and the spinel structure, and had made studies of a number of other crystals. Other investigators, too, had already become active.

The hexagonal closest-packed arrangement was discovered by A. W. Hull in 1917, when he carried out his powder studies of magnesium, and at the same time he discovered the body-centered arrangement, in his investigation of lithium and potassium metals. The rutile and anatase structures were discovered by Vegard in 1916; the third structure for titanium dioxide, brookite, was not determined until 1928. The nickel arsenide arrangement was determined by Aminoff in 1923. The first layer structure was discovered in 1922: the cadmium iodide structure, by Bozorth.

An important consequence of the determination of the structures of the alkali halogenide crystals and the recognition that they could be described in a reasonable way as a packing of cations and anions was the development of a simple theory of ionic crystals, especially by E. Madelung, W. Kossel, M. Born, and F. Haber.

Most of the structures of binary compounds were seen to involve a packing of 4, 6 or 8 non-metal atoms about the metal atoms, at the corners of a tetrahedron, an octahedron, or a cube, respectively. Molybdenite (1923) was found to constitute an exception: in it the six sulfur atoms that surround a molybdenum atom are arranged at the corners of a trigonal prism, with axial ratio close to unity. This coordination polyhedron is, however, rare. The great majority of structures of inorganic substances that have been determined during recent decades can be described in terms of the three simple coordination polyhedra, the tetrahedron, octahedron, and cube. Some others, which occur less often, are described in the following section.

As more and more structures were determined, the possibility of systematizing them through principles arrived at in part by induction and in part by deduction from the laws of electrostatics was recognized. V. M. Goldschmidt made two great contributions in the second decade of the X-ray diffraction period. One was to add greatly to the body of structural information by synthesizing a large number of binary compounds and determining their structures. The other was to classify substances on the basis of structure type and interatomic distances. W. L. Bragg had attempted to formulate a system of atomic radii. His efforts were extended by Goldschmidt, who divided substances into two classes, ionic and atomic (covalent, including metals), and formulated two sets of radii, ionic radii and atomic radii. He then developed a system of classification and explanation of crystal types in relation to the sizes of atoms and ions.

This method of interpreting structural information has been highly developed and found to be of much value. A description of some of the structural principles is given in the following section.

Structure determinations have now been made for hundreds of binary compounds, and the structural chemist is tempted to feel that he understands them. He is, in fact, often able to predict in a reasonably reliable way not only the structure of a substance but also the interatomic distances, usually to within about 0.05 Å. But it is difficult to make reliable predictions for substances more complicated than binary compounds, and even the binary compounds sometimes offer puzzles.

For example, in 1924 Dickinson and Friauf found that lead monoxide forms tetragonal crystals involving layers in which the oxygen atoms lie in one plane and the lead atoms lie in adjacent planes, above and below the oxygen plane. In most layer structures the metal atoms are at the center of the layer, and the nonmetal atoms are on the two sides of the layer. No very convincing explanation of the stability of the observed structure for lead oxide, rather than, for example, the sphalerite structure, has been given.

The structural principles for crystals such as the silicate minerals in which oxygen is the principal non-metal have been well worked out. The corresponding compounds of sulfur and its congeners are as yet not thoroughly understood. The sphalerite and wurtzite structures are, of course, simple ones that can be described in terms of tetrahedral coordination of atoms about one another. Chalcopyrite ($CuFeS_2$), enargite (Cu_3AsS_4), and many other sulfides have structures that represent superstructures of sphalerite and wurtzite, or are capable of similar interpretation in terms of covalent bonds arranged either tetrahedrally or otherwise in accordance with a reasonable electronic structure. But many of these substances have surprising and unexplained structures. An example is sulvanite, Cu_3VS_4. This cubic crystal has a structure in which the sulfur atoms are arranged in the same way as in sphalerite, and the copper atoms occupy the positions occupied by three of the four zinc atoms, per unit cube, in sphalerite; but the vanadium atom, instead of occupying the fourth tetrahedral position, is in another place, where it is still surrounded by four sulfur atoms but also is close to six copper atoms. Each sulfur atom, instead of being surrounded by three copper atoms and one vanadium atom in a tetrahedral arrangement (as, in enargite, it is surrounded by three copper atoms and one arsenic atom) is, instead, in a one-sided relationship with its environment: it forms three bonds with copper atoms, at tetrahedral angles, and a fourth bond, with vanadium, in the same direction, midway among the three bonds to copper.

Many sulfide structures, especially structures of sulfide minerals, have now been determined. It is possible that the amount of structural information available is already great enough to permit the induction of a significant set of structural principles, but there is also the possibility that the interactions in these crystals are so complex that additional structure determinations will be needed before this problem can be attacked with success. I think that it is likely that the next ten years will see the formulation of a sound set of structural principles for the metal sulfides, comparable in its significance to the set of principles that was formulated over 30 years ago for the oxides.

9.2. Inorganic Complexes

The first inorganic complex to have its structure determined by X-ray diffraction was the carbonate ion. When W. H. and W. L. Bragg carried out their study of calcite and isomorphous minerals in 1914 they found that the carbonate ion is planar and that the three oxygen atoms are equivalently related to the carbon atom. This information required that the chemists abandon the conventional valence-bond structure for the carbonate ion, in which one of the oxygen atoms is attached to the carbon atom by a double bond and the other two are attached by single bonds. During the following fifteen years two proposals about the structure were made: one, that the carbon atom foms only three bonds, and has only three electron pairs in its valence shell, and the second, that the carbon atom forms four bonds, and shares four electron pairs with the oxygen atoms, but with the structure a resonance hybrid of the three possible conventional structures, such as to make the three oxygen atoms equivalent. The discovery of the configuration of the carbonate ion in calcite thus contributed in an important way to the development of chemical structure theory during the period between 1914 and 1934.

The discovery that pyrite contains two sulfur atoms only 2.08 Å apart, also made in 1914, did not have much influence on the development of chemical structure theory, in part because the disulfide complex is so simple as not to introduce any serious theoretical problems and in part because in the decade following 1914 there was not great understanding of the significance of interatomic distances, and the sulfur-sulfur distance was at that time not necessarily interpreted as corresponding to a single covalent bond between the two sulfur atoms. On the other hand, the next complex ion found by X-ray diffraction, the dichloroiodide ion in the rhombohedral crystal cesium dichloroiodide that was studied by Wyckoff and reported in his first published paper in 1920, represented a significant contribution to the body of information underlying chemical structure theory. Wyckoff found that the dichloroiodide ion is linear, with the two chlorine atoms equidistant from the iodine atom. The number of valence electrons is such that no electronic structure of the classical type (with shared electron pairs for the bonds and with no atom having a larger number of electrons associated with it than the number for the next noble gas) can be assigned to the complex. Even now, after more than forty years, there is some uncertainty about the best way of describing the electronic structure of the dichloroiodide ions and of the many other complex ions formed by the halogens with one another that have been investigated in recent years, especially by Rose Slater and O. Hassel. There is no doubt that an extension of existing valence theory will be needed to encompass these complexes. The linear sequences of iodine atoms along the axis of a helical shell formed by starch molecules, as discovered by Rundle, also come in this category.

In 1921 Dickinson and, independently, Bijvoet and Kolkmeijer reported the structures of sodium chlorate and sodium bromate. The pyramidal configurations of the ions, in agreement with the theory of electronic structure that had been developed five years earlier by G. N. Lewis, provided strong support for this theory.

The years 1921 and 1922 were important ones for the Werner coordination theory of inorganic complexes. In 1921 Wyckoff and Posnjak reported their structure determination of ammonium hexachloroplatinate, verifying the octahedral arrangement of the six chlorine atoms about the platinum atom in the hexachloroplatinate ion that had been postulated by Werner twenty years before. The octahedral configuration was also verified within a year for the hexachlorostannate ion in the potassium and ammonium salts by Dickinson, for the hexafluorosilicate in the ammonium salt by Bozorth, for the hexachloroplatinate ion in the potassium salt by Scherrer and Stoll, and for the nickel hexammoniate cation in the chloride, bromide, iodide, and nitrate by Wyckoff. Wyckoff in 1922 also showed that the hexahydrated zinc ion in zinc bromate hexahydrate has the octahedral configuration.

The surprising discovery by Werner, on the basis of studies of isomerism, that the tetraligated complexes of bipositive palladium and platinum have a square planar configuration was also verified by X-ray diffraction by Dickinson in 1922, through his determination of the structures of K_2PtCl_4, K_2PdCl_4, and $(NH_4)_2PdCl_4$. Dickinson found that the bond lengths in these square planar complexes are the same as in the corresponding octahedral complexes of the metal atoms with oxidation number greater by $+ 2$. Ten years later, on the basis of theoretical considerations, it was pointed out (Pauling, 1931) that certain diamagnetic crystalline compounds of nickel should contain square planar complexes, and this prediction was soon verified by X-ray diffraction studies of many crystals.

The square planar configuration about bipositive copper in $K_2CuCl_4 \cdot 2H_2O$ was reported in 1927 by Hendricks and Dickinson, and a similar configuration was found in cupric chloride dihydrate by Harker. In these crystals the copper atom forms bonds that are indicated by the interatomic distances to be strong with two chlorine atoms and two oxygen atoms of water molecules, at the corner of a planar rhomb, and also forms two weaker bonds with chlorine atoms at the two remaining corners of an octahedron. This arrangement of four strong bonds and two weak bonds about the copper atom still constitutes something of a challenge to theoretical chemists, thirty years after its discovery.

During recent years some hundreds of crystals containing complexes have been investigated by X-ray diffraction, often with surprising and unusually interesting results. An outstanding example showing the power of the X-ray diffraction technique was the study of duodecitungstophosphoric acid, $H_3PW_{12}O_{40} \cdot 29H_2O$ by Keggin in 1934. Many complexes in which the ligancy of the metal atom is not four or six have also been investigated. An interesting example is the molybdenum octacyanide ion in $K_4Mo(CN)_8 \cdot 2H_2O$, the structure of which was determined by Hoard and Nordsieck in 1939.

9.3. Inorganic Molecular Crystals

The first inorganic crystal to be found to contain discrete molecules was tin tetraiodide, the structure of which was reported by Dickinson in 1923 (cubic, 8 SnI_4 per unit cube). In the same year Dickinson and A. L. Raymond reported the structure of the first organic compound, hexamethylenetramine, and Bozorth reported his determination of the structure of pentoxides of arsenic and antimony, and showed that the crystals contain discrete molecules As_4O_6 and Sb_4O_6.

9.4. Metals and Other Crystals involving Metal-metal Bonds

Although Barlow and Pope correctly assigned copper, silver, and gold to the cubic closest-packed arrangement and magnesium and beryllium to the hexagonal closest-packed arrangement, this assignment had little impact on science because their evidence was not convincing. A great part of the present understanding of the nature of metals and alloys is the result of X-ray investigations, beginning with the determination of the structure of copper by W. L. Bragg in 1914. Most of the metals were found to have simple structures, cubic or hexagonal closest packing or the cubic body centered arrangement; but the discovery of the structure of white tin by Bijl and Kolkmeijer in 1919, in which the atoms have ligancy six, was followed in a few years by the determination of the complex structures of alpha manganese and beta manganese by Bradley and, independently, Westgren and Phragmén, in 1925.

The first intermetallic compounds to be investigated were found to have simple structures. Then in 1927 Friauf determined the structures of $MgCu_2$ and $MgZn_2$ (the Friauf phases), and Bradley as well as Westgren and Phragmén investigated the gamma alloys. In all of these complex structures, including alpha manganese and beta manganese, the smaller atoms are surrounded by larger atoms at the corners of an icosahedron, and the larger atoms have ligancy greater than twelve. Many other structures of the icosahedral type, such as the structures of the sigma alloys, have been investigated in recent years. The nature of the interatomic interactions leading to the stability of these very complex structures has not yet been thoroughly elucidated, but it seems likely that the coordination polyhedra whose faces consist only of triangles are more stable than those involving squares, which occur in the simple structures, because the repulsion of atoms separated by the diagonal of a square leads to instability. In metals and intermetallic compounds, as well as other substances, we may hope that a thorough analysis of the extensive information about interatomic distances will in the course of time provide a much greater understanding of the structures than exists now.

The extreme complexity and variety in the structure of chemical substances has been brought to light largely by X-ray techniques. An interesting early example is silver subfloride, Ag_2F, investigated by Ott and Seyfarth in 1928. This crystal was found to have pairs of hexagonal layers of silver atoms with single layers of fluorine atoms interleaved between them. The bonds between the two silver layers in juxtaposition are closely similar to those between silver atoms in metallic silver, whereas the bonds between silver and fluorine seem to be essentially as in silver fluoride, AgF. Since then many other crystals in which there are metal-metal bonds as well as metal-nonmetal bonds have been investigated.

One of the first complexes to be studied in which evidence was found for a metal-metal bond was the ditungsto-enneachloride ion, $W_2Cl_9^{---}$, found in the crystal $K_3W_2Cl_9$, by Brosset in 1935. The complex has the configuration of two WCl_6 octahedra sharing a face, and the two tungsten atoms are only 2.41 Å apart, much less than in metallic tungsten (2.74 Å). This fact provides evidence that there is a strong tungsten-tungsten bond in this complex. Many other complexes have been found through X-ray investigation of the crystals to have metal atoms bonded together. Interesting examples of these complexes are $(Mo_6Cl_8)^{++++}$ (Brosset, 1945), and $(Nb_6Cl_{12})^{++}$, $(Ta_6Cl_{12})^{++}$, and $(Ta_6Cr_{12})^{++}$ (P. A. Vaughan and coworkers, 1950). In these complexes the metal atoms lie at the corners of an octahedron, with the halogen atoms out from the centers of the faces or the edges of the octahedron. The interatomic distances for the molybdenum complex are close to the single-bond value and for the other complexes close to the value corresponding to bond number 2/3.

The carbonyl complexes provide an interesting problem in structural chemistry that has been in large part solved by the methods of X-ray diffraction of crystals and electron diffraction of gas molecules. The first significant step in the solution of this problem was the investigation of crystals of diiron-enneacarbonyl, $Fe_2(CO)_9$, by R. Brill in 1927. In this molecule each iron atom is surrounded octahedrally by carbon atoms, and the octahedra share a face. The three bridging carbon atoms are at distances from the iron atom corresponding to the formation of two single bonds with these atoms (and a double bond with the attached oxygen atom), whereas the other six carbon atoms form essentially double bonds with the adjacent iron atom. (The discovery that the interatomic distances in carbonyls correspond to double bonds between carbon and metal was made for nickel tetracarbonyl by L. O. Brockway and P. C. Cross in 1935.) The two iron atoms in diiron-enneacarbonyl are 2.46 Å apart, corresponding to the formation of a bond between them. During recent years many complex metal carbonyls and related substances have been investigated, and found to involve metal-metal bonds, bridging carbonyl groups, and other carbonyl groups attached to metal atoms by double bonds, as in diiron-enneacarbonyl.

The amount of accurate and detailed structural information about complexes has now become very great, exceeding the bounds of chemical theory. It is likely that the extensive information about interatomic distances and bond angles in these complexes will be used at some time in the future as the basis for the development of a more penetrating theory of electronic structure and chemical bonding in these complexes than is now available. Other powerful methods of studying complexes, such as electron spin magnetic resonance and nuclear magnetic resonance and the Mössbauer effect, have become available and are also providing information that will need to be taken into consideration in the formulation of a more powerful theory of complexes.

9.5. Other Problems

The discovery of antiferromagnetism and ferrimagnetism has stimulated structural investigations of oxides and other crystals with these properties. In magnetite, Fe_3O_4, for example, it has been found by neutron diffraction that the octahedrally coordinated metal atoms have their magnetic moments pointed in one direction and the tetrahedrally coordinated metal atoms have the opposite orientation of their magnetic moments. In manganous fluoride, which has the rutile structure, the magnetic moments of the manganese atoms have parallel orientation for a string of octahedra that share edges with one another and the opposite orientation for the surrounding strings of octahedra. These and other magnetic superstructures that have been discovered by neutron diffraction indicate that in these crystals the bonds between the metal atoms and the non-metal atoms are one-electron bonds, rather than the usually assumed electron-pair bonds with some partial ionic character.

The hydrogen bond is a structural feature that has received recognition largely through the X-ray diffraction studies of crystals. Although the discovery of the hydrogen bond by Latimer and Rodebush in 1920 was not based to any significant extent on crystal-structure results, a large part of the knowledge now existing about the properties of this bond has resulted from X-ray studies. Among the interesting results of these studies is the recognition of the existence of hydrogen-bonded clathrate compounds, by Powell and his coworkers, and in particular the clathrate hydrates, such as chlorine hydrate. The importance of the hydrogen bond in living organisms, as in protein molecules and deoxyribonucleic acid molecules, has been recognized in large part because of the knowledge about these bonds obtained through X-ray studies.

The boranes and metallocenes provide interesting examples of substances with structures not compatible with simple valence-bond theory. In ferrocene, for example, the iron atom is equidistant from ten carbon atoms, in the two cyclopentadienyl rings. Both the observed interatomic distance and theoretical considerations indicate strongly that the iron-carbon bonds are not single covalent bonds, but are instead fractional bonds. No really satisfactory theory of substances of this sort has been developed as yet. I think that we may expect that during the next few years much progress will be made in the development of a more powerful theory of the chemical bond than exists at present, and also in the interpretation of the experimentally determined bond lengths, and that the method of X-ray diffraction, in collaboration with other methods of investigating the structure of molecules and crystals, will lead to additional great progress in our understanding of the nature of molecules and crystals.

It is, of course, impossible to answer the question as to how the field of inorganic structures would have developed if the discovery of X-ray diffraction had been delayed. The almost complete lack of progress in the decades before 1913 and the great progress thereafter show clearly that the development of this field resulted from Laue's brilliant discovery and the Braggs' effective application of it. Despite the great progress that has been made, there are many problems in structural inorganic chemistry still awaiting solution, and X-ray diffraction will no doubt contribute to their solution during the coming years.

CHAPTER 10

Problems of Organic Structures

by J. Monteath Robertson

In organic chemistry there are two basic and complementary problems, analysis and synthesis. By analysis the chemist attempts to find the number, kind and spatial arrangement of the atoms in his molecule; only then is he in a position to attempt its synthesis out of the elements. By the labours of thousands of chemists for more than a hundred years the vast edifice of modern organic chemistry has been built. This knowledge has been achieved by painstaking analyses and syntheses and by logical processes of deduction and induction from a few relatively simple valency concepts. But in spite of this vast progress an almost infinite field still lies ahead, as chemistry advances towards the more complex molecules associated with living matter. Some of these structures have only been elucidated by the work of large teams of chemists over a dozen years or more, and many others still remain unsolved.

This whole situation is now changing dramatically. Within the last few years it has become possible to determine the structures of organic molecules containing up to 100 atoms or more by the methods of X-ray crystal analysis. Provided that suitable crystalline derivatives can be prepared (and we certainly know enough chemistry to be able to do this) then the structures can be determined accurately in almost every case without fail, even if the chemical formula is quite unknown. Furthermore, this can now be done in a relatively short time. Last December we obtained a suitable crystalline derivative of a complex curare alkaloid. By mid February we had determined the structure completely in three dimensions with an accuracy sufficient to give the bond lengths to within a few hundredths of an Ångström unit. These advances are now radically changing the whole outlook in organic chemistry and are enabling the chemist to come more quickly to grips with his other fundamental problems of mechanism, biogenesis and synthesis. How these very great advances in the science of the

X-ray analysis of organic structures have come about during the last fifty years is the subject of this article.

1912–1922

It may perhaps be convenient to paragraph this account very roughly into the five decades that have elapsed since Laue's original discovery of X-ray diffraction in 1912. During the first decade very little was accomplished in the field of organic molecular structures. In the second decade (1922–1932) the analytical methods were mainly one-dimensional, and considerable advances were made in the structures of the long chain compounds, for example. The third decade (1932–1942) saw the extensive development of two-dimensional methods and the solution of numerous aromatic structures. In the fourth decade (1942–1952) three-dimensional methods were introduced, and it is perhaps not too far fetched to say that during the last decade (1952–1962), with the introduction of still more refined techniques which can determine the thermal as well as the positional parameters of the atoms, a kind of four-dimensional analysis is now possible. This is, however, a very superficial summary and other more important and more fundamental advances were made during these different periods.

Apart from some early observations on crystalline waxes [1*] and a few other compounds, organic structures were not really attempted during this first period. Reference should be made, however, to the successful determination of the fundamental mineral structures diamond [2] and graphite [3] which proved to be the prototypes for the atomic arrangements present in aliphatic and in aromatic compounds respectively. The significance of the measurements of the basic interatomic distances in these structures became clear much later.

1922–1932

This period saw the beginnings of the first concerted attack on the structures of molecular crystals, as opposed to the elementary and simple ionic crystals that had been the main object of previous study. In the early twenties W. H. Bragg [4] published an extensive account

* The notes to this Chapter begin at page 171.

of his X-ray measurements on a number of fairly complex organic crystals, including napthalene, anthracene, acenaphthene, α- and β-naphthol, and benzoic acid. This work did not succeed in determining any of these structures at all fully but it produced strong evidence that the 'molecule' of the organic chemist did have a real existence in the crystal, and that the rings of atoms postulated in the naphthalene and anthracene structures, for example, must have dimensions closely similar to the carbon rings already known to exist in diamond and graphite. The arrangement and general orientation of the molecules in the naphthalene and anthracene crystals were correctly deduced, and this was substantiated later by a more detailed survey [5] of the intensities of the spectra. These important results initiated the X-ray study of organic molecular crystals. However, when attempts were made to go further and postulate detailed atomic structures on the basis of cell dimensions and perhaps a few intensity observations, the results were in general disappointing and even misleading. With a few very notable exceptions, some of which are described below, it can be stated that nearly all the work on organic crystal structures published before 1932 is quite unreliable in so far as it attempts to describe detailed atomic arrangements.

The most notable exception, and indeed the very first organic crystal structure to be completely and accurately determined, was hexamethylene tetramine (I). Dickinson and Raymond's [6] remarkable

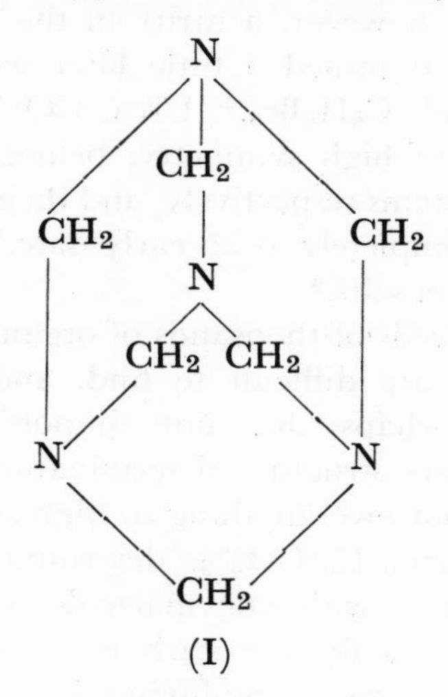

(I)

determination of this structure owed its success to the fact that the crystals are cubic, with only two molecules of $C_6H_{12}N_4$ per unit cell, and that the carbon and nitrogen positions can be completely defined by only two parameters. These were accurately determined, giving a carbon-nitrogen distance of 1.44 Å and a molecule of the shape shown in Fig. 10(1) with the four nitrogen atoms tetrahedrally and the six carbon atoms octahedrally grouped about a common centre. The chemical formula (I) was thus definitely proved.

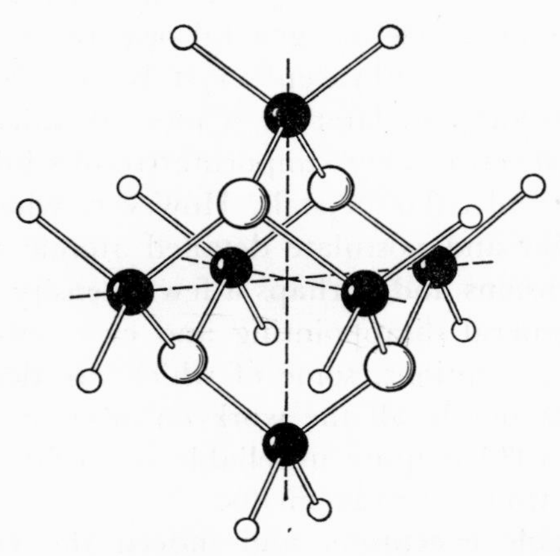

Fig. 10(1). Molecule of hexamethylene tetramine (Dickinson and Raymond, 1923).

Cubic crystals are, however, a rarity in the field of organic structures. Two others determined a little later were the benzene hexahalides, $C_6H_6Cl_6$ and $C_6H_6Br_6$.[7] Urea, $CO(NH_2)_2$, and thiourea, $SC(NH_2)_2$, are also of high symmetry, belonging to the tetragonal and orthorhombic systems respectively, and their structures were again determined rather completely at an early date,[8] as were a number of substituted ammonium salts.[9]

Amongst the hundreds of thousands of organic crystals such highly symmetric examples are difficult to find, and progress was correspondingly slow. Perhaps the most important and outstanding example of a complete structure determination during this decade, and one which did not owe anything to high space-group symmetry, was hexamethylbenzene, $C_6(CH_3)_6$, determined by K. Lonsdale.[10] The system is triclinic, with one molecule in the unit cell, so 36 ndependent parameters for the carbon atoms have to be found. There is, however, one great simplifying feature. The good cleavage plane (001) is found to give an exceptionally strong reflection of which all the higher orders fall off uniformly in intensity in the same manner as those from the basal plane of graphite. The conclusion is inescapable. The molecule must be planar and lie exactly in this plane. Detailed analysis carried out on the basis of a regular hexagonal model led to excellent agreements and showed that the planar benzene ring was of about the same size as the graphite carbon ring (C—C = = 1.42 Å) and that the methyl groups, also in this plane, lay at a slightly greater distance. These results were of fundamental importance and provided a key to the structures of all aromatic molecules.

This chapter would not be complete without a reference to the brilliant work of Muller, Shearer and others on the structures of long chain compounds. Just as graphite and hexamethylbenzene provided a key to the structures of aromatic compounds, so diamond and the long chain hydrocarbons provided an equally essential key to understanding the structures of aliphatic compounds. These two X-ray results now verified in a purely physical manner the two fundamental valency postulates or organic chemistry.

The early work on the long chain compounds was not complete in the crystallographic sense. Single crystals were extremely difficult to obtain but very accurate information about the geometry and dimensions of the zig-zag chains was deduced, particularly by a study of very high-order reflections.[11] In another notable and very early study by Shearer [12] the Fourier series method was employed to fix the positions of the carbonyl group in a series of unknown long chain ketones and to determine the number of carbon atoms. This is probably the first application of the X-ray method to the determination of chemically unknown organic structures, and one of the first practical applications of the Fourier series method.

1932–1942

In all the structures described above there has been some simplifying feature. The long chain compounds present what is mainly a one-dimensional problem. High space-group symmetry with many atoms in special positions account for the success in other cases, while hexamethylbenzene is almost unique in that all the atoms lie in one crystallographic plane.

During the decade that we now describe there was an enormous development both in the power of the analytical methods and in fundamental theory. The Fourier series method was developed [13] and the computational problems simplified [14] so that it could be applied as a useful and practical tool in two-dimensional work, and as a powerful method of refinement. Thus, if a model of the molecule could be postulated from a knowledge of the chemical structure, and its orientation in the crystal described in terms of not too many parameters, then by trial and error, guided by optical and magnetic properties, a rough approximation to the structure could sometimes be found. This could then be refined by Fourier series methods until a highly accurate picture of the molecule was obtained, from which bond lengths to within a few hundredths of an Ångström unit might be estimated.

In this way the structures of naphthalene, anthracene, and a large number of other molecules,[15] which had been looked at provisionally in the 1920's, were now accurately determined for the first time. Results of great theoretical significance were obtained. Although the chemist already knew the structural formulas, the accurate measurements of bond length now available supported in detail the more exact treatment of valency theory that was also now emerging from a study of quantum mechanics.[16] The organic structures that were determined with considerable precision during this period are too numerous to mention in detail. The structural formulas of the chemist were now placed on an absolute scale and on an exact metrical basis. Many stereochemical problems were solved and the first detailed pictures of molecular arrangement and hydrogen bonding were obtained.[17]

In spite of these great advances, however, only in very rare cases was it possible to solve the structures of molecules to which the chemist had not already been able to assign a structural formula. One exceptional case of great importance occurred as early as 1932 when Bernal,[18] by a few simple X-ray measurements, played a great part in the chemical elucidation of the correct skeleton formula for the sterols.[19] Exact solutions for such structures were, however, out of the question. It must be recalled that in the field of organic chemistry we are dealing with molecules that may contain up to 50 or more atoms in addition to hydrogen, and all of very similar scattering power. Without some knowledge of the structural formula to limit the possibilities it is clearly impossible to set up trial structures containing any reasonable number of parameters.

To solve chemically unknown or partially known molecular structures a more fundamental approach was necessary. The foundations of such an approach, which is only now coming to full fruition, were laid during the 1930's. The Patterson vector method [20] was discovered in 1934 and the Harker synthesis [21] in 1936. This approach does not attempt to solve the phase problem but instead it portrays in an elegant manner all the information that can be obtained from a knowledge of the magnitudes of the structure factors, introduced as F^2. The result is a map not of the atomic positions, but showing the superposition of all the interatomic vectors in the crystal. This direct information is immensely valuable and has been used, and is being used, in almost every crystal structure investigation undertaken since that time. For the complex organic structures with which we are now concerned, however, this is not enough. The number of vectors goes up with the square of the number of atoms, and when this number comes to a dozen or more, the chance of being able to resolve the separate vector peaks is very small indeed.

The methods that have proved effective in solving very complex and chemically undetermined organic structures were evolved soon after Patterson's discovery, and they represent a chemical rather than a physical or mathematical approach to the solution of the phase problem. They are known as the isomorphous substitution and heavy atom methods, and they were first developed and applied to organic structures in the case of the phthalocyanines.[22] The problem is to determine the unknown phases of the resultant waves which are scattered by all the atoms in the crystal. Only the amplitude can be measured. A chemical experiment is now performed and an extra atom, preferably of higher scattering power than the others, is inserted at a known point in the structure, or at a point that can subsequently be determined by a simple application of the Patterson method. The effect of this extra contribution to the resultant amplitudes is observed, and from the changes that occur many of the original phases may be deduced. Phase differences which cannot be measured are thus transformed into amplitude differences which can be measured.

An earlier observation on isomorphous substitution was made by J. M. Cork [23] when studying the inorganic alum crystals, but the full significance of the effect was not appreciated, and indeed the crystal structure concerned was not solved, at that time. The phthalocyanine structures represent a beautifully simple and complete example. The electron density map representing the principal projection of nickel phthalocyanine is shown in Fig. 10(2). The central nickel atom can be removed or replaced by another atom without disturbing the remainder of the structure. Hence the phases of nearly all the reflections could be determined by the isomorphous substitution method, and most of them by the heavy atom method alone, which involves the use of only this derivative. As a result this analysis was accomplished without the need of any chemical assumption, and indeed without postulating the existence of separate atoms in the molecule at all.

Almost unlimited possibilities for the analysis of complex structures were thus opened up. Very few molecules exhibit such high symmetry or afford such perfect isomorphous substitution as the phthalocyanines.

Fig. 10(2). Electron density projection for nickel phthalocyanine. Contour lines at unit electron density intervals, except on the central nickel atom where the interval is 5 e.Å^{-2} per line (Robertson and Woodward, 1937).

But the more general heavy atom method has a wider application. It should be realized that the organic chemist is generally interested in determining the structure of some molecular species, rather than the structure of a particular crystal. And from a given organic compound it is nearly always possible to prepare a series of derivatives, containing, for example, the halogen atoms chlorine, bromine, or iodine. Again, if the compound is an acid, then a series of heavy metal salts may be prepared. If the structure of such a derivative, obtained by simple addition or substitution reactions, can be determined, then we have effectively determined the structure of the parent compound.

The prospects of being able to make a great contribution to organic chemistry by the X-ray method thus appeared to be extremely bright, and were much discussed.[24] Many complex crystals had been looked at in a preliminary way, and it was suggested that if a mercury atom could be substituted for zinc in the insulin crystal [25] then X-ray analysis of a structure even of this complexity might become feasible. This prediction is now being realized for some protein crystals, but in 1939 it was still too early, for reasons which we shall presently discuss.

In spite of the advances that we have described, the structures that were solved exactly during this period still represent a somewhat haphazard collection. It was easy to suggest systematic surveys of structures, for example, in the field of the sugars, or carbohydrates generally, or the amino-acids, but extremely difficult to carry them out. When attempted, usually little more than unit cell measurements could be collected. One extremely valuable survey of this kind was carried out for the steroids, mainly belonging to the cholesterol and ergosterol series, and published in a monumental work.[26]

Exact structures for molecules of this complexity, however, were not yet systematically possible. For exact analysis it was a case of finding a molecule that would fit the methods and techniques available, rather than of applying these techniques to a specified series of compounds. The limiting factor was really the effective restriction of the methods of analysis, and especially the powerful Fourier method, to two-dimensional work. To obtain precision it was thus a necessary condition that it should be possible to resolve the atoms in some projection of the structure, and the separation required for resolution, with the usual X-ray wave-lengths and reasonably good organic crystals, is about 0.6 Å.

It was no accident that during this period the most effective and important studies were carried out on aromatic compounds with planar molecules. If the molecule is planar, and the unit cell is not too large, there is always a good chance of being able to resolve the atoms in some projection. On the other hand, if the molecule has a complicated three-dimensional shape, then many of the atoms will overlap in any projection because these separation distances will in general be much less than the minimum required for resolution. So even if the phase problem can be solved, by some of the methods outlined above, a solution capable of precise interpretation may not be reached in two dimensions. Another decade and more had to elapse before the full power of the X-ray method in three dimensions was revealed.

Before leaving this period reference should be made to the important X-ray work that was carried out on a number of macromolecular structures. In fibre structures the crystallinity is generally low and the chemistry complex. Detailed and fully quantitative X-ray studies can hardly be expected. On the other hand the importance of fibre structures is outstanding because they are almost universal constituents of living matter. Biochemical structures do not come within the scope of this chapter, but some brief references will be made because the line of demarcation is hard to draw.

Synthetic fibres were becoming prominent at this time, and the structure of polyethylene, $[—CH_2—CH_2—]_n$, the simplest of them all, received detailed study.[27] It was possible to measure the zig-zag angle, 112°, and the carbon-carbon distance, 1.53 Å, with some accuracy. Polyvinyl alcohol, $[—CHOH.CH_2—]_n$, was also studied at this time.[28]

In the field of natural fibre structures a fairly reliable model for cellulose was deduced by a combination of chemical and X-ray evidence.[29] The isoprene polymers, rubber and gutta-percha, were also studied in some detail [30] and shown to be *cis*- and *trans*-modifications of the polyisoprene chain.

The fundamental researches of Astbury on the fibrous proteins also commenced during this period, but these definitely belong to the biochemical field and will not be described here.

1942–1952

A slow start was made with X-ray crystal analysis during this period because of the interruption of the war, but before the end of the period quite spectacular progress had been achieved, culminating in the complete elucidation of such important structures as the alkaloid strychnine and the antibiotic penicillin. Although this chapter is not immediately concerned with the development of the technique of crystal structure analysis, we must again make some reference to such developments in order to understand the sequence of events.

In the field of organic structure determination by X-ray methods there are perhaps two main objectives. One is concerned with the ever more precise measurement of bond lengths, angles, and ultimately the complete electron distribution in comparatively simple molecules. The other is concerned with the elucidation of the more complex or more difficult structures that the organic chemist has not yet been able to master completely by the classical methods of his science. For both these objectives the power of the analytical methods at the disposal of the crystallographer is always the limiting factor.

In the period now before us there were many important advances that greatly increased the power of the analytical tools. Of the many methods available for the refinement of atomic parameters, the introduction of the method of least squares,[31] differential synthesis,[32] and the more complete development and discussion of the old method of difference synthesis,[33] were perhaps the most important. In more fundamental theory, very important relations between the structure factors were discovered in terms of inequalities,[34] the interpretation and solution of vector distributions were extensively studied,[35] and many valiant attempts were made to solve the phase problem.[36]

All these developments led to important results; but of greatest immediate practical importance was the simple yet prodigious problem of computation. We have seen that in the 1930's two-dimensional methods of analysis were supreme. This was because it was about the limit of what was practicable, with the hand and desk methods of computation then available. However, there was now an urgent need to move into three dimensions. To appreciate the problems involved, let us consider some simple examples. The kind of calculations required may be typified by the summation of a Fourier series. The calculations of structure factors, vector distributions, etc. are similar in kind, and these operations have usually to be carried through many times in the course of one analysis. Now, to evaluate a two-dimensional electron distribution of the kind shown in Fig. 10–2 for the phthalocyanines, a series containing about 200 terms (structure factors) has to be evaluated and summed at at least 1500 points over the area. This involves dealing with a total of about 300 000 quantities which have to be evaluated and added in groups. Many ways of shortening the work were devised in the 1930's, but the computational burden was still heavy, remembering that this kind of operation had usually to be repeated many times before the analysis was completed.

The three-dimensional analysis of a moderately complex structure, however, may require a series of at least 2000 terms which has to be evaluated at a minimum of perhaps 50,000 points throughout the volume of the asymmetric unit. Some 10^7 quantities are now involved, and life is too short for a task of this kind to be performed by the methods hitherto available.

All these problems have now been solved by the invention of electronic digital computers, but in the period we are now discussing these machines had not yet been invented. Although development work started on them in the 1940's they did not become at all generally available until later in the 1950's. One supremely important event which did take place much earlier was the development of the magnificent electronic analogue computer (XRAC), designed and built by R. Pepinsky and his colleagues.[37] Although the output is essentially two-dimensional, this machine can be adapted for three-dimensional work and it is capable of producing sections or projections, like the one illustrated in Fig. 10(2), fully and accurately drawn out on the oscilloscope. As soon as the amplitudes and phases are set up on a series of rheostat dials, the summations are performed instantaneously. This machine played and is still playing a great part in speeding up structural work. Another important development on the computational side in the 1940's was the utilisation of punched card computing and tabulating equipment (Hollerith or IBM) which was now becoming commercially available.[38] Most of the three-dimensional analyses during this period that we now describe were aided by some of these developments, but the work was often still laborious and heroic efforts were needed to achieve success.

It should be mentioned again, as in previous sections of this chapter, that nothing like a complete account of the achievements in organic structure determination can be attempted here. A very complete account and references will be found in the seven large volumes of *Structure Reports* that cover this decade. The feature of the period was undoubtedly the introduction of three-dimensional analysis and the improved methods of refinement that became available. In the simpler molecules this led to much greater accuracy and reliable estimates of probable errors. Whereas formerly bond lengths could be estimated to within a few hundredths of an Ångström unit, they could now, in the most favourable cases, be estimated to within a few thousandths. In the complex structures it became possible towards the end of the period for the crystallographer to contribute actively to the solution of structures that were still to a large extent chemically undetermined.

We begin with an account of some simple structures, and a good example to illustrate progress is one to which we have already referred, the aromatic hydrocarbons naphthalene and anthracene. These were now subjected to the full power of three-dimensional analysis,[39] and the earlier conclusions were verified but made much more accurate. It was possible to establish experimentally the bond-length variations in the aromatic rings that had already been predicted theoretically, and the hydrogen atoms were quite clearly defined for the first time. In

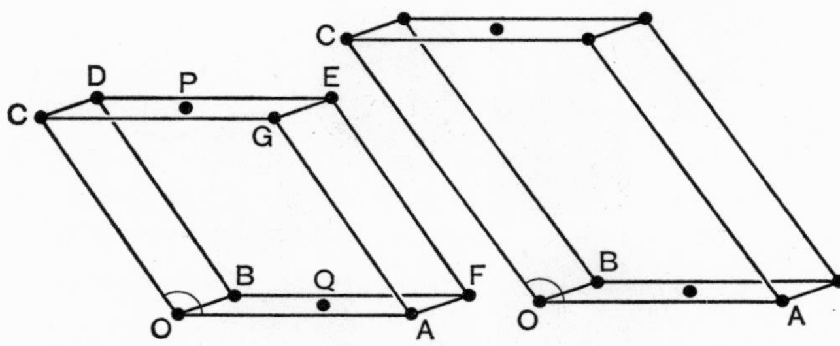

Fig. 10(3a). Unit cells of naphthalene and anthracene, showing positions of molecules (W. H. Bragg, 1922).

Fig. 10(3) we show (a) the results of W. H. Bragg's 1922 investigation in which the positions of the molecules in the unit cells were correctly defined and a clue as to their orientation was discovered, (b) the results of the two-dimensional 1933 investigation showing an electron density projection of the anthracene molecule in which some but not all of the carbon atoms are separately resolved, and (c) a section of the electron density distribution evaluated in the mean plane of the molecule, from the 1950 three-dimensional analysis. The increasing power of our methods is strikingly revealed by these pictures. In the last the hydrogen atoms can be seen, and it is possible to measure the carbon-carbon bond lengths with great

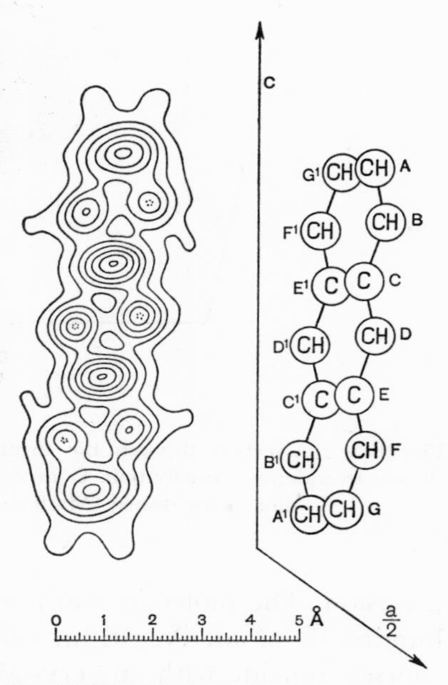

Fig. 10(3b). Electron density projection of anthracene molecule (J. M. Robertson, 1933).

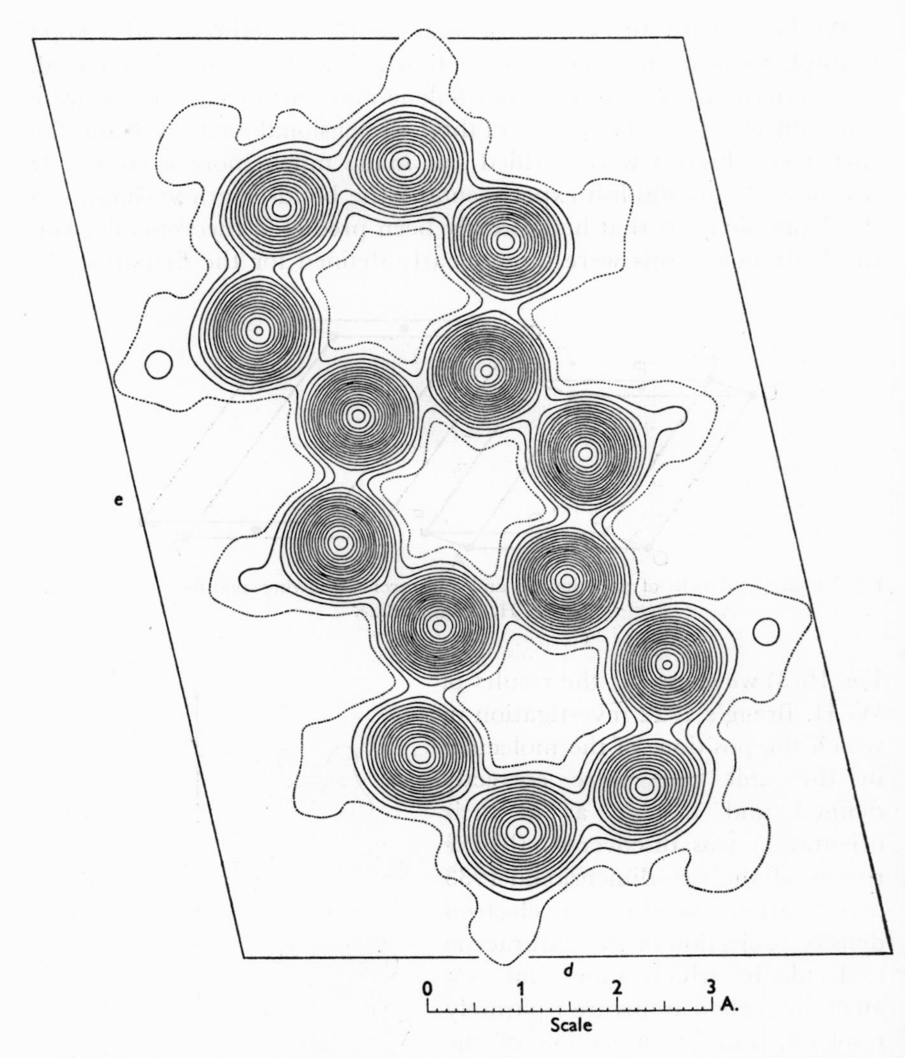

Fig. 10(3c). A section through the mean plane of the anthracene molecule, showing the electron density distribution. Contour levels at intervals of $\frac{1}{2}$ e.Å^{-3}, the half-electron line being dotted (Mathieson, Robertson and Sinclair, 1950).

precision. The molecule was found to be planar to within about one hundredth of an Ångström unit, although this plane does not, of course, coincide with any crystallographic plane. However, this is not quite the end of the story, and in the next decade we shall describe a still more accurate refinement of these structures.

Accurate and detailed structures for many other simple molecules were determined during this period. One important and extremely early three-dimensional analysis for the terpenoid geranylamine hydrochloride [40] may be noted, while the stereochemistry of various camphor derivatives [41] was elucidated by the isomorphous substitution method.

Further study of the ismorphous substitution method in the asymmetric case now led to one of the greatest triumphs of X-ray analysis during this decade. When the replaceable atom site is not coincident with a centre of symmetry, the phase angle can still be determined, but there is an ambiguity of sign. Bijvoet [42] was able to solve this ambiguity by the introduction of a 'phase lag' in scattering, which occurs if the atom used can itself be excited by the incident radiation. The condition for this is that the radiation should have a wave-length near the absorption edge for the atom in question. A phase change on scattering induced in this way invalidates Friedel's law, so that intensity differences between (hkl) and ($\overline{hkl}$) reflections occur. Bijvoet realized that this effect might be used to determine whether a structure was left-handed or right-handed, that is, to solve the chemical problem of absolute configuration. This was achieved in the case of the rubidium isomorph of Rochelle salt, $RbNaC_4H_4O_6 \cdot 4H_2O$, by the use of $ZrK\alpha$ radiation, and it was shown for the first time that the conventional configuration for the *D*-tartaric acid molecule as proposed by Emil Fischer did in fact correspond to reality. Further very important applications of the anomalous scattering method for phase determination in general were developed later by Okaya and Pepinsky.[43]

While it is impossible even to mention the numerous important structures that were determined during these years, attention should be drawn to one very significant feature of general development which now arose from the increasing power of the X-ray method. This was the beginning of really systematic surveys of related compounds, in place of the earlier rather haphazard development of the subject. Formerly it was usually necessary to search for a structure which because of some simplifying feature was likely to yield to the available methods of analysis with a reasonable expenditure of time and effort. Now it began to be possible to select a chemically interesting and important structure and attempt its detailed analysis with a reasonable prospect of success.

The best example of this is found in the beautiful series of strikingly accurate studies on amino acids and simple peptides carried out mainly in the Pasadena school. The structures of glycine,[44] alanine,[45] serine,[46] threonine,[47] *N*-acetylglycine,[48] β-glycylglycine,[49] diketopiperazine[50] and many other compounds were determined. These results were of great importance in formulating possible structures for the proteins. The bond lengths and angles obtained were used to derive the dimensions of the polypeptide chain, and they form the basis for the helical models of protein structure proposed by Pauling [51] that have proved to be so important.

Significant advances were also now made in the very difficult field of carbohydrate structures by the analysis of sucrose,[52] α-D-glucose,[53] and certain fructose derivatives.[54]

The problem of molecular compounds, in which two types of molecule appear to be united without any obvious attractive forces operating between them, had been a puzzle to chemists for generations. This general problem was now very completely studied and partially solved by X-ray crystallographic studies. The work of H. M. Powell [55] and his collaborators on the quinol cage structures, or clathrates, is the most outstanding example in this field. Such structures are exceedingly complex, and often disordered, but the foundations of a new kind of chemistry wer firmly laid by this work.

Finally, in the field of chemically unknown or incompletely determined structures we must first refer to the important and early determinations of the structure of cholesteryl iodide by Carlisle and Crowfoot [56] and of a calciferol derivative (calciferyl-4-iodo-5-nitrobenzoate, $C_{35}H_{46}NO_4I$) by Crowfoot and Dunitz.[57] These were probably the first applications of the heavy atom method to such complex and asymmetric structures. Although the chemistry was known, the exact configurations and stereochemistry were now fully determined for the first time. The same can be said of the complex alkaloid strychnine, determined independently from the hydrobromide by J. H. Robertson and Beevers [58] and from the isomorphous sulphate and selenate by Bokhoven, Schoone and Bijvoet.[59] A very great deal of fundamental chemical work had been carried out on this structure and most of it was known before the X-ray work began. The X-ray work, however, carried out by the heavy atom and by the isomorphous substitution methods, was quite independent of chemical theory, and not only verified the structure in every detail but also provided a complete picture of the stereochemistry.

The power of the X-ray method in aiding the organic chemist to elucidate a difficult and completely unknown structure was perhaps first most conclusively demonstrated in the work on penicillin during the war years, when computing methods were still primitive and laborious. The formula for benzylpenicillin (II)

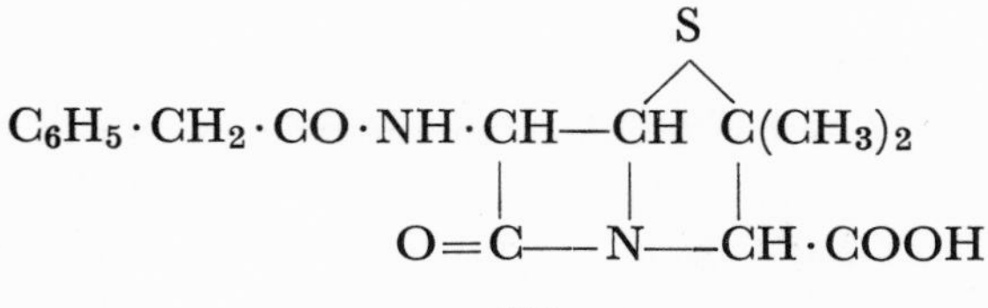

$$
\begin{array}{l}
\qquad\qquad\qquad\qquad\qquad\qquad\quad S \\
C_6H_5\cdot CH_2\cdot CO\cdot NH\cdot CH{-}CH\quad C(CH_3)_2 \\
\qquad\qquad\qquad\qquad\quad | \qquad\quad | \qquad | \\
\qquad\qquad\qquad O{=}C{-}N{-}CH\cdot COOH
\end{array}
$$

(II)

may not now appear to be unduly complicated, but the chemistry is extremely difficult and unusual. The analysis was finally achieved through the rubidium, potassium and sodium salts, which are not all isomorphous.[60] A feature of the work was the close collaboration at every stage between the crystallographers and the chemists. The final result establishes the intricate spatial arrangement of the atoms in full detail, and, as is well known, has led to profound advances in the fields of chemistry and medicine.

1952–1962

Any adequate description of the development of organic structure analysis during the last ten years is extremely difficult because the work cannot yet be seen in perspective. It seems likely, however, that this will be remembered as the period when the discoveries and developments of earlier years first came to full fruition, and when the real impact of X-ray analysis on the science of organic chemistry was revealed.

We have already stressed the importance of computational work and the magnitude of the problems involved when full three-dimensional surveys of complex structures are attempted. For years this had been the bottle-neck. The outstanding structure determinations that we have already described were nearly all performed with great difficulty and were only completed, often inadequately completed, after the most strenuous efforts. The development and general availability of fast electronic digital computers has now radically changed the whole outlook in X-ray crystal analysis, as it has in many other branches of science. It is, of course, easy to exaggerate this aspect. The computer has not solved the phase problem. But if phase determination can be effected by the use of some complex heavy atom derivative, or by a series of such derivatives, then complexity of the structure or lack of resolution are no longer the limiting factors. Indeed, complete three-dimensional analyses can now be effected with less effort than was required for comparable two-dimensional analyses in the 1930's and 1940's.

The impact of these last ten years on our science can be vividly seen by comparing two important volumes on *Computing Methods and the Phase Problem in X-Ray Crystal Analysis*.[61] In the first of these, written in 1950, Mina Rees, from the Office of Naval Research, pointed out in her introduction that high speed digital computers were 'still just over the horizon'. The book is concerned with many aspects of the phase problem and with the description of many analogue computers, including the extremely important X-RAC and S-FAC machines. On the other hand, the 1960 volume is largely written in a language that would not have been understood in 1950, because a major portion of it deals with programmes and the programming of all the varied operations of crystal analysis for over a dozen different kinds of high speed electronic digital computer that are now available in nearly every country of the world.

These great advances in technique have naturally led to an ever-increasing number of significant organic structure determinations, and it becomes more difficult than ever to present anything like an adequate picture of what has been achieved. Once again, however, we may draw attention to advances in two main directions; the more intensive and more accurate refinement of relatively simple structures on the one hand, and the determination of complex and chemically unknown structures on the other.

As an example of the first, we may return once more to our old friends naphthalene and anthracene. The electron density map shown in Fig. 10(3c) is a very direct representation of the experimental results, which can only be altered if more accurate or more complete intensity measurements become available. In deriving this map no assumptions were made regarding the existence of hydrogen atoms or regarding the exact form of the scattering curve for carbon. From this point, however, it is possible to carry out much further refinement, and try to interpret the results in terms of the thermal vibrations of the atoms and the distribution of electronic charge density.

Elaborate refinements to this end have now been carried out by Cruickshank.[62] Assuming a theoretical form factor for carbon[63] it is possible to evaluate the anisotropic thermal motion. For this a symmetric tensor with six independent components is required for each atom to characterize its vibrations. These thermal parameters and the three positional parameters for each atom may be refined simultaneously. Elaborate and lengthy calculations are thus required, which

Cruickshank has been able to programme for the electronic computer. Without this aid it has been estimated that these detailed refinements of the naphthalene and anthracene structures would have taken 100 man-years to complete if they had been attempted in the 1940's.

From the results of this detailed study of thermal motion the molecule is shown to behave very largely like a rigid body, oscillating with amplitudes of several degrees about its centre point. Care was taken to evaluate the thermal parameters from the intensities of the high-order reflections only, because for these the X-ray scattering is due almost entirely to the inner electrons of the atoms (K shell), which are not involved in chemical bonding. Now, in order to obtain a picture of the bonding electrons, the set of 14 vibrating but non-bonded carbon atoms and the 10 hydrogen atoms, as defined by their appropriate wave functions, were subtracted away from the experimentally observed electron distribution (Fig. 10(3c)). The result is perhaps hardly conclusive as yet, because the experimental measurements are not sufficiently accurate, but there is a fairly definite indication of a slight channelling of the residual electrons along the bonds and around the aromatic rings. This example may serve to indicate the kind of detailed picture of molecular structure that can be expected in the future when measurements become more precise and our powers of interpretation are increased.

A vast number of other accurate determinations of molecular structure have now been made, which cannot be enumerated in detail. The work of J. C. Speakman on acid salts, particularly sodium hydrogen diacetate,[64] of J. Iball and his co-workers on further condensed ring hydrocarbons, of W. Cochran and others on salicylic acid and various purines and pyrimidines, of R. E. Marsh and others on various benzene derivatives and other structures, and of D. C. Phillips on acridine, are all well known. In some of these cases, with accurate Geiger-counter intensity measurements, bond lengths and electron densities have been measured with standard deviations as low as 0.004 Å and 0.1 e.$Å^{-3}$ respectively, as in the measurement of a tartrate structure by A. J. van Bommel and J. M. Bijvoet.[65] Further progress in this field of high-precision work clearly depends on accurate counter measurements, combined perhaps with low temperature studies, as in the work of Hirshfeld and Schmidt on α-phenazine.[66]

In the field of complex molecules the most outstanding example during this decade has undoubtedly been the complete solution of the structure of vitamin B_{12} by D. C. Hodgkin, J. G. White and their many collaborators.[67] Furthermore, this feat was accomplished during the early years of the period, and before computing methods had nearly reached their present state of high efficiency. As a complete structure determination it can still be considered the crowning triumph of X-ray crystallographic analysis, both in respect of the chemical and biological importance of the results and the vast complexity of the structure.

The formula, $C_{63}H_{88}O_{14}N_{14}PCo$ together with about 24 molecules of water, shows that, even without counting hydrogen, there are about 350 positional parameters to determine. The cobalt atom is far too light for anything like complete phase determination. Nevertheless, with this as a starting point, and with great determination and skill, involving what can only be described as gifted intuition at some points, the complete structure was finally elucidated. When the work first started in 1948 there was no information about the chemical nature of the vitamin. However, as in the case of penicillin, chemical work and degradative studies went on in close collaboration with the crystallographic work, and during the next eight years the chemistry of various

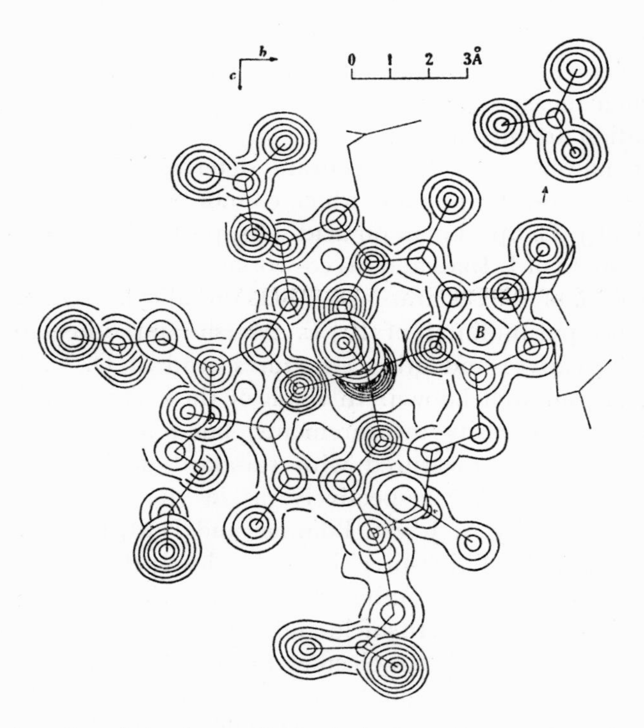

Fig. 10(4a). A portion of the electron distribution in the wet B_{12} crystal, showing the nucleus and most of its side chains (D.C. Hodgkin *et al.*).

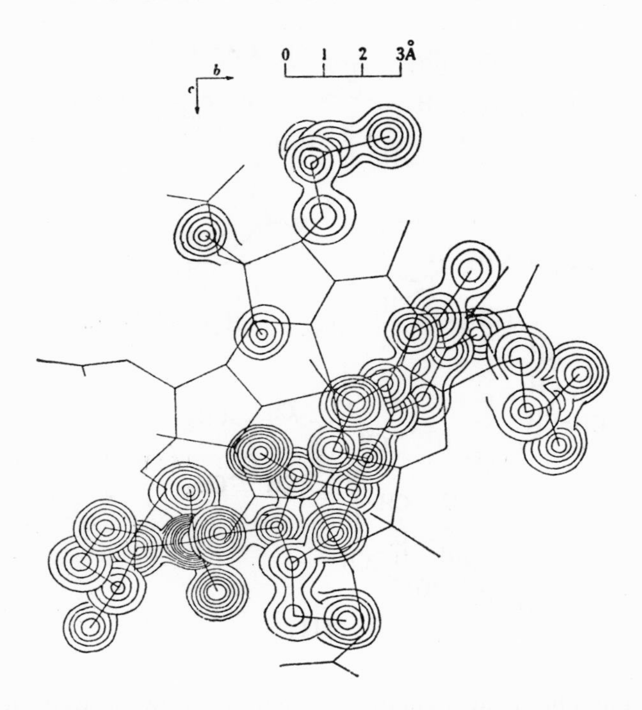

Fig. 10(4b). The remainder of the B_{12} structure showing the benziminoazole, ribose, phosphate and propanolamine groups, with some of the remaining side chains (D.C. Hodgkin *et al.*).

groups within the molecule was determined. A nucleotide-like group consisting of a benziminoazole ribose phosphate, a propanolamine residue and various amides and a cyanide group were discovered. There was also a large fragment containing cobalt which could be some kind of porphyrin-like nucleus. However, there was no evidence as to how all these groups might be linked together.

The X-ray work proceeded by gradual stages and quite early there was direct X-ray evidence for some kind of a planar porphyrin-like unit. Additional atoms were gradually introduced into the phasing calculations, and the structure slowly became more clearly resolved. Both the wet and the air-dried crystals were examined, and a lot of help was obtained from the study of other derivatives, especially a seleno-cyanate, $B_{12}SeCN$, which had a second heavy atom, and a hexacarboxylic acid obtained by Todd and his co-workers by degradation. Finally, the whole structure was obtained with remarkable precision, as shown by the superimposed sections of the three-dimensional electron density distribution reproduced in Fig. 10(4a) and

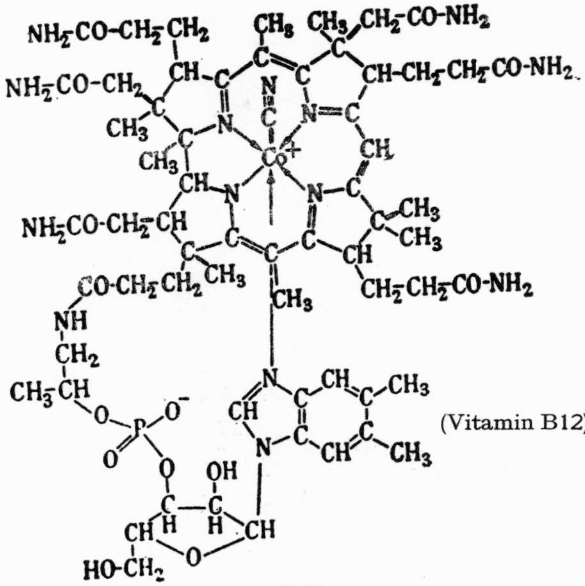

(Vitamin B12)

(III)

10(4b). The chemical structure, which can now be written as (III), contains many surprising and unusual features, such as the direct linking of two of the five-membered rings on the central nucleus. The whole arrangement is extremely compact and is found to conform very precisely to various stereochemical rules that have emerged from the study of many simpler organic systems.

Since the solution of the B_{12} structure, the analysis of complex and chemically unsolved natural product structures has gone on apace. Many important contributions have come from Australia by the work of A. McL. Mathieson and his team, in both the alkaloid and terpenoid fields, from Canada by the work of Maria Przybylska, from the laboratories of Ray Pepinsky, W. N. Lipscomb and others in the United States, from Oxford and Cardiff in England. Another important recent development has been the accurate determination of the spatial arrangement of the atoms in large and medium ring aliphatic compounds by J. D. Dunitz and his co-workers in Zurich.

A very large number of important natural product structures has recently been solved by the Glasgow school, using three-dimensional analysis with phase determination based on a heavy atom derivate. These include the terpenoids isoclovene, β-caryophyllene alcohol, and geigerin, the bitter principles limonin (IV),[68] clerodin (V),[69] and cedrelone, the alkaloids calycanthine (VI),[70] echitamine (VII)[71] and macusine A, and many other similar investigations are now in progress. Owing to the complicated crystal structure, with two molecules in the asymmetric crystal unit, the solution of the limonin structure

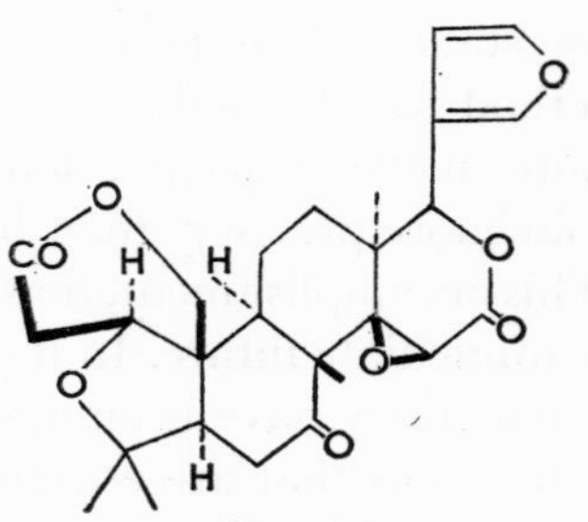

(limonin)
(IV)

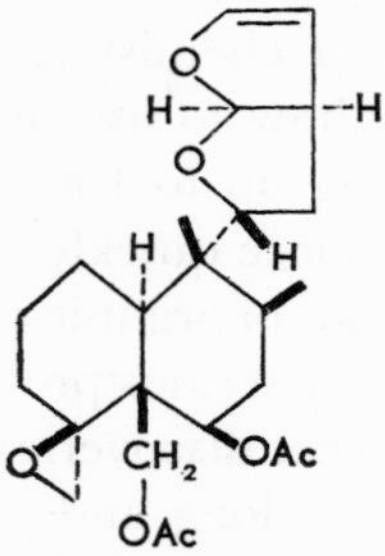

(clerodin)
(V)

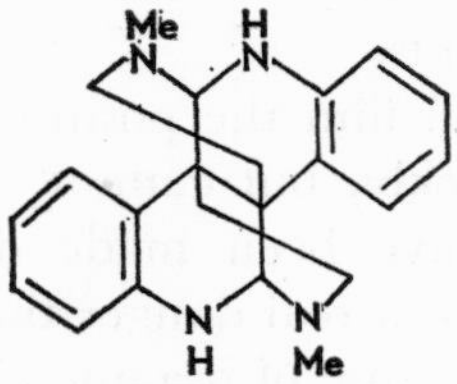

(calycanthine)
(VI)

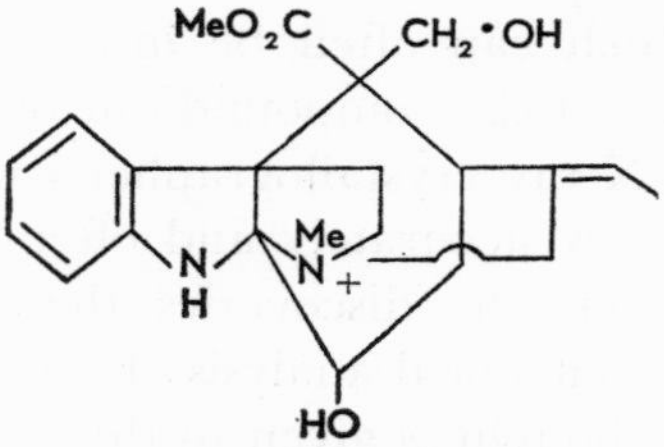

(echitamine)
(VII)

Conclusion

involved finding the positions of 76 atoms other than hydrogen. The structure now found (IV) shows it to be a triterpenoid of the euphol type, and this has solved one of the long-standing problems in chemistry, since this bitter principle of citrus fruits was first discovered in 1841. The other structures shown below have also cleared up many chemical problems, because for years there has been controversy about their precise formulas.

Having presented a brief sketch of present day work in organic structure determination by the X-ray method, it is now necessary to enquire about the future and ask what all this is going to lead to. In the field of simple and even moderately complex structures we may expect still higher precision, not only in bond length measurements but in the direction of mapping out the entire electron distribution. Such very complete physical pictures of molecules must be the background against which fundamental valency theory will develop, and we can perhaps expect the science of chemistry now to move fairly rapidly from an empirical to a more strictly theoretical basis.

In the case of those complex structures on the edge of the ever growing field of chemistry, we can now state with some confidence that we know how to use the X-ray method to solve unknown structures containing up to 100 atoms or more in the molecule. This can often be done more quickly and always far more precisely than by the classical methods of organic chemistry. This accomplishment, however, now raises a very serious problem for the immediate future. In the past many of the great discoveries of organic chemistry have been made in the course of the long and patient investigations that are required in the elucidation of natural product structures. While solving a structure the chemist does far more than merely find the relative positions of the atoms in space. He makes many discoveries and learns a lot of chemistry, which can often be utilised, for example, in effecting a total synthesis of the compound out of its elements.

The X-ray crystallographer can now tell him the positions of the atoms very accurately and often very quickly, but cannot enlighten him about the discoveries that might have been made during a detailed chemical analysis. There is perhaps a real danger that unless serious thought is given to this matter the cause of organic chemistry may not be advanced by this work. It could even be retarded. It is clear that a lot of re-thinking is required, and that, given the structural formula, the chemist must learn to devise rapid and decisive experiments that will enable a full and effective understanding of the chemistry involved to be gained.

Perhaps we solve some problems only to create others. In the long run, however, it is quite certain that a tremendous advance has now been made, and that in another ten years' time organic chemistry will be a very different subject from what it is to-day.

References

1. W. Friedrich, *Physik. Z.*, 1913, *14*, 317; M. de Broglie and E. Friedel, *Compt. rend.*, 1923, *176*, 738; S. H. Piper and E. N. Grindley, *Proc. Phys. Soc.* (London), 1923, *35*, 269; *36*, 31; A. Muller, *J. Chem. Soc.*, 1923, *123*, 2043.
2. W. H. Bragg and W. L. Bragg, *Nature*, 1913, *91*, 557; Proc. Roy. Soc. (London), 1913, A*89*, 277.
3. P. P. Ewald, *Sitzber. math.-physik. Klasse bayer. Akad. Wiss. München*, 1914, p. 325; P. Debye and P. Scherrer, *Physik. Z.*, 1916, *17*, 277; 1917, *18*, 291; A. W. Hull, Phys. Rev., 1917, *10*, 661; J. D. Bernal, *Proc. Roy. Soc.* (London), 1924, A*106*, 749; O. Hassel and H. Mark, *Z. Physik.*, 1924, *25*, 317; C. Mauguin, *J. Phys.*, 1925, *6*, 38.
4. W. H. Bragg, *Proc. Phys. Soc.* (London), 1921, *34*, 33; 1922, *35*, 167.
5. W. H. Bragg, *Z. Krist.*, 1928, *66*, 22.
6. R. G. Dickinson and A. L. Raymond, *J. Amer. Chem. Soc.*, 1923, *45*, 22; G. W. Gonell and H. Mark, *Z. physik. Chem.*, 1923, *107*, 181.
7. S. B. Hendricks and C. Bilicke, *J. Amer. Chem. Soc.*, 1926, *48*, 3007; R. G. Dickinson and C. Bilicke, *ibid.*, 1928, *50*, 764.
8. K. Becker and W. Jancke, *Z. physik. Chem.*, 1921, *99*, 242, 267; H. Mark and K. Weissenberg, *Z. Physik*, 1923, *16*, 1; S. B. Hendricks, *J. Amer. Chem. Soc.*, 1928, *50*, 2455; R. W. G. Wyckoff, *Z. Krist.*, 1930, *75*, 529; 1932, *81*, 102; R. W. G. Wyckoff and R. B. Corey, *ibid.*, 1934, *89*, 462; P. Vaughan and J. Donohue, *Acta Cryst.*, 1952, *5*, 530.
9. S. B. Hendricks, *Z. Krist.*, 1928, *67*, 106, 119, 465, 274; *68*, 189; R. W. G. Wyckoff, *ibid.*, 1928, *67*, 91, 550; *68*, 231.
10. K. Lonsdale, *Nature*, 1928, *122*, 810; *Proc. Roy. Soc.* (London), 1929, A*123*, 494; *Trans. Faraday Soc.*, 1929, *25*, 352.
11. A. Muller, *Proc. Roy. Soc.* (London), 1928, A*120*, 437.
12. G. Shearer, *Proc. Roy. Soc.* (London), 1925, A*108*, 655; R. Robinson, *Nature*, 1925, *116*, 45.
13. W. H. Bragg, *Trans. Roy. Soc.* (London), 1915, A*215*, 253; P. P. Ewald, *Z. Krist.*, 1921, *56*, 129; W. Duane, *Proc. Nat. Acad. Sci. U.S.*, 1925, *11*, 489; R. J. Havighurst, *ibid.*, 1925, *11*, 502; A. H. Compton, *X-Rays and Electrons*, New York: Van Nostrand and Co., 1926; A. H. Compton and S. K. Allison, *X-Rays in Theory and Experiment*, New York: Van Nostrand and Co., 1935; W. L. Bragg, *Proc. Roy. Soc.* (London), 1929, A*123*, 537.
14. C. A. Beevers and H. Lipson, *Phil. Mag.*, 1934, *17*, 855; *Nature*, 1936, *17*, 825; *Proc. Phys. Soc.* (London), 1936, *48*, 772; C. A. Beevers, *Acta Cryst.*, 1952, *5*, 670; J. M. Robertson, *Phil. Mag.*, 1936, *21*, 176; A. L. Patterson, *ibid.*, 1936, *22*, 753; A. L. Patterson and G. Tunell, *Amer. Mineral.*, 1942, *27*, 655; J. M. Robertson, *J. Sci. Instruments*, 1948, *25*, 28; *idem*, *Nature*, 1936, *138*, 683.
15. J. M. Robertson, *Proc. Roy. Soc.* (London), 1933, A*140*, 79; A*141*, 594; A*142*, 659, 674.
16. L. Pauling, *The Nature of the Chemical Bond*, Ithaca, N.Y.: Cornell University Press, 1939.
17. J. M. Robertson, *Proc. Roy. Soc.* (London), 1936, A*157*, 79; J. M. Robertson and A. R. Ubbelohde, *ibid.*, 1938, A*167*, 122.
18. J. D. Bernal, *Chemistry & Industry*, 1932, *51*, 259, 466; *Nature*, 1932, *129*, 277, 721.
19. O. Rosenheim and H. King, *Chemistry & Industry*, 1932, *51*, 464, 954; *Nature*, 1932, *130*, 513; H. Wieland and E. Dane, *Z. physiol. Chem.*, 1932, *210*, 268.
20. A. L. Patterson, *Phys. Rev.*, 1934, *46*, 372; *Z. Krist.*, 1935, *90*, 517.
21. D. Harker, *J. Chem. Phys.*, 1936, *4*, 381.
22. J. M. Robertson, *J. Chem. Soc.*, 1935, p. 615; 1936, p. 1195; J. M. Robertson and I. Woodward, *ibid.*, 1937, p. 219; *idem*, *ibid.*, 1940, p. 36.
23. J. M. Cork, *Phil. Mag.*, 1927, *4*, 688.
24. D. M. Crowfoot (D. C. Hodgkin), *Chem. Soc. Ann. Repts.*, 1936, *33*, 214.
25. J. M. Robertson, *Nature*, 1939, *143*, 75.
26. J. D. Bernal, D. Crowfoot, and I. Fankuchen, *Trans. Roy. Soc.* (London), 1940, A*239*, 135.
27. C. W. Bunn, *Trans. Faraday Soc.*, 1939, *35*, 482.

28. R. C. L. Mooney, *J. Amer. Chem. Soc.*, 1941, *63*, 2828.
29. M. Polanyi, *Naturwisssenschaften*, 1921, *9*, 228; K. H. Meyer and H. Mark, *Z. physik. Chem.*, 1929, *B2*, 115; E. Sauter, *ibid.*, 1937, *B35*, 83; K. H. Meyer and L. Misch, *Helv. Chim. Acta*, 1937, *20*, 232; S. T. Gross and G. L. Clark, *Z. Krist.*, 1938, *99*, 357.
30. J. R. Katz, *Naturwissenschaften*, 1925, *13*, 411; *Kolloid-Z.*, 1925, *36*, 300; *37*, 19; K. H. Meyer and H. Mark, *Ber.*, 1928, *61*, 1939; C. S. Fuller, *Ind. Eng. Chem.*, 1936, *28*, 907; C. W. Bunn, *Proc. Roy. Soc.* (London), 1942, A*180*, 40, 67, 82.
31. E. W. Hughes, *J. Amer. Chem. Soc.*, 1941, *63*, 1737.
32. A. D. Booth, *Trans. Faraday Soc.*, 1946, *42*, 444, 617.
33. A. D. Booth, *Nature*, 1948, *161*, 765; W. Cochran, *Acta Cryst.*, 1951, *4*, 408.
34. D. Harker and J. S. Kasper, *Acta Cryst.*, 1948, *1*, 70; J. Gillis, *Nature*, 1947, *160*, 866; *Acta Cryst.*, 1948, *1*, 76, 174; D. Sayre, *ibid.*, 1952, *5*, 60; W. Cochran, *ibid.*, 1952, *5*, 65; W. H. Zachariasen, *ibid.*, 1952, *5*, 68.
35. J. M. Robertson, *Nature*, 1943, *152*, 411; M. J. Buerger, *Acta Cryst.*, 1950, *3*, 87; *idem Proc. Nat. Acad. Sci. U.S.*, 1950, *36*, 376, 738.
36. Computing Methods and the Phase Problem in X-ray Crystal Analysis, State College, Pa., 1952; J. Karle and H. Hauptman, *Acta Cryst.*, 1950, *3*, 181.
37. R. Pepinsky, *J. Applied Phys.*, 1947, *18*, 601.
38. L. J. Comrie, G. B. Hey and H. G. Hudson, *J.R. Statist. Soc. Suppl.*, 1937, *4*, 210; P. A. Shaffer, V. Schomaker, and L. Pauling, *J. Chem. Phys.*, 1946, *14*, 648, 659; J. Donohue and V. Schomaker, *Acta Cryst.*, 1949, *2*, 344; E. G. Cox and G. A. Jeffrey, *ibid.*, 1949, *2*, 341; E. G. Cox, L. Gross, and G. A. Jeffrey, *ibid.*, 1949, *2*, 351; M. D. Grems and J. S. Kasper, *ibid.*, 1949, *2*, 347.
39. A. McL. Mathieson, J. M. Robertson, and V. C. Sinclair, *Acta Cryst.*, 1950, *3*, 245, 251.
40. G. A. Jeffrey, *Proc. Roy. Soc.* (London), 1945, A*183*, 388.
41. E. H. Wiebenga and C. J. Krom, *Rec. trav. chim.*, 1946, *65*, 663.
42. J. M. Bijvoet, *Koninkl. Nederland. Akad. Wetenschap.*, 1949, *52*, 313; A. F. Peerdeman, A. J. van Bommel, and J. M. Bijvoet, *ibid.*, 1951, *54*, 3; D. Coster, K. S. Knol, and J. A. Prins, *Z. Physik*, 1930, *63*, 345.
43. Y. Okaya, Y. Saito and R. Pepinsky, *Phys. Rev.*, 1955, *98*, 1857; R. Pepinsky and Y. Okaya *Proc. Nat. Acad. Sci. U.S.*, 1956, *42*, 286.
44. G. Albrecht and R. B. Corey, *J. Amer. Chem. Soc.*, 1939, *61*, 1087.
45. H. A. Levy and R. B. Corey, *J. Amer. Chem. Soc.*, 1941, *63*, 2095; J. Donohue, *ibid.*, 1950, *72*, 949.
46. D. P. Shoemaker R. E. Barieau, J. Donohue, and C. S. Lu, *Acta Cryst.*, 1953, *6*, 241.
47. D. P. Shoemaker, J. Donohue, V. Schomaker, and T. B. Corey, *J. Amer. Chem. Soc* 1950, *72*, 2328.
48. G. B. Carpenter and J. Donohue, *J. Amer. Chem. Soc.*, 1950, *72*, 2315.
49. E. W. Hughes and W. J. Moore, *J. Amer. Chem. Soc.*, 1949, *71*, 2618.
50. R. B. Corey, *J. Amer. Chem. Soc.*, 1938, *70*, 1598.
51. L. Pauling and R. B. Corey, *J. Amer. Chem. Soc.*, 1950, *72*, 5349; L. Pauling, R. B. Corey, and H. R. Branson, *Proc. Nat. Acad. Sci. U.S.*, 1951, *37*, 205; L. Pauling and R. B. Corey, *ibid.*, 1951, *37*, 235, 241, 251, 256, 261, 272, 282, 729; *ibid.*, 1952, *38*, 86.
52. C. A. Beevers and W. Cochran, *Proc. Roy. Soc.* (London), 1947, A*190*, 257.
53. T. R. R. McDonald and C. A. Beevers, *Acta Cryst.*, 1950, *3*, 394.
54. P. F. Eiland and R. Pepinsky, *Acta Cryst.*, 1950, *3*, 160.
55. H. M. Powell and P. Riesz, *Nature*, 1948, *161*, 52; H. M. Powell, *J. Chem. Soc.*, 1950, pp. 298, 300, 468.
56. C. H. Carlisle and D. Crowfoot, *Proc. Roy. Soc.* (London), 1945, A*184*, 64.
57. D. Crowfoot and J. D. Dunitz, *Nature*, 1948, *162*, 608.
58. J. H. Robertson and C. A. Beevers, *Acta Cryst.*, 1951, *4*, 270.
59. C. Bokhoven, J. C. Schoone, and J. M. Bijvoet, *Acta Cryst.*, 1951, *4*, 275.
60. D. Crowfoot, C. W. Bunn, B. W. Rogers-Low, and A. Turner-Jones, *The Chemistry of Penicillin*, Princeton: Princeton University Press, 1949, p. 310; G. J. Pitt, *Acta Cryst.*, 1952, *5*, 770.

61. (i) Report of a conference held at The Pennsylvania State College, 1950. Edited by Ray Pepinsky. State College, Pa. 1952.
(ii) Report of a conference held at Glasgow, 1960. Edited by Ray Pepinsky, J. M. Robertson and J. C. Speakman. Pergamon Press, London. 1961.
62. D. W. J. Cruickshank, *Acta Cryst.*, 1956, *9*, 747, 915.
63. R. McWeeny, *Acta Cryst.*, 1951, *4*, 513.
64. J. C. Speakman and H. H. Mills, *J. Chem. Soc.*, 1961, 1164.
65. A. J. van Bommel and J. M. Bijvoet, *Acta Cryst.*, 1958, *11*, 61.
66. F. L. Hirshfeld and G. M. J. Schmidt, *J. Chem. Phys.*, 1957, *26*, 923.
67. D. C. Hodgkin, J. Kamper, J. Lindsey, M. MacKay, J. Pickworth, J. H. Robertson, C. B. Shoemaker, J. G. White, R. J. Prosen, and K. N. Trueblood, *Proc. Roy. Soc.* (London), 1957, A*242*, 228.
68. S. Arnott, A. W. Davie, J. M. Robertson, G. A. Sim, and D. G. Watson, *Experientia*, 1960, *16*, 49.
69. G. A. Sim, T. A. Hamor, I. C. Paul, and J. M. Robertson, *Proc. Chem. Soc.*, 1961, p. 75.
70. T. A. Hamor, J. M. Robertson, H. N. Shrivastava, and J. V. Silverton, *Proc. Chem. Soc.*, 1960, p. 78.
71. J. A. Hamilton, T. A. Hamor, J. M. Robertson, and G. A. Sim, *Proc. Chem. Soc.*, 1961, p. 63.

CHAPTER 11

The Growing Field of Mineral Structures

by F. Laves

11.1. General Remarks

Minerals are substances of either 'inorganic' or 'metallic' character. From a purely crystallographic point of view it would appear unscientific to prefer minerals to other substances as objects for structure determination, as minerals are only a chance selection of possible compounds. Thus, it was to be expected that major progress in '*crystallography*' would be made by people, who did not care whether the object of investigation was a mineral or not. The fact that many mineralogists of influence were hesitant to consider structural work on non-minerals as belonging to the realm of mineralogy may be in part responsible for the present situation, in which crystallography can no longer be considered a part of mineralogy, as it was before 1912. Crystallography is now a science in its own right with many facets and roots, mineralogy being only one of many others, among them mathematics, physics, chemistry and biology. Previously it was desirable for physicists, chemists and biologists to know mineralogy, for it included an education in crystallography. This situation is well demonstrated by some sentences von Laue (1943) wrote commemorating the 100th birthday of P. von Groth who was Professor of Mineralogy and Crystallography at Munich at the time when von Laue conceived his idea:

'Mit Entdeckungen ist es in der Physik—diese im weitesten Sinne verstanden—eine eigene Sache. In dieser 300 Jahre alten Wissenschaft ist jeder Entdecker ein Erbe, ein Erbe des geistigen Gutes welches viele Generationen geschaffen und gemehrt haben. Paul von Groth hat sich neben vielen anderen grossen Verdiensten auch das erworben, dass er die mehr als 100-jährige, anderswo in Deutschland kaum noch vorhandene Tradition der Raumgitterhypothese in München durch seine Lehrtätigkeit lebendig erhalten und so eine der Voraussetzungen geschaffen hat, ohne welche die Auffindung der Röntgenstrahlinterferenzen rein Glückssache, ihre Deutung ganz unmöglich gewesen wäre.'

Today crystallography is nearly equally needed by mineralogy, physics and chemistry; conversely mineralogists, physicists and chemists join in the effort to promote the new science of modern crystallography. From this point of view the last fifty years have been very fateful ones for mineralogy in two respects as far as crystallography is concerned. Mineralogy lost the field of crystallography as its 'dominated colony', but it gained from the new possibilities of solving problems of central interest to mineralogy.

Before going into details of a structural character it should be mentioned that mineralogy and petrography obtained great advantage from Laue's discovery through using the powder method, invented by the physicists Debye and Scherrer in 1916, for the identification of minerals. Thus it became possible to investigate natural processes of mineral formation, even in those cases where the products were badly crystallized or microcrystalline. In addition, in many cases it turned out that substances which were thought to be different minerals proved to be identical from a structural point of view and vice versa. This is impressively shown in the book by Strunz *Mineralogische Tabellen* (1959, 3rd edition). Another application of the powder method of steadily increasing importance is the identification, or at least the characterization, of substances produced in experiments carried out to compare the conditions under which minerals can be formed in the laboratory with those under which minerals may have formed in nature.

Two highlights in this respect may be mentioned: (a) the first artificial production of diamond in 1955; (b) the production of a previously unknown high-pressure form of SiO_2 by Coes in 1953, later named coesite. The last mentioned discovery is of a unique mineralogic and cosmic interest: after the material was produced in the laboratory it was looked for in nature and was found as a mineral on the earth's surface where giant meteorites had collided with the earth. The crystal structure of coesite was determined by Zoltai and Buerger in 1959.

Other important problems are questions of nomenclature and of the classification of minerals. Crystal structure, structural types and the rules of crystal chemistry became leading principles. P. Niggli's textbooks of mineralogy (1920, 1924) are milestones in this respect; they educated and fascinated a generation of mineralogists.

11.2. Highlights of Structure Determination

Some highlights in determining the important structure types occurring in minerals shall be mentioned now, regardless of whether the determinations were done by physicists, chemists or mineralogists.

1913: The first structure determinations (W. H. and W. L. Bragg) were done on the minerals zincblende, diamond and NaCl. It still appears miraculous to the writer how fortunate and ingenious the Braggs were in chosing these 'easy' substances, considering how many much more complex structures could have been picked.

1914: Copper (W. L. Bragg); CaF_2 (W. L. Bragg); FeS_2 (W. L. Bragg).

1915: Spinel (W. H. Bragg, independently Nishikawa); calcite (W. H. Bragg).

1916: Graphite (Debye and Scherrer; independently Hull in 1917); rutile and anatase (Vegard).

1919: $Mg(OH)_2 = CdJ_2$-type (Aminoff).

1920: Wurtzite (W. L. Bragg).

1923: NiAs (Aminoff).

1924: The first silicate, garnet (Menzer).

1925/1926: SiO_2 – structures (W. H. Bragg and Gibbs; Wyckoff; Seljakow, Strebinski and Krasnikow).

1923–1926: V. M. Goldschmidt's famous rules of Crystal Chemistry and the determination of the size of atoms and ions. The main rule may be quoted here in the original formulation (*Geochemische Verteilungsgesetze der Elemente*, VII, pg. 9):
'Die Kristallstruktur eines Stoffes ist bedingt durch Grösse und Polarisationseigenschaften seiner Komponenten; als Komponenten sind Atome (respective Ionen) und Atomgruppen zu bezeichnen.'
The rules were drawn from the results of a large amount of experimental work gained in cooperation with his coworkers, mainly Zachariasen and Barth. The initial purpose of the work was to get information on the rules which govern the behaviour of matter with respect to isomorphism, isomorphous exchange, polymorphism and morphotropism. This would be a basis for investigating the petrological problem of the rules which govern the distribution of the elements within the earth and in its crust. The influences of V. M. Goldschmidt's work on the development of mineralogy and crystallography cannot be overestimated.

1926–1930: Bragg and coworkers (mainly Warren, West and Taylor), Zachariasen, and Pauling established the principles of silicate structures and their dependence on the Si/O ratio (Si_4O_{16} leading to tetrahedral groups; Si_4O_{12} leading to rings or chains; Si_4O_{11} leading to bands; Si_4O_{10} leading to sheets; Si_4O_8 leading to frameworks).

1928: Machatschki recognized that in silicate structures Al can replace Si; he was able to predict important features of the feldspar structures.

1929: Pauling formulated some very valuable rules for the crystal chemistry of compounds of predominantly ionic character. Whereas these rules are covered in principle by those given by V. M. Goldschmidt (1926), some of them formulate additional and very important aspects in a precise way. Two are quoted here in the original:
'IInd rule, electrostatic valence principle: In a stable coordination structure the electric charge of each anion tends to compensate the strength of the electrostatic valence bonds reaching to it from the cations at the centers of the polyhedron of which it forms a corner; that is, for each anion $\zeta = \Sigma_i s_i$.' [ζ = charge of the anion; $s = z/\nu$; z = charge of the cation; ν = coordination number of the cation.]
'IIIrd rule: The presence of shared edges, and particularly of shared faces, in a coordinated structure decreases its stability; this effect is large for cations with large valence and small coordination number, and is especially large in case the radius ratio approaches the lower limit of stability of the polyhedron.' [The limit of stability had been worked out already by V. M. Goldschmidt in 1926].

1928: In agreement with the rules mentioned before Pauling and Sturdivant succeeded in an ingenious way in determining the crystal structure of brookite, the orthorhombic modification of TiO_2. In addition they were able to give reasons for the sequence of stability of the three TiO_2-modifications (rutile—brookite—anatase).

1933: Stibnite, Sb_2S_3, and bismuthinite, Bi_2S_3 (W. Hofmann). General rules regarding the crystal chemistry of sulpho salts were given later by Hellner (1958).

1933: W. H. Taylor's determination of the structure of feldspars (refinements are still in progress in the Cavendish Laboratory, Cambridge, England) may be considered as a most importtna step in structural mineralogy, which brought the first period in this field to a close. An impressive account of this period has been given by W. L. Bragg (1937) in a book *Atomic structures of minerals*. Additional ideas have been put forward by Belov (1960).

This first period was one in which a large number of key structure types were worked out; these still form the foundations and the framework of the edifice of structural mineralogy as far as structural types are concerned. It is a large edifice with much room for many research workers to put walls and windows in the right places and to decorate the rooms with the beauty of further research results in structural mineralogy.

In one respect this 'edifice' seems to be different from those accomodating the results of structure research on substances produced artificially. In structural mineralogy the architectural style is restricted by the fairly small number of building stones offered in nature. In the fields of inorganic and organic chemistry, however, the human spirit may continuously produce new materials under varying conditions differing from those in nature. Thus in these other fields the style may be changed to adjust itself to new developments in building materials from year to year.

This simile should not be taken as indicating that in the writer's opinion structural mineralogy is doomed to reach a state of stagnation. The opposite is the case due to the fact that no crystal is perfect. The degree of perfection (respectively imperfection) can differ both in different mineral species and in different specimens of the same species. The kind and degree of imperfection may be rather significant for the question of mineral and rock genesis. Therefore, the study of mineral imperfections is not only fascinating as a research object of crystallography, but the results of such studies will be important for general mineralogy and petrology.

Some examples may be discussed in this respect following a historical order.

11.3. Point Defect

Minerals are usually formed from solutions (magmatic or hydrous) containing many elements which are not needed for the mineral formation, but which become incorporated in the crystal. The question of how this incorporation can occur (and mostly does occur) was already answered by Vegard in 1917. He found random distribution of different kinds of atoms on identical lattice points and a change of lattice constants nearly proportional to the amount of incorporated material ('Vegard's law'). Deviations of this 'Law' are of special interest nowadays. V. M. Goldschmidt (1926, pg. 83) investigated the question of how different the sizes of atoms or ions may be so that substitution can still occur. He gave the empirical rule:

> 'Isomorphe Mischbarkeit in erheblichem Ausmasse und bei Temperaturen, welche nicht sehr nahe dem Schmelzniveau liegen, tritt ein, wenn die Radien der betreffenden Bausteine um nicht mehr als etwa 15% (in Prozenten des kleinsten Radius) voneinander verschieden sind.' (For more details on this statement see the original paper.)

Hume-Rothery and coworkers (1934) showed that a similar rule holds for solid solutions in metals.

Another kind of more or less randomly distributed point defect (now called 'Leerstelle' or 'vacancy') was first described by Laves (1930) to explain the 'excess' of sulphur usually found in natural 'FeS'. He proposed the formula $Fe_{n-1}S_n$ instead of Fe_nS_{n+1}. Barth and Posnjak (1931) introduced the concept of 'variate atom equipoints' in discussing the imperfections in spinel and other structures, and Hägg (1933) distinguished between 'substitution, addition and subtraction mixed crystals'.

11.4. Feldspars

This concept led Barth (1934) to a hypothesis which turned out to be one of the most important ideas in the advance of mineralogy and petrology, if feldspars are considered as important minerals. He explained the different symmetries of potassium or K-feldspars (optically monoclinic sanidine and optically triclinic microcline) on the assumption that Al and Si may be disordered (as 'variate atom equipoints') in sanidine and ordered in microcline. This new hypothesis contradicted an older one advanced by Mallard (1876), who considered optically monoclinic K-feldspar to be the same 'phase' as triclinic microcline; the true symmetry of the latter would not be recognizable optically owing to submicroscopical twinning. The decision between these hypotheses was not possible, however, as long as the triclinic character of microcline had not been established by X-rays. In those days (Barth, 1928) no differences in geometry between the two lattices had been observed. The problem was taken up again in 1950 and 1952 by Laves. It could be shown that (a) many microclines have a small but significant deviation from monoclinic geometry, $\alpha = 90.5°$ and $\gamma = 92.5°$, later called 'maximum microcline'; (b) microcline is usually twinned polysynthetically after a combination of two twin laws, which fact is understandable if microcline was once monoclinic; (c) some optically monoclinic K-feldspars show the same kind of microcline twinning when X-rayed; (d) there are states with deviations from monoclinic symmetry which lie between those of maximum microcline and $\alpha = \gamma = 90°$, later called 'intermediate microcline'; (e) many hkl and $h\bar{k}l$ reflections show significant differences in intensity, the differences being larger for maximum microcline than for material with intermediate deviations from monoclinic geometry; (f) there are optically monoclinic K-feldspars with sharp reflections in monoclinic positions plus diffuse tails which have shapes and intensity distributions which can best be explained by the submicroscopical twinning of domains of triclinic character with atomic arrangements approaching those of intermediate or maximum microcline. Such material was called 'orthoclase' to distinguish it from sanidine which does not show these tails and which appears to be truly monoclinic—not only optically but also when investigated by X-rays; (g) in addition to the tails mentioned in (f) many 'orthoclases' show diffuse reflections in positions not consistent with the feldspar lattice proposed by Taylor (1933). A qualitative model explaining these reflections was given in 1961 (Laves and Goldsmith; J. V. Smith).

With these observations (a—f) it could be proved that there are at least two K-feldspar modifications: monoclinic sanidine and triclinic microcline, thus supporting Barth's (1934) hypothesis; that there are optically monoclinic K-feldspars with a structural appearance supporting Mallard's hypothesis (1876); and that there are K-feldspar states, which may not be considered as stable phases in the sense of Gibbs, which are neither sanidine nor microcline but something in between. Such material was called common or normal orthoclase or just 'orthoclase'. There is still much discussion nowadays between crystallographers and mineralogists on the significance and characterization of such K-feldspar states which can be considered neither as sanidine nor as microcline as a consequence of structural imperfections which show themselves in X-ray photographs as described above in (f) and (g). The present situation is expressed in a paper given by Laves and Goldsmith at a feldspar symposium sponsored by the

International Mineralogical Association in Copenhagen in 1960 entitled 'Polymorphism, order, disorder, diffusion and confusion in the feldspars'. Other papers and discussions on the subject by R. B. Ferguson, J. B. Jones, W. S. Mackenzie, A. S. Marfunin, H. D. Megaw, J. V. Smith, W. H. Taylor are compiled in the Proceedings of this meeting (*Cursillos y Conferencias* Vol. *8*, September 1961, Instituto 'Lucas Mallada', Madrid), which in addition contains many other important papers by Megaw and coworkers on structural features of the plagioclases which will not be discussed here in detail. The papers of this Conference show in an impressive way the lively activity of crystallographers and mineralogists in using X-rays for the solution of the problems and puzzles which nature has offered through the production of feldspars. Much has been done but more is needed in order to understand the complexities involved.

The feldspars are treated here in great detail for three reasons: (1) The problems involved are typical of the difficulties mineralogy has to put up with. These difficulties are such that it appears to be impossible to reproduce in the laboratory the conditions under which some minerals were formed, because we may vary in the laboratory the pressure and temperature and the chemical compositions of our reaction products but we cannot compete with the time nature had available. All attempts to synthesize the ordered low-temperature modifications of K-feldspar (microcline) and Na-feldspar (albite) failed. The products have been always disordered high-temperature modifications. On the other hand it is easy to produce disordered high-temperature modifications by heating the ordered low-temperature modifications to near the melting point. (2) The feldspars offer an excellent example of the central question in mineralogy and crystallography: *What is a mineral, what is a single crystal*, a question which is not only one of semantics but involves interesting problems of the nature of the condensed state currently discussed. (3) Besides quartz the feldspars are the most frequent minerals as together they make up approximately 65% of the earth's crust.

Returning to Barth's Al/Si order/disorder hypothesis the question arises of how can it be proved? The scattering powers of Al and Si are so similar that it is impossible to decide by X-rays whether a particular site is occupied by Al or Si except perhaps by methods not yet easily feasable. For this reason Taylor gave in 1933 only the sites where Al *or* Si are located. Therefore, other methods had to be tried for information on the Al/Si order and disorder. They are summed up in short. (a) By diffusion experiments near the melting point the Na of an albite could be replaced by K leading to microcline (Laves, 1951). This indicated that the alkali ions are not responsible for the differences between the high and low-temperature forms and that the differences must lie in the $AlSi_3O_8$ framework. It was proved further that the Al/Si distribution of albite and microcline are identical. Very accurate structure determinations of albite and maximum microcline are now at hand (unpublished work by S. W. Bailey, personal communication) which are in line with this. (b) It can be shown theoretically that mechanical twinning cannot take place if the Al/Si distribution is ordered but it may take place if it is disordered. Experiments to twin albite mechanically failed (Mügge and Heide, 1931) even under extreme conditions, but mechanical twinning can easily be performed at room temperature (Laves, 1952) with Na-rich feldspars if they are in a high-temperature state. (c) Results of crystal chemistry gained from substances other than feldspar suggested that the size of $(Al,Si)O_4$ tetrahedra should be approximately proportional to the probability of an Al sitting in the tetrahedra (J. V. Smith, 1954). A refined structure determination of a halfway intermediate microcline (Bailey and Taylor, 1955) revealed significant differences in size between the four tetrahedral sites. (These results should be compared with those of a structure determination of an 'orthoclase' carried out by Jones and Taylor in 1961.) The same was true in a refined structure determination of albite (Ferguson, Traill and Taylor, 1958). The refined structure determination of $NaAlSi_3O_8$ in a high-temperature state, carried out with the same accuracy by the same authors did not reveal any differences in the sizes of the tetrahedra. (d) The difference in the sharpness of infrared absorption bands between low and high-temperature forms strongly indicates differences in the degree of Al/Si order (Laves and Hafner, 1956 and 1957). (e) Finally, measurements of the nuclear magnetic resonance of $KAlSi_3O_8$ and $NaAlSi_3O_8$ in the low and high-temperature state leave no doubt, that the Al/Si order is practically complete in microcline and albite, whereas it is practically zero in the corresponding high-temperature states (Brun, Hafner, Hartmann and Laves, 1960).

The account given above shows that sometimes the combination of different approaches—field evidence, good ideas (Barth), accurate structure determinations (Taylor and coworkers) and physical methods other than X-ray investigations—is necessary if problems are to be solved which are of pertinent interest to mineralogy and petrology.

11.5. One-dimensional Disorder (Stacking Faults) and Polytypism

In the above section some diffuse reflections were mentioned which occur in X-ray photographs of feldspars; these have not yet been explained rigorously by mathematical treatment. Thus, there is still much to be done in the future. However, somewhat similar effects have been observed in other minerals which could be treated mathematically in a more rigorous way: stacking faults. The first observation of this kind was published in 1927 by Mauguin who observed 'streaks' corresponding to rows in reciprocal space in X-ray photographs of micaceous minerals. Similar effects were observed by Nieuwenkamp and Laves (1935) in X-ray photographs of maucherite, Ni_3As_2, and of SiO_2. Hendricks and Teller (1942) added much to the theoretical interpretation of these effects in the case of the mica minerals and A. J. C. Wilson (1942) in the case of metals and alloys. Jagodzinski and Laves (1949) introduced the concept of 'eindimensionale Fehlordnung' (one-dimensional disorder) having observed similar effects in single crystals of graphite and ZnS. The most general theoretical treatment of the observed X-ray effects has been given by Jagodzinski (1949); he considered the influence of the position of a given layer on the position of other layers up to four layers away. As unidirectional stress can change the order-disorder relations (for example, it is possible experimentally to produce 'rhombohedral graphite' domains within a single crystal of hexagonal graphite) applications of such investigations to the geological history of mica and other sheet minerals in rocks may become important for future studies of rock metamorphism.

An important example for such polymorphism or polytypism which is chemically simple and has mineralogical interest is 'tridymite', a 'polymorph' of SiO_2. If we consider the tridymite structure as it is given in textbooks it can be described in some analogy to the hexagonal close-packed structure with a sequence of layers ABAB... If we call this a 2-layer sequence, structural work has shown that no tridymites are known yet which have such an ideal 2-layer sequence. Mistakes occur which necessitate the consideration of an additional C position in such a way that ABC-sequences must be present. Buerger and Lukesh (1942) observed natural tridymites with n-layer sequences with $n = 10$ or 20. Flörke (1955) found the same sequences in tridymites produced synthetically; other kinds of stacking disorder within single crystals were found by him too. Whereas some correlations between layer sequence and impurities could be worked out, the reasons for the preference for 10- and 20-layer sequences are still obscure. Several authors discussed similar structural anomalies observed in SiC where n-layer sequences up to $n = 594$ and more are reported in the literature (Mitchell, 1958). For a discussion of the reasons for such large periodicities perpendicular to the layer plane a paper by Jagodzinski and Arnold (1960) may be quoted. Jagodzinski rejects the opinions advanced by such authors as Bhidde, Verma and Mitchell suggesting that the polytypes are caused by screw dislocations and proposes that the influence of entropy terms may be more important. However, his concluding sentence reads 'A great deal of effort is needed to solve these problems'.

11.6. What is a Mineral? What is a Crystal?

The writer as a professor of mineralogy has frequently been asked 'What is mineralogy?' An answer like 'it is the science which deals with minerals' can easily be given; some people however, are not content with such an answer and want to know 'What are minerals?' At this point it is better to leave the room, for there is no satisfactory answer to such a question. Here are the problems.

Looking into textbooks of mineralogy one finds (neglecting such 'minerals' as amber and others) that a mineral is an inorganic substance produced by natural processes, implying in this context that processes in which men are involved are considered unnatural. Furthermore 'homogeneity' is one of the properties which belongs to a mineral.

However, what does 'homogeneous' mean? This question could not be answered in a satisfactory way before 1912, nor can it be answered today as we shall see. Thus it may appear that the new tool available since Laue's discovery has not helped us to answer this basic question of mineralogy. In a strict sense this is true. However, in many special cases X-rays can tell us that substances previously thought to be homogeneous are actually mixtures of different substances; in other cases X-rays have revealed structural features in 'minerals' which are fascinating and which add considerably to the liveliness of the present day research in structural mineralogy.

One example may be given for illustration. Moonstone, a gemstone with beautiful blue schiller belonging to the feldspar group, can be considered as 'really' being homogeneous when in a high-temperature state, i.e. for example when formed under magmatic conditions.

However, when we pick up such a feldspar in nature it usually has a low temperature and may have changed considerably in its atomic arrangement during the millions of years which may have elapsed since it was formed.

Neglecting minor impurities always present in each 'pure' substance the chemical composition of the feldspars can be expressed as a mixture of three 'molecules' Or = $KAlSi_3O_8$, Ab = $NaAlSi_3O_8$, An = $CaAl_2Si_2O_8$. Moonstone, an alkali feldspar, is a mixture of the molecules Or and Ab and usually appears 'homogeneous' when investigated with a microscope.

X-ray photographs, however, give results which can be interpreted as a single-crystal pattern of 'orthoclase' or sanidine ($KAlSi_3O_8$) plus oriented reflections, more or less diffuse, in positions which correspond approximately to those positions which a pure but twinned $NaAlSi_3O_8$ would show when X-rayed alone (either in the low-temperature form albite or in the high-temperature form analbite or in an intermediate state between albite and analbite).

The conclusion may be that moonstone is not a mineral but a mixture of minerals. Considering that the K and Na-rich areas cannot be discriminated by microscopic investigation but that their existence can be shown by X-rays, and considering that the size and shape of these areas changes from specimen to specimen (as revealed by the varying diffuseness and position of the Na-feldspar reflections) the question 'mineral' or 'not-mineral' finally boils down to a question similar to 'has a Texas Grapefruit a yellow or an orange colour?'

Thus, trying to answer the question 'What is a mineral' leads to the question 'What does homogeneous mean?' Through trying to answer *this* question, theoretical problems of high importance arise which are of general interest not only for mineralogy but also for the theory of the condensed state in its broadest sense. The problems involved have not yet been solved, but X-ray investigations of minerals offer excellent examples for future discussion of the kinetics and thermodynamics of solid-state reactions.

As one of these examples moonstone has been mentioned already. For further discussion it is chosen again for several reasons. It belongs to the group of feldspars which are a major component of the earth (appr. 65%). Thus, it occurs in many localities and there is a good opportunity to study and to compare samples of different origin on the one hand, and to compare the structural states (there is no *one* state characteristic of *the* moonstone) of moonstones with the structural states of other feldspars which are chemically alike.

As a result of such investigations we now know that feldspars with moonstone composition can exist in an infinite number of different states. These states lie between two extremes. (1) An ideal homogeneous mineral in which the alkali sites of the crystal structure are randomly occupied by K and Na ions and in which structural sites exist which are occupied nearly at random by Al and Si ions. (2) An ideal mixture (called perthite) of two ideal homogeneous minerals, namely microcline and albite, both having structural sites which are occupied either by Al or by Si ions. In microcline the alkali sites are occupied only by K ions, in albite only by Na ions. Thus, after the crystallization of a feldspar with moonstone composition processes may take place concurrently during geological times within a morphological unit: (a) separation into areas of different K/Na ratio, (b) change of the original Al-Si disorder into states of higher order, (c) enlargement of the Na-rich and K-rich areas by a process of 're-crystallization' into areas of a visible size.

Whether the state, stable at room temperature (microcline plus albite), is fully reached by these processes or only partly approached may depend on several factors, some among them of geological character. Thus a morphological unit of *one* typical mineral (sanidine) may change continuously into a material of *two* typical minerals (microcline and albite) which can easily be separared mechanically. At which stage does such a morphological unit cease to be *one* mineral and become a mixture of *two* minerals? There is no satisfactory answer to this question.

Another feldspar example may be chosen to elucidate the question 'What is a single crystal?' Na-feldspar in its most disordered state (monalbite) has monoclinic symmetry like sanidine. However the monoclinic states are stable only at high temperatures; at low temperatures a triclinic form is stable. Two kinds of phase transformation can occur leading to 'crystals' that appear to be monoclinic when investigated optically. One transformation (diffusive in the sense of Laves, 1952) is due to a change in Al/Si distribution needing geological times at low temperature, whereas the other transformation (displacive in the sense of Buerger, 1948) takes place without a change in Al/Si distribution and cannot be suppressed either by quenching or rapid heating. In both cases 'microcline twinning' is produced which is frequently on a submicroscopical scale and then detectable only by means of X-rays. As a matter of fact the displacive transformation of monalbite into analbite (the disordered triclinic form of $NaAlSi_3O_8$ in contrast to the ordered triclinic form, albite) has hitherto always led to an optically monoclinic material which is submicroscopically twinned.

Before 1912 such material would have been called in good faith a single crystal. Trouble now arises because we have X-rays. Again we may expect a continuous series of states between two extreme cases (and we find it in nature in the case of microcline): (1) a real single crystal in which the 'twinning' is expressed by the symmetry of a space group somewhat different from the one the transformed phase has; (2) a morphological unit in which the twinned domains are so large that they can be 'seen' and which can be cut into small pieces of real single crystal character. At which stage does such a morphological unit cease to be *one* single crystal and becomes a twinned aggregate of *more* than one single crystal? There is no satisfactory answer to this question.

The difficulties involved in answering the questions 'What is a mineral?' and 'What is a crystal?' are closely related to the question 'What is a phase?' A discussion of this last question is not only important for mineralogists but for all people who are interested in the condensed state. A very valuable contribution to this problem has been given by Jagodzinski (1959).

11.7. Concluding Remarks

The preceding sections have shown how large the variation of the structural state of even *one* mineral can be as revealed by X-rays. Laue's discovery in 1912 provided the tools for recognizing this variation and for using it in attempts to reconstruct the conditions under which minerals in their various structural states were formed in nature. In some cases reproduction can be achieved by experimental synthesis in the laboratory; in other cases where the times involved are available to nature but not to men, intelligent thinking is needed to explain the facts offered by nature.

Unfortunately, only a few topics of structural mineralogy could be discussed in enough detail to give an impression of the development of such problems during the last 50 years and the present-day state of their solution. Many other topics might have been chosen equally well, but the selection given here should illustrate not only the impact Laue's discovery had on the growing field of structural mineralogy, but should also show the importance of this field for problems of general mineralogy, petrology, geology, chemistry and physics.

References

References of papers cited in the text (see also *Strukturbericht* and *Structure Reports* for additional references):

S. W. Bailey and W. H. Taylor (1955), The structure of a triclinic potassium feldspar; *Acta Cryst. 8*, 621–632.

T. F. W. Barth (1928), Die Symmetrie der Kalifeldspäte; *Fortschr. d. Min. 13*, 31–35; (1934) Polymorphic phenomena and crystal structure; *Am. J. Sci. 27*, 273–286.

T. F. W. Barth and E. Posnjak (1931), The spinel structure: an example of variate atom equipoints; *J. Wash. Acad. Sci. 21*, 255–258.

N. V. Belov (1960, short survey in English), Chapter B of the Crystal Chemistry of Silicates; *Fortschr. d. Mineralogie 38*, 4–6.

W. L. Bragg (1937), *Atomic structure of minerals*; Cornell University Press, Ithaca, N.Y.; 292 p.

E. Brun, St. Hafner, P. Hartmann, F. Laves und H. H. Staub (1960), Magnetische Kernresonanz zur Beobachtung des Al, Si-Ordnungs/Unordnungsgrades in einigen Feldspäten; *Z. Kristallogr. 113*, 65–76.

M. J. Buerger (1948), The role of temperature in mineralogy; *Am. Mineralogist 33*, 101–121.

M. J. Buerger and J. Lukesh (1942), The tridymite problem; *Science 95*, 20–21.

F. P. Bundy, H. T. Hall, H. M. Strong and R. H. Wentorf (1955), Man-made diamonds; *Nature 176*, 51–55.

L. Coes (1953), A new dense crystalline silica; *Science 118*, 131.

R. B. Ferguson, R. J. Traill and W. H. Taylor (1958), The crystal structure of low-temperature and high-temperature albites; *Acta Cryst. 11*, 331–348.

O. W. Flörke (1955), Strukturanomalien bei Tridymit und Cristobalit; *Ber. D. Keram. Ges. 32*, 369–381.

General Electric Research Laboratory (Marshall, Bundy, Hall, Strong, Nerad) 1955: *Man-made diamonds*; Research information services, The Knolls, Schenectady, New York, 25 p.

V. M. Goldschmidt (1926), Geochemische Verteilungsgesetze, VII: Die Gesetze der Krystallochemie (nach Untersuchungen gemeinsam mit T. Barth, G. Lunde, W. Zachariasen); *Skrifter Norsk. Vid. Akademie*, Oslo, *Mat. Nat. Kl.* 1926. No. 2; See in addition VIII: Untersuchungen über Bau und Eigenschaften von Krystallen, l.c. No. 8.

St. Hafner und F. Laves (1956, 1957), Ordnung/Unordnung und Ultrarotabsorption I, (Al, Si)-Verteilung in Feldspäten; II, Zur Struktur von Orthoklas und Adular; *Z. Kristallogr. 107*, 196–201 and *109*, 204–225.

G. Hägg und I. Sucksdorff (1933), Die Kristallstruktur von Troilit und Magnetkies; *Z. phs. Chem.* B, *22*, 444–452.

E. Hellner (1958), A structural scheme for the sulfide minerals; *J. Geology 66*, 503–525.

St. Hendricks and E. Teller (1942), X-ray interference in partially ordered layer lattices; *J. Chem. Physics 10*, 147–167.

W. Hume-Rothery, G. W. Mabbott and K. M. Channel-Evans (1934), The freezing points, melting points, and solid solubility limits of the alloys of silver and copper with the elements of the B sub-groups; *Phil. Trans. R. S.* London, A, *233*, 1–97.

H. Jagodzinski (1949), Eindimensionale Fehlordnung in Kristallen und ihr Einfluss auf die Röntgeninterferenzen; *Acta Cryst. 2*, 201–207; 208–214; 298–304; (1959) Struktur- und Phasenbegriff in Mischkristallen; in *Beiträge zur Physik und Chemie des 20. Jahrhunderts*, Vieweg-Verlag, Braunschweig, p. 188–209.

H. Jagodzinski and H. Arnold (1960), Anomalous silicon carbide structure, in *Silicon Carbide*, Pergamon Press, London, p. 136–146.

H. Jagodzinski and F. Laves (1948), Eindimensional fehlgeordnete Kristallgitter; *Schweiz. Min. Petr. Mitt. 28*, 456–467.

J. B. Jones and W. H. Taylor (1961), The structure of orthoclase; *Acta Cryst. 14*, 443–446.

M. v. Laue (1943), Zu P. v. Groths 100. Geburtstage; *Z. Kristallogr. 105*, 81.

F. Laves (1930), Die Bau-Zusammenhänge innerhalb der Kristallstrukturen; *Z. Kristallogr. 73*, 202–265, 275–324; (1950) The lattice and twinning of microcline and other potash feldspars; *J. Geology 58*, 548–571; (1951) Artificial preparation of microcline; *J. Geology 59*, 511–512; (1952) Phase relations of the alkali feldspars; *J. Geology 60*, 436–450, 549–574; see in addition Al/Si Verteilungen, Phasentransformationen und Namen der Alkalifeldspäte; *Z. Kristallogr. 113* (1960) 265–296; (1952) Ueber den Einfluss von Ordnung und Unordnung auf mechanische Zwillingsbildung; *Naturwissenschaften 39*, 546.—Mechanische Zwillingsbildung in Feldspäten in Abhängigkeit von Ordnung-Unordnung der Si/Al-Verteilung innerhalb des $(Si, Al)_4O_8$—Gerüstes; l.c. 546–547.

F. Laves and J. R. Goldsmith (1961), Polymorphism, order, disorder, diffusion and confusion in the feldspars; *Cursillos y Conferencias*, Madrid, *8*, 71–80.

F. Machatschki (1928), Zur Frage der Struktur und Konstitution der Feldspäte (Gleichzeitig vorläufige Mitteilung über die Prinzipien des Baues der Silikate); *Zentralbl. f. Min.* 1928, 97–100.

F. Mallard (1876), Explications des phénomènes optiques anomaux, que présentent un grand nombre de substances cristallisées; *Annales des Mines 10*, 187–240; reviewed in *Z. Kristallogr. 1* (1877) 309–320.

Ch. Mauguin (1928), Les rayons X ne donnent pas toujours la véritable maille des cristaux; *C. R. Paris 187*, 303–304.

R. S. Mitchell (1957), A correlation between theoretical screw dislocations and the known polytypes of silicon carbide; *Z. Kristallogr. 109*, 1–28.

O. Mügge und F. Heide (1931), Einfache Schiebungen am Anorthit; *Neues Jahrb. f. Min.*, Beilage Band A *64*, 163–170.

W. Nieuwenkamp (1935), Zweidimensionale Cristobalitkristalle; *Z. Kristallogr. 90*, 370–380; see in addition l.c. 273–278 and 279–282.

P. Niggli (1920, 1924; 1st and 2nd edition), *Lehrbuch der Mineralogie*; Verlag Borntraeger, Berlin.

L. Pauling (1929), The principles determining the structure of complex ionic crystals; *Am. Chem. Soc. 51*, 1010–1026.

L. Pauling and J. H. Sturdivant (1928), The crystal structure of brookite; *Z. Kristallogr. 68*, 239–256.

J. V. Smith (1954), A review of the Al—O and Si—O distances; *Acta Cryst. 7*, 479–483.

J. V. Smith and W. S. MacKenzie (1961), Atomic, chemical and physical factors that control the stability of alkali feldspars; *Cursillos y Conferencias*, Madrid, *8*, 39–52.

H. Strunz (1941, 1949, 1957, 1st–3d edition), *Mineralogische Tabellen*; Akademische Verlagsgesellschaft, Leipzig, 448 p.

W. H. Taylor (1933), The structure of sanidine and other feldspars; *Z. Kristallogr. 85*, 425–442.

L. Vegard und H. Schjelderup (1917), Die Konstitution der Mischkristalle; *Phys. Zeitschrift 18*, 93–96. See in addition L. Vegard (1921): Die Konstitution der Mischkristalle und die Raumerfüllung der Atome; *Vidensk. Skrifter, Mat. Nat. Kl.*, Kristiania 1921. No. 6, 36 p.

A. J. C. Wilson (1942), Imperfection in the structure of cobalt, II. Mathematical treatment of proposed structure; *Proc. R. Soc. London A 180*, 277–285.

T. Zoltai and M. J. Buerger (1959), The crystal structure of coesite, the dense, high-pressure form of silica; *Z. Kristallogr. 111*, 129–141.

CHAPTER 12

Applications of X-ray Diffraction to Metallurgical Science

*by W. Hume-Rothery**

12.1

At the time of the discovery of X-ray diffraction, the knowledge of the structure of metals was limited to what could be revealed by optical microscopy. From the occasional occurrence of metallic crystals with well defined plane faces, it was recognized that the structure of metals was essentially crystalline. Little but speculation existed as to the actual atomic arrangement, although the geometrical theory of space-groups and space-lattices had been laid down long before the detailed atomic arrangement could be determined. The general principles of metallic phase diagrams had been established by Roozeboom and others, and the experimental work of Heycock and Neville (1897) had shown how the limits of the different phase fields could be established to a high degree of accuracy, even in very complicated systems. At the same time the German School under Tammann had produced a rapid survey of a number of metallic equilibrium diagrams, but the underlying structures remained a mystery. The application of microscopical methods to the study of steels had resulted in the recognition of a number of 'constituents,' but confusion often existed as to whether these were distinct phases with definite crystal structures, or were mixtures of phases on a scale too fine to be resolved by optical methods. The general position was, therefore, one in which further progress depended on the discovery of some method by which the detailed atomic arrangement in metals could be revealed.

* The following review has been written at short notice at the request of the Editor. The time available did not permit any systematic searching for references, and the paper should be regarded as a brief commentary on those applications of X-ray diffraction to metallurgical problems which happen to have attracted the author's attention. At the outset, therefore, an apology must be offered to the many investigators to whose work no reference is made, even though their contributions to the science were considerable.

12.2

The importance of the discovery of the diffraction of X-rays by crystals was recognized from the outset, but its application was delayed for several years by the First World War. Early work by Davey, Hull, and others led to the determination of the crystal structures of many of the more common metals and, owing to the difficulty of obtaining single crystals of metals, the Debye-Scherrer, or powder method of crystal analysis was specially attractive—indeed it was devised by A. W. Hull indepently of, and nearly simultaneously with Debye and Scherrer for the purpose of structure determinations of metals. The original discovery of this method was made in 1916, and from 1920 onwards cameras of various types began to make their appearance, and their application to many metallurgical problems resulted in spectacular advances in the decade 1920–30. It was soon established that the majority of the metals crystallized in one of the three typical metallic structures shown in Fig. 12–2(1). These are the face-centred cubic, close-packed hexagonal, and body-centred cubic structures. Allotropy is common, particularly among the transition metals, and in nearly all such cases the body-centred cubic modification

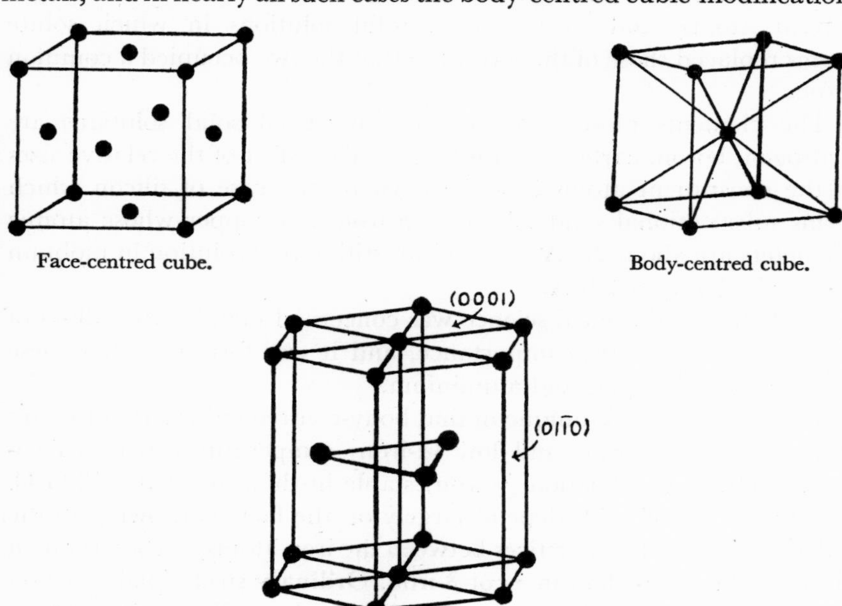

Face-centred cube.

Body-centred cube.

The close-packed hexagonal structure with axial ratio 1.633.

Fig. 12–2(1). The three typical metallic structures.

is the form stable at the highest temperatures. This is because the relatively open nature of this structure results in abnormally large amplitudes of thermal vibrations in certain directions relative to the crystal axes. The corresponding entropy term in the expression $G = U - TS + PV$ is thus large at high temperatures, and tends to lower the value of the free energy, G, and so to make the body-centred cubic structures more stable.

12.3

The older work on the constitution of alloys had led to the conclusion that many metals formed solid solutions* in one another. This was confirmed experimentally by the work of Bain, and of Owen and Preston, who showed that the diffraction patterns from a solid solution resembled that of the parent metal, but that the formation of a solid solution was accompanied by a change in the lattice spacing. Combination of lattice spacing measurements and density determinations enabled a distinction to be made between interstitial solid solutions in which small solute atoms entered the interstices between the larger solvent atoms, and substitutional solid solutions in which solute atoms replaced those of the solvent so that the two occupied a common lattice.

The elements most concerned in interstitial solid solutions are hydrogen, boron, carbon, and nitrogen. The effect of the relative sizes of the constituent atoms is well shown by the case of silicon which forms substitutional solid solutions in iron and copper whose atomic diameters are about 2.5 Å, but an interstitial solid solution in niobium of atomic diameter 2.9 Å.

In 1920, metallurgical science was concerned mainly with alloys of the metals of industrial importance, and it was fortunate that these included iron, copper, and aluminium.

The element iron is unique in that body-centred cubic structures are stable at high (δ-iron) and low (α-iron) temperatures, with a face-centred cubic modification (γ-iron) stable in the range 910°–1390°C. The properties of steels depend largely on the fact that, owing to the different sizes of the interstices between the iron atoms, carbon is much more soluble in γ- than in α- or δ-iron. Ordinary steels (plain carbon steels) may be regarded as impure iron-carbon alloys with carbon

* The term mixed-crystal was also used.

contents in the range 0.1–1.5% carbon by weight together with small quantities of other elements—when relatively large proportions of other metals are deliberately added, the steels are known as alloy steels. The methods of optical microscopy had shown that in plain carbon steels most of the carbon is present in the form of a metastable carbide of iron, Fe_3C or cementite, and a suspicion existed that other carbides might be formed when hardened or 'quenched' steels were tempered, whilst carbides of other metals were known to be present in some alloy steels. The application of X-ray methods to the problems of these steels began with the X-ray diffraction work of Hägg, and of Westgren and Phragmén. The solid solution of carbon in γ-iron was shown to be of the interstitial type and, on slow cooling, the γ-iron changes to α-iron whilst the carbon atoms form cementite whose unit cell was determined as early as 1922 by Westgren and Phragmén. With rapid cooling, the $\gamma \rightarrow \alpha$ change takes place too quickly to permit the formation of cementite, and the carbon atoms remain trapped in the interstices of α-iron, whose unit cell is distorted to a body-centred tetragonal unit with the carbon atoms in what are called the octahedral interstices. The tetragonal nature of this phase, martensite, was established in 1926, by Fink and Campbell and, independently in 1927 by Seljakow, Kurdjumow, and Goodtzow, but it was not until much later that the details of these structures were determined. The writer remembers a meeting of the Iron and Steel Institute in 1940 at which the late Dr. Hatfield complained that the crystallographers could not say where the carbon atoms were in cementite. Whereupon Sir Lawrence Bragg pulled an invisible string, and Dr. (now Professor) Lipson rushed to the platform and produced the answer.

In the years between the two World Wars X-ray methods established the structures of many of the carbides which were of importance in connection with ferrous metallurgy.

In this work many ingenious methods have been used. In some cases, if steels are treated with suitable acids—either directly or electrolytically—the iron matrix can be dissolved, and the carbides left as an insoluble residue which can be submitted to chemical and X-ray analysis. This method has been of great value, but has the obvious disadvantage that the carbide itself may be attacked by the acid. Methods were devised by K. H. Jack, in which nitrides of iron were first prepared, and were then treated with carbon monoxide which resulted in the formation of an iron percarbide, $Fe_{20}C_8$; this had originally been obtained by Hägg, and its composition has at different times been regarded as Fe_2C and $Fe_{20}C_9$. It is a metastable carbide which is of interest because it is formed as an intermediate product in the tempering of martensite.

Carbon is the most important of the interstitial elements in steels but nitrogen, whose atomic diameter is slightly less than that of carbon, is able to form an analogous series of alloys. Nitrogen, like carbon, is more soluble in γ-iron than in α-iron, and nitrogen martensites can be obtained with the same general characteristics as carbon martensites. The smaller atomic diameter of nitrogen has the interesting effect of permitting the formation of hexagonal iron nitrides, in which the iron atoms occupy a close-packed hexagonal structure with nitrogen atoms in the octahedral interstices. The very beautiful X-ray work of Jack has led to the discovery of a whole series of iron-nitrides, the characteristic of which is that the nitrogen atoms arrange themselves in the octahedral interstices of the metallic framework, in such a way that the nitrogen atoms keep as far away from one another as possible.

If these nitrides are treated with carbon monoxide, some of the nitrogen is replaced by carbon, with the formation of hexagonal carbo-nitrides of composition ranging from $Fe_2(C,N)$ to $Fe_3(C,N)$. It is not possible in this way to obtain a pure hexagonal iron carbide, but hexagonal or ε iron carbides are formed as intermediate products in the tempering of some steels, and may be regarded as the end products of the carbo-nitrides which are derived from the nitrides. The work on iron-nitrogen alloys has been of outstanding value because, apart from its direct bearing on the problem of nitrogen in steels, it has led to a greater understanding of the carbides of iron. All these carbides are metastable when compared with the stable system iron-graphite, and according to Jack their free energies are in the order ε carbide $>$ iron percarbide $>$ cementite.

The work on iron-nitrogen alloys has also led to the discovery of two nitrides of great structural interest. The first of these is γ-iron nitride, Fe_4N, whose structure is shown in Fig. 12–3(1). Here the iron atoms occupy a normal face-centred cubic lattice with a nitrogen atom at the centre of each unit cube. The nitrogen atoms thus occupy octahedral interstices and arrange themselves so that, when a particular interstice is filled, both the nearest and second nearest octahedral interstices are unoccupied. A further iron nitride, Fe_8N, is formed as an unstable intermediate constituent in the tempering of nitrogen martensite, and has the remarkable structure shown in Fig. 12–3(2). This may be regarded as a highly distorted martensitic structure in which only 1 in 24 of the interstices are occupied by nitrogen atoms but, in spite of this extreme dilution, the arrangement is perfectly regular, and examination of the diagram will show the fascinating way in which the larger and shorter distances between the iron atoms compensate for each other. In ordinary carbon and nitrogen martensite, the proportion of interstitial atoms is smaller than in Fe_8N, but Fig. 12–3(2) gives some idea of the local distortion which will surround each interstitial solute atom.

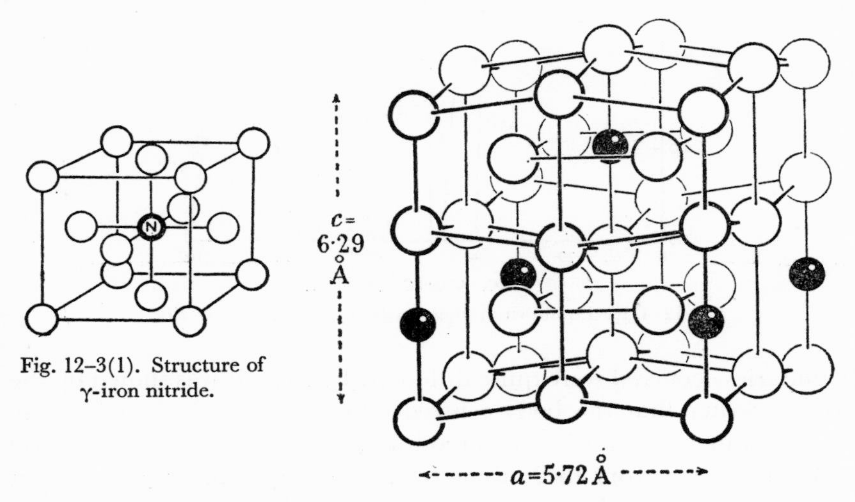

Fig. 12–3(1). Structure of γ-iron nitride.

Fig. 12–3(2). The structure of tempered nitrogen-martensite (α″). O, iron atom; ●, nitrogen atom. From K. H. Jack, *Proc. Roy. Soc.* A *208*, 218 (1951).

The methods of X-ray diffraction have led to great advances in the knowledge of the constituents present in steels, and of the processes occurring during the various heat treatments to which industrial alloys are submitted. The information gained in this way is now being supplemented by the methods of electron microscopy and electron diffraction, and the present time is one of great activity.

12.4

In the study of copper alloys the combination of X-ray diffraction methods with those of the older techniques led to the foundation of the first general theory of alloy structures. In alloys such as those of copper-zinc, copper-aluminium, and copper-tin it was obvious that the copper-rich parts of the equilibrium diagrams were of the same general form (Fig.12–4(1)) although the compositions at which the various phase

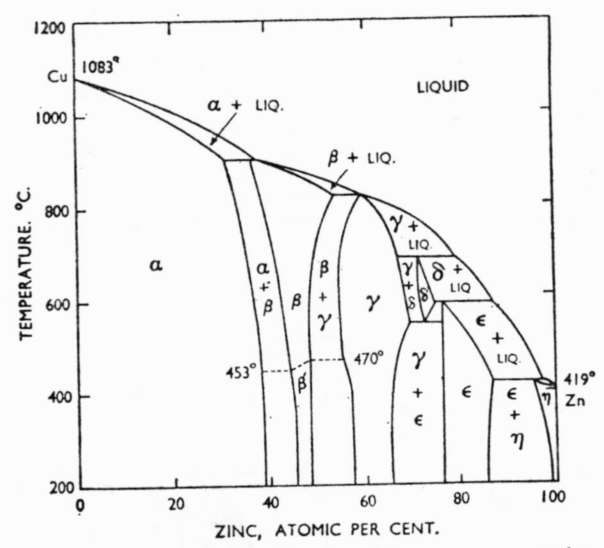

Fig. 12–4(1). Equilibrium diagram of the system copper-zinc.

boundaries occurred were quite different. In 1926, it was shown by the present writer that the body-centred cubic β-solid solutions in these alloys, although of variable composition, occurred at roughly a ratio of 3 valency electrons to 2 atoms (e.g. CuZn, Cu_3Al, Cu_5Sn), and it was suggested that, in some of these intermediate phases, structures might be characterised by constant electron/atom ratios. This idea was taken up enthusiastically by Westgren and Bradley, and it was established that electron/atom ratios, or *electron concentrations*, of 3/2, 21/13, and 7/4 were characteristic of structures of the body-centred cubic, γ-brass, and close-packed hexagonal types. It was, further, shown that ternary and quaternary alloys with these structures could be obtained, provided that the atoms were mixed together so that the characteristic electron concentration was preserved. Intermediate phases of this type were known as *electron compounds*, and it was by a combination of X-ray diffraction methods with the older techniques that the principles underlying this type of inter-metallic phase were revealed.

12.5

The industrial applications of aluminium alloys led to the development of three distinct types of X-ray diffraction work. The element is of relatively low melting point, and the methods of optical microscopy showed that when worked aluminium alloys were heated to comparatively low temperatures, recrystallization occurred in which an entirely new set of crystal grains was produced. The mechanical properties of the alloys depended, among other things, on the grain size, and under some conditions crystals of dimensions of the order 1 cm might be formed. A systematic study of these effects by C. F. Elam (now Mrs. Tipper) led to an understanding of the conditions necessary for the formation of large single crystals, and this in its turn led to methods for the production of single crystals of dimensions of the order $5'' \times \frac{1}{2}''$ diameter. These were then submitted to deformation in simple tension or compression, and remarkable results were obtained in which the deformation of the test piece was clearly related to the orientation of the crystal relative to the axis of the specimen (Fig. 12–5(1)). In the work of H. C. H. Carpenter, G. I. Taylor, and C. F. Elam, X-ray diffraction methods were used to establish the crystal orientation, and in this way the mechanisms of deformation of single crystals of metals were determined. Similar methods for the study of slip deformation were devised by Schmidt and Boas, and by many others; they are now almost a routine operation, and it is difficult for later workers to realize the immense advance which was made in the years 1920–30, and the excitement created by those who first watched single crystals undergoing plastic deformation.

A further characteristic of some aluminium alloys is that, after being heated at a suitable temperature and rapidly cooled, they may be relatively soft, but that they may gradually harden or 'age' on standing at room temperatures—in other alloys 'ageing' may require heating to a low temperature. It was soon realized that such changes were concerned with the precipitation of an intermetallic compound from solid solution, but the interesting fact emerged that the greatest hardness might be found before visible precipitation could be detected. In 1939–40 the very beautiful X-ray diffraction work of A. Guinier and of G. D. Preston led to the discovery that, in alloys where the ultimate precipitate was the θ or $CuAl_2$ phase, the mechanism of precipitation was such that copper atoms first assembled together on (100) planes of the aluminium lattice, with the formation of what are now called Guinier-Preston or G. P. Zones. The next stage is the formation of a metastable θ' phase which, in turn, changes into the final precipitate of $CuAl_2$. It is now recognized that in many cases of age-hardening, the process involves the formation of unstable intermediate structures, and that it is these, rather than the stable precipitate, which produce the greatest hardening effect. Although highly skilled microscopical work might often reveal or suggest the formation of intermediate phases,* it was only by X-ray diffraction that their nature could be established. A wide variety of methods has been used varying from the straightforward identification of precipitating phases by means of their diffraction patterns, to the low-angle scattering techniques, first advanced by Guinier in 1939, in which the size of precipitated particles is estimated from scattering effects analogous to those by which a halo round the moon is produced from the scattering of light by small crystals of ice, or drops of water.

12.6

The interest in the deformation of single crystals of metals in tension or compression led naturally to an examination of the behaviour of metal crystals in the industrial operations of rolling, drawing, swaging etc. Many metals and alloys were examined, and it was soon established that most processes of the working of metals produced material with a pronounced preferred orientation of the crystal, with the result that the mechanical properties might vary greatly in different directions relative to that of the deformation. These preferred orientations are shown by means of 'pole figures' of which Fig. 12–6(1) is an example.

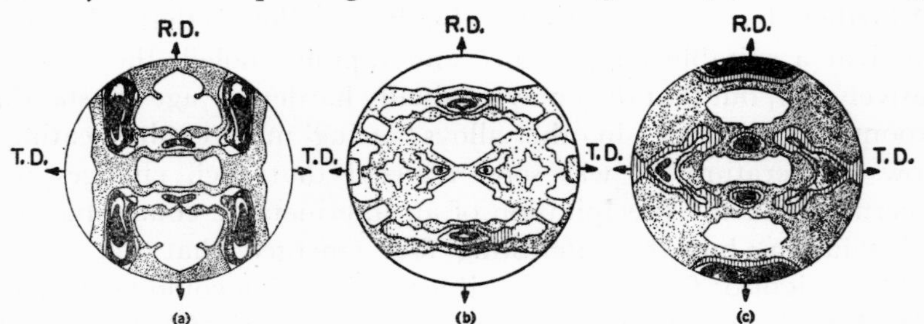

Fig. 12–6(1). Pole figures for aluminium sheet cold-rolled to 96.4 per cent reduction in thickness. (*a*) (100). (*b*) (110). (*c*) (111). R.D. = rolling direction; T.D. = transverse-direction (Grewen and Wassermann, *Acta Met.*). From A. Taylor, *X-ray Metallography*, New York, John Wiley & Sons.

From the problem of preferred orientation, it was a natural step to consider the structure of deformed metal itself. In powder X-ray diffraction films of cold worked metals, the diffraction lines are diffuse, and in single-crystal photographs some of the spots may be elongated

* M. L. V. Gayler was able to reveal the existence of G. P. zones by very careful etching of the alloys and high magnification work by optical microscopy.

(asterism, Fig.12–6(2)). There was acute controversy as to whether these effects were due to a reduction of the particle size of the worked metal, (crystal fragmentation), or to a distortion or bending of the crystals. From the fact that there was a limit to the extent to which diffraction lines were broadened, it was argued correctly that there was a limit to the extent to which fragmentation occurred, but the attempts to estimate particle size from line broadening were incorrect and it was soon realized that most of the broadening resulted from distortion and bending of the crystal. This position was reached by the beginning of the First World War, and further progress was delayed both by the War and by the fact that nearly all work was carried out with films. The later development of counter X-ray sets greatly increased the experimental possibilities, since line profiles could be measured to a much higher degree of accuracy, and permitted more complex problems to be attacked. The structures of deformed metal have been the subject of much work, and interpreted from the standpoint of the dislocation theory referred to below.

12.7

In the work just described, the deformation of the crystal structure is produced by intentional deformation of the metal. Another problem of even greater importance was that of the degree of perfection of ordinary unworked metallic crystals. The investigations referred to above revealed the mode of deformation of single crystals, but calculation showed that the strength of these was very much less than would be expected on theoretical grounds. Measurement of the intensities of diffracted X-rays, and of the range of angles over which a reflection was observed, led to the conclusion that, whilst a diamond crystal was reasonably perfect, metallic crystals consisted of small units which were slightly out of alignment with one another. Various theories of mosaic crystals were advanced, some of which were incorrect, but the evidence for lack of perfection was conclusive, and in 1934 the concept of dislocations was advanced independently by G. I. Taylor, Orowan, and Polanyi. In the years 1945–60 spectacular advances were made in the theory of dislocations, and X-ray diffraction methods have played their part in this work, and it is now realized that the mechanical and many other properties of metals depend as much on the imperfections as on the perfect or ideal crystal structures. In this work the development of micro-beam techniques has been of great value, since these permit very small areas to be submitted to the X-ray beam, and differences between small adjacent areas can be examined. It may, for example, be possible to show that two adjacent areas of what appears to be a 'single crystal' have slightly different orientations, and in studies of a quite different nature, the method may permit the identification of individual particles in a microsection. Another method for revealing irregularities in the structure of what may at first seem to be single crystals is due to Berg and Barrett, and the underlying principle is indicated in Fig. 12–7(1). In this figure a beam of X-rays proceeding from the left is shown striking the surface of a polycrystalline specimen on which rests a photographic plate tilted at a small angle. Occasional crystals will then satisfy the conditions for X-ray reflection, and the reflected beams will pass through the photographic plate as shown. If the crystals are perfectly uniform in structure, each will give rise to a perfectly uniform image, but slight variations in orientation will produce corresponding variations in the intensity of the reflected beam.

12.8

The tremendous increase in metallurgical research work which began about 1920 led to increasing recognition that, even where an industrial alloy was not in a state of thermodynamic equilibrium, a knowledge of the relevant equilibrium diagram was of immense value. In the years 1920–40 a considerable proportion of metallurgical research work was concerned with the establishment of these diagrams. X-ray methods were applied to this problem, and it was shown that in many cases an accurate determination of phase boundaries is possible. In this work specimens of poly-phase alloys are annealed to equilibrium, and the lattice spacings of the different phases are measured accurately, and from the results the compositions of the phases in equilibrium can be deduced if the lattice spacing/composition relations are known for the phases concerned. Accurate lattice spacing measurements were thus of great importance, and A. J. Bradley first showed how the errors of the ordinary Debye-Scherrer method could be greatly reduced by the use of extrapolation methods. The original $\cos^2\theta$ plot of Bradley has now been replaced by that of the more general Nelson-Riley function, but it was the pioneer work of Bradley which was responsible for the first real step towards increased accuracy. Focussing cameras and back-reflection cameras have also been em-

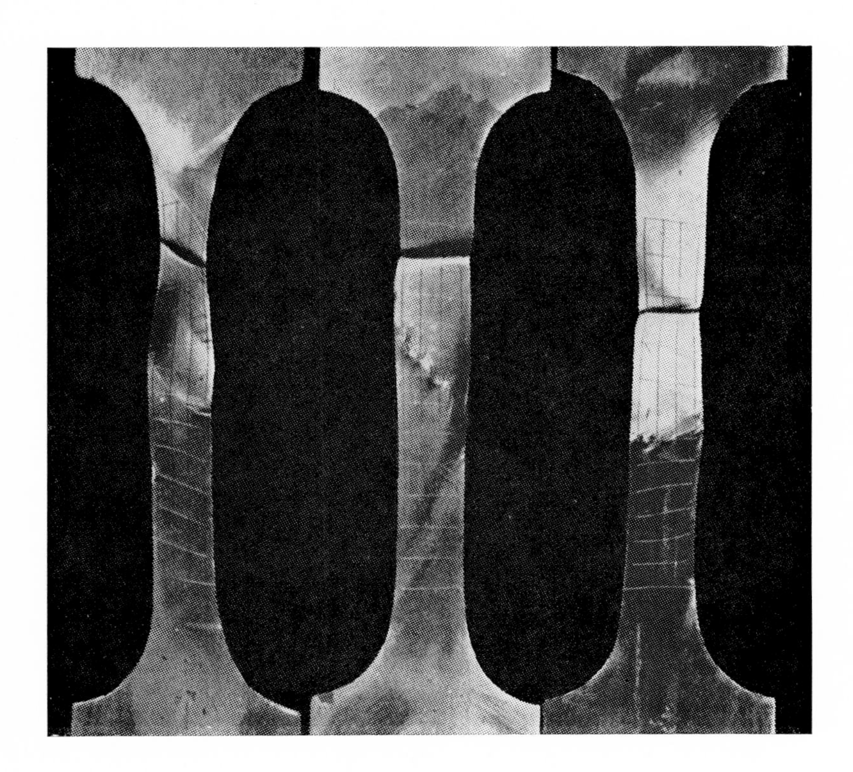

Fig. 12–5(1). From Carpenter and Elam, *Proc. Roy. Soc.* A *100*, Plate 11 (1922).

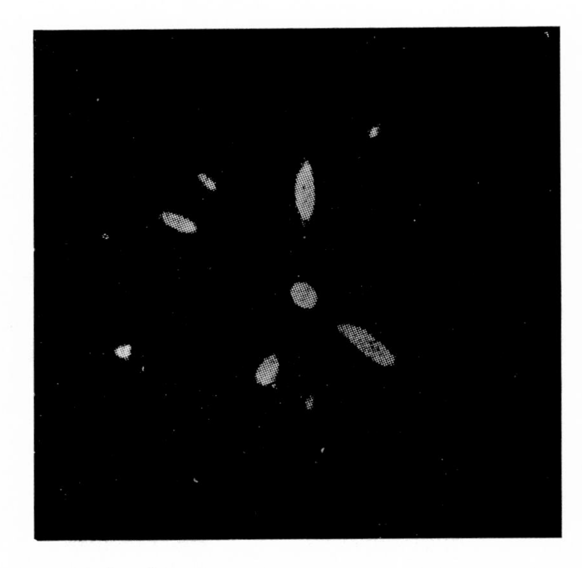

Fig. 12–6(2). Laue diffraction photograph of deformed single crystal of aluminium.

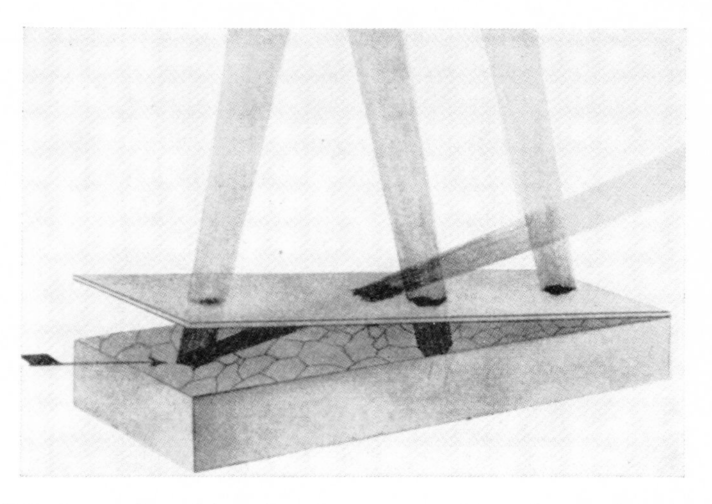

Fig. 12–7(1). Images formed by reflected rays penetrating a photographic plate placed above a polycrystalline sample. Primary beam indicated by arrow. From C, S. Barrett, *The Structure of Metals and Alloys*, New York, McGraw-Hill Book Company, Inc.

ployed, and it is not difficult for an individual worker to obtain lattice spacing results which are consistent to 1 part in 50 000. The absolute accuracy of these measurements is, however, a matter of much greater difficulty. A test was recently carried out by the International Union of Crystallography in which samples of three standard substances were sent to 16 different laboratories for a determination of the lattice spacings. The results submitted showed a scatter of the order 1 part in 10 000, although all the laboratories concerned considered their values to be far more accurate than this. The cause of the differences has not yet been established, but it is clear that the real accuracy is not as great as had been imagined.

With the advent of counter diffractometers, claims have been made to determine lattice spacings to a greater degree of accuracy, but it has not yet been shown how close an agreement can be obtained between results from different laboratories, and tests of this kind would be very welcome.

For most problems of equilibrium diagram work, the lattice spacings can be measured to a degree of accuracy which is as great as is justified by the other factors concerned. There are, however, many experimental difficulties, and it was largely owing to the work of E. A. Owen that techniques were developed which were satisfactory. The development of these methods led in turn to an improvement in the accuracy of results obtained by the older methods of thermal and microscopical analysis, and the whole standard of equilibrium diagram work was raised. Much argument and violent controversy took place between the supporters of the old and the new techniques, and extreme opinions were expressed which varied from dogmatic assertion that the X-ray methods could deal only with diagrams for which the answers were already almost known, to the equally extreme view that equilibrium diagrams should be determined almost entirely by X-ray methods, with microscopy and thermal analysis for purposes of confirmation only. The conclusion eventually reached was that the old and new methods were best used in conjunction, and many papers now appear in which each method is used for the part of the diagram for which it is best suited.

12.9

In the equilibrium diagram determinations referred to above, most of the work was carried out on specimens quenched from the temperature at which equilibrium was being investigated. This meant, naturally, that the methods became unreliable when transformation in the solid phases took place so rapidly that changes occurred during the quenching process. Much interest therefore attached to the possibility of accurately controlled high-temperature X-ray diffraction work. Crude apparatus for this purpose had long been available, the most notable achievement being that of Westgren who in 1921 showed that γ-iron had a face-centred cubic structure. In this work an electric current was used to heat an iron wire in an ordinary Debye-Scherrer camera, and the temperature estimated roughly by optical pyrometry. In 1930–40 a number of more accurate cameras were developed, notably by Bradley and Jay, Hume-Rothery and Reynolds, and by A. J. C. Wilson, and some of the ideas of these early instruments have been incorporated into cameras which are now available commercially.

The difficulties of accurate high-temperature X-ray diffraction work are often very great. Powders of many alloys can be heated in thin-walled sealed tubes of silica, and in this way experiments at temperatures up to 1000°C may be carried out under conditions in which the composition of the specimen is reasonably controlled. Unfortunately many alloys react with glass or silica, and no alternative containing tube is yet available, whilst temperatures of 1000°–1100°C. are the highest that can be used with silica capillaries. For pure metals the method of Westgren referred to above can be used, and R. Ross in the author's laboratory has taken successful diffraction films from tungsten wire at temperatures up to 3200°C. Pure metals in the form of wire specimens can also be heated in platinum wound furnaces, and Basinski and Christian described a pressurized camera for the purpose. The apparatus was used successfully to determine the structures of the high-temperature allotropic forms of manganese (δ-Mn b.c.cube; γ-Mn f.c.cube), and to determine the lattice expansion of iron at temperatures above the A4 point (1390°C). For alloys, the control of compositions at very high temperatures becomes increasingly difficult because preferential evaporation of one constituent is often serious, and the composition cannot be controlled.

12.10

For many theoretical purposes the properties and structures of metals at low temperatures are of greater interest than those at high. Many devices have been used in order to obtain X-ray diffraction photographs from specimens at low temperatures. Of these, one of the most simple is to allow a cooling liquid (e.g. liquid air) to stream over the specimen, and the author remembers a visit to the Royal Institution where he saw the figure of Dr. (now Professor Dame Kathleen) Lonsdale appearing through a cloud of mist, like a glorified spectre of the Brocken, while her assistant pumped liquid air over a crystal. More refined apparatus of this kind has been used with liquid helium, whilst an alternative method is for a metallic specimen to be fixed in a holder of copper the other end of which is immersed in liquid helium. Several cameras have been described in which the specimen is cooled by a stream of gas. Cameras of the focussing type have also been used with the crystalline powder stuck to the arc of the camera the other side of which was cooled by liquid air.

One of the more spectacular results of low-temperature X-ray diffraction work was the discovery by C. S. Barrett in 1947 that, at low temperatures, the body-centred cubic structures of lithium was transformed into close packed modification; sodium was shown later to undergo analogous changes.—Work by Lonsdale and others on diffuse streaks has led to an estimation of the temperature variation of the root-mean square value of the amplitudes of the thermal vibrations.

12.11

In the simple X-ray work described in 12.3 and 12.4, information on solid solutions was gained from the positions of the diffraction spots or lines which, when the specimen is annealed to equilibrium, are almost as sharp as those from a pure metal. It was appreciated from the outset that the sharpness of the lines did not mean that the lattice of the solvent metal had been uniformly expanded or contracted. In a dilute solid solution each solute atom is surrounded by a distorted region in which the solvent atoms are pushed outwards if there is lattice expansion, or collapse inwards if there is a contraction. The positions of the diffracted spots or lines correspond with a mean value of the lattice spacing, and it is these mean values which are used in the phase-boundary determinations referred to in 12.9 above. A little more information can be gained from the intensities of the diffraction lines, but much more valuable knowledge has been obtained by studies of the variation of the general background scattering. In this connection the work of Averbach and Warren has been outstanding, and has provided direct evidence for the existence of irregular interatomic distances in solid solutions. The problem has, however, proved too difficult for accurate solution, and at present the results are qualitative only.

12.12

The existence of the regions of distortion referred to above naturally lead us to expect that, in most cases, the lowest free energy will result if solute atoms tend to avoid each other, since the existence of two or more solute atoms as next-door neighbours will accentuate the lattice strain energy. Such 'short-range order' effects had long been imagined to exist, but it was not until the analysis of general background scattering had been carried out that direct evidence was obtained (Fig. 12–12(1)). It has now been established that, in nearly all solid solutions, there is a short-range order, in the sense that solute atoms tend to avoid each other to an extent greater than would correspond with a random distribution. In a few cases (e.g. Al-Zn) the reverse effect is found and clustering of like atoms occurs—these are sometimes systems in which a miscibility gap occurs in the solid solution at low temperatures.

It is clear that with a sufficient difference between the two kinds of atom, and with a suitable ratio of solvent/solute atoms, we might expect the ordering process to be carried a stage further, and for a long-range ordered structure to exist in which the strain energy was reduced by the two kinds of atom taking up regular positions in the lattice. Such structures are known as 'superlattices', and they were first established in 1927 by the outstanding work of Johansson and Linde in Sweden on copper-gold alloys. Both copper and gold crystallize in the

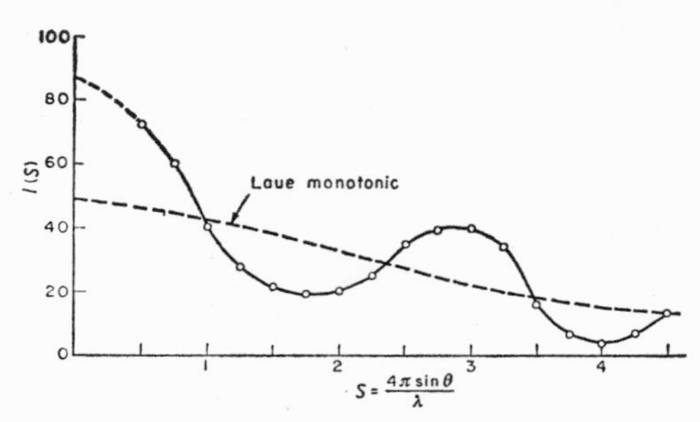

Fig. 12–12(1). Diffuse scattering in electron units for aluminium–30 atomic % zinc alloy at 400°C. Temperature-diffuse and Compton modified scattering subtracted. CoKα radiation at 22 kV monochromated with bent fluorite crystal.

face-centred cubic structure, and at high temperatures the two metals form continuous solid solutions. It had long been known that on cooling alloys of compositions in the region of 25 and 50 atomic % of

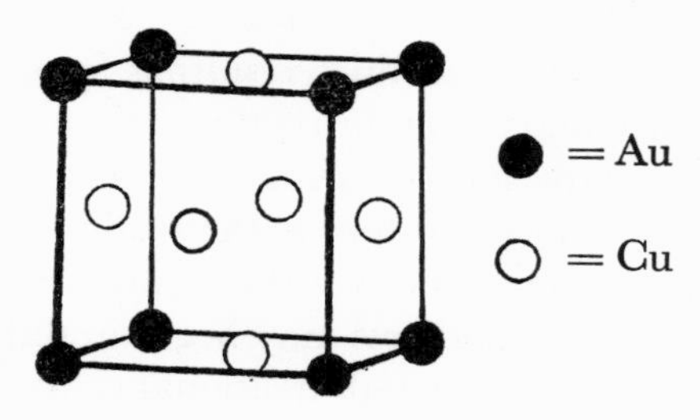

Fig. 12–12(2). The Cu_3Au superlattice structure.

gold, some kind of transformations occur at low temperatures, and there was much discussion as to whether definite intermetallic compounds were crystallizing out from the solid solution. The X-ray diffraction work of Johansson and Linde showed that for Cu_3Au, the process occurring was a rearrangement of the copper and gold atoms on the face-centred cubic lattice to form the structure of Fig. 12–12(2). If the distinction between the two kinds of atom is ignored, this is a simple face-centred cube, but the regular arrangement of the atoms means that some of the forbidden reflections for the face-centred cube occur in the diffraction patterns of the superlattice. This is the simplest kind of superlattice, and the X-ray diffraction pattern contains all the lines characteristic of the random face-centred cube, together with extra lines resulting from the ordered arrangement of the two kinds of atom. Fig. 12–12(3) shows one of the superlattice structures of the equi-atomic

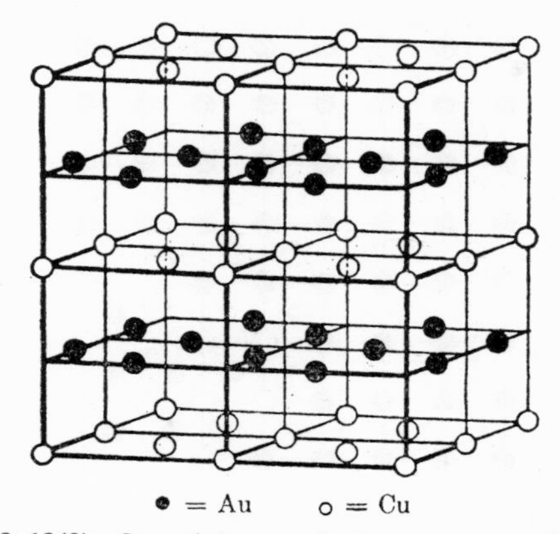

Fig. 12–12(3). One of the superlattice structures of CuAu.

CuAu alloy, and here the separation of the gold and copper atoms into layers results in a slight distortion of the structure from cubic to tetragonal, and the diffraction pattern shows a splitting of the lines of the underlying face-centred cubic structure (except those for which $h = k = l$) together with extra lines resulting from the regular arrangement of the two kinds of atom relative to one another.

If the formation of superlattices is regarded as a rearrangement of the two kinds of atom in order to relieve the strain energy present in a random solid solution, there is clearly the possibility that in different parts of a crystal the ordering process may begin in different ways. Fig. 12–12(4) shows this effect for a two-dimensional centred square lattice, and in such a case the crystal is said to contain anti-phase domains. The face-centred cubic structure may be regarded as resulting from the interpenetration of four simple cubic lattices, and an examination of Fig. 12–12(2) will show that the Cu_3Au patterns in a given crystal can be formed in four different ways, according to which simple cubic lattice is occupied by the gold atoms. In such a case an anti-phase structure can be formed which is very stable for the same reason that a foam structure is stabilized. This was confirmed by the very beautiful

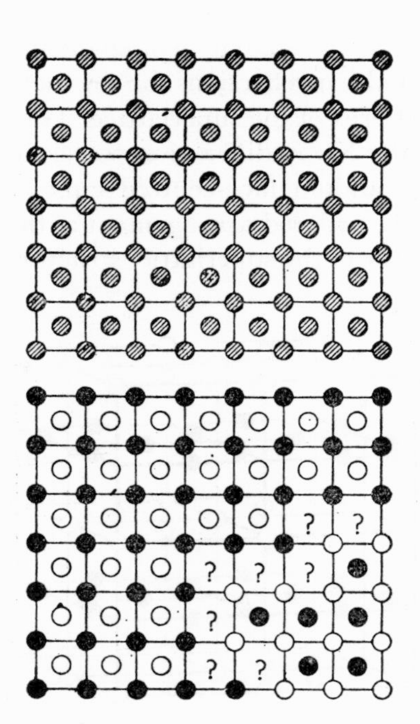

Fig. 12–12(4). Formation of anti-phase domains in a two-dimensional centred square lattice.

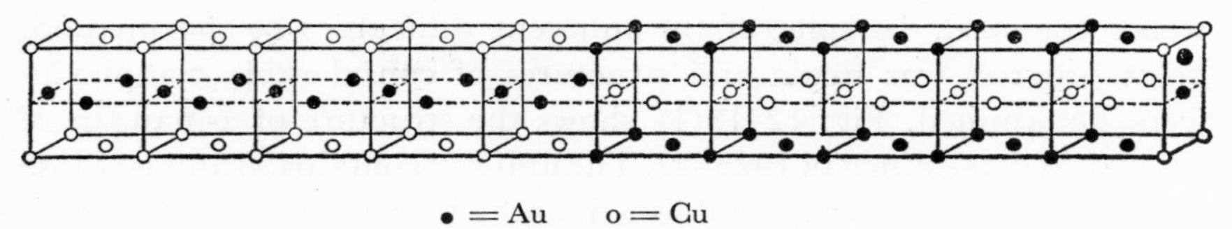

Fig. 12–12(5). One unit cell of the orthorhombic superlattice of CuAu. From C. S. Barrett, *The Structure of Metals and Alloys*, New York, McGraw-Hill Company, Inc.

experimental work of Sykes in which growth of anti-phase domains was accompanied by an increase in the sharpness of the superlattice diffraction lines which were diffuse when the domains were small.

The anti-phase domains referred to above are due to the ordering of the atoms proceeding independently from a number of centres. A more remarkable phenomenon is the formation of the second CuAu superlattice whose structure is shown in Fig. 12–12(5). Here the copper and gold atoms again form alternate layers, but the relative positions of these change at regular intervals so that the unit cell consists of 10 of the simple face-centred cubic cells. The reason for this regular repetition is not known.

12.13

In the early days of the Science, most of the structrual work on metal crystals was carried out by powder methods, whose limitations are now

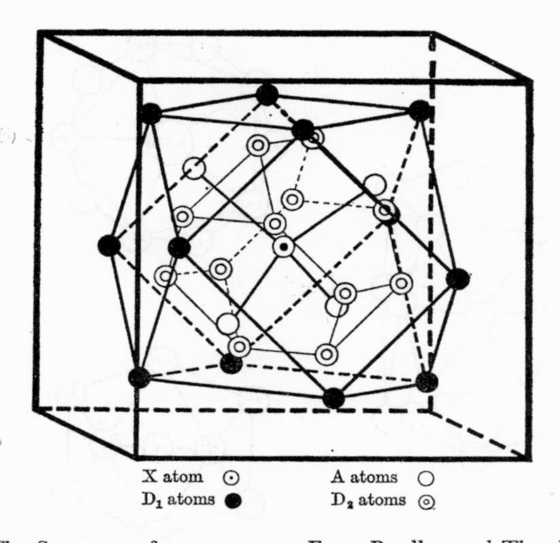

Fig. 12–13(1). The Structure of α-manganese. From Bradley and Thewlis, *Proc. Roy. Soc.* A *115* (1927).

well recognized. In spite of the inherent difficulties of the method, great progress was made and structures of considerable complexity were determined. Fig. 12–13(1) shows the structure of α-manganese which was determined in 1925–27. The unit cell contains 58 atoms whose positions were accurately established, and when the investigator, A. J. (later Dr.) Bradley, brought the work, as a young student, to Sir Lawrence Bragg, the latter at first could not believe that the problem had been solved. At the same period Bradley, by powder methods alone, solved the structure of γ-brass (Cu_5Zn_8), and showed the differences between the detailed atomic arrangements in Cu_5Zn_8 and Cu_5Cd_8. In view of its early date, this work was one of the great achievements of its time.

Taken as a whole, those concerned with the structure of alloys were strangely reluctant to adopt single-crystal methods of analysis, and there is no doubt that more rapid progress would have been made if a greater proportion of time had been devoted to the preparation of single crystals. This defect has now been recognized, and single-crystal work is now as common in metallurgical studies as in other sciences where complicated structures are concerned. Structures of very great complexity have been determined in many countries, and no attempt will be made here to summarize this work. One of the most successful investigations of this kind is the determination of the structures of the intermetallic compounds of aluminium with the

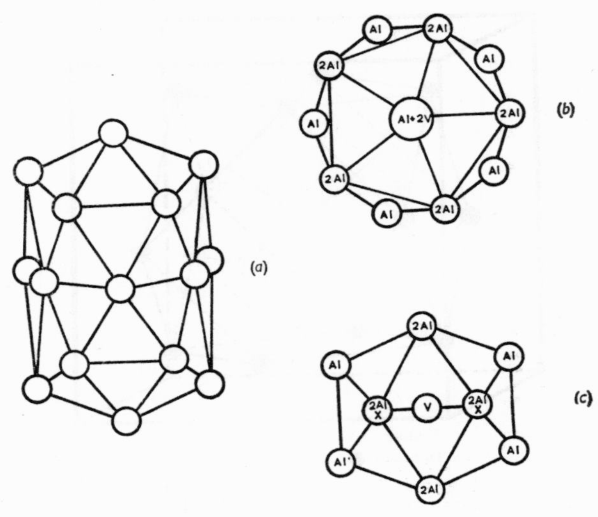

Fig. 12–13(2). Coordination polyhedra in the structure of α′ (V-Al).

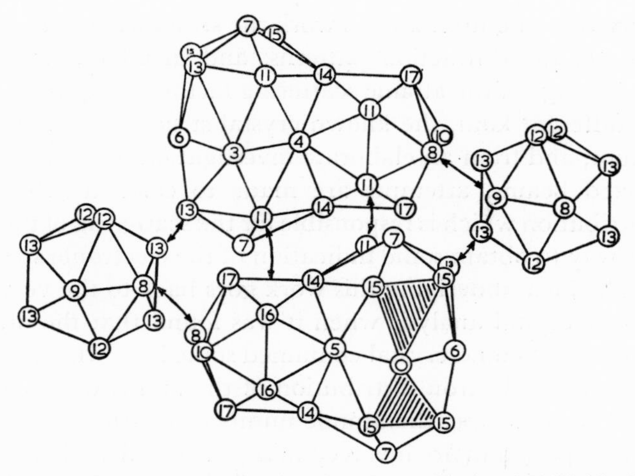

Fig. 12–13(3). Schematic diagram showing the structural units of α′ (V-Al). The numbers refer to the structurally non-equivalent types of Al-atoms.

transition metals, which has proceeded for many years at Cambridge under the general direction of Dr. W. H. Taylor. Many of these structures are extremely complicated, and mention may be made of the α′ vanadium-aluminium phase of ideal composition V_7Al_{45} investigated by P. J. Brown. The unit cell is monoclinic and contains 90 aluminium and 14 vanadium atoms. The structure may be described in terms of the coordination polyhedra around the vanadium atoms which may be divided into three classes denoted V_0, V_1, and V_2. The polyhedra round the V_0 atoms are nearly regular icosahedra, and those round the V_1 atoms are slightly distorted icosahedra. Each pair of V_2 atoms which are in contact form the nucleus of the two interpenetrating icosahedra shown in Fig. 12–13(2). Each of these complexes shares one pair of adjacent faces with a pair of faces of one of the V_1 icosahedra, and thus form the large groups of atoms indicated schematically in Fig. 12–13(3) in which the numbers refer to the different structural types of aluminium atoms. The successful determination of this structure represents metallurgical X-ray crystallography at its highest standard and, with the increasing use of electronic computers, even more complex problems will be solved.

12.14

Nearly all of the work referred to in the preceding section has been concerned with the determination of the positions of atoms in the different crystal structures. In this work the structure is determined by the analysis of the diffraction patterns, and in all but the simplest cases, a knowledge of the atomic scattering factors is required. In work of a quite different kind, the known crystal structure is assumed as a starting point, and from an elaborate investigation of the intensities of the diffracted beams, attempts are made to calculate the electron density distribution which is responsible for the scattering of the X-rays, and in this way to obtain some indication of the electronic structure of the solid. The idea underlying this work goes back to the very earliest days of X-ray crystal analysis when it was found that the diffraction patterns from a diamond crystal contained some lines which would not be expected if the electron distribution of the atoms were spherically symmetrical. In this case, the atomic number of carbon is 6, and 4 of the 6 electrons per atom are involved in simple covalent bonding, and the occurrence of the 'forbidden' reflections is readily understood. Attempts to investigate the electronic structure of metals in this way were made by Russian workers under Ageev, but it is now recognized that too many approximations were involved for the measured electron distribution of the bonding electron to have been significant. All such work is of great difficulty because, in all but the lightest elements, most of the scattering is due to the inner electrons of the atoms, and the slight effect of the bonding electrons can be determined only after numerous corrections have been made, the effects of which are of the same order as the experimental errors. The difficulties and dangers of this work are well shown by the fact that one pair of American authors obtained results for iron which were in apparent agreement with the Pauling view that about two of the $3d$ electrons per atom were atomic, and about 6 electrons per atom were concerned in the metallic bonding. This work indicated a marked and improbable difference between iron and cobalt, and was not confirmed by later and more detailed determination by other American authors whose results were described at the Crystallographic Congress held at Cambridge in 1960. There is, however, little doubt that this kind of work is greatly needed and that, if the experimental difficulties could be overcome, the results would add greatly to our knowledge of the electronic structure of metals.

12.15

It may be claimed fairly that in 50 years the results of X-ray diffraction work have revolutionized metallurgical science. At the present time every metallurgical research laboratory of note has one or more X-ray diffraction units. On the industrial side these may be used for a variety of purposes ranging from the analysis of the structures of minerals and refractories at the one end to the determination of the preferred orientation of the finished product, and the details of the age-hardening or tempering processes. On the scientific side, the first simple steps have been taken, but the possibilities are endless. It must be emphasized that there is at present only a most elementary understanding of the factors which control the formation of the different types of crystal structure of alloys. There is still great need for the systematic determination of the at present unknown structures of innumerable intermetallic phases, and for the careful study of the results obtained. The problems of the structures of liquids, and of the detailed atomic arrangements in what are usually called random solid solutions are essentially unsolved, and little or nothing is yet known of the electron distribution of the bonding electrons in metals.

The present author is only too conscious of the limitations of this review, but he has felt it better not to attempt to describe subjects with which he is not competent to deal. The language difficulty has prevented more detailed reference to Russian work, and it is to be hoped that some other author may be encouraged to review the main progress in metallurgical X-ray diffraction which has resulted from work in Russian laboratories.

CHAPTER 13

Problems of Biochemical Structures

by Ralph W. G. Wyckoff

The distinction between organic and biological chemistry becomes progressively less sharp as chemists learn to establish the molecular configurations and to synthesize more and more of the constituents of living matter. During the past generation such complicated natural products as sterols, alkaloids and antibiotics have been synthesized, as well as many derivatives which nature has not had occasion to make. From the standpoint of X-ray diffraction analysis the line between the organic and the biological has also become increasingly blurred as our more powerful tools have been able to establish molecular structures too difficult for the unaided chemistry of today. Nevertheless it remains worthwhile to consider as a group apart those native products we have not yet learned to prepare in the laboratory; and this is the field to be treated in the present chapter.

The extreme complexity of living matter made it inevitable that knowledge of what it is should lag far behind our understanding of the inanimate world. Biochemistry, and now our X-ray techniques are giving a rapidly increasing insight into the molecular composition of plants and animals—but this is not in itself enough. In all but the simplest manifestations of life, elaborateness of organization is as striking as complexity of chemical composition, and knowledge of this organization is an essential part of any understanding we may acquire of the mechanism of life. It is a basic problem of the biophysics now emerging to ascertain the details of this organization, which often involves an order in molecular arrangement that X-ray diffraction can effectively interpret. We can no longer summarize in a few pages what is now known of this natural order; instead we shall seek to review the growth of our appreciation of the value of X-ray methods and the appearance of other methods that have come to supplement the information X-rays can give.

Biology in its development over the last century has shown how cells are organized into tissues and how in the higher animals these tissues are associated into the organs whose corporate life becomes that of living individuals. Until the advent of X-ray diffraction it was generally considered that, except for teeth and bones, these ingredients of living matter were amorphous, gel-like colloids and that consequently structures showing order in arrangement were scarcely to be expected at and below the cellular level. Discovery of how rare was the amorphous condition even among colloids was one of the important early consequences of X-ray diffraction.

As a result, many of the fibrous colloids of biology were examined during the 1920's and found to give diffraction effects pointing to a partial crystallinity or a definite but sub-crystalline order in particle arrangement. This was true of such widely diverse substances as animal hair and birds' feathers, tendons and other connective tissues of animals, and cellulosic fibers from all sorts of plants. In general the diffraction phenomena were too fragmentary to provide a satisfactory picture of the kinds of order responsible for them but they were the basis for a continuing study of these natural products. It was during this period that Herzog fostered the application of X-ray methods in his institute for cellulose research, that Sponsler suggested that cellulose was a parallel packing of polysaccharide crystallites and that somewhat later Meyer and Mark gave a first crystalline interpretation to the X-ray pattern of cellulose. At this time, too, Astbury began the studies centered around hair and wool which, proceeding uninterruptedly over the next thirty years, have contributed so much of the foundation upon which an intimate knowledge of protein structure is now being built. Katz' observation that rubber yields an oriented crystalline diffraction pattern only after stretching demonstrated how X-rays could reveal reversible molecular rearrangements; this experiment was a prototype for many of the most rewarding subsequent studies of fibrous structures.

Such exploratory examinations of natural fibers were actively continued and broadened throughout the 1930's. Especially significant for the future were Astbury's demonstration of the alpha, beta and super-contracted states of the keratin-like fibrous proteins and the way they pass from one to another. He pictured this in terms of the coiling of peptide chains which forecast the spiral models we have now come to consider as underlying all protein structures.

The work of this period also developed a clear distinction between fibrous and globular proteins and the different types of molecules responsible for each. As they occur in nature the fibrous proteins are insoluble whereas the proteins we designate as globular function in solution. The latter can often be crystallized in the laboratory and these crystals have the same regularities in external form and the same optical characteristics as the crystals of simple substances. The ultracentrifugal investigations initiated by Svedberg in the middle 1920's determined their molecular weights and led to the discovery that their enormous molecules were more or less globular in shape and had the same uniformity in size that prevails among small molecules. When the still larger virus proteins came to be recognized the ultracentrifuge demonstrated their molecular uniformity and supplied a new way to purify them as well as other substances too unstable to be handled without damage by the existing chemical procedures.

Several attempts to obtain diffraction patterns from crystals of globular proteins were made early in the 1930's. At first these met with mediocre success, largely because of the enormous amounts of water in the crystals studied and the ease with which this water and the crystalline structure are lost. With time, however, conditions were established for the preparation of photographs from substantially undamaged single crystals. Their patterns contained thousands of reflections and were evidently caused by the same kind of three-dimensional regularity that prevails in chemically simple crystals. They indicated the experimental basis for an ultimate determination of atomic arrangement while at the same time emphasizing the magnitude of the task. Perutz began his lifelong study of crystalline hemoglobin at this time. The success now being achieved is overwhelmingly due to his skill and perseverance in what must often have seemed a hopeless task, as well as to W. L. Bragg's scientific and material support over the years. This success is a triumph of persistence and faith which is doubly noteworthy now when research projects which do not offer an assured and quick result are rarely able to survive even when they can be begun.

Though the fibrous proteins dealt with during the pre-war years did not yield the many X-ray data obtainable from globular proteins, they did provide much knowledge upon which striking post-war advances have been based. Chemical work suggested that most fibrous solids could not be put into solution and then reconstituted the way a globular protein can be repeatedly dissolved and recrystallized; and ultracentrifugation of solutions obtained from wool, cellulose and other fibrous materials revealed non-uniform particles that could best be interpreted as molecular fragments. X-ray experiments made at this time on reconstituted collagen were chiefly important in showing that such was not necessarily the case. Precipitates from dissolved tendon gave the same X-ray patterns and thus were essentially the same as the native product. The preparation of molecularly ordered structures in the laboratory as suggested by these and analogous experiments with cellulose has since led to a wide extension of our knowledge of fibrous substances.

After the war the X-ray study of the biologically important proteins experienced the same stimulus as did other phases of crystallography. There have been a number of reasons for this, not the least of which has been a more general appreciation of the importance of crystallography itself. In its wake has come increased financial support for the many new workers who have turned to this field. Other factors that have contributed to the current rapid growth of our knowledge of biological structures have been the introduction of improved techniques of X-ray experimentation, of computers able to deal with the great masses of data these improved methods have made possible and of several new and complementary sources of information such as chromatographic amino acid analysis and electron microscopy. These methods have contributed very differently to the understanding we have of the structures of various biological solids and it is important to see what these contributions have been if we are to appreciate how our understanding has expanded. The order we find in such solids is of different kinds and our methods are not equally effective in dealing with them. There is either (1) a true crystallinity which expresses itself as a three-dimensional regularity in molecular arrangement, (2) a para-crystallinity consisting of one- or two-dimensionally repeated molecular packings such as are shown by the fibrous proteins or (3) repetitive atomic distributions within individual molecules. In general X-ray diffraction is most effective where the order is greatest while the electron microscope, for instance, has been most helpful where the ordering is less than crystalline. Thus it is the electron microscope rather than X-ray diffraction which has revealed the laminations of the lipoidal sheats of myelinated nerves and the chloroplasts of plants. The contributions of X-ray diffraction and of electron microscopy have been more equal for the nucleic acids. Neither they nor the nucleoproteins are known to furnish naturally occurring ordered solids but in the laboratory nucleic acids sometimes give X-ray fibre diagrams analogous to those obtained from the fibrous proteins. Both the ultracentrifuge and the electron microscope show these acids and proteins to have thread-like molecules of indefinite lengths and we may expect their solids to be parallel aggregates of these filaments. Chemical analysis shows that all nucleic acids consist of linear associations of a few nucleotides somewhat as the polypeptide chains of proteins are strings of amino acid residues. Following the success of Pauling and Corey's helical structure for proteins, Watson and Crick have proposed an analogous linking of nucleotides as the basis for nucleic acids; and such a chain is compatible with the limited X-ray data existing preparations can supply. The visibility of individual filamentous molecules of nucleic acids under the electron microscope makes it certain that combined X-ray and microscopic studies will be increasingly fruitful as we learn to grow specimens with better molecular ordering.

Cellulose is in some respects a most promising and in other respects one of the most disappointing of natural fibers. Its promise lies in the richness of the X-ray data it often yields. It is disappointing in the sense that we are more dependent on natural products for experimental material than is the case with some of the fibrous proteins; the crystallinity of derivatives made in the laboratory is not yet sufficient to provide all the data required for a thoroughgoing determination of structure. Natural cellulose, however, illustrates in an interesting fashion the different kinds of order that have been mentioned. Its elementary particles as seen under the electron microscope are indefinitely long fibrils that commonly have a diameter of ca. 200 Å. Each of these is made up, in part at least, of minute crystallites whose atomic arrangement is responsible for the sharp X-ray reflections we observe. Since the original proposal of structure by Meyer and Mark, these reflections have been made the experimental basis for atomic arrangements assigned to a number of forms and derivatives of cellulose but we need many more data to make these determinations complete. In the secondary walls of plants the elementary fibrils are in parallel arrays. Some information about them was early gained from X-ray fiber diagrams but by portraying each particle separately, the electron microscope has superceded diffraction as the tool for investigating such arrays. It not only has shown the fibrillar arrangement but has told much about how cellular activity creates and orients the fibrils themselves.

Most has been learned about the order that prevails in protein solids. Though we can sometimes see the molecular arrangement in protein crystals with the electron microscope, the developing power of X-ray methods to establish complete atomic arrangements makes them preeminent. With the paracrystalline fibrous proteins the X-ray data are less complete while the electron microscope has more to show about their molecular architecture.

The early X-ray work of Astbury and others resulted in a natural classification of the fibrous proteins. To one group belong the keratinous proteins such as those which constitute the skin and horny structures of mammals, the feathers of birds, the scales of reptiles and the flagella of some bacteria. Mammalian keratin normally gives what Astbury called the alpha-keratin X-ray pattern. These keratins elongate on stretching and then give the different, beta-keratin, pattern which is also produced by avian and reptilian keratin. Beta-keratin will return to its original alpha form when tension is released (under suitable conditions) and all keratin can be made to shrink to a supercontracted state still more shortened than the alpha. Recently there has been wide interest in showing that this behaviour of keratin and related fibrous structures is compatible with the helical polypeptide model of Pauling and Corey. With the electron microscope one can see the elementary fibrils of such keratinous solids as wool, but the imperfect order of their arrangement automatically restricts the information this instrument can give.

The collagenous proteins are very different from keratin both in X-ray fiber diagram and in physical and chemical characteristics. They are not, like keratin, extensible and they yield solutions from which solids showing high degrees of molecular ordering can be precipitated. Sometimes this order is the same as that observed in nature, under other conditions it is equally good but different. X-ray diffraction has supplied information about this order as it exists in tendon and also about repetitions within the separate fibrils but during the post-war years the application of the electron microscope has contributed far more to our knowledge of both the intra- and inter-fibrillar order of these several polymorphous forms of collagen. An interpretation of what has been learned is now becoming feasible with the help of knowledge of the amino acid sequences in the polypeptide chains from which the molecular filaments are built up. The chromatographic analysis has shown that along the polypeptide chains of collagen every third residue is glycine and that hydroxyproline is its second most numerous amino acid constituent; and we can in a general way picture how reactions involving mainly this residue could result in the several ordered molecular associations we observe. Making use of these amino acid distributions a molecular structure involving three intertwined Pauling and Corey polypeptide helices has recently been proposed which is compatible with what is now known about collagen. Such semi-speculative molecular models as this, like the analogous Watson and Crick model for nucleic acid, are of great assistance in thinking about these exceedingly complex ingredients of living matter and in correlating their properties. They are not firmly established by existing data but the cumulative weight of the comparisons to be made between such models and experiment will with time give a clearer indication of their validity even though the prospect of obtaining enough X-ray data for complete structure determinations is at present very poor.

Muscle is a fibrous tissue whose order, even more elaborate than that of collagen, cannot be inferred from its X-ray diffractions but is being revealed in detail by the electron microscope. More than one kind of filamentous molecule can thus be seen in intact muscle, distributed regularly with respect to one another in an almost crystalline array. Muscular contraction patently involves interaction between these ordered elements of an especially complex 'multicomponent' crystalline structure, and electron microscopy has been giving mounting insight into how muscles work. As would be expected from the different sized molecular filaments seen in muscle, several fibrous proteins can be obtained from it; and after separation from one another they furnish ordered, almost crystalline solids whose study is usefully supplementing what can be seen in muscle itself.

The intimate knowledge now being acquired of the crystal structures of hemoglobin and myoglobin and the corresponding advances being made with other globular proteins, notably ribonuclease, is, as already indicated, the culmination of a generation's development of X-ray methods and the first step towards an entirely new understanding of proteins as the substrate of all life. To appreciate the significance of these determinations of structure we need to take note of the stages through which their analysis has proceeded and the new techniques that have made each advance possible. The pre-war studies of hemoglobin demonstrated its essential crystallinity and the availability of the experimental evidence required for a deeper penetration into its structure. The very richness of these data, however, brought new and formidable problems. Thus the precise measurement and identification of the more than 20000 reflections that can be obtained from the crystals of a protein required X-ray equipment that did not exist before the war and demanded a mass of routine measurement never before encountered in the course of crystallographic work. Methods of manual computation were completely unable to deal with the data thus produced and accordingly advance in this field could only follow the development and availability of high speed computers.

The early determination of the dimensions of the unit cell set certain limitations on the shape of the hemoglobin and myoglobin molecules but progress towards a direct deduction of structure hinged on being able to find the correct signs for measured structure factors. This involved using the heavy metal isomorphous replacement technique that had already led to complete structures for several complicated organic crystals. It required in this case the introduction into the protein molecule of mercury or other heavy atoms without damage to the molecule as a whole or a change in the molecular distribution within the crystal. Much difficult chemical work was necessary to accomplish this but the success that has been achieved in making these metal substitutions augures well for future advances involving other crystalline proteins. In the case of hemoglobin and myoglobin, analysis of the data from the native protein and its derivatives led to knowledge of an increasing number of signs and thus to Fourier summations that pointed to important atomic concentrations within the volumes occupied by individual molecules. The more detailed step of passing from such general, preliminary ideas of molecular shape to a picture of the way the various polypeptide chains are distributed and interconnected has been made by taking advantage of two other sources of information. One of these, to which we have already referred, was complete chemical analysis to establish the amino acid residues present in these substances and their sequence in the polypeptide chains; the other was use of the alpha helix of Pauling and Corey to assign configurations to these chains. In this way probable structures could be limited and direct attacks continued. The interplay of these procedures has led to structures first for the chemically simpler myoglobin by Kendrew and his co-workers and now for hemoglobin. Though they do not yet define the exact positions of all the thousands of atoms in a unit cell they already give very probable pictures of the spatial distribution of the various amino acid residues in the molecules and undoubtedly soon will define these atomic distributions more closely. Probably before long we will have a correspondingly detailed knowledge of ribonuclease and perhaps of other globular protein crystals.

The scientific importance of these results for an understanding of proteins is obvious but they are equally important as an incontrovertible justification of the large amounts of work and money they have entailed. They make it clear that other proteins can be similarly analyzed and at a fraction of the cost in time and money that have gone into these first structures. Such protein analyses will always require programmes having assured support for a group of workers over a number of years but we may now be sure that, given this support, the job can be done. It seems inevitable that the analyses of myoglobin and hemoglobin are but the first steps in an intensive study of protein structure which can scarcely fail to advance in unpredictable and truly revolutionary ways our knowledge of the mechanism of life.

CHAPTER 14

X-ray Diffraction and its Impact on Physics

by Dame Kathleen Lonsdale

14.1. Introduction

The intention in this chapter is to present, in the light of a backward look by one physicist who was privileged to work with Sir William Bragg from 1922 until his death in 1942, the overall significance of a half-century of development in the field of X-ray diffraction, *as it affected the study of physics.*

It will be apparent to those knowledgable in this field that some branches or crystal physics have received little or no mention here and that the treatment given to others is very unequal. This is deliberate. To write an adequate account of crystal physics would require not one chapter, but several volumes. Moreover, the subject considered is really not crystal physics at all. The early workers in the field of X-ray diffraction were not at first interested in the physics of crystals as such but in the fundamental problems of physics itself: properties of matter, heat, light, sound, electricity and magnetism. Crystals and the diffraction of X-rays by crystals were simply tools with which to investigate such phenomena as cohesion, rigidity, elasticity, changes of state, expansion, the nature and properties of radiation and its interaction with matter. Examples are given to illustrate the way in which this enormous project was tackled and the way in which it opened up what to most physicists was a new field: the study of the solid state as such. Some crystal properties had, of course, been known for many years, but they were not understood until the structure of the crystals became known. Other crystal properties were revealed during the course of the studies made. Examples of both kinds of investigation are given here. Later on, the subject became divided. Structure analysis tended to become tied to chemistry and mineralogy; solid state physics established itself as fundamental to many very important industrial developments.

It would have been tidy to have been able to divide the chapter under precise sub-headings. Some attempt at sub-headings has been made, but the subject, in analogy to most real crystals, is an untidy mosaic. The studies were interlocked from the onset and any attempt to sort them out chronologically here would mean losing the very flavour of research, part of the attraction of which is that one never knows what may turn up next! The task of presenting a really balanced account of the whole subject must be left to the historian, who would doubtless place a very different emphasis on the personalities involved. The fact that the writer was trained in a British school of research at the time when many of the most fascinating discoveries were being made has certainly biased this account. To remedy this in part, a selection of books and major articles has been listed and references given in the text to these more readily available books rather than to original papers.

14.2. Early Ideas

The early research workers in the field of X-ray diffraction from von Laue onwards were all physicists. Some of them were theoretical physicists. Their purely physical training was, indeed, a source of some embarrassment to some of the experimentalists among them, for while all chemists know some physics, few physicists include much chemistry in their training, or mineralogy either, for the matter of that. Yet mineralogical and chemical knowledge were necessary when it came to the proper selection of substances for X-ray structure analysis. The idea that the chemist and mineralogist might learn something from these studies came later.

The initial preoccupation of physicists in the early experiments was certainly not without cause. X-ray diffraction was an exercise in advanced optics, it gave information about the nature of X-rays and about the structure, sizes and behaviour of the atoms in solids and it gave promise of providing the solution to the problem so forcefully stated by Isaac Newton when he wrote (*Opticks* 3rd ed. pg. 364 and 369):

'The Parts of all homogeneal hard Bodies which fully touch one another, stick together very strongly. And for explaining how this may be, some have invented hooked Atoms, which is begging the Question; and others tell us that Bodies are glued together by rest, that is, by an occult Quality, or rather by nothing; and others, that they stick together by conspiring Motions, that is, by relative rest amongst themselves. I had rather infer from their Cohesion, that their Particles attract one another by some force, which in immediate Contact is exceeding strong, at small distances performs the chymical Operations above mention'd, and reaches not far from the Particles with any sensible Effect.—There are therefore Agents in Nature able to make the Particles of Bodies stick together by very strong Attractions. And it is the Business of experimental Philosophy to find them out.'

The same problem was expressed in other words by Sir Oliver Lodge who said in the course of a lecture that the most extraordinary thing about a poker is the fact that if it is lifted up by one end, the other end comes up too. Until the structure of solids began to be understood there was very little hope of understanding the phenomena of rigidity and cohesion, or indeed many other properties of solid matter which have been known empirically both to workmen and to natural philosophers for centuries and have been used without being explained. A blacksmith does not have to know physics in order to fashion a horseshoe by heating and hammering. But once the facts of structure and of physical properties are related, it becomes possible to 'tailor' new types of solid to have certain properties that may be desired, as well as to frame and test physical theories.

Solid state physics, or crystal physics, has therefore assumed great importance during the last half-century, but the impact of X-ray diffraction on physics is much wider than would be covered by those terms alone.

14.3. Importance of the Ionization Spectrometer

The impact began with the building, by W. H. Bragg, of an extremely powerful new tool, the *X-ray ionization spectrometer*, a forerunner of the counter diffractometers of today. Of this instrument his son wrote:

'The X-ray spectrometer opened up a new world. It proved to be a far more powerful method of analysing crystal structure than the Laue photographs which I had used. One could examine the various faces of the crystal in succession and by noting the angles at which and the intensity with which they reflected the X-rays one could deduce the way in which the atoms were arranged in sheets parallel to these faces. The intersections of these sheets pinned down the positions of the atoms in space. On the other hand, a suitable crystal face could be used to determine the wave-lengths of the characteristic X-rays coming from different elements as sources. A 'pure' beam of monochromatic X-rays could be selected by reflection from a crystal and its absorption in various substances measured. It was like discovering an alluvial gold field with nuggets lying all round waiting to be picked up.'

The power of this instrument was brought forcibly home to the present writer on receiving delivery of a new X-ray Coolidge tube in or about 1925. A few diffraction-angle measurements with a crystal of rocksalt showed, within a matter of minutes, that the target was not of molybdenum, which had been ordered, but of rhodium (Mo$K\alpha_1$ 0.708 Å, Rh$K\alpha_1$ 0.612 Å). To have found this with certainty by absorption measurements alone would have been quite difficult and would have taken very much longer. In fact, although the discovery of the characteristic K and L radiations (with its subsequent influence on structural atomic theory) was made by Barkla and Sadler in 1908 on the basis of their differential absorptions, yet it was the ionization spectrometer, providing as it did an easy means of isolating X-ray lines from their continuous background, that enabled all *the physical phenomena associated with radiation in this short-wave region* to be intensively studied by many investigators, especially wave-length, absorption, dispersion, anomalous dispersion, fluorescence, interference, polarization and scattering, both coherent and incoherent. Thus this instrument, which provided a simple but accurate method of measuring intensity, could be used to measure *the falling-off of atomic scattering power* with increasing angle of deflection (or decreasing wave-length of incident radiation) not only for solids but also for gases.

14.4. Structure of the Atom

This in turn helped to throw light on *the structure of the atom* itself. Bohr proposed, in 1913, a model of the He atom in which the two electrons revolve in the same orbit at opposite ends of a diameter. But this led to a scattering curve at variance with that observed. Debye in 1915 and J. J. Thomson in 1916 independently solved the problem of the scattering of X-rays by atoms in which there were groups of electrons at fixed distances from each other. But again the calculations did not agree with the facts.

Ultimately it was found that the Schrödinger-type atoms did give reasonable agreement with experimental determinations of scattering fall-off, although no very exact calculations of electron distributions in real atoms are possible, especially when the atoms are in a state of anisotropic combination in molecules and crystals (valence state). The long list of methods and of references given in Volume III of the *International Tables for X-ray Crystallography*, Section 3.3, supplies evidence of the large amount of work carried out in connection with the calculation of atomic scattering factors. In recent years it has even been possible, from the scattering of neutrons by unpaired electrons in the magnetic elements, to calculate the distribution of these particular 3d electrons alone.

The scattering of X-rays by atoms in the forward direction was directly proportional to the number of extra-nuclear electrons; to the atomic number, that is, of the scattering element if not ionized, or to the total number of extra-nuclear electrons in the ion. But this quantity could not be measured directly, since the measured scattering curve began at a value of $(\sin \theta)/\lambda$ corresponding to the largest observable value of the crystal spacing, whereas $(\sin \theta)/\lambda = 0$ would correspond with $d = \infty$. Hence, even for the lightest atoms, it was not possible to deduce the state of ionization directly from the X-ray scattering curves. It was possible to do so indirectly by *a study of interatomic distances*, and hence to classify the various types of chemical binding. This gave the theoretical physicists data from which to begin to answer Newton's question: that is, to find out something, on a quantitative basis, of the forces that hold the atoms in various kinds of solids in their relative positions: both preventing them from coalescing into one dense agglomeration and from flying apart altogether.

14.5. Structure of the Nucleus

The intensity of *scattering of neutrons by the atomic nuclei* does not bear any simple relationship to their positions in the periodic table or to their chemical properties. It has had to be measured empirically for each element; and this has been done using apparatus which is a development of the early Bragg spectrometer, together with structure analysis techniques which are very similar in principle (though not in detail) to those used in the early days of X-ray structure analysis. The data thus obtained, when properly understood, will give information about *the structure of the nucleus* itself, its energy levels and the forces which hold together its component parts, forces of an altogether different nature from those that are important outside the nucleus.

14.6. The Nature of Radiation

In the chapter on X-ray Spectroscopy it is being shown how, using the X-ray spectrometer, *the wave-lengths and frequencies of X-rays* could be measured with extreme accuracy. When X-rays are scattered by any substance the diffraction spectra have a practically unchanged wave-length and frequency. But there is also a *photo-electric effect*: swiftly moving electrons are ejected at the expense of the energy of part of the incident radiation. In the course of a lecture on 'Photo-electricity' (which was intended as an introduction to a course of lectures on 'Wave Mechanics' to be given by Dr. E. Schrödinger at the Royal Institution in March 1928, with a final summing-up by Professor Whittaker), W. H. Bragg spoke of the difficulty he had experienced in accepting the idea of the undulatory theory of X-rays, until that theory was confirmed by X-ray diffraction experiments. He referred to the famous paper (*Ann. d. Phys.*, *17*, 145 (1905)) in which Einstein applied the idea of the Planck 'quantum' to the explanation of photoelectricity, thus resuscitating a corpuscular theory of light which would have somehow to be reconciled with the known phenomena of reflection, refraction and diffraction. Bragg then demonstrated some of the experiments that he had carried out in Australia in 1908, with R. Kleeman and J. P. V. Madsen, which showed that γ-rays falling on a carbon plate excited β-rays, a phenomenon parallel with the photo-electric effect for visible light. He went on to say 'The parallelism was not then so obvious as it is now, and for my own part I could not then believe in its existence. Our experiments fitted in perfectly with a corpuscular hypothesis of the nature of the γ-rays; and I supposed that the undulatory theory of light was unshakeable. *I should, of course, have thought otherwise if I had been aware of Einstein's paper*, to which I have already referred; *but it is easy to miss a single reference when one is in a very isolated laboratory*, and, as I said before, there are few if any allusions to the paper in the current literature of the years immediately following its appearance.' (My emphasis). All workers in isolated laboratories will know this situation, and will sympathize.

It was in fact not until 1912 that Einstein's photoelectric equation $E_{kin} = h\nu - \omega_0$, where ω_0 is the work done in pulling an electron out of an atom, was satisfactorily experimentally tested (see Compton and Allison (1935) pg. 45) and not until 1916 that Millikan showed that h was indeed Planck's constant. Meanwhile, the corpuscular and undulatory theories of X-rays had both been amply demonstrated and the quantum-mechanics theory of electromagnetic radiation became in a sense inevitable once the phenomenon of X-ray diffraction had been proved.

This theory led L. de Broglie (*Phil. Mag.*, *47*, 446 (1924)) directly to the inference that moving electrons and other 'particles' should also behave like waves and be diffracted by crystal gratings. In the lecture that has just been referred to W. H. Bragg spoke of the successful *electron diffraction* experiments carried out by Davisson and Germer in America and G. P. Thomson in Scotland and announced, not without real satisfaction, that the lectures by Schrödinger and Whittaker would be followed soon after Easter (1928) by one given by Professor Thomson on this newly-discovered phenomenon. This must certainly have been a historic series.

The Bragg spectrometer was used by A. H. Compton in 1926 to examine the weak scattered radiation that accompanied the X-ray diffraction spectra. He was able to show that this scattered radiation differed in wavelength from that of the incident beam and that this $\Delta\lambda$ depended on the angle of scattering only and not on the original wave-length or on the nature of the scattering material. This *Compton effect* was capable of a very simple explanation in terms of a quantum theory of X-rays.

The excitation of characteristic fluorescence radiation from a secondary radiator had been known before 1912, but the Bragg spectrometer made *X-ray spectroscopy* a precise study, and Moseley used a crystal to carry out a systematic photographic study of the X-ray spectra of the different elements. (Moseley's two papers in the *Phil. Mag.*, *26*, 1024 (1913) and *27*, 703 (1914) are a model of conciseness that should be studied by every research worker.)

14.7. X-ray Absorption and Related Phenomena

That the *absorption of X-rays by matter* is an extremely complex phenomenon was known to the physicists who first studied X-ray diffraction. They did not, however, realize the complexities that would arise in the quantitative application of the laws of absorption to the case of crystals. These have been slowly discovered by subsequent workers, to many of whom the diminution in intensity of the primary and diffracted beams, as they pass through crystalline material which may be of unknown texture or purity and of irregular shape or composition, is a major practical problem. Not least among the difficulties encountered is that of variable use of such words as *absorption*, *attenuation* and *extinction.* True absorption must involve the conversion of energy from the absorbed radiation into some other form of energy, whether heat, ionization, the recoil of bound electrons or the re-emission of radiation of different wave-length. But the primary beam intensity will be diminished also by scattering of radiation of unchanged wave-length and if a single crystal is used this effect will be a directional one; it will be particularly marked when selective reflection occurs. Using his ionization spectrometer W. H. Bragg was able to measure for diamond the abrupt diminution of the transmitted beam that occurs when a rotating crystal turns through a reflecting position. With a divergent beam of X-radiation and a stationary single crystal the same phenomenon is shown photographically as a pattern of white lines on a dark background.

The measured loss of intensity of a beam after transmission through matter is sometimes called its attenuation. This naturally depends upon the state of the scattering material whereas true absorption is independent of state. A recent tendency to write of 'attenuation coefficients' is wholly to be deplored, as it adds confusion to an already confused subject. Even the use of powdered material and monochromatic radiation does not permit an exact measurement of true absorption to be made, since even from powdered material the incident beam, after transmission, will in practice have lost energy by diffraction as well as by absorption. It is better to use an amorphous substance or a single crystal in a non-reflecting position. Measurements of absorption coefficients for low atomic-number elements are available for a wide range of wave-lengths, but for elements above $Z = 30$ the experimental data are scanty and not very consistent. This is unfortunate, for data are badly needed on which to build a reliable theory.

The dependence of true absorption on wave-length of radiation and on atomic number of the scatterer is therefore not accurately known either from the experimental or from the theoretical point of view. The *positions of absorption edges* were easily measured using the ionization spectrometer or photographic techniques; and the relationship between absorption edge and fluorescent radiation wave-lengths was readily established. But the absorption on each side of an edge is difficult to measure accurately.

In crystal analysis the effect of absorption depends upon the crystal shape and the composition of the primary beam. Earlier workers as a rule used short wave-lengths and their intensity measurements were not so precise, nor their structures so complex that absorption made much difference to their results. But they were interested in it as a physical phenomenon. (For example, see W. H. Bragg and S. E. Pierce, *Phil. Mag.*, *14*, 626 (1914)).

The early workers realized also that X-rays must travel at slightly different speeds in different media, and they determined the *refractive index* by several methods, including that of measuring the apparent wave-length of the X-radiation when diffracted in various orders from sugar and other crystals (Stenström 1919, see Compton and Allison (1935) pg. 280; R. W. James (1948) pg. 168 for a number of references). It was realized, however, that in the neighbourhood of an absorption edge the refractive index would have anomalous values, and that the atomic scattering factor would also show anomalous dispersion effects, including an imaginary term. This will be referred to again later. The point now being stressed is that *the discovery that a crystal could act as a diffraction grating meant that an immense variety of natural gratings was available for experimentation on various aspects of X-ray optics.*

Moreover, the reason why the metals and the alkali halides have been used as diffraction gratings for such a wide range of investigations is not due to interest just in those substances as such. Their structures are simple enough to be amenable to mathematical calculation, and it is this that makes them important from the point of view of physics.

14.8. Crystal Dynamics

Nowadays X-ray crystallographers, particularly if their basic training is in chemistry, study *the dynamics of crystals*; but only because they *must.* In order to refine the structure of an organic compound to the point where accurate bond lengths and angles are obtainable it is necessary to find and to make allowances for the anisotropic atomic thermal vibrations. To the early research workers, as also to the crystallographer-physicist of today, the interest is in the dynamics itself, and in its relation to a host of independent physical properties. Humphrey Davy had suggested that melting involves an increase in the heat motion of atoms, a suggestion which seems obvious to us now, but only because we understand, so much better than he could, the final breakdown of crystal architecture involved in the process of fusion. As X-ray diffraction has shown, the reason why glasses soften, become plastic and finally liquify *without* any definite melting-point, is that their structures do not have the regularly repeating periodic pattern of a crystalline solid; and there is therefore in glasses no such directed system of long-range order to be broken down by the disturbing influence of increasing and varying local disorder.

The historical paper by Born and Karman on the theory of lattice vibrations preceded, by a few months, the discovery of X-ray diffraction by crystals. Indeed, we are told by Ewald that the question that first exercised the minds of some of the theoretical physicists of those early days, when the possibility of crystal diffraction was discussed, was whether the disorder produced by thermal vibration might not be so large as to prevent the observation of diffraction effects altogether. It has occasionally happened in scientific history that a discovery has been postponed for years because theoretical prediction has discouraged experimental workers from undertaking a difficult investigation likely to give a negative result. The Hall effect provides one such example. Fortunately, neither the doubts in this case nor an initial failure, were allowed to inhibit the crucial X-ray diffraction experiments, which showed that for diamond there was diffraction right round to back-reflection regions, but that for copper sulphate, zinc blende, rocksalt, etc. the intensity was so much affected, presumably by *thermal vibration of the individual atoms*, that no spots were observable at large angles of deviation.

In subsequent studies by Debye and others, notably I. Waller, (summarized by P. P. Ewald, *Handbuch der Physik*, *24*, 270 (1933)) the effect that thermal vibration would have on the intensities of the diffraction maxima was calculated, and expressed as an exponential factor, $I_T = I_R e^{-M}$ where I_T is the intensity at a temperature T, I_R the corresponding intensity for a structure undisturbed by vibration or disorder, M a function dependent upon temperature, structure and the forces between the atoms. M could be expressed in many ways. It could be related to the *elastic constants*, to the *specific heat*, to the possible existence of crystal vibration even at the absolute zero of the temperature scale, when thermal vibration had disappeared (*zero-point energy*), and so on. Subsequent experiments, by R. W. James and others, were directed not just to the confirmation of the temperature factor as a correction to intensity measurements (which could have been and was also achieved empirically) but as a means of checking these important physical theories, which however have never been extended quantitatively to more complicated and less symmetrical structures or to temperatures near to the melting point, where anharmonic vibrations become increasingly important. Nor is there any really comprehensive theoretical study of the *thermal expansion* of the less simple structures, although certain empirical relationships with molecular form and arrangement are obvious. It is, for example, an observed fact that whereas the expansion normal to the layer planes of graphite is 28 times that of diamond at room temperatures, it is only a fraction of that found for aromatic compounds such as anthracene or naphthalene, where again the expansion increases as the molecular size decreases. There is a contraction with increase of temperature within the layer planes of graphite and of hexagonal BN, up to fairly high temperatures.

That the thermal expansion of face-centred cubic elements is related to the adiabatic compressibility at 0°K and to the molar specific heat was shown by Grüneisen before the discovery of X-ray diffraction and the fine structure of solid matter. Grüneisen and Goens subsequently extended this theory to hexagonal elements such as zinc and cadmium and explained the negative expansion at low temperatures normal to the *c* axis. The theory does not agree with experiment for body-centred cubic elements such as W, Li, Na, Pd, Ta, nor for most polyatomic compounds. A recent critical review of this entire field has been given by R. S. Krishnan (1958) together with much experimental data for thermal expansion and its variation with temperature.

It is clear that X-ray diffraction provides a method of measuring thermal expansion of the unit-cell parameters, as distinct from that of the macroscopic material. It is perhaps relevant here to point out that thermal expansion is one of many deceptively simple phenomena for which accurate measurements are obtainable, but for which not even the powerful new methods of X-ray diffraction and structure analysis have provided sufficient basic information for the formulation of a really comprehensive theory.

14.9. Crystal Texture

In other cases, it was the study of the X-ray diffraction intensities themselves that drew attention to *unexpected* physical phenomena. For example, physicists were interested in the shape of the curve showing intensity versus deflection-angle in the neighbourhood of a diffraction maximum, and in the resolving power of the scattering mechanism. Darwin tackled these as theoretical problems and so, later, but independently, did Ewald. Naturally they assumed a perfectly regular arrangement of scatterers at the points of a space lattice, and to simplify the calculations, crystal absorption was neglected. If anything, therefore, the observed intensities might have been expected to be less than those calculated. In fact, they were often many times *greater*; and they could be increased still more by roughening the surface of the crystal if the diffraction effects were being observed by 'reflection' and not by 'transmission'. The *primary and secondary extinctions* that were introduced by Darwin as the consequence of crystal perfection, and the need for a single crystal to approximate to a condition of 'ideal mosaicity' in order to give diffraction intensities that can form the basis of a structure analysis—these are a commonplace to the present-day crystallographer. They may involve him in tiresome corrections. But they were exciting discoveries to the early physicists; and many discussions followed as to what was the real nature of that hypothetical *mosaic structure* that must be postulated at least as a limiting condition, in order to get any sense out of the observed intensities. It goes without saying that there are again many properties of solids and particularly of metals, that depend upon texture as much as, or even more than, upon structure.

For example, if high-purity metal wires are twisted under slight tension beyond their elastic limit they begin to extend rapidly, but in a self-annealing metal such as lead the elongation for a given twist and tension depends on the rate of twisting (T. Lonsdale, *Phil. Mag.*, *8*, 703 (1929); *11*, 1169, 1187 (1931) gives reference to earlier work). The classic investigations by G. I. Taylor (*Phil. Trans. Roy. Soc.*, A *230*, 323 (1931) (with H. Quinney); *Proc. Roy. Soc.*, A *145*, 362 (1934)) extended this study to metal *tubes*, and thus he obtained data which could be compared with theoretical calculations and he could show that some kind of structural misfit or weakness must be present to account for the experimental results. (For work on NaCl see also A. Joffé, 1928.)

'The Theoretical Strength of Materials and their Practical Weakness' formed the subject of a lecture to the Society of Engineers given by Sir William Bragg in 1935, but even then he was not able to give a complete or quantitative answer to the question as to what kind of imperfection in fact exists which reduces the tensile strength of rocksalt two-hundred-fold from its theoretical value, although he rightly pointed out that no theoretical value could have been calculated until the fine details of its structure were found by X-ray diffraction. It happened, however, that only a few months before he spoke, a new idea, that of 'dislocations' in crystals, had been put forward by Orowan, by Polanyi and by G. I. Taylor, to explain the phenomenon of plastic deformation in metals and the misfit that must occur at mosaic boundaries; and this idea was developed by W. G. Burgers (one of W. H. Bragg's early research students at the Royal Institution) and others, in particular by F. C. Frank, to give a satisfactory theory of *crystal growth*. It is an interesting thought that a really perfect crystal could not be ruptured except by a force that could pull all its atoms (or atomic groups) apart simultaneously, nor could it grow except by the fortuitous (and, in practice, impossible) circumstance that a whole plane should be deposited in place at once. A perfect crystal has been described as being like Oliver Wendell Holmes' 'deacon's shay' the parts of which were so exactly adjusted that it could not collapse until the day when it disintegrated entirely and the deacon found himself seated on the road in a heap of dust. The postulate of dislocations, since proved by electron microscope pictures and demonstrated in other ways, provides just those necessary points of weakness which allow shear or yield to take place for an applied force far less than would be necessary to separate the material into its component atoms or molecules. In a similar way dislocations in the surface of a crystal provide the growing points where new atoms or molecules can be deposited. As Frank (*Crystal Growth*, pg. 54) has written 'The completely perfect crystal will not grow in any circumstance: the conditions which could cause it to grow would also soon make it imperfect.' In fact it could not even exist except as an abstraction, for a perfect crystal would have to be infinite in all dimensions. Any surface forces would mar its perfection.

The mosaic nature of a single crystal, whether due to dislocations or to some other discontinuity of structure, is not the only kind of 'imperfection' that can affect physical properties. One of the earliest substances studied was quartz. In a sense this was curious, because it was certainly not one of the simpler structures that would easily lend itself to mathematical treatment. Perhaps it was chosen because good crystals were at hand and because its *polymorphism*, its *optical activity* and its *pyro- and piezoelectricity* constituted a challenge that simply could not be resisted. Subsequent studies of α and β quartz led to a complete understanding of the four forms of *twinning* shown by α quartz, two of which are also shown by the high-temperature and more symmetrical polymorph (W. H. Bragg, *Trans. Soc. Glass Techn.*, *9*, 272 (1925)). Again we may emphasize that the modern structural crystallographer is seldom concerned with the *study* of twinning for its own sake (although there are notable exceptions). He learns to detect twins in order to reject them; or if he must use them for structure analysis he does so warily, his object being to find the basic structure of the untwinned material. But twinning is of importance both in its own right and in relation to the *plasticity* of metals. Studies of cold working and of heat treatment bring the subject rather outside the range of the physicist and into that of the metallurgist, although the problem is essentially one of physics; but the understanding, for example, of the flow of metals and even of glaciers depends essentially upon the same principles: an applied force produces strain; the crystals in which the orientation, often achieved by twinning, is such that strain can be most easily relieved by slip, will tend to grow in size, given time, at the expense of those less favourably orientated and therefore retaining higher internal stresses. This explains why the ice crystals at the foot of a glacier may be several centimeters in linear dimensions, although those at the head are only a millimetre or two across.

14.10. Ferroelectricity

In 1945 Ubbelohde and Woodward (*Nature*, *155*, 170) observed a new kind of 'imperfection' associated with a phenomenon that has assumed the greatest possible practical importance in physics. If Rochelle salt, sodium potassium tartrate dihydrate, is studied between $+24.5°C$ and $-20°C$, it is found that reflections from certain $(0kl)$ planes are split into two components of slightly different Bragg angle. The separation is greatest at about the middle of the transition range, 0°C, but is so small that at first it was doubtful whether it was a real effect. It corresponds to the break-up of the orthorhombic crystal into sub-crystalline monoclinic units, all having their a axis in common with that of the parent crystal, and their b and c axes respectively equal to those of the parent crystal but with an angle of $90° \pm 2'$ replacing the former 90°. The new b and c axes are randomly $\pm 2'$ from their former positions, giving four new orientations all present simultaneously and all disappearing again, with a coalescence of the split reflections, when the crystal is heated or cooled to a temperature outside the above range. Now this is just the temperature range within which Rochelle salt had been known, since 1921, to exhibit its *ferroelectric properties*, that is a dielectric constant along a about 400 times as large as that along b, with a dielectric hysteresis which implied that some kind of polarization or electric displacement persisted when the applied field was removed or gradually reduced to zero. Similar very large and anomalous changes take place in the piezoelectric modulus along the a axis. That these phenomena are related to cooperative movements of hydrogen atoms was suspected on structural grounds, but was definitely proved much later, by Pepinsky and others, using neutron diffraction measurements. These showed that certain reflections which changed markedly in intensity within the relevant temperature range, were due to hydrogens only.

In the structurally simpler case of KH_2PO_4 and related compounds, the movements of the hydrogen atoms have been demonstrated by Fourier analyses based on neutron diffraction experiments at temperatures within and outside the ferroelectric temperature region. But there is a second class of compounds, of which barium titanate, $BaTiO_3$, is a typical example, which shows an even larger polarization effect, due to the ease with which a highly-polarizable ion can shift out of a symmetrical position, if the structure is already at the limit of stability and there is a change of temperature. The extraordinary behaviour of such compounds is admirably described in a number of recent textbooks and papers and it rapidly became a matter of much practical importance.*

In the case of ferroelectrics the observed changes of texture which accompany the structural second-order transitions are themselves a very subsidiary effect. The cause of the ferroelectricity is a response of the *structure* to an external stimulus in the shape of a change of temperature and an imposed electric field.

14.11. Fluorescence; Electrical and Thermal Conductivity

Other physical properties of great practical importance are due to *structural imperfections* of various kinds. In some cases the early workers were aware of the application without being aware of the cause, and even perhaps without being sufficiently curious about it to make it an object of intensive study. Every X-ray worker used a *fluorescent screen.* It was often regarded as being a distinct advantage if the material (ZnS or Zn_2SiO_4, usually) were phosphorescent, because then the screen could be inserted into a narrow gap and the luminescence

* It even became a matter of political consequence. From 1952 to 1954 or longer, every physicist or crystallographer wishing to travel from Europe to the U.S.A. was liable, on applying for a visa, to be asked if he or she knew anything about (a) nuclear physics or (b) ferro-electric crystals. The writer of this article was so startled at this unexpected question (b) that she asked what was the reason for it, only to receive the reply 'I haven't the slightest idea. I only know that if you say "Yes" you go through one hoop and if you say "No" you go through another. Since you have said "Yes" your papers have to go to Washington for further investigation and I am afraid that you will have to wait some months for a visa.' It was granted in the end, however, and by this time a little thought had provided the reason for the question.

observed when it was withdrawn. It had long been noticed that many crystals become fluorescent when they are bathed in an X-ray beam (Röntgen 1896). None of the early crystallographers seem to deal with these interesting facts. Only in recent times have the details of the dependence of fluorescence and phosphorescence on the presence of impurities in the crystal structure been opened up to quantitative interpretation on lines very similar to the energy band theory of semiconductors which attracted so much interest and intense study because of its important applications in connection with rectifiers and transistors.

Some of these forms of 'structural imperfection' have already been mentioned. The more important of these, from the point of view of *electrical and thermal conductivity* or resistivity, are the presence of free or partially-free electrons, the existence of 'positive holes' caused by the removal of an electron from a previously neutral region, the effects of lattice vibrations, of impurities (interstitial or substitutional), and of a wide variety of lattice defects such as vacant sites, atoms or ions moved into interstitial positions, dislocations and the presence of foreign high-energy particles (nucleons). This is no place for a detailed description of the mechanism of the processes of conduction in a metal or of semiconduction in, say, germanium, cuprous oxide or indium antimonide, but perhaps it may be mentioned that the theory is based on an extension of the idea of energy levels of free atoms to that of atoms in close combination. The solid has filled, partially-filled, or empty energy bands. If under the action of an applied voltage the electrons remain in the filled energy levels of the solid the substance is an insulator. If the energy gap separating the highest filled level and the next unfilled (conduction) band is small (only of the order of the atomic thermal energy) the substance is an intrinsic semi-conductor. If the two levels overlap the substance is a conductor. If an imperfection such as the presence of impurity atoms causes electron jumps either from the impurity atom to the conduction band, or from the highest filled band to the impurity, then there is impurity semiconduction respectively of the excess (n) or defect (p) types. The 'impurity' may be an atom of higher or lower valency than that of the matrix; or there may be a deviation from stoichiometric composition in the shape of excess or defect anions or cations. X-ray diffraction methods may be used in some cases to distinguish these alternatives. From a knowledge of the crystal structure the energy levels or energy gaps may be calculated or estimated. Without it, the subject would be wholly empirical.

Sometimes the conductivity is of an ionic type. In a remarkable experiment by A. Joffé it was shown that the conductivity of a quartz crystal at 200°C could be reduced sixty-fold, and that of an alum crystal nearly fifty-fold, by passing an electric current through it for a long time, to expel foreign ions. Further purification of the alum by aqueous solution and recrystallization decreased the conductivity another 100-fold. Such ionic conductivity is strongly texture-sensitive, and is a kind of diffusion. But there is an ionic conductivity which is diffusion of another kind and is texture-insensitive but strongly temperature-sensitive, and which is also very sensitive to the size of and charge on the moving ions. A striking example of this ionic conductivity is that of AgI in its various forms. A few degrees below the transition from cubic to hexagonal at 146°C the electrical conductivity is very small; just above it the conductivity jumps 4000-fold and increases steadily up to the melting-point 552°C. Within the range 146°C to 552°C the iodine atoms occupy fixed positions at the corners and centres of the cubic unit cell, but the silver atoms behave almost like an interstitial liquid (L. W. Strock, *Z. phys. Chem.*, B *25*, 441 (1934), B *31*, 132 (1935); *Z. Kristallogr.*, *93*, 285 (1936)).

14.12. Atomic and Ionic Radii

This question of structure transitions and of mixed crystal and defect structures brings us back to one of the most fundamental of all the contributions that X-ray structure analysis has made to physics. It has provided a method of measuring atomic and ionic radii. This subject, because of its importance in connection with *bond lengths* and because of the influence of the radii upon the state of *co-ordination* (number of atoms or ions of one kind surrounding one of another kind) is often thought of as a branch of crystal chemistry. But from our point of view the structure and weight of the elements and the atomic number are all of physical no less than chemical interest; and the atomic radius is one of the factors controlling many very important physical properties of solids, including the possibility of *polymorphism*.

The original experiments with the Bragg ionization spectrometer gave very exactly the angles at which the diffraction spectra occurred when monochromatic radiation was used and hence, knowing the wave-length, the unit-cell dimensions could be calculated (from $n\lambda = 2d \sin \theta$); or knowing the unit-cell dimensions (from $a^3\rho = nM/N$) the wave-length could be found. Once the intensity distribution had been used to determine the structure, it was possible to decide the interatomic distances. The actual atomic (ionic) radii could only be found when some other technique, such as refractivity (Wasastjerna 1922), had been used to estimate the radius of at least one atom or ion. The discovery was then made that in the metallic state the apparent atomic diameter was quite different from what it was in an ionic compound and that even in ionic compounds the diameter varied slightly with the type of structure. This in turn gave valuable information about the types of binding in different types of structure, information that could be correlated with hardness, cleavage, polymorphism, melting-point, electrical, optical and magnetic properties. It also provided information about the effective diameters of the electron orbitals, when the atom was in different valence states.

Measurement of the unit cell dimensions combined with a knowledge of the molecular weight and number of molecules in the unit cell gives an 'X-ray density' which need not be quite the same as the macroscopic density, although it often is. Density is a property which enters into many physical theories; and an X-ray analysis can often give an 'X-ray density' in circumstances where it would be impossible to measure the macroscopic density; for example because the crystal is too small or too volatile, or because it is an inclusion or exists only as a transitory stage in a solid-state reaction.

In order to be sure that the values found are meaningful, it is of course necessary that the structure should be relatively free from defects. For instance, the density of a diamond single crystal can be reduced by about 8 per cent by means of neutron irradiation damage, the unit-cell parameter increasing by a corresponding amount. The damage, in fact, produces an overall increase of the unit-cell size and not an unequal spread of unit-cell dimensions. By heating the diamond to a temperature a little lower than that required to graphitize it, it can be annealed. The true carbon-carbon distance is that found in a diamond free from both impurity and defect. But the change in shape of the space taken up by a carbon atom in diamond and in graphite is a real indication of the change in electronic condition.

Caesium undergoes a *transformation on compression*, from a body-centred to a face-centred cubic structure. Since the latter is close-packed, one would not expect any further change to be possible. In fact there is a further abrupt reduction of 11 per cent in the volume of caesium at a pressure of 45000 kg/cm^2 (P. W. Bridgman, *Amer. Acad. Arts and Sci.*, *76*, 55 (1948)) and this is supposed to be due to a transition of the valency electron from the 6s to the 5d level. At normal pressure and temperature the atoms of caesium undergo very large thermal vibrations indeed, having a Debye coefficient,

$$B = (\ln I_R - \ln I_T)\lambda^2/2 \sin^2 \theta,$$

of the order of 10, as compared with 2 for lead, 1 for antimony and 0.5 for gold. This will increase their apparent size; but a *discontinuous* change of volume is a sure sign of a transition, either structural or electronic.

Strictly speaking, *thermal vibration* (and perhaps zero-point energy) should be allowed for, or measurements should be made of unit-cell dimensions at low temperatures. Apart from the few measurements made by the physicists in the 1920's, however, there are hardly any accurate data for thermal vibrations even in binary compounds. This is partly because of the real difficulty of getting sufficiently accurate measurements of intensities, partly because of the problem of allowing for binding effects in the atomic scattering factor, and partly because there were and are so many other interesting things to be done.

The analysis of caesium antimonide by Jack and Wachtel (*Proc. Roy. Soc.*, A *239*, 46 (1957)) has shown that the two kinds of atoms in such a structure may sometimes have vibration amplitudes which are as different from one another when combined as they are when measured for the separate elements. The Debye characteristic temperatures calculated from vibration amplitude measurements may be compared with those obtained from specific heat, elastic constants, electrical resistance, etc., thus extending our knowledge of the relationships between these properties for an increasing range of simple substances.

14.13. Magnetic Properties of Crystals

The atomic radii in the solid elements show a periodic variation according to their place in the Periodic Table. There is, however, only a small variation in the sizes of the transition elements V-Zn, for which the 3d sub-shell is filled up after 4s is already completed. Fe, Co, Ni have long been known to be *ferromagnetic*. Other transition elements, such as Cr, Mn, Cu are not normally ferromagnetic. For this to be the case, a condition, found empirically, relating the radius of the 3d shell and the interatomic distance in the crystal, must be satisfied; and this is rather critical. Certain combinations of atoms, such as the Heusler alloys $AlMnCu_2$ and $SnMnCu_2$, are ferromagnetic when annealed, but not when quenched from high temperatures, because it is only in the annealed state that the critical distance rule is satisfied and an internal magnetic field can exist. There are also non-ferromagnetic forms of iron (face-centred cubic, f.c.c.) and nickel (hexagonal close packed, h.c.p.) having structures which do not satisfy that rule; and at high temperatures the thermal vibrations are large enough to prevent, in all substances, the co-operative alignment of spins that is required to establish the 'molecular field'. The transition from ferromagnetism to paramagnetism at the 'Curie point' is accompanied by discontinuous changes in specific heat and other properties.

There are other compounds, however, in which the spins of 3d electrons in separate atoms interlock to keep each other polarized but in anti-parallel positions, through an intermediate ion such as oxygen or sulphur. Such substances are called '*antiferromagnetic*'. The idea that such an arrangement might exist was put forward by Van Vleck in 1941, and applied by Néel in 1948 to explain a second-order transition shown by a specific heat discontinuity in the ferrites. The magnetic order-disorder transformation which occurs at high temperature produces very little change in unit-cell dimensions and is therefore not easily studied by X-ray diffraction, nor does antiferromagnetism result in any overall macroscopic magnetic moment which can respond to an external field. But since neutrons possess a magnetic spin there is a strong interaction between neutrons and magnetic atoms; and the existence of an ordered arrangement of spins in a structure, whether ferromagnetic, antiferromagnetic or a mixture of the two (*ferrimagnetic*), results in strong neutron diffraction peaks from which the direction of alignment of spins can be deduced and which disappear when the magnetism changes into paramagnetism (at the Curie or Néel point).

Moreover, there is now considerable work going on, mainly in Eindhoven, by Verwey, Gorter, Braun et al., on *the tailoring of new magnetic materials* which are strongly anisotropic, the magnetic axes being strongly linked with the crystal axes, and which have hysteresis curves of various desired shapes. This work is possible because of previous knowledge of the spinel structure of these materials and of the modification by substituting some ions by various percentages of others.

Magnetite and the spinel group of crystals captured W. H. Bragg's interest as early as 1915 (two letters to *Nature*). Even earlier than this, he had tried to use X-ray diffraction to explain the ferromagnetism of iron but had not succeeded. As is mentioned above, it was not the X-ray but the neutron diffraction method that did succeed in demonstrating the alignment of spins in magnetic substances.

In 1927 W. H. Bragg gave a lecture at the Royal Institution on 'Tyndall's Experiments on Magne-crystallic Action' and he followed this, at the time of the Faraday celebrations in 1931, with a note to *Nature* on 'Faraday's First Successful Experiment on Diamagnetism'. He did no experiments himself in this field but he was keenly interested in those of the Indian schools; especially as K. Banerjee had been able to correct Bragg's early anthracene structure by reference to its crystalline *diamagnetic anisotropy*; and then K. S. Krishnan who with his colleagues examined the diamagnetic susceptibilities of many aromatic crystals, was able to show that the orientation of the molecule could be fairly accurately deduced from such measurements in many cases.

If the crystal structure were already known, however, the magnetic susceptibilities of individual molecules could be calculated and related to their dimensions and electronic structure. Lonsdale was able to use diamagnetic data to measure the effective radii of the σ and π orbitals in benzene derivatives and in metal-free phthalocyanine and to prove conclusively that the π orbitals are of molecular dimensions.

14.14. Electrical Properties of Crystals

Sir William Bragg spoke of some of these results in a lecture at the Royal Institution in March, 1936. In a second lecture 'The Electric Properties of Crystals (II)' in May 1936, he went on to say:

'The recently acquired and more intimate knowledge of the manner in which the molecule is built into the solid allows a further insight into the problems of the dielectric constant and of the electric properties of the crystal. These problems have been widely studied of late because of their extreme importance in industry, especially in the transmission of electricity at high voltages. They are of remarkable interest also for their applications in radio telephony, where great use is being made of the piezo-electric property in the production of oscillators of known high frequency, and in the construction of various types of microphone.'

This lecture was illustrated by experiments showing how the polarity of resorcinol crystals was explained by the arrangement of molecular dipoles and how this could be varied by heat (*pyro-electric effect*). A model of Rochelle salt with flexible molecular linkages was used to exhibit a possible form of *piezo-electric action*; and various illustrations were arranged by the National Physical Laboratory and the Building Research Station to show the many uses of the piezo-electric properties of quartz. The phenomenon of pyro-electricity seems to have been known for many centuries B.C. That of piezo-electricity was discovered in 1880 by Pierre and Jacques Curie, who were also aware of its relationship to the crystal symmetry. But although an interesting physical property, little use was made of it until, during World War I, Langevin used quartz plates, excited electrically, to emit and receive high-frequency (h-f) sound waves under water. Langevin was thus the originator of the *science of ultra-sonics*, now used extensively in the measurement of *elastic properties* and for many other scientific and industrial purposes. As Cady (*Piezo-electricity*, 1946, pg. 8) points out, however:

'Voigt came very near to being the originator of the piezo resonator. In the *Lehrbuch* he gave the differential equations for elastic vibrations in crystals, without, however, mentioning the bearing of the piezoelectric effect on such vibrations. He mentioned the use of h-f in the measurement of dielectric constants, recognizing the fact that anomalous results are to be expected at frequencies of molecular resonance. What he did not foresee was that similar anomalies would be found with all vibrating piezoelectric crystals whenever the applied frequency coincided with that of a normal vibrational mode of the entire crystal specimen. It was the electronic generator of h-f alternating current, supplanting the induction coil of Voigt's day, that paved the way for the advent of the piezo resonator.'

Communication, entertainment, adult education and propaganda have been almost entirely transformed during the past forty years because of this single development of an obscure and somewhat mysterious crystal property.

'Crystal sets' were used as receivers of wireless signals from 1913 on (one of the first in Gt. Britain was tested by the writer's elder brother in that year). These used zincite (ZnO) and bornite (Cu_5FeS_4) in conjunction. The writer possessed one of the original 'cat's whisker' sets. The crystal was silica (though others were possible), the resistance of which varied for some unknown reason with the direction of the applied voltage. The 'cat's whisker' probe was a fine copper wire, but not so fine as modern 'whiskers' of tin or cadmium! The signals received through head-phones were very weak and unreliable. Over 30 years later the original *transistors* were made at the Bell Telephone Laboratories by using two cat's whiskers, very close together, on a very pure crystal of germanium. Modern transistors are small, light and, unlike the original crystal sets, they are most reliable.

Nowadays, of course, quartz plates and other piezo-resonators and oscillators are cut with very precise orientations, to avoid disturbing effects of coupling between different modes of vibration and to avoid also the effect of temperature change upon frequency. Orientation can be accurately controlled by X-ray diffraction measurements. One of the more glamorous uses of piezoelectricity is in the *quartz clock*, where a vibrating quartz plate or ring replaces the pendulum and gives a time piece more constant than any other, being unaffected by slight daily variations in the acceleration of gravity or fluctations in the earth's rotation. The quartz clock has introduced a new order of precision into astronomical measurements.

In a simple compound such as ZnS the development of a piezoelectric moment normal to a (111) plate means either that compression has moved the Zn lattice relative to the S lattice, or that there is a change in the polarization of the bonds along the direction of the compression. The question that early physicists had to answer was: 'Does the positive charge develop on a Zn face or on a S face?' ZnS, which is of course hemihedral, has an equal number of larger shiny faces, and smaller rough ones. Ordinarily there would be no difference in the intensities of X-ray spectra diffracted from the (111) or from the opposite ($\bar{1}\bar{1}\bar{1}$) planes, even in the absence of a centre of symmetry (Friedel's rule). But the existence of anomalous dispersion in the neighbourhood of an absorption edge made it possible to distinguish the two. Coster, Knol and Prins used radiation from a gold target to study the (111), ($\bar{1}\bar{1}\bar{1}$) reflections from a plate cut from a well-formed crystal of ZnS. The wave-length of the ZnK absorption edge lies between Au$L\alpha_1$ and α_2. They found a difference in the (111), ($\bar{1}\bar{1}\bar{1}$) intensities which indicated that the atoms in the larger shiny faces are S, those in the smaller rough faces are Zn and that on pressure being applied the S faces become positively, the Zn negatively charged. Von Hippel has explained this by a change of the tetrahedral bonding angles, without change of bond length; and with a consequent movement of charge. Actual calculation of the piezoelectric field or of ferroelectric effects for a given strain is, however, a problem of an altogether different magnitude, involving a knowledge of the atomic dynamics as well as of structure and structural changes. Brave attempts have been made, particularly by the Indian schools, to relate piezoelectric and other physical properties to structure and to calculate electrical, elastic, thermal and optical properties at least for some of the simpler compounds.

14.15. Optical Properties of Crystals; Absolute Configuration

The fact that quartz was found in right- or left-handed forms and that these rotated the plane of vibration of polarized light in clockwise and anticlockwise directions, respectively, was certainly one of the properties that led to its intensive study in the early days of X-ray structure analysis. A relationship with the spiral arrangement of atoms in the structure was at once expected.

It has been seen elsewhere that the discovery of X-ray diffraction by crystals grew out of Ewald's previous interest in the *double refraction* to be expected from hypothetical three-dimensional arrangements of atoms on a crystal lattice. It was not long before Born (1915) used the newly-determined structures of simple crystals to calculate various physical properties, including *refractive indices*. But it was necessary to know the dispersion curve of the ordinary ray in order to calculate that of the extraordinary ray and the optical rotatory power of quartz. These calculations, however, he made with considerable accuracy.

Recently it has proved possible to make quantitative calculations of the *optical activity* (S. Chandrasekhar, *Proc. Indian Acad. Sci.*, A*37*, 468 (1953); A*39*, 290 (1954)), and of its variation with temperature, by relating these effects to the characteristic frequencies of the substance and other physical properties. But a simple quantitative theory in terms of structure alone is not yet available.

The particular case of tartaric acid, which rotates the plane of polarization in one direction in the liquid state and in the other direction in the solid state was naturally something of a challenge to the early physicists and this problem was tackled by W. T. Astbury in 1922. But it was far too difficult a structure analysis to be solved at that time. This was accomplished by Beevers and Hughes in 1941. What became of even greater general interest later on was the question as to whether it was possible, by X-ray methods alone, not merely to differentiate between a right- and a left-handed screw but to determine the absolute configuration of an enantiomorph in any particular case. The solution was of importance to chemists because of its bearing on the Fischer convention. The determination of absolute configuration by Bijvoet et al. (*Proc. K. Akad. Wet. Ned.*, *54* B, 16 (1951); *Nature* (Lond.), *168*, 271 (1951)) which was a brilliant extension of the experiments of Coster, Knol and Prins referred to above, showed that light travels more slowly when the atomic spiral and that representing the electric field along the light beam wind in the same sense.

The quantitative determination of optical refraction indices by a method much simpler than that of Born was made by W. L. Bragg, in 1924, for calcite and aragonite, two polymorphs of $CaCO_3$ (*Proc. Roy. Soc.*, A *105*, 370). He assumed that each atom becomes an electric dipole under the influence of an electric field and he used the ionic refractivities as a measure of the polarizability of the ions (calculated theoretically in 1922 by Wasastjerna, and used by him in 1923 to give the first accurate estimates of ionic radii, see above). The mutual influence of nearest and next-nearest neighbours had to be taken into account in order to obtain a good approximation to the experimental data. The method was also successfully applied to other carbonates and nitrates, including the four polymorphs of NH_4NO_3. A qualitative extension of this work to crystal structures containing ionic groups or molecules of various shapes and arrangements and in various states of rotation (for example, simulating spherical or cylindrical symmetry) has allowed birefringence and optical sign to be used, just as diamagnetic anisotropy can similarly be used, as a valuable adjunct to crystal structure analysis.

14.16. Thermo-optical Properties

A change of temperature can change the birefringence. Mitscherlich (best known for his studies of isomorphism early in the 19th century) showed that when a gypsum plate, positive biaxial at room temperatures, is heated, it becomes uniaxial at 90°C and then biaxial above that temperature but with its optic axial plane rotated through a right angle. Tutton (Vol. II, 1922) records the case of $CsSeO_4$ for which, over a 0°–250°C range of temperature each of the three axes of the optical ellipsoid in turn becomes the acute bisectrix. Naturally the temperature at which such changes take place depend upon the wave-length of the light employed.

In an excellent review of this subject, R. S. Krishnan (1958) points out that the thermo-optical behaviour of crystals depends upon their structures, thermal expansions and temperature variations of absorption frequency.

The absorption of light, whether in the ultraviolet and visible or in the infrared regions, was studied long before the discovery of X-ray diffraction. However, knowledge of crystal structure has shown that the colour of a crystal and particularly any pleochroism, or variation of colour with direction of vibration of light, is closely linked with structure. In organic crystals absorption in the visible region is greatest when the light is vibrating parallel or nearly parallel to such double bonds as –C=C– and –N=N–. By the use of polarized infrared radiation also, the orientation of particular atomic bonds having known characteristic frequencies can be found. These are applications of chemical rather than of physical interest and indeed, the use of solid phase techniques is limited. The study of the Raman spectra of crystals does give information, however, about lattice and molecular (or group) vibrations.

It is only recently that enough experimental data for the elastic properties of crystals have been available to test relationships between these and the strength of atomic bindings; although for such simple structures as the alkali halides a reasonably good theory was available in 1922.

This brief and inadequate review of a tremendous field does less than justice to the crucial part in it played by that great physicist, Max von Laue. Attention is therefore especially drawn to the article by P. P. Ewald published in the *Biographical Memoirs of Fellows of the Royal Society*, *6*, 135 (1960).

References

(additional to those in the text)

Bacon G. E. (1955) *Neutron Diffraction*, Clarendon Press, Oxford.

Bhagavantam S. and Venkatarayudu T. (1948) *Theory of Groups and its Application to Physical Problems*, Andhra University, Waltair, India.

Boas W. and Mackenzie J. K. (1950) Anisotropy in metals, *Proc. Met. Phys.*, *2*, 90.

Cady W. G. (1946) *Piezoelectricity*, McGraw-Hill, New York (Full bibliography).

Compton A. H. and Allison S. K. (1960) *X-rays in Theory and Experiment*, Van Nostran d, New York.

D'Eye R. W. M. and Wait E. (1960) *X-ray Powder Photography in Inorganic Chemistry*, Butterworth's, London (gives examples showing how to distinguish different kinds of defects).

Dunlop W. C. Jr. (1957) *An Introduction to Semiconductors*, Wiley, New York.

Evans R. C. (1946) *An Introduction to Crystal Chemistry*, University Press, Cambridge. (This gives information about physical properties in relation to structure.)

Ewald P. P., Pöschl Th. and Prandtl L. (1932) *The Physics of Solids and Fluids*, Blackie, London.

Flügge S. (Ed.) (1957) *Encyclopaedia of Physics*: *26* (1) *Crystal Optics—Diffraction. 30 X-rays. 32 Structural Research.*

Frank F. C. *Crystal Growth*, Butterworth's, London 1949, repr. 1959.

Hartshorne N. H. and Stuart A. (1960) *Crystals and the Polarizing Microscope*. 3rd ed. Arnold, London.

Herrmann K. and Krumacher A. H. (1932) Röntgenuntersuchungen an kristallinischen Flüssigkeiten IV. *Z. Kristallogr.*, *81*, 317 (19 intermediate states between solid and liquid, and diffraction effects.)

International Tables for X-ray Crystallography (3 Vols.) *Vol. III* (1962) *Physical and Chemical Tables*. Kynoch Press, Birmingham, England.
James R. W. (1958) *The Optical Principles of the Diffraction of X-rays*. 5th ed. Bell, London.
Joffé A. (1928) *The Physics of Crystals*. McGraw-Hill, New York.
Kittel C. (1956) *Introduction to Solid State Physics* 2nd ed. Wiley, New York.
Krishnan R. S. (1958) *Progress in Crystal Physics, I.* Viswanathan, Madras. (Many references.)
Landolt-Börnstein (1955) I (4) *Kristalle*. Springer, Berlin.
Megaw H. D. (1957) *Ferroelectricity in Crystals*. Methuen, London.
Mott N. F. and Gurney R. W. (1953) *Electronic Processes in Ionic Crystals*. University Press, Oxford.
Nye J. F. (1957) *Physical Properties of Crystals*. Clarendon Press, Oxford.
Roberts J. K. (1947) *Heat and Thermodynamics*. Blackie, London.
Schmid E. and Boas W. (1950) *Plasticity of Crystals* (English ed.) Hughes, London.
Seitz F. (1940) *Modern Theory of Solids*. McGraw-Hill, New York.
Shockley W. (Ed.) (1952) *Imperfections in Nearly Perfect Crystals*. Wiley, New York.
Shubnikov A. V. (1958) *Osnovy Optichiskoy Kristallografie*. Acad. Sci. USSR, Moscow.
Wood R. W. (1948) *Physical Optics*. Macmillan, London.
Wooster W. A. (1938, repr. 1949) *A Textbook on Crystal Physics*. University Press, Cambridge, England.
Idem (1953) Physical Properties and Atomic Arrangements in Crystals. *Rep. Prog. Phys., 16*, 62. Phys. Soc. London.
Voigt W. (1910, repr. with addl. appendix 1928) *Lehrbuch der Kristallphysik*. Teubner, Leipzig.
Wasastjerna J. A. (1922) *Soc. Sci. Fen. Comm. Phys. Math. 1*, 38; ibid. *6*, 18, 19, 21; (1923) *Z. phys. Chem. 101*, 193.
Zwikker C. (1954) *Physical Properties of Solid Materials*. Pergamon Press, London.

CHAPTER 15

Dynamical X-ray Optics; Electron and Neutron Diffraction

P. P. Ewald

15.1. Inadequacy of the Kinematical Theory

Laue's original theory of X-ray diffraction was characterized in Ch. 6 as a kinematical theory. Just as in mechanics the sub-section of kinematics deals with the analysis and superposition of given motions, regardless of any forces producing them, so this theory only considers the result of superposing given elementary wavelets which are derived from an incident or primary wave by its successive scattering on the atoms. No further scattering is considered which the secondary rays might produce on their passage through the crystal, and the omission of this reaction entails immediately an infringement of the law of conservation of energy. Each secondary ray, generated by 'constructive interference' of the wavelets, transports energy in its own direction, and this energy flow must obviously be taken out of that of the primary ray. On the assumption of the kinematical theory, however, the strength of the primary ray is not affected by the scattering, so that energy is not withdrawn from that source. For this reason it cannot be expected that the kinematical theory renders correct account of the intensities to be found in the secondary rays.

This deficiency of the kinematical theory was not apparent from Laue's first experiments, because the primary beam seemed to have such an overwhelmingly large intensity. With a scattering action weak of the first order, the reaction on the primary beam would be small of the second order, and indeed, the kinematical theory may be safely applied if, as in a fine powder, the individual crystal is so small that its total scattered intensity remains a very small fraction of the intensity incident on it. Only with the use of X-ray tubes giving strong monochromatic X-rays did it become certain that a diffracted (or 'reflected') ray was not as weak as it appeared to be from the Laue diagram. It was recognized that the central spot of these diagrams contains the entire spectrum of the incident radiation and that, referred to the same spectral range that is contained in the secondary ray, this and the primary ray are of similar intensity.

15.2. Darwin's Theory

Comparable intensity of incident and reflected ray is assumed in C. G. Darwin's very early dynamical theory (1914).

Considering first a single plane of atoms, Darwin calculates a reflection coefficient for it by the classical considerations by which J. J. Thomson determined the scattering by a single electron. This reflection coefficient is then applied to the back and forth reflections of the two beams, primary and reflected, between the equally spaced atomic planes of the crystal. The theory has many points in common with that of light falling on a plane-parallel glass plate, for instance in the Lummer-Gehrcke interferometer. Its result is a curve for the reflected and transmitted amplitudes as functions of the angle of incidence on the crystal plate. In the symmetrical Bragg case, i.e. when the reflecting planes are parallel to the crystal surface, total reflection occurs over a small angular range of incidence of the order of 10 seconds of arc, next to the Bragg angle; here the crystal acts as a no-pass filter to the incident ray. In the adjoining angular regions the reflectivity drops rapidly to zero. The centre of the region is shifted slightly from the position indicated by Bragg's Law $n\lambda = 2d \sin \theta$ as usually applied, namely by taking for λ the X-ray wave-length in free space, which we will now call λ_0. Instead, λ should be understood as the wave-length in the crystal, that is λ_0/μ, where μ is the refractive index for X-rays in the body. Normally, μ is smaller than 1, $\mu = 1 - \varepsilon$, where ε is a positive quantity of the order of 10^{-6}, so that the wave-length and phase velocity are greater in the body than in empty space by this amount. Measurable deviations from Bragg's Law result because the refractive index varies from one order to another; they were noticed when the spectroscopists derived different wave-lengths for the same line reflected in different orders by applying the uncorrected Bragg Law.

In the symmetrical Laue case, i.e. when the reflecting planes are at right angles to the surface of the crystal plate, the primary and secondary ray emerge at the underside of the plate equally inclined. Again there exists an angular range of some seconds of arc in which reflection occurs; but instead of reflectivity 1 as in total reflection, the maximum cannot exceed $\frac{1}{2}$ and occurs exactly under the uncorrected Bragg angle of incidence.

Darwin's paper is a masterpiece of physical insight into what is essential for explaining the features of X-ray diffraction by perfect crystals, and all later work, experimental and theoretical, has vindicated it. It grew out of the reflection idea current in England in 1913/14 which was not fully accepted in Germany. It could not convince the German physicists for two reasons. Thinking in terms of diffraction rather than reflection, they were conscious of the fact that a single plane of atoms, owing to its periodic structure, would produce a whole pencil of cross-grating spectra besides the specular reflection which was exclusively taken into consideration by Darwin. The second reason was the fact that in the Laue arrangement, then currently used in Germany, the incidence was generally taken along a direction of symmetry and therefore not a single reflected beam but all the symmetrically related ones were produced simultaneously. Darwin's treatment was not at all well adapted to deal with this case. It thus came that Ewald developed a dynamical theory along more general lines without fully recognizing the significance of Darwin's work, except when in the last stage the results of both theories in the case of only two rays turned out to be the same.

15.3. Ewald's Dynamical Theory (1917)

The construction of the directions of the 'strong' diffracted rays by means of the sphere of reflection shows that if this sphere passes through n points of the reciprocal lattice, nothing would be changed if each of these points in turn were assumed to be the origin of this lattice, and the wave-vector pointing to it were that of the primary ray. In other words, given the wave-length λ_0 which determines the radius of the sphere of reflection, and the centre of the sphere at the 'tie-point' T, the n plane waves form an inseparable unit which we call the X-optical field. If for the moment we considered only one of the component waves of this field to travel through the crystal alone, it would, in doing so, promptly generate the other $(n - 1)$ waves. There is, at this stage, no meaning attached to considering one of them as the 'primary' wave. The problem then arises of setting up the condition for the entire X-optical field to travel through the crystal in a self-consistent way. This means that at any atom the field should produce a scattered wavelet of exactly the amplitude and phase that are needed in order to build up the field as a sum over all the wavelets. As a result of this fundamental condition the tie-point T cannot be chosen arbitrarily; it has to lie on a surface in reciprocal space, the 'Surface of Dispersion'. Its particular shape extends only through that very small region of this space where the n chosen waves remain the only strong component waves of the X-optical field. This condition is not critical in so far as, with the tie-point moving farther away, the interaction between some of the waves becomes so weak that it becomes unnecessary to take it into account. In some cases, however, a new 'strong' wave-component of the field might turn up which should be considered.

Once a tie-point T is chosen within this region and on any part of the surface of dispersion, the wave-vectors which go from T to the n lattice points of the reciprocal lattice are precisely determined regarding both length and direction, and they change with a shift of T on the surface of dispersion. This means that for each dynamically possible position of T the component waves have well-determined directions and phase velocities or refractive indices, but there is not a refractive index *of the medium* which would determine the velocity of all the waves.—The surface of dispersion contains one further important information, namely the relative amplitudes of all the component waves of the field. These depend on the position of T; in fact, the amplitude of a component wave h is proportional to $(K_h^2-K_o^2)^{-1}$, where $\mathbf{K}_h$ is its wave-vector T→h and $\mathbf{K}_o$ the wave constant of the elementary wavelets. Since this amplitude is extremely sensitive to small changes in the length of $\mathbf{K}_h$, the restriction of T to the surface of dispersion is the means of adjusting the amplitudes of the component waves of the field so that it becomes self-consistent.

It can be shown generally that in a region of reciprocal space giving rise to n strong waves the surface of dispersion consists of 2n sheets, the factor 2 corresponding to the transversality of the waves. Thus if for one wave-vector, say $\mathbf{K}_1$, the direction is prescribed, 2n different lengths of this vector are dynamically possible with tie-points T^1 to T^{2n}, whereas the wave-vectors leading to the other points of the reciprocal lattice have each not only 2n different lengths but also slightly different directions. The differences in length and directions are minute since the spread of the tie-points is over a distance about 10^{-5} to 10^{-6} of the length of the wave-vectors themselves, yet they are the all important feature because of the dependence of the amplitude ratios on the shifts of the tie-points. Needless to say, each possible field can be given an arbitrary overall amplitude attached to the corresponding tie-point.

Fortunately the knowledge of the dynamically possible fields in the interior of the unbounded crystal—the 'meso-fields'—is sufficient for the construction of the dynamically balanced state of a crystal bounded by a plane surface on which the 'incident' wave falls. The way this comes about is the following.

The field in the unbounded crystal, which is represented by the tie-point T, is the sum of the spherical wavelets issuing from the atoms or dipoles filling *all* space. If the summation is limited to the wavelets issuing from the lower half-space only (the 'half-crystal') the same field,—the meso-field—,is again generated inside the crystal, but there are additional waves created both inside and outside it ('epi-waves'). This is a result only of breaking off the summation. All epiwaves have phase velocity c as in free space. The internal epiwaves stand in a very close relation to those mesowaves which move from the surface towards the interior of the halfcrystal; in fact, each of these mesowaves is accompanied by an epiwave of the same amplitude (but of opposite phase) and a direction differing only by a refractive correction corresponding to its phase velocity c or wave-vector of length k_0. The relation of the external epiwaves is of the same kind with respect to those mesowaves which travel from the interior of the half-crystal towards its surface. These waves exist only outside the crystal, and they are there the continuation (except again for a slight refraction) of the mesowaves running up against the surface.

Clearly the internal epiwaves, as well as the incident wave which has to be superimposed, disturb the self-consistent régime which found its expression in the surface of dispersion. The mere fact that these waves have free space velocity c in the interior of the body shows that they do not belong there. If their elimination could be achieved by the superposition of self-consistent fields of suitable strengths, then a dynamically balanced state would remain over throughout the crystal. This is exactly what can be achieved with 2n conditions of annihilation of unacceptable waves and 2n overall amplitudes that can be chosen for the wave-fields represented by the 2n tie-points.

In this solution, in each of the main directions of diffraction **h** a bundle of waves progresses with very nearly equal wave-vectors $T^1\mathbf{h}$, $T^2\mathbf{h}$, . . $T^{2n}\mathbf{h}$. These lead to multiple beats between the constituent waves, the spatial period of the beats being determined by the differences T^2—T^1, T^3—T^1, . . which are very small compared to the length of, say, $\mathbf{K}_1 = T^1\mathbf{h}$, so that the change of amplitude in any one of the directions of diffraction takes place slowly. Through these beats the energy flow is shifted from one direction of diffraction to the others, and possibly back again; for this reason this type of solution was called the pendulum type solution (Pendellösung). It will be noticed that the beat periods, which are given by the above differences T^2—T^1 etc. are the same for all orders of diffraction, i.e. for all values h. At the surface the energy flux has of course at first the direction of the incident wave and the diffracted waves moving towards the interior start with amplitude zero; they pick up amplitude as they advance into the interior, while the primary wave is diminished. The details of the energy transfer, the beat period, the maximum amplitude reached by each diffracted wave, and, in the case of a plane parallel crystal slab of thickness D, the distribution of amplitudes among the diffracted rays emerging from the plate—all this depends on the exact direction of the incident wave and its wave-length and can best be visualized with the help of the surface of dispersion.

15.4. Laue's Form of the Dynamical Theory

Before discussing the simplest applications of this theory, namely to the cases of a solitary ray and of a primary and one secondary ray, the form should be considered which Laue gave to this theory in 1931. The main innovation lies in the model of the crystal, which to Laue is a three-dimensionally periodic dielectric, i.e. locally polarizable, medium. Laue does not concentrate the polarizability at certain points of the cell, as was the case of the dipoles of Ewald's theory, but considers a continuous distribution. M. Kohler, in 1935, derived Laue's assumption by a perturbation method from a wavemechanical model of the crystal.

Actually the models are not fundamentally different since Laue could assume the scattering power to be localized at certain points, or Ewald could fill the cell with a suitable distribution of dipoles which in the limit form a continuous distribution. The main difference then is one of technique: Laue makes use of the Fourier development of the dielectric constant from the start, whereas with Ewald this appears only at a later stage in form of the structure factor of the polarizability distribution in the cell. A difference of perhaps greater significance occurs when the 'half-crystal' is considered. Its surface is obtained in Laue's theory as a plane cutting through the continuous mass distribution, and boundary conditions are formulated which must be fulfilled at all points of this plane. In Ewald's half-crystal the surface plane is defined by the last point-atoms, but between them it has no physical reality. There is no 'boundary condition'; instead the condition of self-consistency holds at every atom, no matter how deep in the interior it is situated. Both theories seem to give the same results and the choice between them is a matter of liking.

15.5. One and Two Rays in the Dynamical Theory

The first case, that of a solitary X-ray in the crystal, occurs if the sphere of reflection passes close to the origin of the reciprocal lattice *only*, without approaching any other lattice point (Fig.15–5(1)). In contrast to the assumption of the kinematical theory that the phase velocity is c (the value for free space), and the length of the wave-vector $k_o = \nu/c$, we have to allow the solitary wave to travel in the body with a different velocity, q, and wave-vector of length $K_o = \nu/q$,

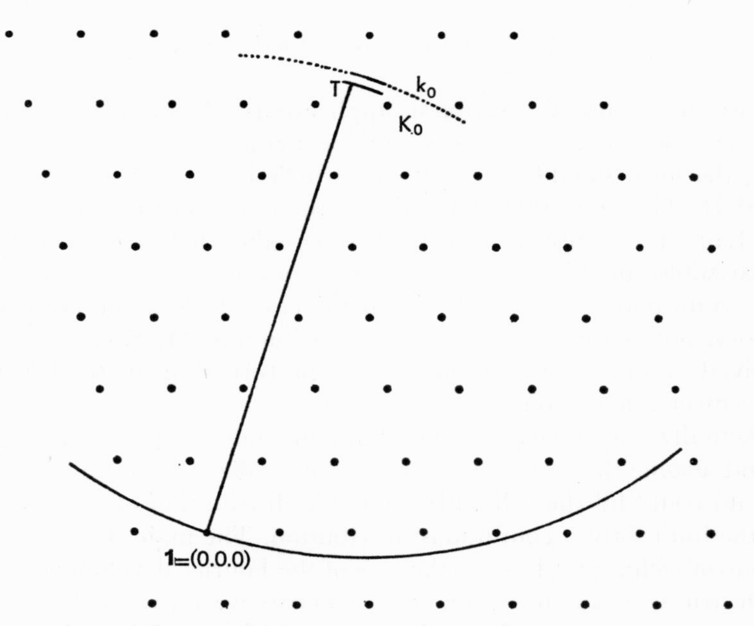

Fig. 15–5(1). Reciprocal lattice and single ray.
Only the origin (000) = **1** lies close to the sphere of reflection. The circle shown dotted in parts, of radius k_0, would be the locus of the tie-point T according to the kinematical theory. The small length of the inner circle, of radius K_0, is the surface of dispersion of the ray as long as it remains solitary.

which is determined by the polarizability of the medium. The 'surface of dispersion' for this solitary wave is then a sphere of radius K_o and is valid for all those directions of propagation of this ray for which no secondary rays are generated. Any point on this surface may be taken as tie-point T about which the sphere of reflection, of radius k_o can be drawn. This no longer goes through the origin of the reciprocal lattice, but passes it at a distance $k_o - K_o$ which in general is of the order of $10^{-5}k_o$, corresponding to a departure of the refractive index from 1 by about 10^{-5}.

Consider now the case of *two* strong wave-components of the optical field. This happens in positions of the tie-point T which are approximately at distance k_o also from a second point, **h**, of the reciprocal lattice. If we draw the surface of dispersion of radius K_o about **h**, as

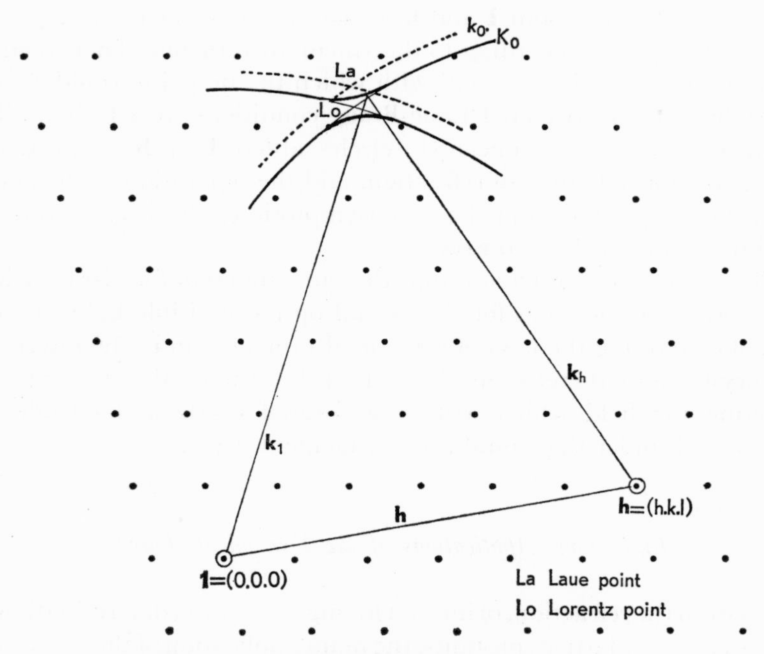

Fig. 15–5(2). Surface of dispersion for two rays.
Dotted circles, radius k_0: locus of tie-point for wave-vectors ending in reciprocal lattice points **1** and **h** according to the kinematical theory. Intersection at Laue point La.

Drawn-out circles: same for dynamical theory assuming two uncoupled rays (radius K_0). Intersection: Lorentz point Lo. The coupling of the two rays produces the splitting-up of the circles and interconnection as shown. This is the true surface of dispersion on which the tie-point has to lie. Shown for one case of polarization only.

N.B. In this and the preceding figure the difference between k_0 and K_0 should be one part in 10^5 or 10^6, instead of one in twenty-five as drawn.

well as about the origin **1** (= 000), the point of intersection of the two spheres, in a given plane, is called the Lorentz point; it is that point where the two rays would co-exist except for their interaction. On account of the latter, however, the surfaces split up and are connected in a way that leaves the Lorentz point the centre but not a point of the surface of dispersion near this point. Fig. 15–5(2) shows the surface for one mode of polarization only, for instance when the electric vectors in both rays are normal to the plane of the rays. (When they lie in this plane, the splitting up of the surface would be slightly less.) In the figure the splitting is shown on a highly magnified scale; actually the minimum distance of the two sheets should be only about 10^{-5} of the lengths of the wave-vectors $\mathbf{K}_1 = T\,\mathbf{1}$ and $\mathbf{K}_h = T\,\mathbf{h}$, respectively. The spheres of radius K_0 are the asymptotic surfaces of the surface of dispersion. The drawing also shows the two (dotted) circles of radius k_0 about **1** and **h** which are the surfaces of dispersion of the solitary rays according to the kinematical theory. Their point of intersection is the 'Laue-point' with which the tie-point would have to coincide if the uncorrected Laue-Bragg conditions were to hold. The distance between the concentric circles about **1** or **h**, respectively, shows the normal effect of refraction, and the separation of the actual surface of dispersion from the two asymptotic circles is a measure of the interaction of the two rays.

The surface of dispersion fulfills the same function for X-rays which the 'normal surface' has for the crystal optics of visible light, namely that of containing the laws of wave-field propagation in the interior of the crystal and thereby offering an easily visualized means of constructing the field which will be generated inside and outside the half-crystal under the stimulus of an incident wave.

15.6. Some Applications of the Dynamical Theory

Apart from the refraction, of interest for high-precision determinations of wave-lengths or lattice constants, the main application of the dynamical theory is to the discussion of the intensities of diffracted rays. The usual test of the theory is performed on the symmetrically reflected ray with the reflecting atomic planes parallel to the crystal surface. In the ideal case of a simple lattice of non-absorbing point atoms, the reflection would be total within a certain angular range of incidence of the order of 5–10″, and drop off quickly outside this range. The total, or integrated, intensity which is taken out of a beam of wider opening, or by turning the crystal through the reflecting position, is proportional to this angular range which varies with the angle of diffraction and is different for the two cases of polarization. The existence of a base in the cell makes the range of total reflection proportional to $|F_h|$, not to $|F_h|^2$ as in the kinematical theory. Absorption in the crystal modifies still further the expression for the integrated intensity.

Within the resolving power of the experimental methods of Bergen Davis, Compton and Allison, Coster and Prins, L. Parratt, J. A. Bearden, M. Renninger and other specialists, the predictions of the dynamical theory concerning the range, shape and intensity of the reflection curve have been substantially confirmed on selected crystals. Renninger showed that even rocksalt, from which the standard data of mosaic crystal reflectivity were taken, gives rise to the totally different reflection curve of the perfect crystal provided only a few square millimeters of the cleaved surface of a crystal grown from the melt are

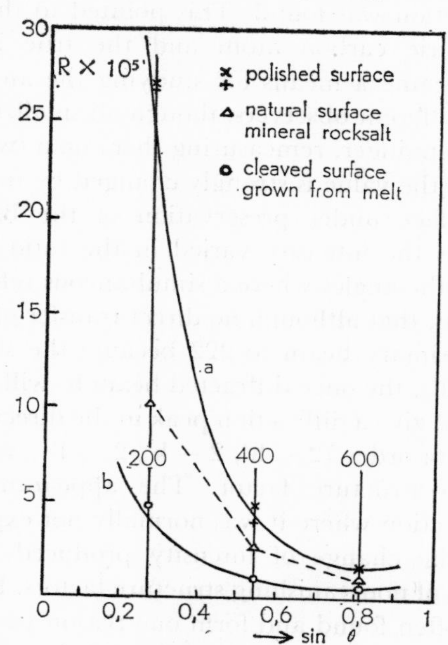

Fig. 15–6(1). Reflectivity of rocksalt for CuK-radiation.

a theoretical curve for mosaic crystal;
b same for perfect crystal;
×+ natural crystal, standard treatment (ground and polished);
△ natural crystal, untouched cleavage;
○ crystal grown from melt, untouched cleavage, small area.
After M. Renninger, *Zs. f. Krist.*, A 1934, *99*, pg. 368.

used for the measurement. The curves showing the reflection for the perfect and the ideal mosaic crystal show the great increase in intensity of the mosaic crystal over the perfect one (Darwin's discovery), and the measured points show that either of these curves can be very closely approached, and that a crystal taken from a salt mine may give values anywhere between the two extremes (Fig. 15–6(1)). For this reason the accurate determination of Fourier coefficients from measured intensities requires a preceding study of the perfection of the individual crystals on which the measurements are being made.

The dynamical theory is also essential in accounting for the changes of intensity of a reflection of order **h** when there occurs, *simultaneously*, a reflection of order **h**′ on a different set of planes. The first example on which this was studied was the 'forbidden' reflection 222 of diamond. The fact that this reflection was missing gave W. H. and W. L. Bragg the clue to the structure of diamond in 1913, but on close inspection a trace of the reflection was found. This pointed to the existence of a non-centrosymmetric carbon atom and the true intensity of this reflection thus became a means for studying the atomic shape. The values found by different observers, though all small, stood not in good agreement and Renninger, remeasuring them on a twocrystal spectrometer, found that the value is strongly changed by turning the crystal in its reflecting face under preservation of the Bragg angle. For different azimuths the intensity varied in the ratio of about 1 : 10, showing peaks at the angles where a simultaneous reflection occurred. The explanation is, that although no direct transfer of energy can take place from the primary beam to 222 because the structure factor is zero (or very small), the once diffracted beam **h**′ will act like a second primary beam and give a diffraction peak in the direction of 222 which is now, however, of order (2 — h′, 2 — k′, 2 — l′), and therefore may have a non-zero structure factor. The appearance of diffracted intensity in a direction where it was normally not expected was easier to detect than the change of intensity produced by simultaneous reflection in cases of non-vanishing structure factors. But these have by now been quite often found and form one reason why the 'reliability index R' (see Chapter 7) cannot be reduced below a minimum value.

A very remarkable confirmation and extension of the dynamical theory began in 1948 with the first observations by G. Borrmann of what is now called the *Borrmann Effect*. This is the surprising fact that an absorbing, near perfect crystal suddenly becomes transparent when a secondary ray is split off. The effect is very sensitive to imperfections and has been most successfully observed in selected germanium, silicon and calcite crystals. There remains, even in the position of greatest transparency, a considerable absorption, but the reduction of intensity is by many powers of ten less than that expected by applying the ordinary absorption coefficient which indeed may prevent any measurable penetration of the crystal by X-rays outside the Bragg reflection position. The angular opening of the transmitted ray is very small, and A. Authier in Paris, by placing a slit between two crystals showing the Borrmann effect, managed to isolate a monochromatic beam of an angular opening of 1/50″. The emerging beam is strongly polarized through the Borrmann effect, so that a pair of suitably cut germanium crystal plates has been used by H. Cole as a polarizer.

M. v. Laue gave the surprising explanation of the effect by discussing the results given by the dynamical theory after the introduction of a complex dielectric constant; the real part of this gives the refraction, and its imaginary part absorption. It turns out that just as the normal refractive index of a solitary wave suffers changes when the coupling with another ray sets in, so also does the absorption index: that sheet of the surface of absorption which comes closest to the 'Laue point' (see above) gives rise to the greatest reduction of absorption, while on the sheets which are pushed away from the Laue point absorption is enhanced. In travelling through a thick crystal only

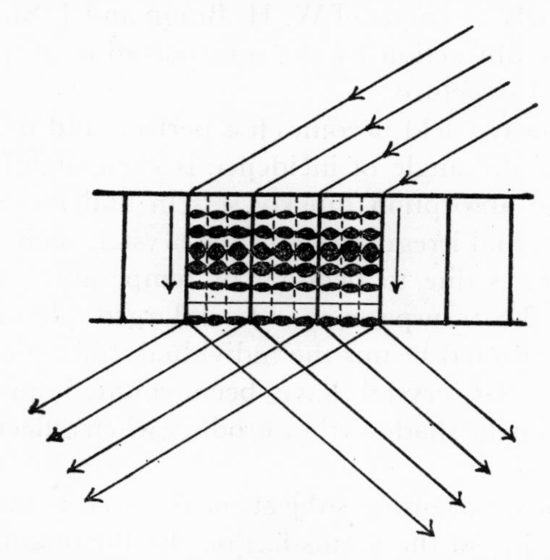

Fig. 15–6(2). Schematic drawing of wave-field in Borrmann effect. Between the reflecting planes (full lines) a standing-wave pattern is progressing downwards. The nodal lines between the atomic planes are shown dotted. The wave-field splits up at the exit surface.

the least absorbed field survives, and this is represented by the tie-points lying closest to the Laue point. In the case of two rays, and four sheets of the surface of dispersion, the closest approach occurs for that case of polarization where the coupling between the rays is strongest, that is when the electric vector is normal to the plane of the rays. This explains the polarizing action of the effect.

In analysing the field inside the crystal it is seen that at the peak position of transparency in the symmetrical Laue case it consists of a wave travelling downwards along the reflecting net planes of atoms while across these planes the field has a standing-wave character. The planes of the atomic lattice are interleaved by nodal planes of the field on which the electric vector is permanently zero in the optimum case, or else remains small. If the absorbing atoms, for instance the calcium atoms in calcite, lie on these planes they are permanently situated in zero, or small, electric field and no photoeffect or absorption happens. The remaining absorption, in the case of calcite, would be that of the CO_3 groups, which is only a fraction of that of the heavy atoms.

The study of the optical field in the Borrmann effect led Laue to investigate the energy flow. Since no energy can be transported across the nodal planes of the standing wave, the flow of energy takes places along the reflecting planes. It was shown experimentally by Borrmann that only at the exit surface of the crystal the two wave-components of the internal field separate. In a rather sophisticated way this result vindicates the early attempts of W. H. Bragg and J. Stark (see Ch. 5) to explain X-ray diffraction by the transportation of 'photons' in the pits of the crystal structure.

The standing-wave field becomes less perfect, and moves relative to the net-planes if the angle of incidence is even slightly altered, and with this change absorption quickly sets in and renders the crystal opaque. Any internal irregularities of the crystal, such as dislocations, the internal stresses due to bending or temperature gradients, etc. locally destroy the transparency. On enlarged photographs of the transmitted or reflected beams the individual dislocations in carefully annealed Si and Ge crystal have been counted and their shapes reconstructed from the shadows they produce when reflection takes place on different planes.

One of the most promising subjects of research is the influence the temperature motion of the atoms has on the Borrmann effect.

15.7. Electron and Neutron Diffraction.

That diffraction effects can be obtained with particle streams is due to the fact that particles of mass m, velocity v, and momentum p have to be associated with a wave of wave-length $\lambda = h/p$, where h is Planck's constant $6.62 \cdot 10^{-27}$ erg·sec.

This relation between particle and wave was first used by L. de Broglie in 1924 to show that the wave associated with an electron travelling on the Bohr orbit in an atom would just fit on to the orbit with an integer number of wave-lengths. Out of this idea arose, in 1926, Schrödinger's theory of wave-mechanics. In 1925 W. Elsasser first pronounced the likelihood that a beam of electrons could suffer diffraction in a crystal similar to X-rays, provided it had a velocity producing a suitable wave-length for the crystal lattice. In 1936, four years after the discovery of the neutron by James Chadwick, Elsasser applied the same idea to neutrons and the first experiments for verifying it were made together with H. v. Halban and P. Preiswerk using a radium-beryllium neutron source.—Electron diffraction was first obtained independently by C. J. Davisson and L. H. Germer at the Bell Telephone Laboratories and by G. P. Thomson in England in 1927. Since the first demonstration of these effects, an enormous amount of technical development of the instrumentation, of highly successful research, and of theoretical advancement has taken place in both branches of particle diffraction. Both subjects, closely linked as they are to X-ray diffraction, would deserve full chapters in this book; they are mentioned here as a *pro memoria* rather than as an adequate treatment, and electron diffraction in particular as a subject requiring the dynamical theory in a much higher degree than X-rays.

The construction of reactors towards the end of the last war provided neutron sources of incomparably greater strength than the radioactive ones used earlier. A neutron flux of 10^{13} to 10^{14} neutrons per square centimeter and second made possible all the experimental methods which had been developed for X-rays, although much larger crystals and counters had to be used because of the much smaller interaction of neutrons with matter than of X-rays (ten to twenty times smaller for 'thermal neutrons', that is those which have been slowed down, mainly by collisions with the hydrogen atoms in paraffin, to an average equivalent wave-length of about one Ångström).

For electrons, on the other hand, the interaction with matter is so much stronger than for X-rays that the scattered amplitude for the same wave-length is greater by a factor of the order of 10^4. The 'optical density' ($\mu^2 - 1$), where μ is the refractive index, is therefore much larger for electrons than for X-rays, where it is of the order 10^{-4} to 10^{-6}. For a solitary electron beam $\mu^2 - 1 = \phi/V$, where ϕ is the average electrostatic potential in the crystal (perhaps 6 Volt) and V is the accelerating potential through which the electron gained its energy, namely about 130, 15000 and 40000 Volt for electrons of wavelengths 1, 0.1, and 0.06 Ångström, respectively. Thus the optical density for these three wave-lengths is of the order of 1/20, 1/2000, and 1/6000, respectively. As a consequence of this much stronger interaction the validity of the kinematical theory of diffraction ends much earlier than for X-rays, and the dynamical theory has to be applied. The low-energy electrons, like X-rays of long wave-lengths, are strongly absorbed and wave-lengths of 0.1 Å and below are preferred for electron diffraction, except in surface investigation. This leads to a larger radius of the sphere of reflection, and therefore in the interior of the crystal to an electron-optical field consisting of a great number of component waves of comparable strength, instead of two or three waves which suffice in most cases for the discussion of X-optical fields.

The diffraction of electrons is caused by the periodic electric potential within the crystal. H. Bethe gave the theory in his thesis in 1927 by developing the potential in a Fourier series and expressing the results in terms of the Fourier coefficients—the same procedure which was followed by Laue in 1931 in his form of the X-ray theory. The main physical difference between the two cases lies in the atomic factors: electrons are scattered by the charges of the electrons *and* of the nucleus of each atom, whereas the nucleus does not scatter X-rays.

In the neutron case, the scattering is an interaction between the two heavy particles, the neutron and the nucleus, and electrons play an indirect role by creating magnetic fields. Neutron scattering without change of wave-length ('classical scattering') consists of two parts: one is the short-range neutron-nucleus interaction and the other the result of the magnetic interaction of the neutron spin and the magnetic field of the atom which is due partly to the nuclear spin and partly to the orbital and spin moments of the electrons.

It would lead too far to give a full account of these newer applications of the diffraction principles. In both cases the existence of the experimental and theoretical methods evolved for X-rays has cut the time of development down to a fraction of that spent on X-ray methods. On the other hand, the peculiar features of both new fields have stimulated new developments, especially in theory, from which in turn the X-ray case profits. Never before has the dynamical theory been discussed to such depth and detail as in the great spate of papers that has been coming in the last ten years from the Japanese school of electron diffraction which goes back to S. Kikuchi. At a recent international conference on electron diffraction in Kyoto (September 1961) it became evident that in Japan alone there must be more physicists familiar with this theory than in all other countries taken together. The theory was shown to be essential not only for diffraction experiments as such, but also for the full explanation of the intensity distribution in electron-microscopical images. Apart from Japan, this study has found a home also in Australia (Melbourne), where J. M. Cowley, in collaboration with A. F. Moodie, has developed a new theoretical approach which is often better suited to deal with the strong interaction and the many-wave fields in electron diffraction.

Neutron diffraction opens up new means of crystal structure analysis because the scattering power of atoms for neutrons differs radically from the simpler distribution of atomic factors for X-rays which go roughly with the atomic number or weight. Thus hydrogen gives an amplitude of coherent neutron scattering of nearly the same size as potassium but of opposite phase, and deuterium scatters with nearly twice the amplitude and the same phase as potassium. This strong scattering makes the detection of the hydrogen positions easier with neutrons than with X-rays. In non-magnetic neutron diffraction there is furthermore no dependence of the scattered amplitude on the scattering angle such as leads in the X-ray case to the complication of the atomic form factor and the rapid decrease of intensity in higher orders of diffraction; the size of the nucleus is so small compared to the neutron wave-length that it acts as a true point-scatterer. Since isotopes may have entirely different scattering powers for neutrons (as hydrogen and deuterium) a new type of 'heavy atom method' for crystal structure analysis by the substitution of chemically identical atoms becomes possible.

The greatest innovation due to neutron diffraction is, however, the intimate study of the repartition of the magnetic moments among the atoms of a crystal. It is possible to 'polarize' a neutron beam, that is, to give the spin of the neutrons a parallel orientation by reflection on a single magnetic crystal such as magnetite or an iron-cobalt alloy. Since the amplitude of the magnetic scattering by an atom depends on the angle between its spin or magnetic field and that of the neutron, it is possible to investigate the distribution of magnetic moments in a crystal diffracting the polarized beam of neutrons. This method has opened up a new vista in structure determination which now reaches one level deeper than the knowledge of atomic sites and of electron distribution. Charts have been obtained for the repartition of the unpaired electrons alone (those creating the magnetic field) and of atomic spins. The classical structure theory of Schoenflies and Fedorov has to be enlarged by the discussion of the possible repartition of equal particles carrying a vectorial or even higher order property besides the insignificant property of a scalar mass. A beginning of this was made by N.V. Belov with the study of black-and-white and coloured space groups, but much remains to be done.

Apart from the 'coherent' scattering of electrons, neutrons and X-rays which leads to the formation of diffracted rays in sharply defined directions, there is also an 'incoherent' scattering in which part of the energy is transferred from the incident wave to the crystal where it augments the thermal elastic waves in which the thermal motion of the atoms resides. This incoherent scattering is not confined to the directions of the secondary rays but fills the intermediate directions with a continuous and generally not uniform background of scattered radiation. In the case of X-rays the individual collision (scattering) process between the (X-ray) photon and the (elastic wave) phonon transfers so little energy that the change of wave-length in the scattering process can be neglected. But owing to the large mass of the neutron a much greater energy is involved in the transfer, so that the change of wave-length becomes measurable and permits to sort the scattered neutrons according to their energy loss. This has become an important method for exploring the elastic spectrum of crystals, so fundamental for many of their properties and so hard to find by other methods.

CHAPTER 16

X-ray Spectroscopy

by Manne Siegbahn

16.1. Early History

Before the discovery of the diffraction of X-rays in crystals some very important studies had been made of the quality of these rays by measuring their absorption in different materials. Especially Barkla had been able, by this rather simple method, to find some fundamental properties of the X-rays when they were scattered by different elements. It was known that the radiation leaving an X-ray tube was heterogeneous and dependent on the material of the anti-cathode. After passing a series of e.g. aluminum foils the 'softer' components of the radiation were successively reduced and finally the beam consisted of a rather homogeneous 'hard' radiation. The hardness of this radiation depends on the voltage applied to the X-ray tube. When such a homogeneous X-ray beam hits a plate of some element three kinds of *secondary* radiations are emitted. One part is scattered radiation with the same quality as the incoming one. The other two parts are characteristic of the element of the secondary radiator. One of them is an X-radiation, the other consists of negative particles, electrons. By systematic studies of this *characteristic* X-radiation from different elements in the periodic system Barkla distinguished two series of homogeneous X-radiation which he called the K- and the L-fluorescent radiations. Both these radiations become stepwise harder when the atomic weight of the 'radiator' increases. For the same element the emitted K-radiation is about 300 times as penetrating as the L-radiation. A condition for the appearance of these fluorescent radiations is that the incoming beam is somewhat harder than the characteristic radiation of the element in question.

The emission of the electronic component of the secondary radiation was shown to be closely linked to the K- and L-series.

Several important facts about the characteristic radiations were revealed empirically in these early researches on X-ray spectra, which formed the basis for the more quantitative and detailed studies made possible by using the diffraction of the X-radiation.

As mentioned above, these earlier researches on characteristic X-radiation were carried out by using the fluorescence method with an ordinary X-ray tube as primary source. At this time the X-ray tubes, with cold cathode, mostly had an anti-cathode of platinum or platinized plates of copper, nickel, iron or silver. But in some cases experiments were also made with X-ray tubes of laboratory type having interchangeable anti-cathodes. The characteristic radiation sent out as 'primary' rays had a much higher intensity than could be obtained by secondary excitation. When the diffraction of X-rays was introduced for the analysis of the X-ray spectra the primary method remained for several years the usual method of producing the characteristic radiation from the different elements.

16.2. The Advent of the Diffraction X-ray Spectroscopy

The discovery of the diffraction of X-rays had solved the old controversy regarding the nature of these rays in favour of the hypothesis that they were electromagnetic waves. This meant that the X-ray spectra are of the same character as the optical spectra except that the wave-lengths are far shorter than even in the ultraviolet region of the optical spectroscopic field. We now know that the X-ray spectra have wave-lengths from hundred Ångström to tenths of an Ångström.

In the first stages of the development of X-ray spectroscopy by means of diffraction in crystals, the radiation from ordinary technical X-ray tubes was registered. W. H. and W. L. Bragg (1913) used for the analysis a goniometer, where a rocksalt crystal was mounted on the rotating table. An ionization chamber on a turnable arm served for the registration of the 'reflected' X-ray beam. The distribution of the intensity as a function of the angle of incidence then showed, in two or three orders, three rather broad peaks on a background of the 'white' X-ray spectrum; these corresponded to the L-series of platinum which constituted the anti-cathode.

In France, Maurice de Broglie in the same year (1913) registered on a fixed photographic plate the X-radiation reflected by a rotating crystal. In this way X-ray spectra with sharp and well defined lines of the same general appearance as those of optical spectra were obtained. Besides the emission lines formed by the characteristic radiation of the anti-cathode, these spectral plates also showed two sharp discontinuities at fixed wave-lengths which seemed independent of the tension on the X-ray tube. The interpretation of these sudden changes of the blackening in the plates was easy to give on the basis of Barkla's earlier researches, which had shown that the excitations of the K- and L-series are connected with a strong increase in the absorption of the incoming rays. The two discontinuities in the blackening observed by de Broglie on his spectral plates were due to the K-absorption of Ag and Br in the photographic emulsion.

These very first studies of the characteristic X-radiation by means of crystals showed that the K- and L-series each consisted of several well defined spectral lines placed close together, with a very wide gap between the wave-lengths of the two groups. As just mentioned the 'absorption spectra' appeared on the spectral plates as sharp, well defined 'absorption edges' connected with the K- and L-groups: a single edge for the K- and three edges for the L-series.

16.3. X-ray Spectra and Atomic Structure

It may be considered a fortunate accident that the new X-ray spectroscopic technique was ready for application at a time when the foundations for the understanding of atomic structure were just in their first state of development. Two years earlier Rutherford had presented good evidence for the hypothesis that the atoms are constituted of a minute positively charged nucleus, less than 10^{-12} cm in size, surrounded by a cluster of negative electrons with a total charge equal to the positive charge of the nucleus. Bohr (1913) was the first to apply the quantum theory of Planck to this atomic model and to give a quantitative theory for the simplest of all optical spectra, that of hydrogen, where just one electron is moving around the singly charged nucleus, the proton, in an orbit with dimensions of the order 10^{-8} cm.

In a first systematic study of the K-spectra from the consecutive elements 19 Ca to 29 Zn H. G. J. Moseley in 1913 was able to show that the square root of the frequencies of the K-lines progressed linearly with the atomic number in good agreement with formulas which could be derived from the quantum theory of radiation. In a second paper he measured several lines belonging to the L-series of the elements 40 Zr to 79 Au where the same regularities were found. This result, which is called the 'Moseley law', has been of fundamental importance for the further development of atomic theory. Through it the atomic number Z, which had been earlier introduced in the periodic system of the elements only as an ordinal number, obtained the physical significance of measuring the nuclear charge. One of the consequences was e.g. that X-ray spectra of unknown elements could be interpolated with full certainty and the existence of such elements could be established by registering their X-ray lines. The first successful application of this kind proved the existence of the missing element number 72 (*Hafnium*) which was found in a Zirconium mineral by D. Coster and G. v. Hevesy in 1923. Later W. Noddack, I. Tacke and O. Berg found X-ray lines belonging to the missing element 75 (*Rhenium*) in the spectrum from the mineral columbite.

16.4. Development of X-ray Spectroscopy

These brilliant first results of crystal X-ray spectroscopy confirmed and extended the knowledge of the characteristic X-radiation. They showed that X-ray spectra were a complement to the optical spectra. A condition for further advance was an increase in the precision of the measurements of the wave-lengths both of the emission lines and the absorption discontinuities. This required an increase also of the resolution in the registered spectra. Only by complete and exact measurements of the X-ray spectra could a reliable picture be obtained of the atomic structure, as far as it influences this radiation. Even if it was soon shown that X-ray spectra were somewhat more complicated than the first studies seemed to indicate, they were ever so much simpler and less confusing than the optical spectra which often contain many thousands of spectral lines for a single element.

As one of the first results with the new methods for registering X-ray spectra with increased resolution, it was found that the strongest line in the K-series, designed by Moseley $K\alpha$, was a narrow doublet, and that the fainter line $K\beta$ had a still fainter line on its short wave-length side. The number of lines in the L-spectrum was soon found to be considerably higher than the first registrations had shown.

All methods for determining X-ray wave-lengths using crystals as gratings are based on the Bragg law $n\lambda = 2d\frac{1}{2} \sin \theta$ and therefore λ is obtained by measuring the angle θ between the reflected beam (in the nth order) and the lattice surface. The value of the lattice constant d had been deduced for several crystals by the Braggs. The accuracy of this value was not very high; but as long as the same crystal is used for the registration, the *relative* values of the measured wave-lengths are not influenced by this uncertainty. By measuring the angle of reflection with high accuracy using a double angle method it was possible to increase the precision about a hundred-fold compared with earlier measurements. This of course was important for the numerical testing of the relations between different spectral lines which formed a check on systematics of the atomic structure.

The domain of the X-ray spectra which can be registered using crystals as gratings extends from 20 Å to 0.1 Å, that is over nearly eight octaves, this may be compared to the range of the visible part of the spectrum which is just about one octave.

Voltages corresponding to these X-ray wave-lengths are 600 to 130 000 Volt. The great differences in absorption in this wide region make it necessary to use special spectrometers as well as X-ray tubes and other equipment for each part of the X-ray range. For the greatest part the absorption of the X-radiation in air makes it necessary to use vacuum spectrometers. After these instrumental problems had been solved a rather complete and exact study of all X-ray emission spectra from 11 Na up to 92 U was carried through.

A picture of the K-spectra of seven elements in the region 33 As to

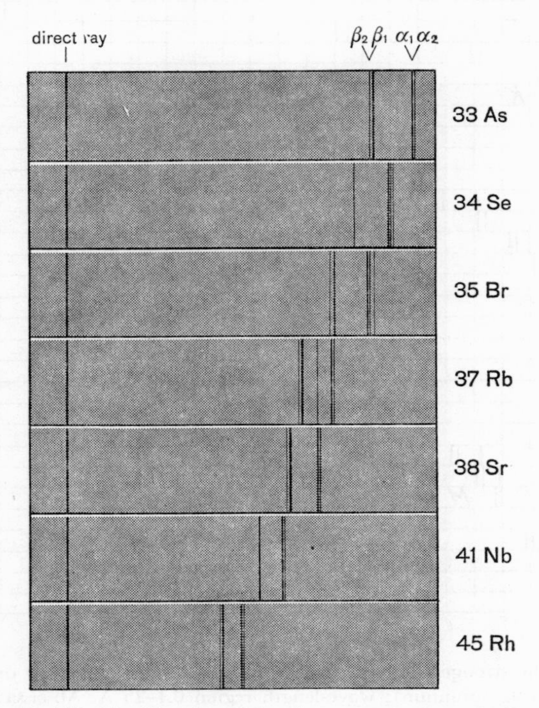

Fig. 16–4(1). The K-series of the elements arsenic to rhodium.

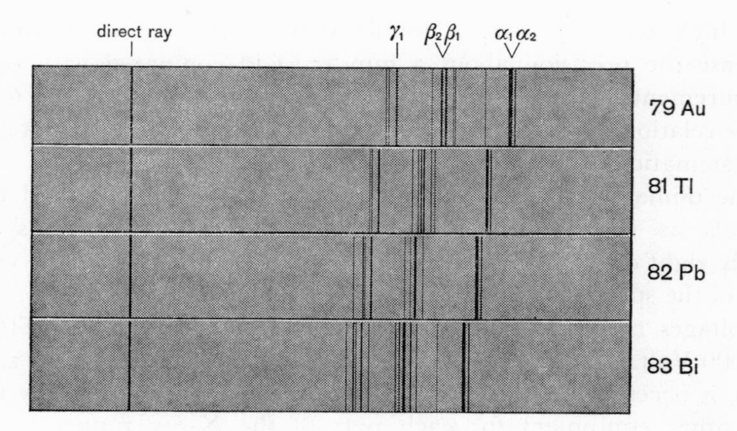

Fig. 16–4(2). The L-series of the elements gold to bismuth.

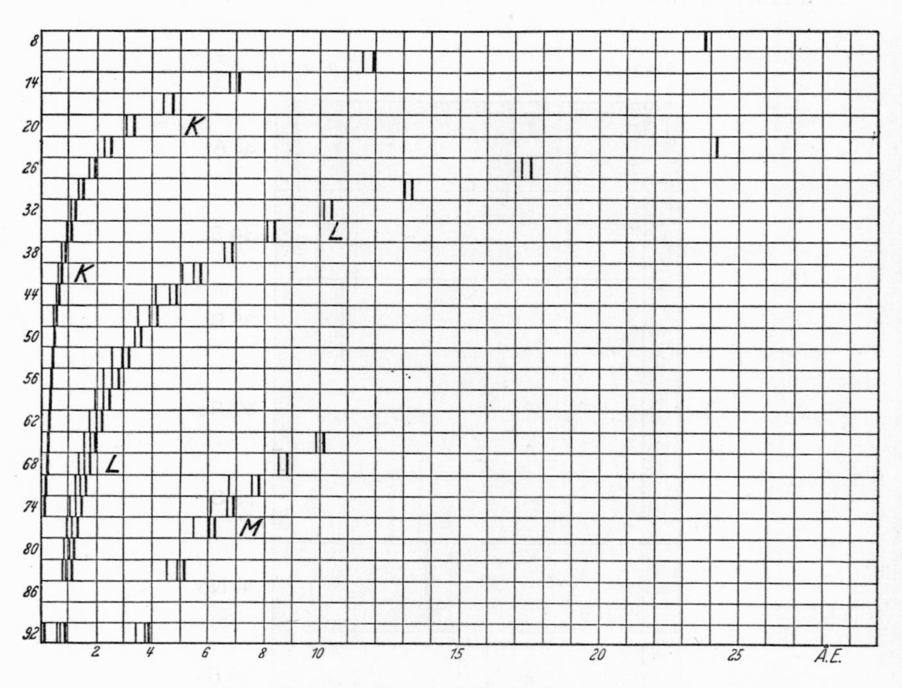

Fig. 16–4(3). The strongest lines of the K-, L-, and M-series of every third element from 8 (oxygen) to 92 (uranium); wave-length region 0.1–25 Å. Abscissa: wave-lengths; ordinate: atomic numbers.

45 Rh is shown in Fig. 16–4(1) and a similar picture of the L-series of the four elements 79 Au, 81 Tl, 82 Pb and 83 Bi is given in Fig. 16–4(2).

A schematic picture of the main lines in the K- and L-series as well as a part of the later discovered M-series is shown in Fig. 16–4(3) covering elements from 8 O to 92 U and the wave length region up to 25 Å.

As already mentioned, characteristic absorption appears as 'absorption edges' in the recordings of X-ray spectra. For the K-series a single absorption edge at a wave-length very nearly equal to that of the shortest emission line was observed. In the region of the L-series three, and in the M-series five edges were found. As with the emission lines the square root of the frequencies of the absorption edges shows a linear relation to the atomic number of the elements as expressed by the Moseley law.

16.5. Emission and Absorption Processes

The very simple structure of the emission and absorption spectra which is nearly unchanged from element to element, apart from a stepwise displacement to shorter wave-lengths with increasing atomic number, facilitated the interpretation of these spectra. Kossel, one of the many prominent members of the Sommerfeld school in Munich shortly after the time when Laue joined the group, suggested the following processes for the X-ray emission: The first step is the excitation of the atom whereby one of its inner electrons is thrown out by absorption of energy provided either by electrons from the cathode of the X-ray tube or by primary X-radiation. When the gap in the electronic configuration is then refilled by one of the outer electrons an X-ray line characteristic of the atom in question is emitted.

This general picture then directly connects the K-, L-, M-, etc. series with electron groups in the atoms called *electron shells*. The refilling of a free electron place in the K-shell gives rise to the K-series whose different spectral lines are due to transitions of electrons from the L-, M- etc. shells, respectively. Similar emission processes occur at the refilling of free places in the L-, M-, etc. electron shells, respectively.

With the exact and complete measurements of the X-ray emission and absorption spectra it was then possible to draw a level diagram connecting the different emission lines with the corresponding absorption limits, as is shown in Fig. 16–5(1) for 30 Zn. For heavier elements

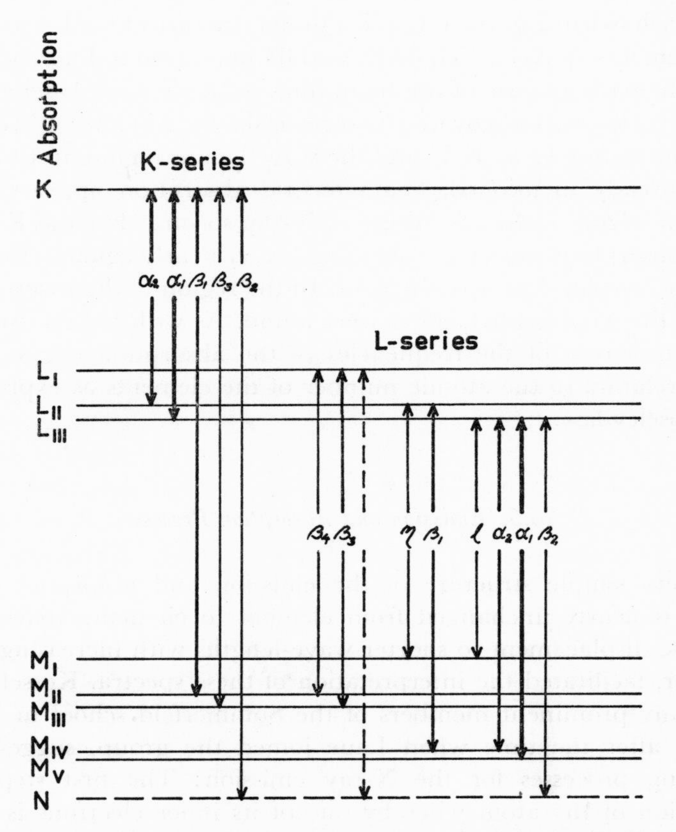

Fig. 16–5(1). Energy levels, absorption and emission lines for element 30 (zinc).

these diagrams extend to higher levels N, O, P whereas for the lighter elements the outer levels successively vanish.

As is seen from the level diagram not all of the possible transitions between the levels are represented by spectral lines, e.g. the transition KL_I, KM_I, L_IM_I etc are not found. These missing lines were eagerly looked for in the earlier studies of the X-ray spectra. As a result of not finding them, 'selection' rules for existing transitions were formulated. The quantum theory of the atom, as it advanced through its stages of development, imbedded the empirically found selection rules in its system of general principles. It is outside this article here to describe these theories.

It may be added that these level diagrams include what may be called the normal spectral lines. But in different regions of elements several lines were observed and called 'satellites' which do not fit in with these 'normal' diagrams. These satellites are found e.g. in the K-series of the lightest elements in the vicinity of the $K\alpha_1$, α_2 lines, also near to the $L\alpha$ of the middle elements and near the M-lines of the heaviest elements. In some cases these lines may be due to transitions in which two electrons are involved.

Commonly the X-ray spectra have been studied when emitted from atoms in solids. In general these spectra are independent of the chemical binding and of the crystal lattice, because the transitions of the electrons occur in the shells which are mainly influenced by the field of the nucleus and of the electrons in the inner shells of the emitting atom. If the outermost shells or the valence electrons are involved in the emission or absorption process it is to be expected that the neighbouring atoms should influence the X-ray spectra. This was found to be the case as soon as the resolution and precision in the study of the X-ray spectra had been sufficiently developed. These effects were first found in the K-absorption spectra of Cl, S, P, Si and later in some cases up to Fe and to still heavier elements. In general these effects give rise to a complicated structure in the region of the absorption edge. Similar influences were also found in several emission spectra as a displacement of lines when the outermost shells were involved in the transition. Again in this case the K-series lines of the lighter elements were first studied. The displacement of e.g. the $K\alpha$ doublet of 16 S is of the same magnitude as the wave-length difference between the two α-lines.

As mentioned, these effects are due to the chemical state of the emitting atom or to the crystal lattice environment to which it is bonded. Inversely it is possible from the spectrogram in such cases to draw some conclusions about the atomic bonding state of the emitting substance if this is unknown.

16.6. The Electronic Shells of the Atom

The Rutherford conception of the atom as consisting of a positively charged minute nucleus surrounded by a cluster of electrons was given a most convincing support by the study of the X-ray spectra. At the same time the collected results of these studies led to a picture of the way in which the electrons in the atoms were distributed. It was shown that if electrons are successively added to a nucleus of increasing charge from a few electrons for light atoms up to the 92th electron of uranium, they will arrange themselves in shells. In conformity with the designation of the spectral groups in the X-ray spectra these shells are called the K-, L-, M-, etc. shells. In the two first elements 1 H, 2 He only the K-shell is populated with electrons: one respectively two. In the next elements 3 Li to 10 Ne the L-shell is built up by adding successively 1 to 8 electrons in this shell. The final number of electrons in the M-shells is 18, in the N-shells 32. Uranium has 18 electrons in its O-shell and the rest (14) of the outer of its 92 electrons divided between the P- and Q-levels.

For a more detailed picture it is necessary to consider, as suggested by S. Goudsmit and G. E. Uhlenbeck, the spin of the electron, and the Pauli Exclusion Principle which says that a certain electronic state, defined by its orbital quantum numbers, cannot be populated by more than two electrons. These two electrons then have their spins in opposite directions.

16.7. X-ray Spectroscopy as an Analytical Tool

Already in the very first stages of X-ray spectroscopy the possibility of using these methods for the purpose of chemical analysis was tried out. The discovery of two new elements, mentioned above, was among the first results. Even though it was soon shown that a very high sensitivity of the analysis could be obtained over a wide range of elements, the method found only a rather limited application for several years. The reason of this was that at that time, in order to obtain sufficient intensity of the spectra, the sample had to be introduced as a target into the tube. This required elaborate vacuum equipment and was otherwise inconvenient.

The subsequent construction of technical X-ray tubes of high power, and the development of new, much more sensitive detectors for X-radiation, like Geiger, proportional, and scintillation counters, made it possible to use the fluorescence method of excitation for the analysis. What is then needed for X-ray spectroscopic analysis is electrical equipment for the running of the X-ray tube, and a crystal spectrometer with a suitable detector. The specimen is placed as close to the exit window of the tube as possible. The window may be made of light material, e.g. lithium glass or beryllium foil, so as to allow also the softer X-rays to come out. The spectrum of the secondary X-radiation emitted by the specimen is generally registered on a recorder. Suitable equipment has now been highly developed and is commercially available using both flat and bent analysing crystals with detectors suited for the different wave-length regions.

The X-ray analytical method with a modern spectrographic outfit allows a more rapid analysis than the wet chemical methods. It is applicable for all elements from 22 Titanium to 92 Uranium and if specimen and spectrometer are placed in an air tight chamber under a pressure of 1 mm Hg also the elements down to 11 Sodium can be analysed. The sensitivity is such that many elements in as low a percentage as 0.001 may be found.

In special cases also the absorption spectra have been used for analytical purposes.

The X-ray analytical methods are non-destructive. They are nowadays very widely applied in such fields as metallurgy, geology, mineralogy, ceramics and for a great number of chemical problems.

A most interesting application first suggested and developed by Castaing in a doctor thesis (1951) is the scanning microanalyzer. In this instrument the electron beam is focussed by two magnetic lenses to an exceedingly small area, about $1\mu^2$, of the specimen. The primary X-radiation sent out from this point enters an X-ray spectrometer, which then shows what elements are present in this small area. Alternatively the secondary electrons may be registered. The instrument, which is evacuated, has a built-in optical microscope to observe what part of the specimen surface is hit by the electron impact. It is also possible to move the specimen for scanning the surface. Another device for this purpose is to let the electron beam sweep over a small area say 0.5×0.5 mm.

The scanning microanalyzer has already found wide and important applications, especially for detailed study of the microcrystalline structure of metals, alloys, minerals and other solids.

16.8. General Conclusions

As may be seen from this short review of the development of X-ray spectroscopy it was early recognized that X-rays are emitted by the electron cloud surrounding the atomic nucleus. This opened the main way to explore the structure of the atom outside the nucleus. The shell structure of the atoms was not only confirmed in general but a detailed quantitative scheme of the energy levels could be gained. Taken together with the information about the outer energy levels as secured by the optical spectra, this provided the present detailed picture of the atom outside the nucleus. But even in the study of the nucleus, now in rapid progress, the knowledge of the energy levels of the electron shells plays an important role. The 'Auger electrons' as well as the Compton effect, which are intimately connected with the study of the phenomena of the nuclear radiation, were both discovered in the course of investigations of the X-ray spectra.

PART V

In Memoriam

My Development as a Physicist

AN AUTOBIOGRAPHY

by Max von Laue *
(1879-1960)

It would seem that my essentially bookish nature was noticed by perceptive adults at an early age. In any case, when I was nine or ten years old, my grandfather, Theodor Zerrenner, who loved me dearly, made me a Christmas present of the comprehensive ten volume edition of Brehm's *Tierleben*. I remember looking at the beautiful illustrations many times and even today I retain a certain general knowledge of the principal divisions of the animal kingdom. This preoccupation, however, came at a time when a boy does not think of his future profession and I never felt any inclination to study biology. Later on my dear Mother may have mentioned Law as a profession, but that too did not last long for other things began to occupy my attention.

It happened in the Tertia of the Wilhelms Gymnasium† in Berlin—we had moved recently to this town (1891) because my father had been transferred there from Posen—that I heard at school, I cannot remember in what context, of the deposition of copper from copper sulfate solution by an electric current. The impression which this first contact with physics made on me was overwhelming. For several days I went about lost in thought and totally useless, so that my worried Mother asked me whether I was ailing. After finding out the cause, she saw to it that I visited the 'Urania' now and then. This was a society to popularize science, in whose house in the Taubenstrasse a large collection of physics apparatus was on show,

* Translated by P. P. Ewald and R. Bethe. The footnotes are those of the translators.—The German original has been re-published in Max von Laue, *Gesammelte Schriften und Vorträge*, Bd. III, pg. V–XXXIV, Friedr. Vieweg & Sohn, Braunschweig 1961. The two orations following here on the autobiography proper formed part of the original publication arranged by Laue in the book by Hans Hartmann, *Schöpfer des neuen Weltbildes* pg. 178–207, Athenäum Verlag, Bonn, 1952.

† The Gymnasium is the classical German secondary school, offering 9 years of Latin and 6 of Greek. Pupils in the 'Tertia' are 12 to 14 years old.

ready for simple experiments: following the printed explanation one had only to press a button in order to see some instructive demonstration. Only very much later, at the Physics Meeting in Koenigsberg in 1930, did I learn that I owed this impressive kind of instruction to Eugen Goldstein, the discoverer of Canal-rays; it was he who had arranged the display of demonstrations. I also listened to several of the lectures given in the 'theatre' of the 'Urania' and visited the astronomical observatory of the society in the Invalidenstrasse—but of these I do not remember any details. When I myself gave a talk in the 'Urania' in 1913 or '14 I recalled the old time affectionately.

We remained in Berlin only one and a quarter years. Then another transfer of my father took us to Strassburg in the Alsace. There I went to the famous 'Protestantische Gymnasium', and had indeed no reason to complain of the change. This institution, belonging to the Alsatian Church of the Augsburg Confession, was headed by a most understanding and well-meaning director, H. Veil, who knew that each individual has the right to follow his own law of development and who, spite of being a classical scholar with strong theological leanings, fully acknowledged the claims of mathematics and science, and more than once sided with me against teachers who would occasionally take offense at my mathematical and scientific bent. Long after leaving school I paid visits to this man, the last time at his place of retirement in Jena, where he died in 1938, high in his eighties.

The teaching faculty, as far as Veil had been able to pick it, also deserved on the whole high praise. There were, it is true, some old members, left over from the time before 1870, who could neither adapt themselves to the new political situation, nor change to another job. They went about rather embittered and were also, therefore, without influence on the students. For them, too, Veil tried to ease the situation, for example by excusing any teacher for whom it meant a conflict of conscience from giving the oration at the official celebration of the Kaiser's birthday. Veil never tolerated political oppression at his school and much less inflicted it.

Among the teachers I remember Professor Erdmann, a theologian, who, besides religious instruction, taught classical languages and German. He was perhaps the wisest of all my instructors. Two of his statements I recall to this day: In the upper forms the subject matter in German was, naturally, History of Literature. Erdmann, who liked to stray slightly from the prescribed course material, came to talk about Richard Wagner and the Ring of the Nibelungen. The 'Siegfried Ethics' of this musical drama threw him into a frenzy of excitement and he predicted terrible consequences if this should ever rule the conduct of man.—His other saying was: 'He who is burning for a great cause will never quite perish.' This I have found true in my own life.

Why do I report all this here where I deal with my development as a physicist? Because this development cannot be separated from my general mental development, least of all in those years when a boy changes into a youth and the foundations for all his later maturing are laid. I doubt also that I would ever have devoted myself completely to pure science had I not acquired at that time that feeling of intimate connection with the language and culture of the Greeks which the humanistic gymnasium and no other type of school produces. The joy of pure knowledge is, with few exceptions, after all acquired only from the Greek classics. If one attempts nowadays to train a scientific generation on a larger scale than in previous decades I have one formula to offer: send the boys to a humanistic gymnasium and let them have a good training in the Classics. And of course teach them also to read and to love the German classical authors. Ludwig Boltzmann, the great Viennese physicist who died in 1906, writes in the preface to his *Popular Essays* (1905): 'Without Schiller there might well exist a man with my nose and beard, but never myself.' This I endorse word for word.

I might mention in this connection that many of the boys who received their secondary schooling under Veil eventually became university teachers. Besides myself there were for instance: the historians Wolfgang Windelband, and Robert Holtzmann (both in Berlin), the physicians Friedrich Holtzmann (Karlsruhe) and Wolfgang H. Veil (Jena), the Lawyer Eduard Kohlrausch (Berlin), the chemist, Walter Madelung (Freiburg/Br.), and the physicists Erwin Madelung and Marianus Czerny (both in Frankfurt). But let me return to my subject.

Among all my teachers the one of greatest importance for me was Professor Göhring who taught mathematics and physics. He was an old bachelor with a thousand idiosyncrasies which would have made any other teacher impossible in front of a class of schoolboys. But in his case they were more than balanced by his mental superiority which fascinated even the worst rowdy in the class—and we were really not model pupils. To play tricks in front of him, impossible! Nobody ever even though of it. For that reason he never had to punish, rarely even to scold. The utmost I remember him saying were these words, addressed to a boy who had made himself a real nuisance: 'Huber grins; yes, Huber grins.' Huber did not grin again for a long time. Youth has a fine appreciation for intellectual greatness and is awed by it—provided it genuinely exists.

From this man I learnt my first mathematics, and I may say that I soon became one of his best students. We were taught the usual high school mathematics—algebra including quadratic equations, geometry according to Euclid and other ancient geometers, logarithms and trigonometry, etc. Much of it I have long since forgotten, and I would have to work up afresh the properties of the Appollonian circle if I needed them. But at this stage it is less the factual knowledge that is important, than the facility one acquires for scientific reasoning. 'Education is what remains after all school learning is forgotten' * is an often quoted saying of I don't know which great man. It was in this spirit that Göhring taught us. He soon recognized my mathematical gift and prodded me on with occasional praise.

My inability to do numerical work formed a strong contrast to my mathematical achievement. This went so far that when in the final school examination a number of mathematical problems had to be solved in a written class exam, the teacher,—it was no longer Göhring—referring to a trigonometry problem with numerical work, said: 'don't even try to do this one, you'd get the wrong answer anyway.' Only at the university and when I needed it for physics did I learn to do numerical work and by conscious attention to it acquired some dependability.

Under Göhring I also studied physics for the first time seriously. I do not think that the few experiments he could show with the inadequate facilities of the Gymnasium were nearly so significant, as was his ability to convey scientific thinking to his students. Besides, it was important that he referred us to suitable books. On his advice I acquired (in October 1896, as a note in the books indicates) the *Vorträge und Reden* by Helmholtz, and I still hear him recommending these two volumes to us in his Thuringian accent which knows no hard consonants: 'These are popular essays, but for a populus that isn't a bit dumb.' I studied these essays and discourses with the greatest avidity, read and re-read what I did not understand the first time, and I widened my horizon considerably through them and later through Helmholtz' *Lehre von der Tonempfindung*. Are not these both in form and content classics of physics literature? I did, however, not limit my reading of the lectures to those on physics. I also read the academic

* 'Bildung ist das was übrig bleibt, wenn man alles Gelernte vergessen hat.'

discourses and revelled in the thought of later becoming a university student. My favorite piece, however, was the magnificent autobiography which Helmholtz gave as a banquet speech on the occasion of his seventieth birthday. I even tried reading his philosophical talks—but with minor success.

In spite of all this, the years in the Secunda and Prima of the Gymnasium would not have had nearly so decisive an influence, had I not found among my schoolmates two of similar inclinations, Otto B. and Hermann F. We were drawn together mainly by our common interests, and were soon known throughout the school as the Mathematical Trefoil. One of us, Otto B., came from the highly intellectual family of a professor at the University of Strassburg. I saw very little of his parents or of his older brothers and sisters. But Otto B. had the extraordinary gift of fully absorbing what he heard from them, and, being interested himself, to pass it on in an interesting way. Also, he read much and with full understanding in the mathematical and scientific books which he found in his home, and he knew how to stimulate us to similar reading. In this way the three of us, by ourselves and with a kind of stealth, got acquainted with differential and integral calculus, browsed in the several volumes of Wüllner's *Lehrbuch der Physik*, and began experimenting on our own as boys will do. The experiments succeeded particularly well after we had managed to buy small induction coils which gave us fairly high tension. Early in 1896 Röntgen's famous *Communications*, which Hermann F. received from a book seller uncle promptly after publication, acquainted us with his great discovery. We managed to procure primitive discharge tubes and tried—in vain—to detect those mysterious X-rays, as they were then called. As sources for the primary current we used chromic acid and Bunsen elements, as the apartments of our parents were not connected to the city's electric power. Many a hole did the required chemicals burn in our clothes! I remember in particular a home-made galvanometer with many windings and an astatic pair of needles with thread suspension. This gave me the proof, which I had not been able to obtain from the experiments at school, that the discharge current of a Leyden jar deflects the magnetic needle.—We also often discussed optical phenomena, in particular, the interference of light and optical diffraction, which at that time were most mysterious to us. The great attraction of these phenomena lies in their being directly perceptible to our senses, without measuring instruments. My strong interest in Optics, which became important later in connection with X-rays, dates from these high school times.

These were stimulating hours which we had together, often in Otto B's attic room in the stately home of his parents in the Goethestrasse. The window offered a view over the University Gardens, far out to the bluish hills of the Black Forest on the horizon. These mountains, and also the Vosges, were often the goal of our excursions, lasting one or several days. Generally, we were not stay-at-homes but cycled through wide stretches of the Rhine plain and beyond, and swam in the little or the large Rhine near Kehl; I have often crossed the Rhine by swimming, with the help, to be sure, of its many gravel banks. No organization was needed to get us out of doors, the freedom accorded us by our elders at home and in school was enough. And this freedom was one of the main attractions of these hikes. What would they have been to us without our independent prior planning, our study of maps and time-tables, estimating our own stamina and, once on the way, the not always easy task of finding our path through the mountains in fog or snow. Our hikes exercised not only our body, but contributed also to our mental maturing.

Unfortunately, the Trefoil did not continue together up to the Abitur. A year earlier Otto B. transferred to another gymnasium in Baden, because of an inherited nervous affliction which progressively retarded him mentally, later prevented him from attending a university after he had passed the Abitur, and finally, in 1904, made him shoot himself. I have never forgotten him, and feel even today the great influence he has had on my work. Had he retained his health, he would have become an outstanding university teacher, an inspiration to his students, and perhaps a great scientist. He was buried in Strassburg.

In March 1898 I passed the final school examination, the Abitur. I had prepared for it to some extent, and felt a kind of serene cheerfulness in contrast with the excitement which took possession of most of my classmates as the date approached. The same thing happened in my later examinations, and at the last one, for the Teachers Certificate, which I took in Göttingen, in 1904, I felt slightly sarcastic at the importance which other candidates and even some of the examiners attached to this business. In the Abitur, I received 'Good' in Religion, Latin, Greek; 'Satisfactory' in German, French and History; 'Excellent' in Mathematics and Physics. The certificate contains a comment on German: 'Laue has come up to the requirements of the curriculum in German and has done good work occasionally. His intellectual potential is higher than his achievement in written or oral expression. The examination composition was satisfactory.' This was all too true. Schiller's moan: 'If the soul speaks, Oh! it's no longer the soul'* I have sympathized with all my life. Even worse, if I had to speak in a foreign language I suffered such torment that I could never give a fluent and correctly pronounced lecture.

Perhaps the reader missed English among the subjects of the Abitur. At that time there were no classes in English given at the German Gymnasium; this I have later felt to be the worst deficiency in my education. I learned English only after leaving school, from scientific journals and books which became more and more indispensable as time went on. I spent months in America and was compelled to use English there. Although I had very little practice in French after the gymnasium, I still speak French better than English. 'Was Hänschen nicht lernt, lernt Hans nimmermehr.'† Undeniably, a certain lack of personal aptitude also plays a part. In the gymnasium in Posen, from Sexta to Tertia, I heard much Polish spoken by my comrades, but I never managed to understand even a single word. In Strassburg I had a similar experience with the Alsatian dialect. Any language I tried required a conscious and determined effort. And for this reason it was a particular blessing for me that at the age when learning was easiest the gymnasium laid so much stress on grammar and expression; this emphasis compensated to a certain degree for my lack of natural ability.

Military service began a few days after the final examination, and brought an interruption of my intellectual training.‡ I could, however, register for the main lecture course in Experimental Physics during the second half of my military year, in the Winter Semester 1898/99. The course was given by Ferdinand Braun and I well remember his brilliant demonstrations and his elegant and often witty presentation. I watched and listened enthusiastically. Sometimes my military duties prevented me from arriving on time, and whenever this happened my entrance in uniform was quite a sensation and caused some unrest in the audience. This led to an amusing incident, characteristic of Braun, which I want to tell, all the more because Braun himself laughed heartily when he heard it from me at a physicist's meeting many years later.

At the end of the semester the students had to 'sign out'; this was a way of certifying their regular attendance. In view of the fact that

* 'Spricht die Seele, so spricht, ach, schon die Seele nicht mehr.'

† 'What Johnny doesn't learn, John never can'.

‡ By reaching the level of Obersecunda in a secondary school the student became an 'Einjährig-Freiwilliger' that is, his future compulsory military service was reduced from two or three years (depending on the branch of service) to one, with the option of receiving a commission as an officer of the reserve after further training.

Braun had several hundred students in his lecture course he was naturally not in a position to know; he therefore sat at a table in an antechamber of the lecture theatre signing his name mechanically and hardly looking up as the students laid their open books before him one after the other. Only when I approached in my uniform did he look up slightly and say with an ironical sigh: 'Yes, that *you* have attended I really *can* certify.

In later years, too, I tried hard to attend lectures regularly. Whenever I was obliged to miss one, I found it difficult to pick up the thread of the lecture, especially in mathematical courses. How other students could miss lectures, for example to fulfill fraternity obligations, was beyond my comprehension. My aim was Science.

Yes,—but what part of Science? This was my greatest problem in the first semesters. That my field lay somewhere among Mathematics, Physics, or Chemistry, of that I was certain since the Secunda. As a start I took many courses in all of these subjects, first in Strassburg, then (beginning in the autumn of 1899) in Göttingen. I attended probably more laboratory courses in Chemistry and Physics, than those who knew their goal and chose the shortest way. In Göttingen, under the influence of Woldemar Voigt, I clearly recognized my destination at last: Theoretical Physics. Next to Voigt's course, the published lectures of Gustav Kirchhoff contributed much to this decision. Otto B. had pointed them out to me when we were still at school and also Prof. Göhring had occasionally spoken of the first volume, the lectures on mechanics. But the decisive impression was my astonishment at seeing how much information about nature can be obtained by mathematical methods. Profoundest reverence for Theory would overcome me when it cast unexpected light on previously obscure facts.

Pure mathematics, too, did not fail to impress me, especially in the brilliant courses of David Hilbert. This man lives in my memory as perhaps the greatest genius that I ever laid eyes on. Mathematics represents the experience of truth at its purest and most immediate; on this rests its value for the formation of the human mind. One of my main delights, even at school, was a well rounded mathematical proof. And yet, mathematics has interested me only when I could find some physical application for it. Otherwise my endeavours in this field appeared to be 'swimming in empty space', or like exerting a force on a non-existing object. Other theoretical physicists differ in this respect and owe great results in physics to their preoccupation with mathematics *per se*. But I had to follow my own natural bent as I said, and come to terms with it.

Important as these and later courses were to me, I learned more effectively from books than from lectures. The spoken word never made as permanent an impression as what I saw before me in black and white. In reading you can stop wherever you like and think about what you have read. In a lecture you are bound to follow the lecturer's train of thought and the least inattention makes you lose the thread. Often enough the lectures only served to spur me on to delve into the appropriate books.

In these semesters, and even long afterward, Maxwell's *Theory of Electricity and Magnetism*, which had won full acknowledgment in Germany only a few years before, gave me particular trouble, and also great pleasure. Finally it happened to me as no doubt to everyone else; a new world opened up before me. What manifold and complex processes can be traced back to Maxwell's simple and beautifully harmonious equations—this insight is one of the most sublime experiences a man can have.

> 'War es ein Gott, der diese Zeichen schrieb
> Die mir das innre Toben stillen,
> Die Kräfte der Natur rings um mich her enthüllen?' *

That is what Boltzmann once quoted from Faust regarding these formulae. My predilection for optics suffered no damage from the strong impression Maxwell's Theory made on me; on the contrary, my admiration increased with the knowledge that optics formed part of the system of electrodynamics.

I remained in Göttingen for four semesters. Then during the winter of 1901–2 I went to the University of Munich. However, in those days a theoretical physicist could not find much in Munich. Boltzmann's chair had not been filled. I remember only Alfred Pringsheim's course on Theory of Functions and W. C. Röntgen's physics laboratory where he talked to me once at some length and was obviously pleased with my knowledge. Normally the assistants conducted the laboratory. But at that time I first became acquainted with the Alps in winter. I had been hiking there during summer vacations together with Hermann F. and Otto B. (who was at that time still physically quite able) and also had some ascensions behind me. But winter in the high

* 'Was it a god who traced these signs
That soothe the raging in my inner self
Revealing Nature's forces that surround me?'

mountains was something quite new and wonderful. It was a pity that skiing did not exist in Germany at the time. We therefore plodded along in the soft snow of the Rofangebirge, sometimes sinking in up to our hips, and when we wanted to push ahead to the Wendelstein after a sleigh descent from the Brünnstein we got into real trouble. We—that means the colleagues of the mathematics club to which I belonged in Munich, as earlier in Göttingen and later in Berlin; these scientific student societies maintained a fine esprit.

I learned to ski on the Feldberg under the tutelage of Paul Drude and Willy Wien, but not until 1906, and by then I was no longer young enough to become a real expert. I did gain enough skill to feel safe skiing cross-country, which after all is a satisfactory way to experience the great joy of this wonderful sport. Until well into World War I, each March found me skiing in Mittenwald with Willy Wien—Paul Drude lost his life in the summer of 1906—and the memory of this great, scientifically inspiring, noble, kind, and at the same time sports-loving savant, is one of the most beautiful of my life.

In the summer of 1902 I moved to Berlin. What drew me there mostly were my old school friends, Hermann F., who was there working for his doctorate in chemistry, and Otto B., who continued to make desperate efforts to complete his studies. Although I arrived at the end of June, the first eight weeks of the term having been swallowed up by a military exercise, I registered without difficulty.

I hurried to Planck's lectures on theoretical optics; I knew of him as the author of a textbook on thermodynamics, and knew also that he had accomplished much in this field. But I knew nothing as yet of his greatest achievement—the discovery, in 1900, of the Law of Radiation and the quantum-theoretical arguments leading to it; these results were not generally accepted at the time and hence they were little known. I could follow the lectures easily despite their advanced stage because I had heard the same material once before with Voigt, and I even managed to draw Planck's attention during the first exercise period by producing a rather elegant solution to the assigned problem. I also attended the 'small course' of O. Lummer on 'Special Problems in Optics', i.e., on interference phenomena, especially in ruled gratings, echelette gratings and plane parallel plates. He himself together with Gehrcke had just at that time introduced plane parallel glass plates in spectroscopy. Lummer's principal association was with the Physikalisch-Technische Reichsanstalt,* and for demon-

* Comparable to the Bureau of Standards in the U.S. and the National Physical Laboratory in Great Britain.

stration purposes, he brought to his lectures in the Physical Institute of the University some of the excellent apparatus of that institution. In these lectures I acquired the feeling for optics which became so very useful to me later on. Toward the end of the semester I went to Planck and asked him for a subject for a doctoral thesis. Referring to the above-mentioned lecture course he gave me 'Theory of Interference Phenomena in Plane Parallel Plates'. I worked on it until the summer of 1903. In July of that year I passed the doctoral examination *magna cum laude* with mathemetics and, as required, philosophy, as minor subjects.

Something should be said about this. Though I have never attended a course in philosophy, I have spent much time and thought on Kantian philosophy. To begin with, I read Kuno Fischer's presentation in his *History of Philosophy* and then I read and re-read Kant's *Critique of Pure Reason* and others of his works, especially those on ethics. The stimulation of this interest dates back to school days and came from Otto B., but only at the University, it would seem to me, did I mature sufficiently to understand philosophy. This transformed my life completely. Since then even physics seems to derive its real dignity only from the fact that it provides an essential resource for philosophy. My conception is that all the sciences must be grouped around philosophy as their common centre and that service to philosophy is their real purpose. Thus, and only thus, can the unity of scientific culture be preserved in the face of ever increasing specialization; without this unity our whole culture would be destined to disintegrate.

In those days the doctoral degree was still conferred in a truly solemn ceremony. Among other things the Dean read an oath which says in part: 'Te solemmiter interrogo, an fide data polliceri et confirmare religiosissime constitueris, te artes honestas pro virili parte tueri, provehere atque ornare velle, non lucri causa neque ad vanam captandam gloriolam, sed quo divinae veritatis lumen latius propagatum effulgeat.' * I have endeavoured to keep this oath.

While working on my thesis I still attended Planck's lectures on thermodynamics and his most impressive lectures on theory of gases and heat radiation. Boltzmann's principle of the relation between entropy and probability, Wien's displacement law with its proof given in final form only by Planck, and then the audacious derivation of the radiation law from the hypothesis of finite energy quanta—they

* 'I ask you solemnly whether by the given oath you undertake to promise and confirm most conscientiously that you will defend in a manly way true science, extend and embellish it, not for gains sake or for attaining a vain shine of glory, but in order that the light of God's truth shine bright and expand.'

Max von Laue (1879–1960).

were revelations to me at the time. There was in addition a magic charm which emanated from the personality of this great man and fascinated every one of his students. All this combined to give me the feeling that the University of Berlin was my proper spiritual home.

The University, not the town. I have always had an aversion to large cities. Therefore, for the continuation of my studies which I considered an absolute necessity, I preferred the typically provincial town of Göttingen and spent four semesters there, once again. I took among other things, Max Abraham's course 'Theory of Electrons', the atomistic confirmation of Maxwell's Theory, and Karl Schwarzschild's Geometrical Optics, although the latter was too highly specialized for my taste. During this time I also passed my state examination for certification to teach in secondary schools. This I did on the side and I was really surprised when I received a 'good' in it.

Since I had chosen chemistry as a subject for examination I had to take an exam in mineralogy as well. Now, I had never studied mineralogy. During my first stay in Göttingen I had made a half-hearted attempt to attend a mineralogy course but had given up very soon. From books I then learned the rudiments of crystallography, that is to say, crystal classes, that was all. The exam was administered by the geologist, Prof. Könen, and I still remember how his amusement grew and grew in the face of my entirely obvious ignorance until he finally broke off the interview. Thanks to my knowledge of chemistry, unusually extensive for a candidate for a state teaching certificate, the committee declared the exam to be successfully passed. The impression that I would never make use of this certificate may have contributed to this decision.

But how did it happen that my training in physics had this obvious deficiency, which by the way, was shared by most physicists in those days, both those older and younger than myself. The reason for this lies probably in the fact that in University teaching crystallography had been pushed almost completely into the lectures on mineralogy where, for the most part, only the descriptive part of mineralogy was stressed. In physics lectures crystals were mentioned generally in connection with optics and occasionally with elasticity, that was all. Even Voigt, who did so much for crystal physics, acted not differently. His fundamental book on crystal physics did not appear until 1910. Perhaps in my case this complete lack of knowledge had the happy consequence that later, in 1912, I could approach a question of crystal physics with complete freshness of mind. Much later, during my years in Frankfurt (1914–18) I filled this gap, but I never achieved the perception of space which marks all those who do crystal physics in their youth. But I did acquire something of that dedication which every crystallographer has for his subject, as it appears with Voigt in an almost touching way in the preface of his book.

In the fall of 1905 Planck offered me the assistantship in his Institute for Theoretical Physics which was about to become vacant. I accepted with pleasure and remained there for more than three years. In addition to supervision of the library my duties consisted in checking the problems which Planck collected each week after his lecture and discussed next day in the problems period after the assistant had looked them over and brought to his atention anything that seemed noteworthy. As a student I had faithfully done these problems. But to correct the work of others was almost more instructive because in this way one could learn what mistakes and misunderstandings were possible and discuss them and many other things with Planck. At this time I also began to do scientific work of my own. It is true I had already published in Göttingen an investigation of the propagation of natural radiation in dispersive media, based on Planck's hypothesis concerning the nature of radiation. However, now, once more following Planck's lead, I attacked the more profound problems of the reversibility of optical reflection and refraction.

Planck's formula for the entropy of a pencil of light rays showed unequivocally that the division of energy from one ray into two geometrically equal rays, e.g. equally long ones, is accompanied by an increase of entropy if, as was then usual, one added their entropies. Then, according to the second law of thermodynamics, the separation of a beam into a reflected and a refracted part should be irreversible. However, a simple argument on optical interference showed that because of their coherence the two beams could be reunited into one, that differs in no way from the original one. This was a profound dilemma. Would one have to abandom the second law for optical processes?

The explanation lay with the already mentioned Boltzmann principle of entropy and probability. It showed clearly that while one could add the entropy of incoherent beams, this was not possible for coherent ones. The entropy of the two beams resulting from reflection and refraction is exactly equal to that of the entering beam.

An hour after leaving Planck's home in the Grunewald following our decisive conversation on this subject, I found myself at the Zoological Gardens without knowing what I wanted there or how I had got there. So overwhelming was this experience.

However, I do not want to enlarge on every phase of my researches. I know from reading the biographies of others that this gets to be rather dull, interesting at most only to a careful historian of science. But I think I may be allowed to mention some high points.

When I returned to Berlin in 1905 I heard Planck talk in one of the first colloquia of the winter term—or was it the very first?—on Einstein's work, which had appeared in September: *On the electrodynamics of moving bodies*. The transformation of space and time performed by the relativity theory, proposed in this paper, appeared strange to me and I was by no means spared the doubts later voiced by others. But the idea took hold of me, especially because Planck soon published several research papers of his own on the subject. Thus in 1907 I could show how Fizeau's famous interference experiment in moving bodies, which until then had been considered the unimpeachable proof for the existence of the luminiferous ether fitted into the new theory which denied the existence of such quasi-matter. What had been taken for granted until then, namely that one can add the velocity of light and the velocity of a body was simply not correct.

I do not know whether it was this work alone or other things as well which resulted in the inquiry made by the publishers, Fr. Vieweg & Sohn, asking whether I would write a monograph on relativity theory. I did, and thus became the author of its first comprehensive presentation. The book was generally well received and eventually reached four editions. I wrote it—I had moved from Berlin to Munich in 1909—in a little boat-house which stood on posts in the water at the shore of the Starnberger See in the ducal gardens of Feldafing and had a marvelous view of the Herzogenstand, Heimgarten, Benediktenwand and the mountains of the Karwendel. I never again had it so good.

At the University of Munich there worked at that time besides Röntgen also Arnold Sommerfeld who had been asked some years earlier to fill Boltzmann's former chair. A powerful inspiration emanated from this incomparable teacher. Among many other things he concerned himself with the theory of X-rays, and developed a theory of the creation of X-rays at the anticathode of the X-ray tube, which proved very useful. It was based completely on the wave theory for these rays, therefore stood in contrast to W. H. Bragg's corpuscular theory which was very popular in England. The basis for making a choice between the two was supplied by Walther and Pohl in Hamburg in their photographs of the diffraction of X-rays by a wedge-shaped slit, which P. P. Koch, Röntgen's chief assistant, was measuring photometrically. Sommerfeld applied the traditional diffraction theory, with the result that he obtained a rough but very useful estimate for the mean wave-length which he published in 1912. Furthermore, Barkla's proof of the polarization of X-rays was discussed in the Sommerfeld colloquium, and so were also the features of the characteristic X-rays of the elements. Thus, one lived there in an atmosphere saturated with problems concerning the specific nature of X-rays.

Röntgen himself began to withdraw at this time and in summer he no longer lived in Munich but in Weilheim, a small town 60 km to the south. He commuted daily by train for his lectures. He still directed his institute, and his assistants and doctoral students produced valuable research. He also still did some work himself, e.g. together with Joffé he investigated certain questions of crystal physics in which he had been interested for some time. He had grown, however, over-cautious in the publication of his results. Again and again new confirmation had to be collected and in this way nothing was really completed. He did not appear for the colloquium and seemed to take little interest in the newer researches. Some theories he favoured with mild ridicule: when the Friedrich-Knipping experiment was interpreted as showing interference, his instinct was for a long time to refuse to accept this interpretation. On the other hand in 1919 he immediately recognized the importance of C. T. R. Wilson's cloud chamber for research on radioactivity and he gave it high praise, as I know from a reliable source. As far as I am concerned I spoke to him only once at any length when, on a train ride to Feldafing, I happened to find the only available seat in a third class compartment opposite His Excellency. At that time I gained the impression that for people like me he could have been easy to talk to, had there been more opportunity to do so.

In 1919 I talked with Röntgen once again informally. Shortly after the Rätezeit in Munich* I went to visit him at his institute but found him ready to leave for the train to Weilheim. So I accompanied him on foot to the Starnberger Bahnhof. But instead of talking about science he expressed his pleasure at all the signs of returning order and observed with a certain delectation the really very delicate system of cracks which surrounded each bullet hole in the glass of display windows. People have often speculated about the reasons which drove this man into virtual retirement after his epoch-making achievement

* A brief period in 1919 when Munich was governed by a kind of Soviet, a council of workers and soldiers.

in 1895/6. Many motives have been suggested, some of them far from flattering to Röntgen. I think they are all wrong. In my opinion, the impact of his discovery was so overwhelming that he, who was 50 at the time, never recovered from it. For—as few people seem to realize—every great intellectual discovery is a heavy burden for the man who makes it. And this was not exactly alleviated when, like many other discoverers, he was forcibly made aware of the less admirable qualities of some of his fellow-men.

The magnitude of his exploit can be recognized particularly if one considers the large number of other physicists, some of them of great renown, who experimented with the same kind of apparatus before Röntgen and did not discover X-rays. For such a break-through into completely unsuspected territory one must have enormous courage and in addition a self-discipline which preserves mental calm and clarity in the midst of the great joy and excitement of the first findings. Many observations had to be made and correlated to make possible for Röntgen to write three treaties like those of the years 1895 to 1897, which exhausted the subject so completely that for almost a decade nothing new could be said about it. With what ingenious care they are written! I know only few accounts of discoveries which do not contain mistakes of one kind or another. Röntgen was correct in every detail.

It became of special importance to me that in Munich the hypothesis of space lattices for the explanation of crystal structure was still alive although elsewhere it was hardly mentioned any more. The reason for this was partly that the university had a collection of models from the times of Leonhardt Sohncke, who had taught in Munich until 1897 and had made considerable contributions to the mathematical theory of space lattices. Mainly, however, the credit must go to Paul von Groth, Professor of Mineralogy, who used the crystal structure models in his lectures. In February 1912 it happened that P. P. Ewald, then working on his doctorate under Sommerfeld, came to visit me in my apartment and asked for help with his problem: to investigate mathematically the action of light waves in a lattice of polarizable atoms. Although I could not help him, I said quite by chance in the course of the discussion that some time one should irradiate crystals with shorter waves, i.e. X-rays. If the atoms really formed a lattice this should produce interference phenomena similar to the light interferences in optical gratings. This was discussed among the younger physicists around Munich, who met every day after lunch at the Cafe Lutz. One of them, Walther Friedrich, who had just finished his doctoral thesis on X-ray scattering under Röntgen, and was now one of Sommerfeld's assistants, offered to test the idea experimentally. The only difficulty was that Sommerfeld did not think much of the idea at first and preferred to have Friedrich do an experiment on the directional distribution of the rays emanating from the anti-cathode. But this difficulty was overcome when Paul Knipping, another of Röntgen's doctoral students, offered his help. And so the experiment on the transmission of X-rays through crystals began around Easter 1912.

Not the very first, but already the second experiment showed a result.* The photograph of the X-rays transmitted through a piece of copper sulfate showed, besides the primary X-rays, a circle of spectra diffracted by the lattice. I was plunged into deep thought as I walked home along Leopoldstrasse just after Friedrich showed me this picture. Not far from my own apartment at Bismarckstrasse 22, just in front of the house at Siegfriedstrasse 10, the idea for a mathematical explanation of the phenomenon came to me. Not long before I had written an article for the *Enzyklopaedie der mathematischen Wissenschaften* in which I had to re-formulate Schwerd's theory of diffraction by an optical grating (1835), so that it would be valid, if iterated, also for a cross-grating. I needed only to write down the same condition a third time, corresponding to the triple periodicity of the space lattice, in order to explain the new discovery. In particular it was thus possible to relate the observed circular pattern of rays to the cones corresponding to the three interference conditions. The decisive day, however, was the one a few weeks later when I could test the theory with the help of another, clearer photograph.†

The theory proved valid far beyond my expectation. This was a surprise because, as had been clear from the beginning, it was only an approximation. Indeed the precision measurements done in Siegbahn's Institute in Uppsala (1920) showed minor deviations; the perfected theory required to explain them one owes to P. P. Ewald. However for the overwhelming majority of the many cases investigated since then, the first 'geometric' theory is perfectly adequate. Although it does not always work out as well as for X-rays, the theory can be used even for the electron interferences in crystals which were discovered in 1927 by C. C. Davisson and L. H. Germer, and by G. P. Thompson.

Standing on the very same spot as Planck, when he spoke for the first time in December 1900 about his theory of radiation and the quantum

* See Fig. 4–4(1).
† See Fig. 4–4(3) and (4).

theory, I presented my discovery on 8 June 1912 to a meeting of the Deutsche Physikalische Gesellschaft in the Institute of Physics of the University of Berlin. Since then a vast body of experimental and theoretical literature has grown up around it, and when in 1941, I wrote a book, *Röntgenstrahlinterferenzen*, to present merely the theoretical aspects of the field, it turned out to be 350 pages long. The first major step beyond my original work, which had appeared in the Bayerische Akademie, was taken by W. H. and W. L. Bragg.

This step, which consisted essentially in the investigation of individual crystal structures was one I could hardly have taken myself. I have always been primarily interested in the great underlying principles in physics—which is why Planck's lectures, emphasizing just these, had fascinated me—and the general principles, concerning X-rays on the one hand and crystals on the other, had after all been established with the help of the experiments of Friedrich and Knipping. The Braggs, father as well as son, had a love for the individual substance and could devote their labour to the structure of crystals from sodium chloride, diamond, and so forth, to the most complicated silicates. Science needs scholars with many different talents and would soon come to a standstill if all scientists were of the same intellectual type.

The history of the discovery of X-ray interference illustrates beautifully the value of scientific hypothesis. Many people irradiated crystals with X-rays before Friedrich and Knipping. However, their observations were limited to the directly transmitted ray which revealed nothing remarkable beyond the weakening produced by the crystal; they missed completely the less strong diffracted rays. It was the theory of the space lattice which provoked the idea of investigating the neigbourhood of the direct ray. Of course the diffracted rays would eventually have been discovered as stronger and stronger X-ray tubes were developed; some accident would have pointed them up. But it is hard to guess when this might have happened, and we can certainly say that the theory of the space lattice was absolutely essential to account for their presence.

As I said before, I don't want to present a complete list of my scientific works, nor do I want to go into the details of my contributions to the mathematical development of X-ray diffraction theory. But I do want to mention a series of publications influenced by my earlier studies in connection with Maxwell's theory. For several years (until 1934) I acted as a consultant on theory to the Physikalisch-

Technische Reichsanstalt and there came into personal and intellectual contact with Walther Meissner who was an old acquaintance from my days as assistant in Planck's problems periods. Meanwhile he had become a member of the Reichsanstalt and Director of its low-temperature section. He was particularly interested in super-conductivity, that strange disappearance of electrical resistance which some metals exhibit when cooled to liquid helium temperatures. It was known that a sufficiently strong magnetic field could reverse this phenomenon. However, the measurements that showed the required field strength had an incomprehensible dependence on its direction with respect to the axis of the wire, which at that time was the only form of superconductor used for experiments. It occurred to me that the super-conducting wire itself distorts the field in such a way that the field actually present at the wire surface is considerably greater than that measured at a distance. My conjecture that actually the same surface magnetic field strength was always needed to destroy the super-conductivity could now be verified quantitavely and applied to other shapes such as super-conducting spheres. I presented this theory at the physics meetings at Bad Nauheim in 1932 when I was awarded the Planck Medal. During the following years W. F. de Haas and his collaborators produced full confirmation of the theory with their measurements at the low-temperature institute at Leyden. Later I worked on the thermodynamics of super-conductivity and, following Fritz and Heinz London's lead, on an extension of Maxwell's theory to include super-conductivity. But not all the conclusions possible from this theory have as yet been checked by experiment; here many fundamental tasks remain for postwar times.

* * *

I wrote the above in 1944 for publication at that time. But the war and immediate post-war conditions interfered and only now (1951) is there a prospect for publication. In the meantime, however, much has happened which after all should also find a place in my autobiography and I have decided to continue the story. I considered for a long time whether this addition still belongs under the title 'My Development as a Physicist', or whether I had better choose for it the title 'Ausklang'

(Diminuendo). However the development of the true scholar ends only with his death, and in any case I have to go far back to make understandable what I have to tell.

As I said earlier in this story, when I left the Gymnasium I had to accommodate myself to military service no matter how much I considered universal military training to be an unjustified encroachment on personal liberty by the state and a waste of some of the most precious years of development. These were the years dominated by Bismarck and consequently very peaceful. Outwardly I did not fare badly after overcoming my initial difficulties and on the insistence of my father I even became a reserve officer and completed all the prescribed exercises. But inwardly my attitude to all things military remained wholly negative. I do not at this point want to go into the non-military circumstances which increased this opposition and finally brought on a severe illness. Suffice it to say that afterward I could request my discharge with a clear conscience. In 1914, at the outbreak of the First World War in which (as I thought at the time and still think) Germany suffered an injustice, I tried to rejoin the army. I even refused a position in Switzerland so that I might share the fate of the German people no matter how hard that would turn out to be. I was, however, rejected; the same old circumstances which I do not wish to discuss, stood in the way. Much as this pained me at the time, it turned out for the best, for when they wanted to put me back into uniform for the Hitler war, I could successfully request deferment by pointing to this previous rejection. Already in 1937 I had sent my only son to America so that he should not be compelled to fight for Hitler.

All my life I was prudent enough to remain, if possible, aloof from all political activities save voting in elections, even though I took a lively interest in politics; I knew my limitations. But who remained in a position to ignore politics after 1932 when absolutely everything was pulled into a political context and evaluated politically? My urgent sense of justice was particularly violated by the lawless capriciousness of National Socialism and my pride as a scholar was hit hard by Nazi interference in the freedom of science and of the universities. Although it could not touch me personally, I had—to put it mildly—an insurmountable aversion for anti-semitism ever since my schooldays. Before 1933 I never gave even a passing thought to the question of the 'race' of a new friend. Never before, not even in 1918–19 had I been therefore in such despair over the fatherland as I was during its death struggle in 1933–34 until it received, on 4 August 1934, the last fatal stab in the back.* Like many others I frequently quoted to myself the verse

> Denk ich an Deutschland in der Nacht
> So bin ich um den Schlaf gebracht.†

Often I asked myself upon awakening whether I had not been dreaming when I remembered the horrors of the preceding day. Unfortunately they were reality, cruel reality.

But I did not suffer my mood to paralyze me. As much as I could, I assisted those directly affected, for instance by warning them in time, especially those colleagues who had lost their positions. There were a few rare cases of colleagues who were able, of course not solely because of my assistance, to survive in Germany through that entire unfortunate period. Much more frequently I smoothed the way for those who emigrated by sending advance reports to the foreign aid organizations about their personalities, their domestic circumstances, their particular abilities and desires. Since the postal censor confiscated such reports, we found safer ways of sending them across the border. Once I even used my automobile to get a colleague who believed himself in danger, across the border into Czech territory. This was easier at the time than seems likely in retrospect.

All this had to be carried on as secretly as possible. But I declared my own attitude publicly as two documents appended here will substantiate. One contains the text of a speech which I gave as chairman of the German Physical Society at a physical society meeting in Würzburg on 18 September 1933, and had printed shortly thereafter in *Physikalische Zeitschrift*. The other is an obituary I wrote for the eminent physical chemist Fritz Haber, whose dismissal and banishment belong among the particularly inglorious achievements of Hitlerism. The obituary appeared in *Naturwissenschaften*, in the spring of 1934. Both earned me reprimands from the Ministry of Culture in Berlin. Apparently the ministry felt obliged to do something for my entertainment.

As professor at the University of Berlin and Acting Director of the Kaiser-Wilhelm-Institut für Physik I was tied to Berlin even after 1939. This situation eased up after the Ministry of Education declared me an emeritus on 1 October 1943, a year before I had reached retirement age, but with my consent. Remaining in Berlin I was exposed to the bombing raids on Berlin until the middle of April 1944. I witnessed

* After Hindenburg died, two days earlier, Hitler concentrated all power in himself.

† Thinking of Germany at night I find my sleep in headlong flight (Heine).

for example, in the unforgettable night of 15 to 16 February 1944, the burning of Otto Hahn's Kaiser-Wilhelm Institut für Chemie. A sea of flame burst forth in terrifying beauty from the framework of the roof and the shattered south wall of this monumental building. When the Kaiser-Wilhelm-Institut für Physik was moved away from Dahlem and bombs had made my house in Zehlendorf, if not uninhabitable, at least unpleasant, I went with the Institute to Hechingen. There at the foot of the Hohenzollern castle my wife and I had a peaceful year, and although fleets of planes would thunder by high over the town, they left it almost unscathed. And thus I recall these times as relatively happy despite the severe hardships of war. Toward the end I must admit the terrors of the Hitler regime came very close to us, as when a good friend was suddenly arrested in the street and we had reasons to fear that he would be shot. Even the preparations for the defense of Hechingen, childish as they were, would have meant great danger if there had actually been a struggle. Fortunately, the feeling prevailed that such hopeless resistance was criminal and it was with a sense of relief, in spite of the military measures that would inevitably accompany such 'conquest', that the inhabitants watched the French and red-Spanish troops enter the town without a struggle on 23 April 1945. All of this could surprise no one.

But it was a surprise when, on the following day, a squad of American and British soldiers appeared, and occupied and searched the Kaiser-Wilhelm-Institut für Physik. As it turned out later, this was part of the American 'Operation Alsos'. The name of this operation was a translation of the Commanding Officer's name, General Groves, into Greek (ἄλσος = grove).

Well-known scientists formed part of this mission, among them my good friend S. Goudsmit. To my amused surprise he suddenly appeared in our home, adorned with a steel helmet. However, he was carrying out his activities in deadly earnest. He was in deep mourning for his parents who had been taken from Holland to one of the concentration camps and there had been 'liquidated'. (Heisenberg and I had intervened as soon as we heard of such a possibility but we had been too late.) Well, as they departed the Americans took away with them a number of nuclear physicists including myself, thereby according me the completely undeserved honor of being accounted a nuclear physicist. With some additions this group increased to ten members. We were taken first to Heidelberg, then to Reims, then to the Paris suburb of Le Vésinet, next to the delightful country residence Faqueval south of Huy in Belgium and finally to Huntingdon in England. We had no complaints about the treatment accorded us and the military diet was most beneficial after the deprivations of the war period. At Huntingdon we had for our use not only the spacious, two-hundred-and-fifty-year-old country house 'Farmhall' but also parts of the adjoining gardens. We had a beautiful view of the park, stretching all the way to the river Ouse; we had English and American newspapers, journals and other literature as well as scientific books, and we had an excellent radio to bring us the many musical offerings of the BBC. Quite often one of the English officers guarding us took us for a ride in the beautiful surroundings of Huntingdon, to the famous cathedrals of Peterborough and Ely and even to London. In London, by the way, lived the only physician and dentist whom we were permitted to consult. We never went to Cambridge even though it was very close by, since presumably we would have been recognized in that university town. Our presence, after all, was a secret of the highest order. For the first four months we were not even allowed contact with our families and later on, when we could write them, we were not to disclose even the country in which we were being held. From these precautions one surmises a certain ambivalence on the part of the victors as to how they should deal with the German physicists whose competence on the one hand they esteemed greatly and yet, on the other hand, feared as potentially dangerous. The situation had its comical aspects but was emotionally oppressive for us. Nevertheless, I was able to write a paper in Huntingdon on the absorption of X-rays under the conditions of interference which later appeared in *Acta Crystallographica.* Frequent colloquia also helped to keep us mentally alert.

Early in 1946 we were allowed to leave England and, after a short intermediate period in Westphalia, most of us found ourselves in Göttingen, where I stayed until April 1951 and where my activities included first once again the post of Acting Director of the Kaiser-Wilhelm-Institut für Physik, and also soon thereafter an adjunct professorship at the University. I wrote a book *The Theory of Superconductivity*; revised my book on electron diffraction (*Materiewellen*) in parts for a new edition, and made several contributions to scientific journals. Also I was able to almost complete the manuscript of a revision of that very first book of mine, *Relativity Theory*, which I mentioned earlier and which had long been out of print. In April of 1951 I accepted an offer to become director of the Institute for Physical Chemistry and Electrochemistry of the Forschungshochschule in Dahlem, the former Kaiser-Wilhelm-Institut founded by Fritz Haber. That one should be offered and accept such a position at the age of 71 is certainly singular. But then, is not all of present day Berlin a singularity?

I have often been asked why I did not emigrate during the Hitler times. There were many reasons for this: for one thing, I did not want to take away any of the few posts available abroad from colleagues who needed them more urgently than I did. Above all, however, I wanted to be immediately available when the collapse of the Third Reich, which I had always foreseen and long hoped for, would make possible cultural reconstruction on the ruins created by that regime. This became the main direction of my activities after 1945. In 1946 it was my privilege to initiate the founding of the 'Deutsche Physikalische Gesellschaft in der Britischen Zone' (German Physical Society in the British Zone),* and in 1950 to participate in the creation of the 'Verband Deutscher Physikalischer Gesellschaften' with which the former affiliated under the name of 'Nordwestdeutsche Physikalische Gesellschaft'. From the end of 1946 to the middle of 1948 I worked on the reunification of the Physikalisch-Technische Reichsanstalt which had been widely dispersed during the war. In 1949 this institution was renamed Physikalisch-Technische Bundesanstalt. In both these enterprises the participation of the British Research Branch was decisive and one cannot be grateful enough to Dr. Ronald Fraser, its director at the time. It was he who procured for the Bundesanstalt its present large tract of land near Braunschweig with its undamaged German air force buildings, which offered undreamed-of possibilities for expansion. However the corresponding American and French organizations also deserve thanks. They appeared on the scene later when it had to be decided that the Bundesanstalt should be a joint enterprise of all three Western occupation zones, and they then acted with sincere goodwill toward German science and economy. Under the circumstances administration of the Bundesanstalt could be taken over by Germany only after the founding of the Deutsche Bundesrepublik; and it has been well supported ever since.

Not only inside Germany did gaps have to be filled after 1945, but also the relations between foreign and German scholars had to be gradually re-normalized. My personal knowledge in this field is limited to physics and crystallography but there, I am convinced, we will succeed within a few years if politics in general continue to be calm. Conditions are rapidly returning to normal as I myself could see during several trips abroad and as is confirmed by all the reports of

* The Allied Control Commission did not authorize organization across the boundaries of the occupation zones into which Germany had been split up.

other Germans. Foreigners, especially Englishmen and Americans, come in growing numbers to German scientific conferences. In this connection I must tell of two incidents which are simply unique.

Once, while we were still imprisoned in Huntingdon, on 2 October, 1945, to be precise, O. Hahn, W. Heisenberg and I were invited to tea by the President of the Royal Institution in London, Sir Henry Dale. One of our English guards took us there by car; punctually at 5 p.m. our host received us at the entrance and led us up the stairs to his private appartments where we were offered the usual afternoon tea. We met there a small group of famous members of the Royal Society, not only physicists, and discussed with them the future of science in the British Zone. To preserve at least the formalities of secrecy that surrounded us, our hostess, Lady Dale, was not permitted to be present, nor, of course, any of the servants. Promptly at 7 our host rose, since a longer visit was not permitted, and accompanied us Germans to the entrance door where the car was already waiting to drive us back to Huntingdon.

For me this tea held a special surprise. I received an oral but completely official invitation to attend a special session of the Royal Society on 9 November 1945, to be held in memory of Röntgen's great discovery, which according to his diary was initiated by an accidental observation on 9 November, 1895. Full of wonder I accepted the invitation expressing of course my doubts that the military authorities would allow me to attend since they were after all holding us in the strictest secrecy. My doubts were completely justified. I did not get the permission. To be consistent, the military authorities could not agree. But consider the implications of this invitation: exactly half a year after the most terrible fighting there had ever been, the Royal Society honours a scholar of the enemy nation with a Commemorative Session and invites another scholar of that nation to attend! I do not believe that any other scientific organization can boast of such an action.

My experiences in London in July 1946 reflect no less glory on the Royal Society and all its scholars. First there was an international conference on crystallography which dealt, as is usual today, mainly with questions of crystal structure and interference phenomena of X-rays and electrons. I was the only German invited and in this way returned to England only four months after the end of my imprisonment there. On the journey there and back I was escorted by a British officer and in London my charming host, a well-known English crystallographer, stayed with me for the first few days. After that I could wander freely in London and its suburbs, and I found that being a German (which I did not wish, nor indeed could conceal) I had no need to fear insults from the inhabitants. I often had to inquire about street directions or means of transportation and always received friendly and correct information. Once in a suburban train I happened to sit opposite a native Englishman, who was using the time on the train to read a German short story, I believe by Storm. Who in Germany would have dared to read an English book in a public conveyance during World War I or the Hitler period?

This is not the place to report on the scientific discussions of that meeting. At the official banquet the chairman made an after-dinner speech in which he remarked in extremely warm words on my attitude during the Hitler period. In my reply I stressed that I was most grateful for his praise but that thousands upon thousands of German men and women were at least equally deserving of it.

Immediately after the conference came the Royal Society's Newton celebration. Actually this should have taken place in 1942 on the anniversary of the day in 1642 when this great scholar died. The only German actually invited was Max Planck, who had long been a Foreign Member of the Royal Society and who came despite his 88 years. But I went with a bachelor member of the Society, who used the invitation he had received by mistake for his non-existing wife to bring me along to the festivities held in the reception rooms of the Royal Society as part of the Newton Celebration. There were gathered together scholars and scientists from all nations, with the possible exception of Russia. And they all came up to me—not only Englishmen, but also Americans, Frenchmen, Poles, Czechs, Hungarians, Scandinavians, Italians, Belgians, Dutchmen, etc.—sometimes without a word, to shake hands. It was not easy to hide how moved I was.

The freedom of science which we lost in 1933 we have not yet regained; not only such things as control commissions for nuclear research stand in the way. Research can also not survive when financial support is reduced below a subsistence level; and this is now the case in many universities. The oppression is worse in the East Zone than in the West Zones. The continuing escape of scholars from there is clear enough evidence for this. Like most other West-Berliners I feel as if I were occupying an outpost against the advance of anti-spiritualism. It is not an easy position, and there are moments when one doubts that this struggle has any hope of success. But then I read the great old Goethe:

Feiger Gedanken
Bängliches Schwanken,
Weibisches Zagen,
Ängstliches Klagen
Wendet kein Elend,
Macht Dich nicht frei.

Allen Gewalten
Zum Trutz sich erhalten,
Nimmer sich beugen,
Kräftig sich zeigen,
Rufet die Arme
Der Götter herbei.*

Appendix

In an address and an obituary note I have tried to illuminate in broader perspective some of the stages in the development of the physical world picture.

First, the *opening address at the physicist's convention in Würzburg, 18 September 1933:*

'When we convene tomorrow in the institute of physics of this university we will meet on historical ground. In this building, toward the end of 1895, Wilhelm Conrad Röntgen discovered the rays named after him. To discuss their importance for physics and the whole range of their application would be superfluous, if not in bad taste. But let us remember the grandeur of the achievement of this man who was the first to see consciously and to raise from the twilight of uncertainty into the full light of sure scientific knowledge what others before him had approached closely and passed by. We know from the Röntgen biography by Glasser that across the ocean there exists a genuine X-ray photograph, taken in 1890, which was recognized as such only after Röntgen's work was published.

'Then, in this place we remember Röntgen's successor Willy Wien who taught and did research in this building for 17 years. His classical theoretical work concerning heat radiation, which, in the form of Wien's displacement law, became an integral part of our science, dates back to his early years under Helmholtz at the Physikalisch-Technische Reichsanstalt. But nowhere else could he devote so many years with so many students and friends to quiet and highly productive experimental research as here. The knowledge of canal rays which was gained here forms no small part of the contents of those magnificent volumes of the *Handbuch* which are just coming out on this subject.

* Timid indecision, cowardly thought,
Unmanly hesitation, anxious lamentation
Bans not misfortune, makes you not free.
Stand up defiant of force!
Never bow under! Prove yourself strong!
This calls the arms of the gods to your side.

I want to mention especially two of Wien's papers of this period because they were truly fundamental: the first quantum theoretical determination of the frequency of X-rays, which gave only an approximate but basically correct value; then the experimental confirmation of the relativity of electromagnetic fields, demanded by relativity theory: light from the particles of canal rays moving in a magnetic field, shows the same spectroscopic Stark effect which is familiar for stationary light sources in an electric field.

'But this same building has also experienced entirely different things: During the World War [I] Wien transformed his institute and, in cooperation with Max Seddig, also the adjoining chemical institute, into a laboratory for the investigation of all the apparatus needed for modern communication technology, and especially for the production of amplifier tubes. The communications system of the army we put into the field in 1914, in contrast to other preparations and equipment, was not at all up to date; in this our enemies were far ahead of us. If we caught up, slowly but very substantially, in the course of the next four years, the physicists who worked in this building deserve a good part of the credit.

'Physics observed a memorial day of special significance on 22 June of this year. It was on this day three hundred years ago that the trial of Galileo before the Inquisition came to an end. The cause for this trial was, as is well known, the theory of Copernicus concerning the movement of the earth and all other planets around the sun, a theory which was in conflict with then accepted views and aroused as much attention and excitement in those days as relativity theory does in our century. Galileo was not its only but its most successful defender, because he was able to support it so convincingly with his own wonderful discoveries of the moons of Jupiter, the phases of Venus and the rotation of the sun around its own axis. Galileo was found guilty. He had to abjure the Copernican system and was condemned to life imprisonment. To be sure, this restriction of his freedom was administered rather gently; he was assigned a house which he could not leave and in which he could not even receive visitors without permission. Actually he was able to publish his principal physical treatise, *Dialogue concerning two New Sciences*, while in custody. Nevertheless, he remained a prisoner for the rest of his life.

'This condemnation has attached to it the well-known legend that Galileo, while renouncing the doctrine of celestial movement and signing the abjuration, said: "And yet it moves". It is a legend, historically unprovable and inconsistent, and yet it is ineradicable from the popular mind. What is the source of this legend's persistence? Most probably the fact that Galileo must have asked himself throughout the proceedings of the trial: "Of what importance is all this? Whether I, whether any human being now affirms it or not, whether political, whether ecclesiastical power is for or against it, this does not change the fact at all. Power may hold up knowledge of it for a while, but still, someday it will break through." And so indeed it did; the triumphal march of Copernican theory was irresitible. The Church itself which had damned Galileo, gave up its opposition with all due formality, even if only after two hundred years.

'Also later there occurred occasional bad times for science, as in Prussia under the otherwise meritorious Friedrich Wilhelm I. But under all oppression the representatives of science could find comfort in the victorious certainty which expresses itself in the simple sentence: *And yet it moves*.'

The obituary note of Fritz Haber (1934, *Naturwiss*. *22*, 97) delineates in a few strokes his importance:

'On 9 December 1928, Fritz Haber's sixtieth birthday, a small group of friends and colleagues assembled in front of the Kaiser-Wilhelm-Institut für physikalische Chemie und Elektrochemie, *his* institute, to plant there the Haber-Linde* while he himself was sojourning in the south. This was the only birthday celebration; however, like other scientific journals, the *Naturwissenschaften* published a special issue with contributions from the most notable researchers in Haber's field. To it let the reader turn to discover to its full extent the loss which chemistry and physics, agriculture and industry both for war and for peace have sustained by Haber's death. But he will discover there too that Haber's greatness only in part finds expression in his own publications. His qualities as director were perhaps even greater—while his collaborators had full freedom to develop their own abilities, his was the all-pervasive spirit in the institute. His sixtieth birthday did not put a stop to his work. Oxydation and explosion processes and how they are influenced by atoms and radicals, auto-oxydation and reduction in solutions, especially in biological processes, are themes and projects of his own researches during the last years. Of work done by others in his institute let me mention Bonhoeffer and Harteck's separation of para- and ortho-hydrogen and their use for the exploration of complicated chemical processes and of magnetic

* A tree, called in English variously Linden- or Lime Tree or Basswood.

properties of substances. Throughout its existence his institute was known far and wide as a home for widely diversified scientific inquiry. On the 2nd of May, 1933, Haber submitted his resignation.

'Themistocles is known to history not as an exile at the court of the Persian king, but as the victor of Salamis. Haber will be known to history as the ingenious inventor of the process of the fixation of nitrogen with hydrogen which is the foundation for the industrial extraction of nitrogen from the atmosphere; as the man who thereby—as was said at the presentation of his Nobel prize—"has created a most important means for advancing agriculture and the welfare of man"; who created bread from air, and excelled triumphantly "in the service of his country and of all mankind".

'And what remains for those who knew him and now mourn him? Memories—"His friends and collaborators know the goodness of his heart, his loyalty, his chivalry and gentleness, his indignation at sordid self-interest and wickedness, and the happy equanimity with which he overlooks minor faults. For those younger than himself he has an untiring kindness, endeavouring constantly to give hope and to point out new approaches; he has no patience for narrow-mindedness and intolerance; while he appreciates what is excellent in other countries and cultures, his heart beats for Germany: for like no other man he helped this country to protect and feed its children in its time of greatest distress."

'Thus wrote Margrete von Wrangell in 1928. Thus we remember Haber. For he was ours.'

*William Henry Bragg**

1862-1942

In the late 'fifties Robert John Bragg, a young man of twenty-five, retired from the sea, where he had been serving as an officer in the merchant navy, and purchased with some moneys that had been left to him the farm called Stoneraise Place at Westward, near Wigton, in Cumberland. Here he settled down to a farmer's life. In 1861 he married Mary Wood, the daughter of the Vicar of the parish of Westward, and the next year, on 2 July 1862, William Henry Bragg, later to be President of the Royal Society, was born.

Fortunately there are available certain notes dealing with his early life which Bragg himself prepared some time before his death.† These contain many vivid little pictures which, besides giving us intimate glimpses of those early days, will, by their manner of telling, recall the man to those who knew him, and give to those who did not something of his direct and unaffected charm of style. Thus, of his mother he says, 'I do not remember my mother very well, as she died when I was barely seven. Just a few scenes remain. I think she must have been a sweet and kind woman. I remember how one day I was sitting on the kitchen table, and she was rolling pastry, and how I suddenly found I could whistle: and how we stared at one another for a quiet moment amazed and proud of the new accomplishment...'

Of the school at Market Harborough, in Leicestershire, to which he went at the age of seven, he writes, 'Uncle William had in 1869 succeeded in re-establishing the old grammar school in Market Harborough. It is a quaint structure raised on wooden pillars. The butter market used to be held underneath it. The newly appointed

* By E. N. da C. Andrade; reprinted by kind permission from Obituary Notices of Fellows of the Royal Society, Vol. 4, 1943, pg. 277–292. The original is followed by 8 pages of bibliography of Sir William Bragg's papers.

† The family has very kindly placed them at my disposal. They were mainly written about 1927, small additions having been made in 1937.

master, Wood by name, was an able man, I believe: and the school grew. I was one of the six boys with which it opened after a long interval. Perhaps it was because of my uncle's connexion with the school that at the end of the first year I was given a scholarship of £8 a year exempting me from fees. At the prize-giving—there were many more than six boys at that time, so that there was quite an assemblage —my name was called out and I went up to the desk to get the scholarship, not knowing what it was: I was puzzled and disappointed to go back empty-handed. The school was quite good and I got on quickly enough: in 1873 I went up for the Oxford Junior Locals and was the youngest boy in England to get through: I got a third class, and was told that I would have done better but that the regulations forbade a higher class to any one who did not pass in Church History; in that I failed, as also in Greek.'

In 1875 he left Market Harborough for King William's College in the Isle of Man. 'The place was a very healthy one and after the first year or two, when the bullying was rather unpleasant, I was happy enough. I stood high in the school and liked my work, especially the mathematics: and fortunately I was fond of all the games and played them rather well. So, though I was a very quiet, almost unsocial boy, who did not mix well with the ordinary schoolboy, being indeed very young for the forms I was in, I got on well enough.' He rose to be head of the school. Early in 1880, when he was seventeen, he went up to try for a scholarship at Trinity College, Cambridge, and was awarded an Exhibition: he was considered rather young and advised to go back to school for a year. The following year he tried again and did not do so well, but was elected to a minor scholarship on the strength of his previous performance. In his notes Bragg puts his 'stagnation' down to a storm of religious emotionalism that swept through the school—the boys were scared of eternal damnation and of hell fire, and very much exercised as to what they should do to be saved. 'It really was a terrible year', says Bragg, who, though essentially a religious man, adds, 'But for many years the Bible was a repelling book, which I shrank from reading.' This from his private notes, but the period evidently left a strong mark on his mind, for in the Riddell Memorial Lecture on 'Science and Faith', given in the year before he died, he says, 'I am sure that I am not the only one to whom when young the literal interpretation of Biblical texts caused years of acute misery and fear'. What he says in this lecture should be in the hands of all those who have to deal with susceptible adolescence.

At Trinity, which he entered in 1881, he enjoyed himself, although he was a somewhat shy and lonely lad. He liked the work and the tennis ('my tennis was fairly good') and the boating; he was keenly alive to the beauty of the place and to his good fortune in belonging to Trinity. In 1882 he obtained a major scholarship, which gave him an added status which he appreciated. In 1884 he was placed third wrangler in the Mathematical Tripos, Part I. 'I had never expected anything so high, not even when I was in my most optimistic mood. I was fairly lifted into a new world. I had a new confidence: I was extraordinarily happy. I can still feel the joy of it! Friends congratulated me: Whitehead (of Harvard now) came and shook me by the hand, saying, "May a fourth wrangler congratulate a third?" He had been fourth the year before. As for the Uncles!'

He continued at Cambridge, taking part three of the Tripos as it then was and attending lectures, including those being given by J. J. Thomson, who had been appointed Cavendish Professor at the end of 1884. It was at the end of 1885 that J. J. Thomson asked him if Sheppard, who had been senior wrangler in Bragg's year, was applying for the professorship in mathematics and physics at Adelaide, which had just become vacant by the resignation of Horace Lamb. Lamb had held the post since the foundation of the university in 1875. Sheppard was not a candidate, but the query put it into Bragg's head that a man of his own age and qualifications might have a chance for what he had till then regarded as a post for a senior man. Accordingly he sent in his application. With two others he was interviewed by the electors, who were the Agent General (Sir Arthur Blyth), J. J. Thomson and Horace Lamb, and in due course learned that he had been chosen for the post. The part played by J. J. Thomson in the events that sent Bragg to Australia was described in a letter which he wrote to J. J. on 17 December 1936, to convey congratulations on the Master's eightieth birthday. '... I must be allowed to add my own personal congratulations. Just fifty-one years ago, I was walking with you along the K.P. on our way to the Cavendish where you were going to lecture and I was going to be one of the audience. You asked a chance question, which sent me off to the telegraph office after the lecture was over and I applied for the Adelaide post which Lamb was vacating. It was the last day of entry; and of course your remark sent me to Australia. Perhaps you were the one who asked a certain Adelaide man—then visiting London—whether the Council of the University of Adelaide was likely to prefer a senior wrangler who occasionally disappeared under the table after dinner to a young man who had so far shown no signs of indulging in the same way. The Adelaide man was Sir Charles Todd, whose daughter I married a few years afterwards.'

Bragg was much elated. The salary of £800 a year, a very respectable income in those days, was much beyond what he had expected at his age, and the thought of going to a new country and being his own master excited him. His relations were delighted, in particular his Uncle William, who had always helped him and was very proud of his academic career, although the parting was a great blow to the old man. Bragg enjoyed the voyage out, and was fond of telling in after life how he spent part of his time on the boat in studying physics. The post was that of professor of mathematics *and* physics, and Bragg was wont to declare that, at the time, he had not studied the latter subject and knew nothing of it. The electors apparently attached little importance to his inexperience in this respect, and supposed that he would pick the subject up as he went along. At any rate, he read Deschanel's *Electricity and Magnetism* while outward bound. Apparently the demands on a professor of physics were not very high in those days.

The life in Australia delighted him from the first. 'Going to Australia was like sunshine and fresh invigorating air,' he wrote. He at once made friends all round, particularly with the Todds. Charles Todd, the head of the family, was Postmaster General and Government Astronomer of South Australia; he was elected a F.R.S. in 1889 and later became a K.C.M.G. In 1889, three years after he landed, Bragg married his daughter Gwendoline. The three children of this very happy marriage were born in Australia—William Lawrence, the present Sir Lawrence Bragg; Robert, who was killed in the Dardanelles during the First World War; and Gwendolen, now Mrs Alban Caroe.

Bragg's simple, modest, unaffected nature, which had rendered him shy in the conventional Cambridge surroundings of those days, expanded in the free, open, good-natured atmosphere of Adelaide society, which, he said, was a revelation to him after the more formal life of Victorian England. Being a personage agreed with him. It is recorded that in his early days in Australia he was one of the least impressive lecturers, but by careful application he developed towards that perfection in the art of exposition which he afterwards attained. His social gifts won him a wide popularity. He helped to lay out the first golf course in South Australia, at Glenelg, played golf, a game at which he was good, painted in water colours and enjoyed himself. Bicycling came into vogue during his Australian days, and he loved bicycle tours and picnics during the long lazy summer vacations by the sea. It was a pleasant life for a man with a zest for the good simple pleasures of the open air.

More important for science, he became interested in experimenting, of which he had had little or no experience in England. Following Röntgen's great discovery in 1895 he set up the first X-ray tube to operate in Adelaide, possibly the first in Australia. He still, however, made no attempt to carry out any original investigation, and no doubt to those around him it might well have appeared that he was destined to lead a pleasant and useful life as a popular teacher and good friend in the Adelaide community, content with his local fame. At the age of forty-one he had amassed a long experience of teaching the fundamentals of physics and had developed a keen critical sense: he was a clear and mature thinker but had produced nothing that could be called research. This maturity and long training in exposition were of immense advantage when he turned to discovery.

In January 1904 the Australian Association for the Advancement of Science met in Dunedin, New Zealand, and Bragg was confronted with the task of giving the presidential address in the section dealing with astronomy, mathematics and physics. At the time the early discoveries in radioactivity were holding the attention of the world of physics and work on the electron was being actively pursued in the Cavendish Laboratory and elsewhere. Exciting discoveries in the new fields were pouring out. In particular, Lenard's second famous paper on the absorption of cathode rays by matter had just appeared. In his address Bragg reviewed the work on radioactivity and on the properties of the electron and, in consequence, he became greatly interested in the question of the penetration of matter by elementary particles. At the time the atom was held to consist either of electrons widely dispersed in a sphere of positive electricity (J. J. Thomson) or of widely spaced 'dynamids' (Lenard), a dynamid consisting of a positive and a negative charge, closely associated. This conception of a neutral pair was one that Bragg used later in another connection. In either picture there would be comparatively strong, highly localized electric forces separated by large spaces in which the forces were feeble. The scattering of the beam of electrons in Lenard's experiments was put down as the effect of a series of random deviations due to passages close to the force centres.

It occurred to Bragg that, if this were so, an alpha particle should, on account of its mass, pass through a thin foil practically undeviated and that, in consequence, the exponential laws of absorption, which held for a beam of electrons, should not apply to a beam of alpha particles. In his own words, taken from his 1904 address,* 'it cannot be correct

* See W. H. Bragg, *Studies in Radioactivity*, London, 1912.

to say that the amount of the radiation which penetrates a distance x is proportional to the expression e^{-ax}: it must rather be proper to say that—

(1) The number of α particles penetrating a given distance does not alter much with the distance until a certain critical value is passed, after which there is a rapid fall.

(2) The energy of the α particles penetrating a given distance gradually decreases as the distance is increased, and dies out at the same critical value.' The passage is quoted, since it shows the typical Bragg clarity and precision, the power to put in the simplest language the essence of a novel problem. Until he had reduced his thoughts to the state where simple and direct expression was possible he was not at ease.

A few months after the address some radium bromide was placed at his disposal, and, assisted by Kleeman, he began his classical researches on the range of the alpha particle and the closely allied questions of the ionization produced by the particle and of the stopping power of substances. Using as detector an electrometer which measured the ionization produced in a thin slab of air between a gauze and a plate, he established the sharply defined range of the alpha particle and the variation of the ionization along the path, obtaining with radium which had come into equilibrium with its products the four ranges, corresponding to the particles from radium, radon, radium A and radium C, according to Rutherford's theory, which was thus confirmed. With Kleeman he found that the 'stopping power' of matter was approximately proportional to the square root of the atomic weight. 'Considering the complexity of the phenomena involved in the absorption of an α particle by matter, it is a matter of interest and also of practical convenience that such a simple rule should hold roughly over the whole range of the elements.' * He corresponded freely with Rutherford about the work on the α particle, whose properties they were pursuing independently along different lines.

He also carried out fundamental work on the ionization produced by α-rays and on the properties and behaviour of the secondary electrons expelled from matter by incident β and γ radiations, electrons to which Bragg applied the name β-rays, or secondary β-rays. In some of the experiments on these β-rays he had the collaboration of Madsen, now Sir John Madsen.

The experiments on the secondary electrons expelled by γ-rays were

* Rutherford, Chadwick and Ellis, *Radiations from Radioactive Substances*, Cambridge, 1930.

important for the development of Bragg's theoretical views. They established that the velocity of the β-rays depended not upon the intensity but upon the 'quality' of the γ-ray, increasing with the penetrating power of the latter, and that the velocity was independent of the atom from which the γ-ray was expelled. These results were shown by Sadler and others to apply to the β-rays produced from matter by incident X-rays. It was largely on experiments of this kind that Bragg based his views that X-rays were corpuscular in nature, views which his own fundamental experiments later showed to require modification. The paradoxical coexistence of particle and wave properties was a favourite theme of Bragg's, to which we shall return.

The fundamental researches carried out, in quick succession, in Adelaide between 1904 and 1908 speedily established his name throughout the world of physics as an original investigator of the first rank. He was elected a Fellow of the Royal Society in 1907, less than three years after the reading of his first original paper, Rutherford being his proposer. It was inevitable that he should be called to a chair in the old world. In 1908 he received the offer of the Cavendish professorship of physics at Leeds, which brought him back to England.

During his twenty-two years in Australia Bragg had identified himself with the life of the community in which he lived and had established himself as a good man, a great teacher and a firm friend. He had many ties with Adelaide, and had thoroughly enjoyed life there. In after days he always spoke with the greatest affection of South Australia.

At Leeds Bragg was, at the beginning, fully occupied with organizing the teaching at the laboratory, and naturally, for a time, he did comparatively little experimental work. He developed and defined his view that X-rays and γ-rays were of a corpuscular nature, the lack of deflection in a magnetic or electric field being explained by the hypothesis that an elementary ray was in the nature of a neutral doublet—'an electron which has assumed a cloak of darknes in the form of sufficient positive electricity to neutralize its charge' — to use Bragg's own typically arresting phrase. Although this picture can no longer be deemed satisfactory it did account for many properties of the radiation which are embodied in the quantum theory. In particular Bragg insisted that many of the experimental facts seemed to show that an elementary X- or γ-ray was a definite and concentrated unit rather than a spreading pulse or wave, as indeed they do. The doublet theory was also successful in explaining in a general way certain aspects of the conversion of cathode rays into X-rays and of the release of electrons by X- and γ-rays. It had its usefulness in concentrating attention on the particle aspect of the radiations, which plays so prominent a part in modern theories.

In the course of his work on secondary radiation Bragg was led to the view that the ionizing effect of the X-rays is an indirect one, due to the secondary electrons released by the primary X-ray. He was the first to insist on this important fact, and he supported his contention by experiments carried out in conjunction with H. J. Porter, which were published in 1911.

Laue's discovery, with Friedrich and Knipping, announced in June 1912, that X-rays could be diffracted by passing through crystals, caused a sensation in the physical world. Bragg's interest was at once captured, and it may be of interest to quote what he wrote in *Nature* in November of that year: 'Dr Tutton suggests that the new experiment may possibly distinguish between the wave and the corpuscular theories of the X-rays. This is no doubt true in one sense. If the experiment helps to prove X-rays and light to be of the same nature, then such a theory as that of the 'neutral pair' is quite inadequate to bear the burden of explaining the facts of all radiation. On the other hand, the properties of X-rays point clearly to a quasi-corpuscular theory, and certain properties of light can be similarly interpreted. The problem then becomes, it seems to me, not to decide between two theories of X-rays, but to find, as I have said elsewhere, one theory which possesses the capacity of both.' The theory of the diffraction spots which Laue obtained by consideration of a three-dimensional grating is somewhat complicated, involving as it does the consideration of interfering wavelets in three-dimensional space. The same year W. L. Bragg gave a much simpler interpretation of the phenomena, by considering the reflection of waves from parallel layers of atoms or diffracting points, each typical set of parallel crystal planes acting as a reflecting surface for radiation whose wavelength fulfilled the Bragg law $n\lambda = 2d \sin \theta$, where d is the distance between parallel crystal planes and θ is the glancing angle, i.e. the complement of the angle of incidence. The more densely populated crystal planes gave, speaking generally, stronger reflections, so that planes with higher indices were not strongly represented. W. H. Bragg at once took up, with his son, the experiments on the reflection of X-rays which this interpretation suggested, and early in 1913 there appeared in the *Proceedings* of the Society the first joint paper, which founded the science of crystal analysis by means of X-rays.

Up to the beginning of the first Great War in 1914 five further classical papers were produced by Bragg, in one of which, that on the structure of diamond, he had the collaboration of his son. Among other subjects dealt with were the general technique of the X-ray spectrometer; the characteristic absorption of the different radiations and its effects; the structure of sulphur and quartz; and the general question of intensities. An investigation with S. E. Peirce led to the Bragg-Peirce law, according to which if we keep to frequencies below the band at which the absorption discontinuity takes place, the absorption coefficient per atom is proportional to the fourth power of the atomic number and to the 5/2 power of the wavelength.* In these early experiments Bragg made use of the ionization chamber to detect and measure the rays. His earlier work had taught him how to overcome the difficulties connected with this type of measurement and he and his son were strikingly successful with the ionization spectrometer. The photographic method had already been used by H. G. J. Moseley in his classical researches, but it was only later that Bragg adopted it.

The work of Bragg and his son Lawrence in the two years 1913, 1914 founded a new branch of science of the greatest importance and significance, the analysis of crystal structure by means of X-rays. If the fundamental discovery of the wave aspect of X-rays, as evidenced by their diffraction in crystals, was due to Laue and his collaborators, it is equally true that the use of X-rays as an instrument for the systematic revelation of the way in which crystals are built was entirely due to the Braggs. This was recognized by the award of the Nobel prize for Physics in 1915 to them jointly 'pour leurs recherches sur les structures des cristaux au moyen des rayons de Roentgen', and a further formal acknowledgment was the appearance in Leipzig, in 1928, of a collected reprint, in German translation, of the early papers, under the title *Die Reflexion von Röntgenstrahlen an Kristallen: grundlegende Untersuchungen in den Jahren* 1913 *und* 1914 *von W. H. Bragg und W. L. Bragg.*

The outbreak of war in 1914 found Bragg hard at work at Leeds, where all his early experiments on crystal structure were carried out. He was a leading figure in the university there, and occupied the office of Pro-Vice-Chancellor. He continued his X-ray work into 1915, publishing, for instance, a paper on the spinel group of crystals. In this year he was appointed Quain Professor of Physics at University College, London, but by then he had become involved in war work.

The Board of Inventions and Research was instituted in July 1915,

* The value of the exponent is usually taken as 3, instead of 5/2, to-day.

for the purpose of giving the Admiralty expert assistance in organizing and encouraging scientific effort in connection with the requirements of the Naval Service, and Bragg was an original member. The submarine menace was becoming acute and the use of acoustic methods for locating underwater craft came up for discussion before the Board, with the result that Bragg was put in charge of research on the detection and measurement of underwater sounds, within the Anti-Submarine Division of the Admiralty. In the first instance he was, in April 1916, made Resident of Research at the Admiralty experimental station at Hawkcraig. After many troubles, largely within the Admiralty (see J. J. Thomson's *Recollections and Reflections*), a laboratory was built for him at Parkeston Quay, Harwich, where Bragg started work in 1917, having under him, among other physicists, A. O. Rankine. The hydrophone or underwater receiver, developed by Bragg and his team, rendered great service in the war against the submarine. The instrument was afterwards simply described by Bragg in his *World of Sound*, a book founded on his first course of Christmas Lectures at the Royal Institution, and in two lectures of which an account is given in *Engineering* for 13 June 1919. During the course of the experiments and research on anti-submarine work principles were established and methods, as well as apparatus, devised which were of great service. It was probably in acknowledgment of his war work, as well as of his scientific eminence, that Bragg was made a C.B.E. in 1917 and was knighted as a K.B.E. in 1920. In the same year, 1920, he was made an Honorary Fellow of Trinity College, Cambridge, a distinction that gave him great pleasure.

The war over, Bragg took up his work as Quain Professor of Physics at University College, London, and promptly started research there. He gathered about him several young research workers, among whom Backhurst may be mentioned, and founded that school of searchers after the secrets of crystal structure which later flourished at the Royal Institution.

In 1921 Shearer and Astbury joined him, and shortly after Müller and Miss Yardley (Mrs Lonsdale) appeared. In those days Bragg was himself actively experimenting with his own hands, as well as directing research: he was seated at his spectrometer whenever he got the chance. At the beginning of his work at University College he still used his first well-tried weapon, the ionization chamber, as detector, but gradually gave it up, for most purposes, in favour of the photographic plate. The equipment at University College was somewhat scanty at first and Bragg and his students, Müller and Shearer in particular, set to work to develop it. Continuously evacuated X-ray tubes, both hot wire and gas filled, were introduced and a self-rectifying gas tube was evolved, which gave useful service for many years. In those days vacuum pumps had not reached their present state of efficiency and hot wire tubes were rather trouble-some. The work was supported by generous grants from the Department of Scientific and Industrial Research, which were more than justified by the results.

The University College period was notable for the first attack on the structure of organic crystals. For much of this work Bragg employed the powder method, whereas hitherto he had worked with single crystals. He embodied his results on naphthalene and naphthalene derivatives in his presidential address to the Physical Society in 1921, having been elected to the presidential chair in the previous year. He worked on the assumption that the benzene or naphthalene ring is an actual structure, which preserves its general form and size from compound to compound and, to the satisfaction of the organic chemist, his results justified this hypothesis. This work was the starting point of the series of investigations on different classes of organic compounds which he afterwards directed at the Royal Institution. He also worked out the probable structure of ice, and at an annual dinner of the Alpine Club exhibited a model, which, being made of soft dental wax, proved itself by wilting as the evening grew warmer.

In 1919 occurred an event, apparently insignificant, that was to have a profound effect on Bragg's life—he gave the Christmas Lectures ('A Course of Lectures adapted to a Juvenile Auditory' as the old phrase runs) at the Royal Institution. The title of the course was 'The World of Sound' and it not only established Bragg's name as a popular lecturer but also showed the originality, the personal qualities, which he could bring to an apparently hackneyed subject. He classified his subject on new lines, the six lectures being entitled 'What is Sound?' 'Sound and Music', 'Sounds of the Town', 'Sounds of the Country', 'Sounds of the Sea', 'Sounds in War'. 'Sounds of the Country' in particular, showed his love for, and careful observation of, nature out of doors. His powers of simple exposition, his personal and persuasive style, his affection for young people—all became known to a wide circle.

It may be worth while to quote the opening sentences from the book that embodied these lectures, as they give some idea of his style. 'All around us are material objects of many kinds, and it is quite difficult to move without shaking some of them more or less. If we walk about on the floor it quivers a little under the fall of our feet; if we put down a cup on the table, we cannot avoid giving a small vibration to the table and the cup. If an animal walks in the forest, it must often shake the leaves or the twigs or the grass, and unless it walks softly with padded feet it shakes the ground. The motions may be very minute, far too small to see, but they are there nevertheless.' The words are simple and colloquial, with no straining after effect, but they are striking and tell us at once that we are in the presence of one who was a lover of nature and a master of simple, direct and captivating exposition.

It was, perhaps, partly as a result of this course of lectures that, when on the death of Sir James Dewar in 1923 the headship of the Royal Institution fell vacant, Bragg was elected to succeed him under the complex of titles Fullerian Professor of Chemistry in the Royal Institution; Director of the Laboratory of the Royal Institution; Superintendent of the House; and Director of the Davy-Faraday Research Laboratory. The Fullerian Professorship of Chemistry is a traditional title which carries somewhat indefinite duties; the Laboratory of the Royal Institution and the Davy-Faraday Research Laboratory are run more or less as a whole, the Managers of the Institution administering the Davy-Faraday funds and the Resident Professor in the Royal Institution being also the Director of the Davy-Faraday Laboratory. The superintendence of the House entails the general responsibility for the maintenance and smooth running of the Institution which the head would naturally be supposed to exercise; the Davy-Faraday Laboratory was founded by Dr. Ludwig Mond under a deed of trust for the carrying out of original research. The whole complicated position is one that could only exist in England, but it works, though whether it works better than a simpler and more unified administration would do has not been proved.

Dewar was over eighty years old at the time of his death and there was much to be done in the way of reorganizing, for instance, the work in the Davy-Faraday Laboratory, which Bragg promptly directed to problems of crystal structure. He brought with him from University College Müller and Shearer, who were mainly responsible for the actual installation of the new type of apparatus needed for the X-ray methods. They also continued the work on long-chain compounds which they had begun at University College, while Shearer further worked on the theory of space groups. The work which Müller carried out on the fatty acids was typical of the tendency towards the investigation of organic structures which the laboratory took under Bragg's direction, while his son at Manchester concentrated on inorganic crystal structure. Among the early workers in the laboratory may be mentioned J. D. Bernal; R. E. Gibbs, who did not pursue organic compounds but worked on the structure of quartz; Miss Yardley, afterwards Mrs. Lonsdale, who became one of the pillars of the laboratory; Miss C. F. Elam (Mrs. Tipper), known for her work on metals; W. T. Astbury, who came from University College and afterwards investigated the crystal structure of the products of the living body, such as hair and horn; and J. M. Robertson, somewhat later, who worked on the crystal structure of anthracene and naphthalene, and applied the methods of Fourier analysis to deduce crystal structure from intensity measurements. There are other distinguished names among those who spent a short time in the laboratory. It speedily became a world-famous centre of research. Bragg was no longer able to spend as much time as formerly at his own researches, but the work of the laboratory was an informed whole which in all its features gave evidence of his wise guidance and prescience.

One of Bragg's first tasks at the Institution was to give the Christmas Lectures. He chose as his title that of the famous poem of Lucretius, 'Concerning the Nature of Things', and talked about atoms, gases, liquids and crystal structure with masterly simplicity and inimitable charm. He always gave the impression that he thoroughly enjoyed addressing the 'Juvenile Auditory' for whom these lectures were traditionally intended. On two other occasions, namely at Christmas 1925 and Christmas 1931, he gave the lectures on 'Old Trades and New Knowledge' and on 'The Universe of Light'. The 1925 discourses gave evidence of his appreciative interest in the industrial aspects of science which he later emphasized by taking 'Craftsmanship and Science' as the subject of his Presidential Address at the British Association at Glasgow in 1928.

Bragg brought to the social life of the Institution a charm and suavity which rapidly won all hearts. Lady Bragg, with a kindheartedness whose sincerity was apparent, and with a graciousness free from all affectation, was an ideal hostess at the mixed gatherings that thronged the Braggs' private apartments after the Friday evening discourses. Their daughter, Gwendolen, an accomplished artist, some of whose drawings appeared in Bragg's *Universe of Light*, lived with them, and Lawrence Bragg was a frequent visitor. The unity of the family was felt outside the walls of their official residence. Lady Bragg's death in 1929 after much suffering was indeed a heavy blow to Bragg and a source of sincere grief to all those who had come to know her at the Royal Institution.

After her mother's death Gwendolen Bragg became her father's

William Henry Bragg (1862–1942).

closest companion, both as the centre of his personal home life and as hostess at the many friendly assemblies, great and small, that characterized the Institution. On her marriage in 1932 to Alban Caroe she gave up residence at the Institution for some eighteen months, but then returned and, with her husband, lived there for the rest of her father's life. Between Bragg and his daughter there existed a warm affection and comradeship which illuminated both their lives. 'We had grand times together', she wrote to a friend.

An essential part of Bragg's family life was the country cottage at Chiddingfold, Surrey, which he purchased in 1923. Hither he would retire for relaxation during holidays and at week-ends, and here he would entertain those working at the Royal Institution in an atmosphere of intimacy which brought them very close to him. They, too, felt that they were part of the family.

Many important changes took place in the Institution under Bragg's rule. In 1929 an extensive reconstruction of the Royal Institution house, which lasted two years, began. It had become clear that the lecture theatre did not comply with the requirements for safety, especially against fire, demanded of a building to-day, and the complete reconstruction of the interior of the theatre and the provision of proper exits involved a number of alterations which were embodied in a comprehensive architectural scheme. Bragg took a leading part in the negotiations and the liberal contributions that were made to the expenses of the undertaking were largerly the results of his persuasive charm. A further reconstruction took place in 1936, in which he was again active. Many notable additions were made to the equipment of the Davy-Faraday laboratory: in particular a giant X-ray generator, of a new type designed by Müller, was built and installed in the Institution. This possessed a rotating hollow target, with water cooling, and dealt with an input of fifty kilowatts. A smaller tube with an input of five kilowatts was also erected. The great intensity obtainable with these tubes very much accelerated work in general by permitting very short exposures and rendered feasible certain experiments which otherwise could not have been successfully carried out. Thus high dispersion was achieved by using much larger distances from object to plate than are practicable with weaker sources, and such experiments as those of Müller on the effect of compression on the lattice spacing of certain organic compounds were made possible.

In spite of the many calls upon his time which his name and fame entailed, and of advancing years, Bragg kept well abreast of physical thought and keenly welcomed all advances. Even when he did not pretend to have mastered all the mathematical intricacies of the latest developments he was able to grasp very quickly the more important implications and to give them a new turn, which not only simplified them but set them in relation to the general advance of science. In 1928, for instance, he arranged for Schrödinger to deliver a course of lectures at the Institution on wave mechanics, then a very recent development. (Incidentally it is said that an enthusiastic, but unmathematical, yachtsman appeared at the first lecture, hoping to learn something about sailing in a rough sea.) He gave an introductory lecture as preparation for the course, in which he returned again to the wave and corpuscle dichotomy. He referred back with pleasure and some pride to his old experiments with Kleeman and Madsen, which have already been mentioned, saying, 'I may say, I think, that in these experiments we were, though unwittingly, carrying out Einstein's suggestion that the corpuscular hypothesis deserved careful exploration ... It is true, however, that I thought of the X-ray and γ-ray problems as distinct from that of light.' He ended on a typical thought, typically expressed, 'When the picture is finally clear there will no doubt be atoms in it, electrons, wave motions, energies, momenta and so on. But have we got them all rightly joined up? Perhaps wave motion belongs to more than the photon or to something else than the photon? We can only wait.' No doubt he was thinking of the experiments of Davisson and Germer and of G. P. Thomson, indicating the wave nature of the electron, of which accounts had just appeared.

On more than one occasion, while at the Royal Institution, Bragg showed his extraordinary power of taking up a new subject, adding something to it, and then laying it down again. A particularly good example is offered by his work on liquid crystals. In connection with the Faraday Society's discussion on this subject, held at the Royal Institution in 1933, he became interested in the phenomena of smectic crystals, and he then showed, in an extremely simple and elegant way, that the static arrangement of a number of equidistant parallel layers, without rigidity, was a set of surfaces formed in successive layers around an ellipse, having as lines of discontinuity the ellipse itself and a hyperbola in a plane perpendicular to it, each curve passing through the focus of the other. In short, he explained the phenomena in terms of the cyclides of Dupin. He never seems to have concerned himself much with liquid crystals after his lecture on the subject at the Royal Institution in November 1933. The case serves to show how he retained his rare geometrical sense and could apply it at will.

Late in his work at the Institution Bragg came to the conclusion that it was desirable to have some fundamental theoretical work proceeding at the same time as the experimental investigations and a systematic mathematical attack on details of the methane spectrum was made by H. A. Jahn, working sometimes in collaboration with W. H. J. Childs. This kind of work was far from anything that Bragg himself ever did, but it had his warm support and encouragement.

Towards the end of his life all his old enthusiasm was aroused by the so-called extra reflections or diffuse spots, which can be observed with powerful X-ray illumination of single crystals. These had been adventitiously observed from time to time, but in 1939 G. D. Preston published a careful study of them in certain simple cases and explained them as due to small crystalline fragments in imperfect alignment. The matter was immediately taken up at the Royal Institution, where the powerful tubes to which reference has already been made allowed the effects to be obtained with relatively short exposures. Mrs. Lonsdale and H. Smith published a detailed study of the spots obtained with both organic and inorganic crystals, in which Bragg took the greatest interest. He was much intrigued by the effects, and was concerned about their explanation. The elaborate mathematical theories which had considered the thermal movements within the crystal as responsible for the reflections did not appeal to him, and he evolved a simpler theory along the lines suggested by Preston, attributing the spots to a lack of regularity in the crystal structure. Discontinuities at the boundaries of the regular fragments of which the crystal is supposed to be composed lead to discontinuities in the phase relationship. Bragg wrote several short papers and notes on the subject, and, in fact, his last paper, written just before his death, and published posthumously, was on the secondary X-ray spectrum (extra reflections) of sylvine.*

By 1930 Bragg had become not only one of the great figures of English science but also something of a national figure. In that year the Royal Society bestowed on him the Copley medal, its senior award. He had received the Rumford medal in 1916. He was an honorary doctor of some sixteen British and foreign universities, and a member of the leading foreign societies. In 1931 he received the Order of Merit. If any great scientific body had a ceremonial lecture to be delivered, Bragg was asked to act, and he generally consented if he found it possible. Whether he enjoyed delivering addresses it is hard to say. Beforehand he appeared to feel some diffidence and certainly the utterance that seemed to come so easily and so spontaneously was the

* A simple account by Mrs. Lonsdale of the extra reflections may be found in *Engineering*, pg. 254, 27 March 1942.

fruit of more labour and thought than the audience often suspected. I think that once on his feet he did enjoy it, as a man enjoys doing anything in which he is supremely competent. Afterwards he received the compliments and congratulations with an artless modest pleasure that became him well. He never lost his zest; if he always maintained the interest of his audience it was, perhaps, because he always maintained his own interest.

In 1935 Bragg, now seventy-three years of age, was elected to the Presidency of the Royal Society. His reputation, his fine presence, his dignity tempered with geniality, his wide knowledge and his ready sense of the right word made him a complete figure in the Chair. His kindness and ease of access commended him to the younger Fellows; his respect for tradition and his historical sense commanded the confidence of the older Fellows; his close connection with Trinity College oiled many wheels. His kindliness led him to welcome certain ambiguous advances from learned bodies in Nazi Germany, and he did his best to further ostensible plans for an understanding between the two countries which, in his goodness of heart, he took at their face value.

The war found him burdened with a variety of heavy duties which he, never sparing himself when the good of the scientific community or of the country at large was in question, allowed to be increased. Soon he was not only President of the Royal Society and responsible for the many duties that attach to the headship of the Royal Institution, but Chairman of the Scientific Committee on Food Policy, Chairman of the Scientific Advisory Committee, and Chairman of the Science Committee of the British Council. He was also a member of the Advisory Council of the Department of Scientific and Industrial Research, and held a number of other appointments, none of them sinecures. Very few men of seventy-seven could have carried out in any fashion the variety of responsible tasks which he discharged with grace, dignity and efficiency. His voice on the wireless service gave pleasure to hundreds of thousands, and he was keenly interested in the teaching of the elements of science to the boys of the Air Training Corps. He even wrote a little book *The Story of Electromagnetism* to help them in their studies. Naturally he did not feel inclined to embark on a campaign for reforming the whole organization of science in the war-time service of the country, but within the cumbersome, complicated and capricious frame set up by the Government through its peace-time servants he rendered great services. He laid down the Presidency of the Royal Society in 1940, and with it some other duties, but the demands on his services outside his duties at the Royal Institution were still many.

With his naturally strong constitution and strong sense of duty he carried on bravely all through the worst period of the war, but his friends were distressed to see how exhausted he was from time to time. He would always make an effort when needed, but it cost him more and more. For some time his heart had been giving him trouble and he endeavoured to avoid physical exertion, while still using his mind like a young man. Reference has already been made to his enthusiastic interest in a new X-ray phenomenon, and as late as December 1941 he was writing in *Nature* about it. On Tuesday, 10 March 1942, the gallant veteran had to take to his bed; two days later he was dead.

Bragg had an astonishing career. Up to the age of forty he never showed any desire to carry out original experiment. He then straightaway embarks upon a perfectly precise and important piece of work and within a few years his name is known wherever physics is seriously studied. He spends some years carrying out a careful series of experiments which can be interpreted to prove the corpuscular nature of X-rays, and he stresses this interpretation. He then himself conclusively demonstrates, by the work with which his name will always be associated, the wave nature of X-rays. He starts life as an extremely shy and retiring youth, never, apparently, quite at home in Cambridge, and in his old age becomes a national figure, at ease in all surroundings, whose personal appeal is known all over England.

Yet there is nothing contradictory in his character. Bragg's nature was simple, straightforward and tenacious—incidentally, of course, he was a man of genius. It would never have occurred to him to embark upon research with the object of publishing a paper or papers to impress other people. He waited until he found something that seemed to him to ask for experiment and in the first case he did not come across this until he had trained both his critical faculties and his powers of exposition in the faithful discharge of his duties as a teacher. This long period of apprenticeship probably had a profound influence on his work. It helps to account for his great objectivity and for his power of formulating a problem simply and directly. Bragg's fundamental ideas, as he expounded them, were accessible to any undergraduate. His greatness was shown in their originality and in the skill and perseverance with which he shaped them to well-recognized ends. He was a very great experimenter who never wasted his time on the trivial or hid difficulties under the graceful veils of mathematical obscurity. His hypotheses were stated clearly as hypotheses, his experimental results as experimental results—the former he was ever willing to modify or to abandon as later discoveries might dictate, the latter he knew to constitute knowledge won for all time. His work, like his personality, was simple yet profound, sincere and compelling.

Bragg preserved all through his fame many of the more admirable characteristics of youth. He was capable of a warm burst of enthusiasm at any new idea or achievement that appealed to him. He was apt to believe that all men were as sincere and diffident about their own achievements as he was about his own, which occasionally led him to lend support where it might, without disadvantage, have been withheld. But how much better this generosity of outlook became him than a coldly critical attitude would have done! Again, he never shrank from asking for information or advice: he was far too big a man to mind admitting gaps in his knowledge. His warm, simple, persuasive utterance, his personal tone in lecturing which made each member of the audience think that the remark was intended for him, was also more reminiscent of the wise elder brother who was sharing with you the pleasure of a discovery, than that of the great sage who was instructing you. But he was a great sage.

Bragg was a man of very strong family feeling, who was never happier than with his children, and, later, his grandchildren. He took a particular pride in the achievements and career of his brilliant son. It was always a delight to his hearers to note the affection that came into his voice when, in lectures, he found occasion to deal with some one or other piece of work which had been carried out by 'my boy'. His unaffected pleasure and surprise at the news of some new high post or distinction awarded to him was very lovable. He was, for instance, obviously overjoyed when Sir Lawrence received his knighthood. The quiet affection that existed between him and his daughter was a source of pleasure to many outside his family circle.

There was nothing narrow about Bragg's interests. Reference has already been made to his love of games, and he was always ready to talk of life in the open air. He had a great affection for the sea and was always at his happiest on an ocean voyage, the first day of which seemed to wipe away his cares. He was a lover of simple music and enjoyed playing the old tunes on his flute, the sound of which, in the old days at the Royal Institution, could often be heard coming from his study at the evening hour. In his younger days he had been a very good draughtsman and painter in water colours, and he knew more than a little about pictures. His range of reading was wide, both among the English classics and the moderns. With all this, he was an exceedingly good conversationalist and a first-rate impromptu speaker. Nature had given him a fine presence. In any company he was an unmistakable figure.

Religion was a strong influence in Bragg's life. He had no strong dogmatic views but he had a simple and genuine piety and was an enemy of unbelief, as he was of loose talk of every kind. He was not comfortable at any discussion, however serious and philosophical, that touched on the weaker side of man's nature. There were many things whose existence he preferred not to acknowledge. In religion, as in other matters, he was tolerant of the views of those who sincerely held views different from his own. Something of his own belief will be found in his address on 'Science and Faith'.

There was, we like to think, something peculiarly British about Bragg. His attitude towards physics was that characteristic of the great experimenters of our land, especially his strong pictorial sense. He was a lover of the traditions, especially those of the great institutions with which he was connected. His lack of pedantry, his gift for popular exposition, his strong feeling for the craftsman in factory and workshop are all characteristics which he shared with Faraday, with Tyndall, with J. J. Thomson. He was an ornament, not only of English science, but of English learning, a great teacher and a good man, whose death came as a personal loss to all those who knew him. With him went an outstanding representative of a great period of English physics.

E. N. da C. Andrade

Shoji Nishikawa

1884–1952

No sooner had the news of the discovery of X-ray crystal diffraction by Laue, Friedrich and Knipping reached Japan, than Torahiko Terada (1878–1935), at that time an assistant professor at the Department of Physics, Faculty of Science, University of Tokyo, carried out his diffraction experiments using single crystals of rocksalt and other minerals. He observed, not only photographically but also visually by use of a fluorescent screen, the movement of the Laue spots as the crystal was gradually turned, and came to the conclusion that the diffraction spots might be explained as coming from simple reflections of X-rays from different netplanes in the crystal,[1] quite independently of W. L. Bragg. One evening, when this original visual observation was being made, Shoji Nishikawa (1884–1952), at that time engaged in the study of radioactivity as a post-graduate student under the guidance of S. Kinoshita at the same Department of Physics, was led by Terada into his room and shown for the first time the diffraction spots moving on the fluorescent screen in the dark. Encouraged by Terada, Nishikawa soon turned to the field of X-ray diffraction, where he made significant contributions to the analysis of crystal structure in the early period of its development.

Nishikawa was born in 1884 in Hachioji near Tokyo, a son of an important dealer in silk textiles, but he was brought up in Tokyo and finished his undergraduate course in physics at the Faculty of Science, University of Tokyo, in 1910. He was known to be very brilliant in his scientific ability and modest and shy in his personal character. After his graduation he became a postgraduate student and, by chance as described above, he devoted his later life mostly to the study of modern crystallography as well as to the education of numerous active research scientists in this field.

In 1913 he published, with his fellow student, S. Ono, his first paper, a diffraction study entitled 'Transmission of X-rays through

Fibrous, Lamellar and Granular Substances'.[2] For the fibrous substances, they took asbestos as the first example and obtained diffraction photographs with the X-ray beam in perpendicular or oblique incidence to the axis of the fiber. These photographs are nothing other than what are now known as fiber patterns, but taken with continuous X-rays. Assuming that the fiber consisted of innumerable minute crystallites in the uniaxial orientation with respect to the fiber axis, they could explain the characteristic features of these patterns on the basis of the reflection law of Bragg and Terada. They also observed fiber diagrams with fibrous gypsum, while they obtained negative results with quartz fibers or glass wool. In a similar manner they observed fiber patterns from organic fibers such as silk, wood, bamboo and *asa* (*Cannabis sativa*—a kind of hemp). Carbonized wood showed no discernible pattern. Similar experiments were extended to lamellar substances such as talc and mica. As for the granular substances, they first took marble, which showed a pattern corresponding to a spotted Debye-Scherrer diagram of later days. They also observed ring patterns produced with finely pulverized rocksalt, quartz, carborundum, etc., as early as 1913. Silica precipitated from its hydroxide showed no ring, while ordinary candle wax showed a pattern of concentric rings more or less similar to that obtained by Friedrich[3] a little earlier. In connection with these various kinds of substances, they investigated the diffraction effect of metals rolled in thin sheets. This investigation was suggested by the experiment of Hupka[4] with a platinum foil. The metals they examined were copper, iron, nickel, zinc, tin, lead, etc. For example, a rolled sheet of copper, when taken with the X-rays in normal incidence to the sheet, gave a pattern with ill-defined pairs of patches in the direction of rolling, which on annealing the sample became sensibly faint, and turned into a spotted ring pattern like that of marble. These experiments with metals were carried out quite independently of Keene.[5] Their first paper, then, contained the germs of later applications of the X-ray diffraction method to metallography and high-polymer science. The second paper by Nishikawa appeared in 1914, and in it the author recognized the possibility of utilizing the fibrous substances as analyzers of X-ray spectra and also found the determination of the netplane spacings.[6] Nishikawa then proceeded to the structure analysis of some crystals of the spinel group and magnetite by means of Laue photographs,[7] independently of similar work by W. H. Bragg.[8] What is significant to the history of the X-ray crystal analysis is his very early realization of the importance of the theory of space groups as a general and logical means of analytical procedure. He learned from Terada of the existence of this theory. In this connection it may be noted that Teiji Takagi, professor of mathematics, had been in Göttingen with D. Hilbert for some time, and Seiji Nakamura, professor of physics, had worked with W. Voigt; both were Terada's colleagues at this University of Tokyo. A collection of books on classical structure theories assembled by Y. Kikuchi, assistant professor of mineralogy at the Faculty of Science, University of Tokyo, was also available.

After publishing two more papers on crystal analysis, one on garnet [9] and the other with K. Hudinuki on nitrates of lead, barium and strontium,[10] Nishikawa and his colleague, Genshichi Asahara (*1891), were sent abroad for further study in 1917 by the newly established Institute of Physical and Chemical Research, Tokyo. They went first to the United States of America, but dislocations caused by the First World War made it difficult for them to find a suitable place for their purpose. After a time, Nishikawa finally visited E. L. Nichols at the Physical Laboratory, Cornell University, where he got acquainted with Wheeler P. Davey, then of General Electric Co., and Ralph W. G. Wyckoff, of the Chemistry Department. As seen in the preface to the first edition of the classical book *The Structure of Crystals* by Wyckoff, Nishikawa's meeting with Wyckoff may be said to have an implicit significance to the history of X-ray crystallography in the United States of America. Asahara, who had been engaged in some classical metallographic studies with W. Campbell at the School of Mines, Columbia University, then joined Nishikawa at Cornell, and there they carried out X-ray studies of metals such as aluminium, cadmium, copper, lead, silver, thallium, tin, zinc, and several kinds of brass, observing the effects of rolling and of annealing after rolling, and also trying to determine the transition points of thallium and tin by the change of X-ray patterns.[11] After the end of the War, Nishikawa crossed the Atlantic to England and stayed there for about half a year at the Department of Physics, University College, London, where W. H. Bragg was continuing X-ray crystallographic work using his famous ionization spectrometer.

In 1920, Nishikawa returned to Tokyo from abroad to organize and lead a research group, the Nishikawa Laboratory as it was called. In this position he was a chief research member of the Institute of Physical and Chemical Research. Using both the ionization spectrometer and diffraction photographs he began X-ray crystallographic investigations with his students Y. Sakisaka (*1893), I. Nitta (*1899), I. Sumoto and others, and, although a serious earthquake in 1923 and another in the next year interrupted the progress of the research work considerably, they recovered gradually and proceeded with their projects. When, in 1924, Nishikawa became professor of physics at the Department of Physics, Faculty of Science, University of Tokyo, he began giving lectures on X-ray physics and X-ray crystallography. Although the excellence of his lectures was well recognized by all students, his low voice, which resulted from his modesty and shyness, was a source of complaint.

During this period he was interested in the analysis of orthorhombic crystals, having made his own systematic absence table for the orthorhombic space groups. With the ionization chamber spectrometer made in the Institute, he measured reflections from a very good single crystal of aragonite, but, when he had nearly finished the structure analysis, he noticed that a paper by W. L. Bragg [12] on the structure of the same crystal had just been published, and he gave up completing the analysis. A similar circumstance occurred a second time to Nishikawa; when he had nearly determined the structure of α-quartz, he again found that a paper on the same subject by W. H. Bragg and R. E. Gibbs [13] had just appeared. Owing to the great distance from other scientific countries it was then very difficult for Japan to get current information from abroad.

His interest in structure and properties of quartz lasted for some time more. Thus he studied with Sumoto the α-β transformation by means of Laue photographs. Sakisaka made a series of experiments concerning the effects of surface treatment on the intensity of reflection, using various crystals, especially quartz.[14] Further, Nishikawa, Sakisaka and Sumoto investigated the effects of thermal strain [15] and mechanical vibration [16] on the intensity of reflection, using various crystals for the former and primarily quartz for the latter. All these subjects arose out of Nishikawa's keen interest in the problems relating to the extinction phenomena and crystal imperfections. This interest of his influenced numerous later studies carried out by his students as will be described in Part VI. Another basic contribution of Nishikawa is his ingenious experimental proof, with K. Matsukawa, that Friedel's law does not hold for the non-centrosymmetric crystal of zincblende; [17] this precedes the studies of D. Coster, K. S. Knol and J. A. Prins [18] and of I. G. Geib and K. Lark-Horowitz. [19] Besides his natural interest in the physical aspects of X-ray crystallography, Nishikawa had not a little interest in its chemical aspects and encouraged I. Nitta, a graduate of chemistry of the University of Tokyo, to undertake structure analyses of organic crystals, this field being rather new at that time. Also he introduced M. Nakaidzumi, at that time a post-graduate student of medicine, to the X-ray spectroscopic chemical analysis of human tissues by use of an ionization chamber spectrometer. Other students he introduced to X-ray crystallography included S. Shimura, a postgraduate student of metallurgy, and Z. Ooe, a student of mineralogy. Later all of the above mentioned became professors at the University of Tokyo and elsewhere.

Soon after the discovery of the electron diffraction by crystals by C. J. Davisson and L. H. Germer [20] and by G. P. Thomson and A. Reid [21], Seishi Kikuchi (*1902) published a series of papers [22] on photographic experiments of electron diffraction. These papers contained the discovery of the so-called Kikuchi-lines, -envelopes and -bands and are of historical importance. All these experiments of great significance were made by Kikuchi in a very short period of time by virtue of the wide experimental experience and the profound theoretical knowledge of his teacher Nishikawa. After these experiments the interest of Nishikawa's students was directed to electron diffraction, and, as will be described in Ch. 23, these students contributed considerably to the progress in the study of electron diffraction by crystals, assisted by the very relevant advice of Nishikawa. Thus he suggested them to use a spider's thread for supporting the powder, cathodic sputtering to prepare thin metallic films for reflection study, and the rotating crystal method with a knife-edge, etc.

In 1934 Nishikawa attended the General Meeting of the International Union of Physics held in London, and on returning to Tokyo next year he began to study with Nakagawa and Sumoto the slowing down of neutrons by paraffin. Further, with two eminent Japanese physicists, H. Nagaoka and Y. Nishina, he became engaged in building up the first cyclotron in the Institute of Physical and Chemical Research. In this field of nuclear physics he led I. Sumoto, T. Yazaki, K. Shinohara, S. Nakagawa, M. Kimura, M. Hatoyama and others on to problems such as the scattering of slow neutrons, artificial radioactivity, electron pair production, etc. up to the beginning of the Second World War. In 1937 he became a member of the Japan Academy. In 1945, immediately after the end of the War, he retired from his professorship at the University of Tokyo. Four years later, after thirty-two years of service he retired from the Institute of Physical and Chemical Research, and Shinohara, who was then a professor of physics at the Department of Physics, Kyushu University, succeeded him in the Institute. At the invitation of P. P. Ewald that Japan adhere to the newly established International Union of Crystallography, he organized in 1950, with T. Ito, S. Miyake, I. Nitta, R. Uyeda, T. Watanabé and others, the Crystallographic Society of Japan, and was elected to be the first President of the Society. At the same time the National Committee for Crystallography was formed within the Japan Science Council, and he was also chosen as the first Chairman of the Committee. The Second General Assembly and Congress of the International Union of Crystallography was held in Stockholm in 1951, and there Japan's adherence was approved. It was regretted that Nishikawa could not join this meeting because of his high blood pressure. He was confined to his bed, and, although he seemed better again, he died suddenly of apoplexy at his home on the 5th of January, 1952. In this way we have lost one of the most eminent scientists of Japan in the field of modern crystallography as well as experimental nuclear physics. By all of his friends and his students he is respected for his modest personality, his wonderful skill in experimental techniques and his profound understanding of theory.

A few lines may be added concerning his family. Soon after he became Professor of Physics at the University of Tokyo, he married Miss Kiku Ayai, a graduate of the Nara Women's Higher Normal School and once teacher of science at a girls' high school; they had four sons and one daughter. The first son, Tetsuji Nishikawa, recently became a professor of physics at the Faculty of Science, University of Tokyo, proving to be a good successor to his father.

I. Nitta

References

1. T. Terada, *Proc. Math. Phys. Soc.* Tokyo, *7*, 60 (1913); *Nature*, *91*, 135 (1913).
2. S. Nishikawa and S. Ono, *Proc. Math. Phys. Soc.* Tokyo, *7*, 131 (1913).
3. W. Friedrich, *Physik. Z.*, *14*, 317 (1913).
4. E. Hupka, *Physik. Z.*, *14*, 623 (1913).
5. H. B. Keene, *Nature*, *91*, 609 (1913).
6. S. Nishikawa, *Proc. Math. Phys. Soc.* Tokyo, *7*, 296 (1914).
7. S. Nishikawa, *Proc. Math. Phys. Soc.* Tokyo, *8*, 199 (1915).
8. W. H. Bragg, *Nature*, *95*, 561 (1915); Phil. Mag., *30*, 305 (1915).
9. S. Nishikawa, *Proc. Math. Phys. Soc.* Tokyo, *9*, 194 (1917).
10. S. Nishikawa and K. Hudinuki, *Proc. Math. Phys. Soc.* Tokyo, *9*, 197 (1917).
11. S. Nishikawa and G. Asahara, *Phys. Rev.*, *15*, 38 (1920).
12. W. L. Bragg, *Proc. Roy. Soc.* London, A*105*, 16 (1924).
13. W. H. Bragg and R. E. Gibbs, *Proc. Roy. Soc.* London, A*109*, 405 (1925).
14. Y. Sakisaka, *Jap. J. Phys. 4*, 171 (1927); *Proc. Mat. Phys. Soc.* Japan, *12*, 189 (1930).
15. Y. Sakisaka and I. Sumoto, *Proc. Math. Phys. Soc.* Japan, *13*, 211 (1931).

16. S. Nishikawa, Y. Sakisaka and I. Sumoto, *Phys. Rev.*, *38*, 1078 (1931).
17. S. Nishikawa and K. Matsukawa, *Proc. Imp. Acad.* Japan, *4*, 96 (1928).
18. D. Coster, K. S. Knol and J. A. Prins, *Z. Physik*, *63*, 345 (1930).
19. I. G. Geib and Lark-Horowitz, *Phys. Rev.*, *42*, 908 (1932).
20. C. J. Davisson and L. H. Germer, *Phys. Rev.*, *30*, 705 (1927); *Proc. Natl. Acad. Sci.*, *14*, 317 (1928).
21. G. P. Thomson and A. Reid, *Nature*, *119*, 890 (1927); G. P. Thomson, *Proc. Roy. Soc.* London, A*117*, 600 (1928); A*119*, 651 (1928).
22. S. Kikuchi, *Proc. Imp. Acad.* Japan, *4*, 271, 275, 354, 471 (1928); S. Nishikawa and S. Kikuchi, *ibid.*, 475 (1928); *Nature*, *122*, 726 (1928); S. Kikuchi, *Jap. J. Phys.*, *5*, 83 (1928).

Charles Mauguin

1878-1958

Charles Mauguin, through his research and teaching, took an outstanding part in the development of modern crystallography in France.

He was born in Provins, a small town of Ile de France, where his father was a baker. There he received the elementary education and passed, at the age of 16, the competitive examination of the Normal Teacher's Training College of Melun. He thus started his educational career by teaching young children to read and to count, at the same time, however, he prepared for the entrance examination at the Ecole normale of Saint-Cloud, where he was admitted in 1902, intending himself to become a teacher in a Normal Teacher's Training College. The instruction given there, extended from mathematics to natural sciences, and was especially suitable for his studious mind, eager for knowledge. He was lucky in being noticed by his professor of chemistry, Simon, who also directed the laboratory of organic chemistry of the 'Ecole Normale Supérieure' of Paris. Therefore Charles Mauguin's first steps into Scientific Research was academical work on organic chemistry; this lead him, in 1910, to the degree of Doctor of Sciences. In his thesis 'Bromine sodium amides and their part in the transposition of Hofmann', he was studying the amides $R{\cdot}CO{\cdot}NH_2$ where he substituted sodium bromide for the two hydrogens; these products are unstable and as they release the sodium bromide, they provide isocyanates OCNR which may be considered as resulting from the oxidation of the first amides. He thus discovered a close link between the amides and the ureides which are of major importance in the chemistry of the living cell; he also thought that, through a process of biochemical oxidation, the same reactions could be reproduced. Charles Mauguin soon discontinued the research in organic chemistry, but kept in view the biological aspect of chemistry during his whole life.

At the same time as he was preparing his thesis, he was attending at the Sorbonne the mathematics classes of Emile Picard, Poincaré, Painlevé, Goursat. But he was particularly struck by the outstanding course of lectures given in 1905 by Pierre Curie on symmetry in physical phenomena. This led him to take a special interest in crystallography, he attented the classes of Fr. Wallerant who was at the head of the laboratory of mineralogy at the Sorbonne. Wallerant made him his assistant and directed him to the study of liquid crystals. The German physicist Otto Lehmann had discovered some organic substances which may possess the same fluidity as real liquids (some are more mobile than water) as well as the birefringence of crystals (several of them are twice as birefringent as calcite). This curious state of matter appears as a turbid phase in a range of temperatures included between those where the isotropic liquid and the solid phase exist. The constitution of this turbid phase was much discussed; some thought it was due to the presence of an insoluble impurity. Mauguin thought that the turbid liquid is constituted by birefringent elements, the orientation of which varies from one point to another, so that it becomes almost opaque in a thick layer. Testing on azoxyanisol and azoxyphenetol which, in the turbid phase, are extremely fluid, he imposed a uniform orientation to the different birefringent elements to make it homogeneous and perfectly transparent. He succeeded in this by acting with agents such as the magnetic field, whose action penetrates through the mass itself of the liquid, or else by applying appropriate surface actions.

The turbid phase of the azoxyanisol, placed between the pole pieces of a magnet, becomes clear almost at once and then resembles optically a very highly birefringent uniaxial crystal, the axis of which is parallel to the direction of the magnetic field. Mauguin achieved a complete optical study of the anisotropic phase and found the classic phenomena of uniaxial crystals, for instance he obtained in the visible spectrum beautifully channelled spectra with several hundred grooves, using a one millimetre thick layer. He thus determined the two indices of refraction. When the current of the electromagnet is cut off, the orientation is lost in a very short time, together with the homogeneity and transparency of the sample.

Contact with solids may force the molecules into a strictly determined orientation; Mauguin showed that a homogeneous layer of azoxyanisol is obtained by simple fusion of the crystals between two plates of glass, provided they are absolutely clean. When observed in the polarizing microscope, the preparation appears as a uniaxial crystal the axis of which is normal to the support. The situation is completely different when the azoxyanisol is melted on a newly cleaved muscovite mica plate. The liquid crystal then takes such an orientation that the optic axis is parallel to the cleavage, namely along a crystallographic direction at about 30° to the symmetry plane. This was at that time (1912) rather unexpected; X-rays have since shown that (010) is a glide-symmetry plane. If the azoxyanisol is melted in a wedge-shaped crevasse obtained by cleavage, the two optical axes in contact with the two surfaces of the wedge, form an angle of 60° with one another. The result is a helicoidal structure which Mauguin studied experimentally in all detail and fully explained theoretically.

This outstanding research was interrupted by the First World War. The results have lost nothing of their interest, and the techniques developed by Mauguin are still used in the research now continuing on these substances which are so important in physicochemistry as well as in biology.

After short periods as Associate Professor, in 1912 at the Faculty of Sciences of Bordeaux, and in 1913 at the Faculty of Nancy, Mauguin came back in 1919 to the Faculty of Sciences of Paris, as Associate Professor under Frederic Wallerant, whose successor as Professor of Mineralogy he was from 1933 to his retirement in 1948.

When he resumed work in a research laboratory after four years of interruption during the war, Mauguin was one of the very few scientists in France who had understood the importance of the discoveries of Laue and the Braggs for the analysis of crystalline structures. Being a chemist, he was immediately aware of their great interest for making accurate statements about the mode of linkage of atoms in the different chemical compounds. He had submitted to the first Solvay Council of Chemistry, in 1922, a report *Electronic Theory of Valence*, where he methodically and clearly set forth work almost unknown in France. He considered X-rays to be the most direct means for finding the nature of the chemical bond. He thereupon followed this new way, built the apparatus, including cameras, and perfected the techniques. He published in 1924 a remarkably well documented book *La structure des cristaux determinée au moyen des rayons X* which played in our country a decisive part in the formation of a new school of crystallographers. His first original publication in this new field is of 1921 and deals with the atomic structure of quartz. In 1923 he published the atomic structure of cinnabar, laid stress on its relationship with galena and showed how the atom of mercury, because it may be approximated by a flattened ellipsoid of revolution, leads to an helicoidal arrangement which gives to cinnabar a strong rotatory power.

In order to stress this interpretation of the rotatory power, he determined the ordering of atoms in calomel, being convinced that in this case an analogy would be found with the structure of rocksalt. This is not so, calomel being a structure of molecules Hg_2Cl_2. He then grappled with the structure of graphite, being unaware that Bernal in London and Mark in Berlin were already studying this problem. Mark was the first to publish his results, but the three authors reached the same conclusions, which did not difter essentially from the result A. W. Hull had given seven years earlier (1917).

In 1926, he published a paper on the use of the reciprocal lattice for the graphical interpretation of Laue and rotation diagrams, which proved to be extremely useful to the X-ray crystallographers.

Mauguin then undertook a fundamental investigation of micas; the result formed the subject of several much quoted publications and numerous papers on silicates were based on them. Mineralogists had always been at a loss to reconcile the extremely diversified chemical composition of micas with their remarkably invariable crystallographic characteristics. This resulted in the distinction of a very large number of mica species merely for chemical reasons. Thus the problem of micas, in the setting of 1925, involved a reconciliation of chemistry and crystallography which, on this subject, were in strong opposition. Mauguin performed the chemical analysis of a great number of micas, and measured their density and lattice constants, the latter with the help of X-rays. This enabled him to determine, purely by experimental means and for each one of the micas, the chemical content of the unit cell. He thus observed that the cell content of non fluoric micas always includes 12 oxygen atoms; for fluoric micas, the sum of oxygens and fluorines is twelve. As for the other chemical constituents, the number of atoms present in the unit cell varies from one mica to another and is mostly fractional, for example 2.12 to 3.69 for silicon, 1.13 to 3.80 for aluminium, etc.... If a fractional number is found, it means that the unit cell is not repeated identically and that X-rays give only the average cell. In such cases oxygen and fluorine which are negative ions and more numerous and bigger than the cations, are determining the structure which is basically the same for every mica.

In a similar way Mauguin studied the problem of chlorites and found that the crystalline fundamental set always includes 18 atoms of oxygen. These results have now become universally accepted and it is not the least merit of Mauguin to have greatly contributed to rendering the chemistry of silicates more understandable.

Mauguin had a strong mathematical background. He spent much of his time on studies of Group Theory and Fourier Transforms. Though he published little on these subjects, his investigations were profitable to his students thanks to seminars and discussions which he held in his laboratory. Still, he published in 1936 in the *Organic Chemistry* of V. Grignard a restatement of the theoretical aspect of the X-ray scattering by a molecule with applications to the study of organic compounds. The same year he also published an important memoir *On the theory of the reflection of X-rays by crystals* where he went back to the theory of Darwin, amplified it and reached general formulas giving the solution to the problem of the reflection of X-rays by a system of parallel reticular planes.

In 1943 he studied the dependence of the sizes and the shapes of the diffraction spots on those of the crystal with a particular reference to crystals of regular octahedral shape.

Charles Mauguin was one of the small group of crystallographers who, under the honorary chairmanship of W. H. Bragg and von Laue, undertook in 1933 to work out International Tables of all the geometrical and physical data apt to facilitate the work of the increasing number of scientists who were busy determining the atomic structures of crystals. He took an active part in the writing of the volume issued in 1935 dealing with the geometrical study of the 230 symmetry groups of Schoenflies-Fedorov. He had already published in 1931, in the *Zeitschrift für Kristallographie*, a report 'Sur le symbolisme des groupes de répétition ou de symétrie des assemblages cristallins'.

The same symbols which appear in this report, slightly modified after collaboration with C. Hermann, are still now in universal use and they have made the name of Mauguin familiar to every crystallographer. It is well known that the main aim of this symbolism is the creation of a direct correspondence between the group of spatial symmetry and the X-ray diagrams.

Having reached the age of retirement in 1948, Mauguin continued up to his last days his life-long habit of studying. Childless, he with Mme Mauguin, who was blind for thirty years, lived a very retired life at the outskirts of Paris, and his time was entirely devoted to study, reading, and the care for his wife. During his last years, he was chiefly interested in atomic physics and chemistry in relation to biology. He was particularly keen on the problem of the origin of life and had assembled on this question a very important documentation. Nevertheless, his last publication which he presented to the Academy of Sciences in March 1952, deals with Astronautics and Relativity and the challenging title is 'Assaulting the time-space'.

He took a keen interest in Botany and specially in mushrooms; he was President of the Mycological Society of France and before Mrs. Mauguin fell ill, one of his great pleasures were walking tours in the mountains and botanising in the woods. During the last years of his life, he gathered a beautiful collection of precious old books about plants.

Mauguin was an easy man to get along with and with his wide erudition, he was held in high esteem both by his students and his colleagues. Though being a very hard worker, he published little, but what much he published are models of clearness and accuracy. I often felt sorry that the results of his deep-probing investigations were not published; but exacting as he was for his students, he was so still more for himself, he would not leave anything unexplained and could not be satisfied with incomplete solutions. This need for clearness is probably his most characteristic feature and made him such a remarkable teacher.

This man, whose bodily needs were rudimentary and whose only luxury was books, was highly preoccupied by the condition and the future of men and by the social responsibility of the scientist. On several occasions he clearly expressed his opinion about the problem of Peace. He was deeply convinced that the material and economic future of mankind depends on scientific progress, and he was somewhat apprehensive of the use that would be made of the discoveries of scientists.

His influence on the growth of our international scientific Unions could have been greater if the infirmity and bad health of Mrs. Mauguin had not prevented him from travelling. He has been on the Advisory Board of *Acta Crystallographica* since it was created in 1947. During the ten last years of his life, he no longer attended the meetings of the French Society of Mineralogy and Crystallography over which he had twice presided, and the only days on which he left his house were Mondays, for attending the meetings of the Academy of Sciences to which he had been elected in 1936. He then used to come right after the meeting to the laboratory of the Sorbonne. There he met some of his former students. He told them of his work and picked up the documents and bibliography that he needed for the problems to which he would give his attention during the week.

He died on 25 April 1958 after having been ill for some weeks, and kept to the last moment his whole lucidity and faith science. Madame Mauguin died five months after him.

J. Wyart

E. S. Fedorov

1853-1919

Among the scientists who in various ways paved the way for Laue's discovery and for Bragg's determination of crystal structures, there is one who occupies the most important place: E. S. Fedorov (1853–1919), the eminent Russian crystallographer. His most outstanding achievement is the derivation of the 230 symmetry space groups which now serve as the mathematical basis of structural analysis.

Already in his first excellent book, *The Elements of Configurations*, Fedorov clearly outlines the idea of this derivation. From his autobiography it is known that he started writing this book in 1879, at the age of 26, prior to his enrollment as a student at the Mining Institute (Gorny Institute).*

During the years 1881–1882 Fedorov gave many of his papers on 'the theory of crystal structures' before the St. Petersburg Mineralogical Society.

In 1883 the finished book was accepted for printing on the recommendation of A. V. Gadolin, member of the Academy, the well-known author who derived the 32 crystal classes. Nevertheless, the book was published only in 1885 in the form of the 21st volume of *Transactions of the Mineralogical Society* (*Zapiski Mineralogicheskogo Obshchestva*).

An important point, underlying the future derivation of space groups, is found on page 240 of this book. Proof of the theorem that 'every real point system is a system of corresponding points of stereohedra' is substantiated by the following specific remark by Fedorov: 'In the past, the definition of real point systems has been taken from Sohncke. But equally, this term is applicable also to an aggregate of other real systems where one is symmetrical with the other... If, for the purpose of differentiating, systems of points with symmetry planes are

* Son of an Army Engineer, Fedorov lost his father at an early age and had to attend a complete course at a military school. At 18 years he was a combat officer in Kiev but resigned from military duty after two years to devote himself to the 'Sturm und Drang' [storm and stress] which was characteristic of the progressive Russian youth of the sixties and seventies [19th century]. He then tried two higher institutes of learning, belonged to the revolutionary underground, vigorously participated in the publishing of the newspaper of the Revolutionary Organization, and spent several years abroad on business, and finally his fascination with crystallography led him in 1880 to the Mining Institute (Gorny Institute).

called real double systems, then it is seen that the theorem just proved is applicable to simple as well as double systems.'[1]

In the following years, E. S. Fedorov systematically developed the studies of symmetry, the outcome of which was his 'Symmetry of real systems of configurations' with the complete derivation of space groups. Separate preprints of this work were published in 1890 and were sent to all his friends, including A. Schoenflies, but the complete 28th volume of the Transactions of the Mineralogical Society, which contained Fedorov's paper, was published in 1891.[2]

E. S. Fedorov himself comments on his work in the following words: 'A complete derivation of real point systems is given and a derivation of the possible forms of crystal structures is outlined. The systems of Sohncke are included among the others only as a special case and are called simple systems. Every group is rigorously determined by an algebraic equation.' [3]

As is generally known, Fedorov's derivation is very closely interrelated with the derivation of the same space groups which were almost simultaneously derived by the German mathematician, A. Schoenflies.

The first two papers by Schoenflies were published in 1888 and Fedorov's comments were as follows: 'The papers by Schoenflies, published in Göttingische Gelehrte Anzeigen, have come to my attention recently. It is with pleasure that I see a repetition of the important underlying features of my theory of crystal structures presented in these papers, although in a less developed form.' [4]

A year later Schoenflies published his paper describing 227 space groups. At that time Federov had already submitted his 'Symmetry of real systems of configurations' for publication. However, the final 230 space groups were not represented yet. As a result of the publication of the paper by Schoenflies, Federov requested that his preliminary Table of the derived 228 groups be recorded in the Minutes of the Meeting of the Mineralogical Society, held on 21 November 1889, and that it be compared with the results obtained by Schoenflies. Referring to the overall similarity of the results of these two derivations Fedorov commented: 'Nevertheless, such concordance is accidental and is dependent on the circumstance that Schoenflies neglected as many possible groups as he repeated in the derivation of identical groups.' [5] From this moment on the authors enter into a lively correspondence which is preserved partially in the Archives of E. S. Fedorov and was published in 1951.[6] A study of this correspondence reveals, step-by-step, the details of the path which lead the authors to one and the same final result.

The following is an excerpt from the book by S. A. Bogomolov [7] which gives a brief résumé of these letters.

'In his first letter of 14 December 1889, Schoenflies acknowledges that the Russian scientist has the priority in time.' * In his third letter, of 29 October 1890, he raises an objection to Fedorov's statement that groups V_8^d and V_9^q are identical [these groups coincide with V_d^3 according to the later classification by Schoenflies, the first edition (page 622) erroneously speaks of V_d^4] and to the omission of a group from each pair (5s) and (6s), (1h) and (2h) †, and could not clarify the later problems completely owing to a difference in classification. In his letter of 10 November 1890, Schoenflies completely acknowledges the validity of Fedorov's statements and also mentions the disagreement in the numbers of groups of the following symmetries:

$$V^h, C_4^v, C_3^v.$$

In a post card, dated 17 November 1890, he states that he is investigating the latter problem. The ninth letter, of 7 January 1891, concerns the problem of group (103a) [in the final compilation in *Zeitschr. f. Kristal.*, Vol. 24, it has the number (93a) given by Fedorov; afterwards Schoenflies called it T_d^6]. Schoenflies does not admit its existence, but in the thirteenth letter, of 17 March 1891, fully admits his error. The truth is that even Fedorov omitted this system in his book; he records it, however, in the list of errata.

This, then, is the way in which our scientists arrived at the whole set of 230 space groups.[8]

The well-known book by A. Schoenflies, *Kristallsysteme und Kristallstruktur*, was published in 1891. The author repeatedly refers to E. S. Fedorov, indicating his priority in many problems concerning theoretical crystallography.

For example: 'Die Notwendigkeit die Sohncke'sche Theorie so auszubilden, wie es durch die reine Strukturtheorie im engeren Sinn geschieht, wurde wohl zuerst von E. Fedorow betont.' [The necessity to expand Sohncke's theory as it is done in the narrower sense of pure structure theory proper was emphasized first by E. Fedorov.]

And further: 'Eine Schrift von Fedorow, welche eine vollständige Ableitung aller Raumgruppen und ihre Beziehung zur Kristallsymmetrie enthält, ist 1890 unter dem Titel *Symmetrie der regelmässigen*

* 'Die Priorität gebe ich Ihnen gern zu' [I gladly acknowledge your priority].

† 5s, 6s (fifth and sixth symmorphous)—Fedorov's classification of groups P_m and C_m; 1h and 2h (first and second hemisymmorphous)—groups P_c and C_c.

Systeme von Figuren in russischer Sprache erschienen.' [A book by Fedorov which contains a complete derivation of all space groups and their relation to crystal symmetry has been published in 1890 in Russian under the title *Symmetry of regular systems of configurations.*]

A comparison of these two papers—the one by E. S. Fedorov and the other by A. Schoenflies—clearly shows two principally different approaches by the two scientists: for Schoenflies it is just an interesting case of representation in the theory of groups, in particular infinite groups, which were being developed at that time; for Fedorov it is a means of studying real systems of configurations, the underlying feature of a crystal.

Fedorov found his results by deriving the only possible 230 types of basic design which underlie all natural crystals; accordingly, the most important part of his book are the 230 diagrams compiled in amazingly compact Plates. In 1894 these diagrams, slightly revised by Fedorov himself, were re-published in *Zeitschrift für Kristallographie* and in 1900 presented by Hilton in his well-known *Mathematical Crystallography* in English and in 1919, with certain revisions, by Niggli in *Geometrische Kristallographie des Diskontinuums.** Fedorov's name is mentioned every time Hilton uses any part of his diagrams, Niggli however, omits it completely. The same omission in the presentation of Fedorov's diagrams appears in the Atlas by Astbury and Yardley (Lonsdale) and then also in the first edition of the *Internationale Tabellen* (1935). This error was corrected at our request only in the revised edition of the *International Tables* (1952).

In 1892 E. S. Fedorov published 'A Comparison of the Crystallographic Results of Mr. Schoenflies with mine'. ('Zusammenstellung der kristallographischen Resultate des Herrn Schoenflies und der meinigen') in *Zeitschrift für Kristallographie.* From this time on he starts an extensive correspondence with the founder of this journal, the well-known P. Groth, who also played an active role in preparing the way for Laue's discovery.

The letters from Groth, preserved in Fedorov's archives, and drafts of his own letters, 30 in number, cover a period of 25 years from 1891 to 1915. These letters show how highly Groth valued the work of the Russian crystallographer and how widely he publicized Fedorov's achievements abroad. The great interest of the German scientist in Fedorov's theory of the structure of crystals deserves special mention. 'I have carefully studied your manuscript *Theory of the Structure of*

* The compact diagrams of E. S. Fedorov, in most instances, show only one quadrant of the elementary cell. Hilton and Niggli have drawn out the quadrants for the complete cell.

Crystals and have prepared it for publication. The results are astonishing. However, I cannot raise any contradictions whatsoever', writes P. Groth in one of his letters, dated 26 February 1902.[9]

Thus, the fundamental geometrical rules of the structure of crystals were established and only the crystals themselves remained to be analyzed, i.e., by determination of the concrete picture of the distribution of atoms. The only way to conduct such an analysis at that time was the optical method and geometrical analysis with the aid of a goniometer. Fedorov became and remained to the very end of his life the most prominent specialist in the field of petrographic (optical) analysis of minerals; at the same time he was also very interested in the goniometric analysis of minerals. At an amazing speed he brought forth inventions which in themselves would have made his name immortal, i.e., the two-circle goniometer and the Fedorov universal stage which are the basic instruments of crystallographers and petrographers the world over.*

Fedorov and the students close to him collected enormous quantities of data from the measurements of all possible crystals; this material was systematized and published only a year after the death of the author as the monumental volume *Das Kristallreich.* From the geometry of crystals Fedorov derives conclusions about their internal structure, an idea taken up by Harker, Donnay, and others 40 years later.

All of this was complicated work requiring long periods of time, a factor to which Fedorov's searching geometrical mind could not reconcile itself. He itemized the 230 groups and established the theory of *Stereohedra* and *planigons*, i.e., division of three- and two-dimensional space into geometrically identical cells which continuously and without gaps fill the space. Despite the importance of the purely geometrical results (for this work Fedorov was elected to the Bavarian Academy), they seem at the present time to be a deviation from the correct path, a deviation caused by the negative attitude of the Petersburg chemists whose authority Fedorov accepted and which led him to study electrolytic dissociation, ions and the coordination nature of matter. Fedorov passively accepted the concept of the molecule as the final stage of matter and considered, in principle, that the ultimate aim of crystallography should be the classification of all possible

* All of Fedorov's outstanding discoveries and inventions, including the derivation of the 230 groups, date to the years 1895–6, a time when he was in great financial difficulties. Although he was one of the best students graduating from the Gorny Institute, he was not permitted to remain there; for many years he worked in the office of the Committee of Geology, making field trips in the summer under rather strenuous conditions, so as to enable him to work more intensely in the winter.

'receptacles' for this finite unit. Desiring to accommodate the exact sciences, namely chemistry, he even attempted to consider the 103 asymmorphous groups, included among the 230 groups, as merely theoretical.

Fedorov acquires world-wide fame which he deserves. He is granted the chair at the St. Peter's (now Timiriazev) Academy in Moscow and triumphantly undertakes the journey to the center of western crystallography, which at that time was located in Munich, to meet Groth. Soon afterwards he combines his professorship in Moscow with a professorship in St. Petersburg, where he travels twice a week during the academic year. In 1905 Fedorov becomes the first elected director of his alma mater, the Gorny Institute (Mining Institute) and permanently moves to St. Petersburg. A large group of scientists gather around him, many from abroad (Barker, England; Duparc, Switzerland, and others). Fedorov is the founder of the *Zapiski Gornogo Instituta* (*Transactions of the Mining Institute*), where his numerous papers are published.

The impression of Laue's discovery on Fedorov is illustrated by his letter to the well known revolutionist and scientist N. A. Morozov (1854–1946) which describes this portentous scientific event.

2 October 1912

Dear Nikolai Aleksandrovich,

You conclude your letter by saying that the human eye shall never see atoms. You wrote this approximately at the time when people saw atoms with their own eyes; if not the atoms themselves, then the photographic images caused by them.

How does this come about? Very simply really. With a dividing machine we can draw a thousand parallel lines on glass within the range of a millimeter; this is a diffraction grating which gives a series of magnificent diffraction spectra and the number of divisions on glass is readily determined from these spectra.

A thin crystalline plate in itself represents two intersecting diffraction gratings where the lines are not a thousandth but a ten-millionth part of a millimeter apart.

Light waves are too coarse for obtaining diffraction spectra. But there are the X-rays with a wavelength millions of times shorter than that of light waves.

Several weeks ago in Röntgen's laboratory such diffraction spectra of atoms were photographed by means of X-rays. Indirectly people were able to see the immediate effect brought about by atoms, that is, in principle, they saw the atoms with their own eyes.

For us crystallographers this discovery is of prime importance because now, for the first time, we can have a clear picture of that on which we have but theoretically placed the structure of crystals and on which the analysis of crystals is based.

I am sure you will be very pleased about this news.

With the very best regards from me and the family.

E. Fedorov

(Archives of the Academy of Sciences USSR, Leningrad, 543, 4, No. 1952). [10]

Already Bragg's earliest work on the analysis of crystal structures by means of X-rays aroused Fedorov's lively response.

In his paper 'The first experimental demonstration of the asymmorphous real system' Fedorov emphasized that the determined structures belong to the systems of true point systems derived by him twenty-two years earlier. This paper starts with the following passages:

'The application of X-rays has enabled W. L. Bragg (and his father) to draw conclusions which are of the utmost importance to the theory of crystal structures. Some of these conclusions are unexpected, at least in the sense that in the points of real systems one expected to find centers of chemical particles, while the experiments of this scientist permit one to draw the conclusion that these are the centers of atoms. As a result, in substances of the simplest chemical composition special real point systems are obtained and the symmetry centers are occupied by separate atoms as though the atoms themselves have a high symmetry.' [11]

And it is stated further that 'the distribution of Fe and S atoms in pyrites confirms the asymmorphous real point group (25).'*

In conclusion Fedorov explains his derivation of the space groups and how the scientific circles of that era underestimated its value.

'Somehow I did not think that I would live to see the day when the distribution of atoms as I predicted it in my papers would actually be

* Group Pa 3 in the contemporary classification; this is one of the 103 which were previously considered as 'imaginary' Fedorov systems, in the same sense as $\sqrt{-1}$ is imaginary. Fedorov is very pleased to acknowledge this fact and soon afterwards, together with Groth, starts campaigning for the ionic-coordination-nature of crystalline substances rendering his imaginary systems more real than the non-imaginary symmorphous ones.

determined. In a letter to Prof. Groth I stated that a detailed analysis of the systems predicted in these papers could, perhaps, be realized only after 100 years.

'In 1891 I submitted the Russian work which predicted the possible atomic distribution or, rather, the laws of such distributions, to the St. Petersburg Academy of Sciences in competing for the Makarev prize of that year; neither did it receive the prize itself, nor was it found worthy of a mention, and was not even included in the lengthy official list of papers submitted on this occasion.' [12]

In addition to this first printed declaration E. S. Fedorov published a series of papers concerning Bragg's conclusions. In them the eminent crystallographer discusses the results obtained, attempting to bring them into harmony with his previously advanced views of crystal structure. The following is as complete a list as possible of these papers:

'On the structure of diamond crystals according to Bragg' (1914).

'The first steps on the path to the determination of the distribution of atoms in crystals' (1915).

'Results of the first stage in the experimental investigation of crystal structures' (1916).

'The basic law of crystal chemistry' (1916).

'The chemical aspect of crystal structures' (1916).

An interesting description of the acquaintance of E. S. Fedorov with Bragg's work can be found in the letters preserved in the Archives and in the now published letters of the well-known British crystallographer T. Barker (1881–1931). In 1908–1909 Barker came to St. Petersburg to study Fedorov's methods of analyzing crystals. E. S. Fedorov warmheartedly greeted the young scientist, installed him in his laboratory, and soon enlisted his services in the compilation of the Tables of Crystal Chemical Analysis *Das Kristallreich* (together with his students B. P. Orelkin, V. I. Sokolov, D. N. Artem'ev). After his return to England Barker continued his work on these Tables in Oxford, assisted Fedorov in proofreading the prints of *The Realm of Crystals*, and enthusiastically spread the word among the British scholars about the achievements of the Russian crystallographer.

His letters to Fedorov contain much interesting information about the scientific events of that era. An amusing misunderstanding which occurred among the British scientists is very vividly described in one of Barker's letters (15 November 1912): 'Sir Oliver Lodge in a lecture before the Chemical Society on recent developments of Natural Science stated some sentences which show that perusal of Tutton's second article had given him the impression that the Röntgen ray work and crystallochemical analysis were the same discovery!!

'On account of this I wrote my article in order to show that the credit of the method of 'crystallochemical analysis' belongs to you alone, and that no one else has taken any part in the matter.' [13]

Barker's letter of 7 January 1914 describes the impression left by Fedorov's paper 'The first experimental demonstration of the asymmorphous real system': *

'I have been busy for the last two months writing the annual report on Crystallography and Mineralogy for the Chemical Society... I have made a special feature in the report of your work on the correct setting—as far as it is possible to write about such a subject for chemists who have little knowledge of crystallography—and the Röntgen-ray work on crystals. The latter is causing somewhat of a sensation in the scientific world here.

'I do not know whether you are acquainted with the recent developments contributed by the two Braggs. If you are interested in the work I will ask them to send you copies of their papers.' [14]

E. S. Fedorov, after having received the reprints from W. L. Bragg, writes as follows: [15]

St. Petersburg, January 15/25, 1914

Dear Mr. Bragg,

Thank you for sending your valued papers. I hope that you will soon receive 'Reguläre Plan- und Raumtheilung' [The regular planar and spatial divisions].

Respectfully,
E. S. Fedorov

The Archives of Fedorov contain two short letters from W. L. Bragg. The first acknowledges the receipt of the monograph 'Reguläre Plan- und Raumtheilung'.

The text of the second letter is as follows: [16]

* This paper was sent to Barker, obviously, in the form of proofs or a manuscript (judging by the dates).

Trinity College, Cambridge
17 February 1914

Dear Sir,

Many thanks for your letter with the suggestions as to ways of depicting a space-arrangement of points. I have considered your method with great interest, because I find it so very difficult to describe a structure [in words] * when I have worked it out properly. If there is already a recognized way of denoting the situations of the atoms in the crystal, I would be very glad if you would tell me of it.

I will try to depict the great lot of crystals I do in your way, though they may be too complicated.

Yours sincerely,
W. L. Bragg

In conclusion we should like to quote the comments on E. S. Fedorov recently made by W. L. Bragg in his letter, dated 14 May 1958, to G. I. Kovan'ko, published in a paper by the latter:

'Fedorov was then to me an almost legendary being who had worked out the 230 crystal classes.

'Few people at that time were interested in crystallography. Such interest as did exist was in the outer forms of crystals, not in their inner structure. When I started analysing crystals with X-rays, I knew nothing at all about their geometry. It was wonderful for us to discover that great men like Fedorov and Barlow, whom I also got to know, had studied the inner geometry of crystals and provided a sure theoretical basis for our work.' [15]

I. I. Shafranovskii and N. V. Belov

References

1. E. S. Fedorov, Nachala Ucheniia O Figurakh, *Zap. Min. Obshch.* (The elements of the study of Configurations. *Trans. Mineral. Soc.*) Vol. 21, p. 240, 1885.
2. E. S. Fedorov, Simmetriia Pravil'nykh Sistem Figur. *Zap. Min. Obshch.* (The symmetry of real systems of configurations. *Trans. Mineral. Soc.*) Vol. 28, pp. 1–146, 1891.
3. E. S. Fedorov, Protokol'naia Zapis'. *Zap. Min. Obshch.* (Official report. *Trans. Mineral. Soc.*) Vol. 26, p. 454, 1890.
4. E. S. Fedorov, Zametka Ob Uspekhakh Teoreticheskoi Kristallografii Za Poslednee Desiatiletie. *Zap. Min. Obshch.* (Report on the progress of theoretical crystallography during the last decade. *Trans. Mineral. Soc.*) Vol. 26, pp. 374–375, 1890.

* [not clearly legible.]

5. E. S. Fedorov, Protokol'naia Zapis'. *Zap. Min. Obshch.* (Official report. *Trans. Mineral. Soc.*) Vol. 26, p. 455, 1890.
6. Correspondence of E. S. Fedorov with A. Schoenflies (Published by G. B. Bokij and I. I. Schafranovskii). *Nauchnoe Nasladenie* (*Scientific Heritage*) Izdaniya Akademii Nauk SSSR (Publishing House of the Academy of Sciences USSR) Vol. 2, pp. 314–343, 1951.
7. S. A. Bogomolov, *Vyvod Pravil'nykh Sistem Po Metodu Fedorova* (*Derivation of real systems by the Fedorov method*) Part I, pp. 9–10, 1932. Leningrad, Kubutch.
8. A. Schoenflies, *Kristallsysteme und Kristallstruktur*. p. 622, Leipzig 1891.
9. Correspondence of E. S. Fedorov with P. Groth (Published by I. I. Shafranovskii, with comments), *Kristallografiia* (*Crystallography*). Sbornik Leningr. Gornogo IN-TA (Collection of Papers, Leningrad Mining Institute) No. 3, p. 219, 1955, Izd. Leningr. Gosuniversitet.
10. *E. S. Fedorov's Manuscripts*, Archives of the Academy of Sciences USSR (Ed. by I. I. Shafranovskii and N. M. Raskin) p. 140, 1957. Publ. Acad. Sci. USSR.
11. E. S. Fedorov, Pervoe Konstatirovanie Opytnym Putem Asimmorfnoi Pravil'noi Sistemy (First experimental demonstration of the asymmorphous real system) *Zap. Gornogo Instituta* (*Transactions of the Mining Institute*) Vol. 5, p. 54, 1914.
12. Ibid. p. 56.
13. Letters of T. Barker and others to E. S. Fedorov. *Kristallografiia* (*Crystallography*) (Sbornik Leningr. Gornogo IN-TA) (Collection of papers, Leningrad Mining Institute) No. 3, p. 239, 1955. Izd. Leningr. Gosuniversitet (Published by the Leningrad State University).
14. Ibid. p. 243.
15. G. N. Kovan'ko, Iz Istorii Strukturnoi Kristallografii. *Zap. Min. Obshch.* (From the history of structural crystallography. *Trans. Mineral. Soc.*) Part 87, No. 6, p. 676, 1958.
16. Letters of T. Barker and others to E. S. Fedorov. *Kristallografiia* (*Crystallography*) (Sbornik Leningr. Gornogo IN-TA) (Collection of papers, Leningrad Mining Institute) Publ. Leningrad, Gosuniversitet, pp. 249–250, No. 3, 1955.

Translated by: A. Werner, Bell Telephone Laboratories, Inc., N.Y.

Artur Schoenflies

1853-1928

Schoenflies was born in the small district town of Landsberg an der Warte, then belonging to Brandenburg, now in Polish territory. He began studying mathematics in Berlin just after the war in 1870 and obtained his Ph. D. in March 1877. His main teacher was E. E. Kummer, famous for his research in geometry. The next six years Schoenflies spent as high-school teacher, the first two in Berlin, the others in Colmar in Alsace. He managed to continue research in this period along the lines begun in his thesis, combining in it geometrical inspection methods with those of analytical, synthetic and projective geometry. The success of his work led to his becoming Privatdozent

(Lecturer) in 1884 and later (1892) Associate Professor of Applied Mathematics in Göttingen. Here he became interested in the geometrical properties of rigid-body motion which had first been studied in this sense by Camille Jordan nearly twenty years earlier. Two theorems established by Schoenflies will serve to illustrate the kind of properties involved: (i) All points of a rigid system which lie on straight lines in three of its positions belong to a (spatial) curve of the third degree; (ii) there exists a curve of the sixth degree whose points lie on circles in any four different positions of the body.—The results of these studies were incorporated in a book *Geometrie der Bewegung in synthetischer Darstellung* (Leipzig 1886) and later again summarized in an article on Kinematics in *Enzyklopaedie der mathematischen Wissenschaften*, Vol. I, section IV, 7, pg. 190–278, written together with M. Grübler.

From problems of motion, Schoenflies' attention was next drawn to those of 'plane configurations' whose study had been begun in 1887 by the Italian geometer Martinetti. A 'configuration', such as n_3, is a system of n straight lines and n points such that each line carries three points and each point is the intersection of three lines. The aim of the theory is to obtain a classification of all possible types of configurations by exploring their geometrical properties.

These by no means easy studies prepared Schoenflies well to deal with the periodic discrete groups of movement, that is, the covering operations of a space group. The earlier work on continuous groups by Jordan (1869) and on periodic discrete groups by Leonhard Sohncke (1879) Schoenflies found to be incomplete because symmetry elements of the second kind, i.e. rotation-reflection and rotation-inversion axes, were not taken into consideration. Their inclusion added 165 groups to the 65 which Sohncke had derived, bringing the total to 230. Whereas Sohncke as well as E. von Fedorov considered that the space groups, in order to be of physical significance, should be restricted by certain pre-conceived physical ideas and were otherwise not 'real', Schoenflies demanded only that the group of covering symmetry operations be geometrically possible. In this separation of pure geometry from physical statements lies the strength of his theory. Within the 'fundamental domain' (the asymmetric unit is the inadequate term now in use for its contents) 'the crystallographer may do whatever he likes', as Schoenflies puts it (*Encycl. d. math. Wiss.* l.c., pg. 468); he has full liberty there, and it is not within the province of the geometrical structure theory to impose any restrictions on the way the *Fundamentalbereich* is filled.

Schoenflies' work on structure theory began with three papers in

Mathematische Annalen (1887 and 1889), and was completed in his book *Kristallsysteme und Kristallstruktur* (Leipzig 1891). At the same time nearly identical results had been developed quite independently in St. Petersburg by E. v. Fedorov. Already earlier the latter had stressed the desirability of considering covering operations of the second kind in his book *Gestaltenlehre* (St. Petersburg 1885) and carried this idea out in the book *Symmetrie der regelmässigen Systeme von Figuren* (St. Petersburg 1890). Here, he obtained the 230 arrangements but, coming from the morphological side of crystallography, he attributed physical significance to the polyhedral fundamental domains (called by him stereohedra) and thus distinguished between the actually possible ('real') and the other 'asymmetric' space groups.

Again, in England, Lord Kelvin discussed close-packed arrangements of equal spheres and their mechanical stability (1889), and W. Barlow derived the regular packings of spheres and their symmetries (1883) and later (1891) extended this to the packing of spheres of two or three different sizes.

There was thus a sudden resurgence of interest in this problem after a stage in which it had lain dormant. Schoenflies' contribution is that of a careful mathematician who does not exceed his competence and therefore, within it, gives a final answer.

Schoenflies became deeply interested in the theory of sets, a subject then fluid and controversial, on which he wrote a famous report in 1914. He was the first author, and in numerous later editions co-author with Walter Nernst, of an introductory calculus for science students: *Einführung in die mathematische Behandlung der Naturwissenschaften*. A summary and critical review of crystallographic structure theories was given by Schoenflies in 1905 in *Enzyklopaedie der mathematischen Wissenschaften*, Vol. V, 7, pg. 437–492 (published 1922!), and in view of the increasing use of structure theory after the First World War he prepared an improved version of his older book under the title *Theorie der Kristallstruktur* (Gebr. Bornträger, Berlin, 1923).

Schoenflies became Professor of Mathematics at the University in Königsberg in 1899 and accepted in 1911 a position at the 'Academy' in Frankfurt which was about to obtain university status. He helped bringing about this transformation and became, in 1914, the first Dean of the Science Faculty. In 1920/21, the year before he retired, he was Rector of the University.—He was married and had five children.

Whoever knew Schoenflies admired his pure and considerate personality, and his modesty. He was beloved as a teacher by the students and as a colleague by the Faculty.

William Thomas Astbury

1898-1961

W. T. Astbury was one of Sir William Bragg's earliest and most devoted disciples. Born of humble parentage—a fact of which he was always proud—in Longton, Stoke-on-Trent, in the pottery district of England on 25 February 1898 he received his secondary training in Longton High School from 1908 to 1916. Scholarships enabled him to go to Cambridge in order to study Chemistry. He was there 1916/17 and 1919/21, war service intervening. In both parts of the Natural Science Tripos he obtained first class, 1920 in Chemistry, Physics and Mineralogy, and 1921 in Physics. He immediately became Demonstrator in Physics at University College, London, under Professor W. H. Bragg who took him along as Assistant to the Royal Institution and the Davy-Faraday Laboratory when he moved there in 1923. Astbury remained at the R.I. for five years and was the soul and activator of the keen group of young workers that Sir William had brought together there (conf. J. D. Bernal and K. Lonsdale in Part VII). The reason for this lay in his unlimited enthusiasm for the new subject of crystal structure analysis, his temperamental approach, and the unexpected and sometimes provocative, but often most helpful turns in his conversation.

In 1928 Astbury came to Leeds on Sir William's recommendation, where basic research on the physics and chemistry of wool was to be started at the University. He became Lecturer in Textile Physics in 1928, Reader in 1937 and in 1945 Professor of a newly established Department and Laboratory of Biomolecular Structure. From the very beginning of his career Astbury stressed the close connection between the chemical and physical changes in wool, as in other fibres. By their X-ray diffraction effects these substances do not provide information of the same precision and in the same profusion as single crystals and this makes it necessary to combine with it all possible evidence that can be gleaned from their physical and chemical behaviour—a discussion that often requires great imagination. It was hereby that Astbury's unsinkable optimism helped him along where more anxious scientists might have feared to tread. The designation of the Department for Astbury was the first of its kind, and Astbury was proud of the name:

Biomolecular Structure; this type of name has since been adopted by departments or laboratories in other universities, British and foreign. As Astbury conceived it, it was to be a place where biological structure and texture on the molecular scale could be attacked in a catch-as-catch-can style, using chemical, physical and biological properties in conjunction with microscopy, electron microscopy, X-ray and electron diffraction and whatever else appeared hopeful. When I last saw Astbury in Leeds in 1959 I found him at the daily laboratory tea party, presiding over about eight co-workers, all of a very mature scientific stature, and leading as challenging and lively a discussion as ever.

Astbury's most important scientific contributions are, according to his own evaluation, his three papers on the Structure of Hair, Wool, and Related Fibres (*Trans. and Proc. Roy. Soc.*, A 1931–35), his studies on the Denaturation of Proteins (1935; poached eggs he used to speak of), and his Diffraction Studies of Bacterial Flagellae (1949, 1955). He wrote a book *Fundamentals of Fibre Structure* (Oxf. Univ. Press 1933), contributed a noteworthy essay on 'The Forms of Biological Molecules' in the presentation volume to d'Arcy Wentworth Thompson *Essays on Growth and Form* (Oxf. Univ. Press 1945), and gave the Croonian Lecture in the Royal Society in 1945 'On the Structure of Biological Fibres and the Problem of Muscle'. Besides, he was a prolific writer of papers in physical, chemical, biological, textile-technological, and general journals, and a great lecturer. The acceptance which his outstanding work found is documented in a long list of medals, prizes, and honorary degrees and memberships he received.

Astbury had undertaken to write for this volume, together with Dame Kathleen Lonsdale, about the early years of X-ray diffraction in England, and especially at the Royal Institution. He began writing, and among his papers a draft page was found which is so characteristic of him, in spite of the severe heart affliction which then already often incapacitated him, that no better description of his buoyant personality could be given than by adding it to this Notice. It will explain better than many words why Astbury was one of the most beloved among the early crystallographers.

P. P. Ewald

Early days at University College, London and the Davy-Faraday Laboratory of the Royal Institution

In sharing out who shall write what in these X-ray diffraction memoirs, the two 'oldest hands' available to recall those early days at U.C.L. and the R.I. were obviously Kathleen Lonsdale and myself, but what was not clear was who should do the actual writing. We could hardly say 'we remember' when it was sometimes only one of us who remembered, and anyway we like to think we have distinctive styles; so we compromised—I mean, she agreed to let me do the job. But it is to be understood, though, that it is a joint effort and unless otherwise stated, any particular memory or anecdote may be privy to either or common to both. For the purposes of this article we are to be considered, like the two unresolved heirs in 'The Gondoliers,' as a single unit.

I am the older hand of the two, in years and crystallographically, because I joined the staff of the Physics Department of University College in 1921 after graduating two years late through war service, while Kathleen (then Yardley) joined in 1922 after graduating two years younger than most people. I had read chemistry and (classical) crystallography at Cambridge besides physics, and I was also married in 1922, so this experienced old man ventured to take the precocious child under his wing, for a very brief start at least. It was a good case of the blind leading the blind (to mix the metaphor a little), but that of course was the fun of the thing. It has been well said that the greatest asset of scientific research is its naivety, and we must have been wonderful examples of that—all of us, not even excluding Sir William at his own high level. And before going any further let me make it crystal clear that Sir William—the Old Man, or Bill Bragg, as we called him behind his back—never 'led' any of us, in the technical sense. That was not his way—if he had a way—and if you were stupid enough you might even claim that you 'led' him, since, especially after we migrated in 1923 to the Davy-Faraday Laboratory of the Royal Institution, as often as not when he popped into your room (not terribly often) it was to ask *you* a question in connection with some lecture that he was due to give. Or you might meet him on the stairs and he would say: 'Hello! How's the family', or some such. He turned up at tea as often as possible, where we rarely talked 'shop' in any case.

Carl H. Hermann

1898-1961

Those of us who knew Carl Hermann deeply appreciate the loss they personally and Crystallographers in general have suffered through his untimely death from a heart attack at the age of 63. Hermann belonged to the small number of fortunate people to whom the intricate geometry of space groups comes, at it were, naturally and without effort. If coordinates of equivalent points were needed, he rarely took the trouble of looking them up in tables because he wrote them down just as quickly 'by inspection', deriving them from the symmetry elements which he saw in his mind. This insight made him the unquestioned editor of the *Internationale Tabellen* when these were first proposed, and was of great value also in the critical attitude which was to be carried through in the first volumes of *Strukturbericht.*

Of Hermann's scientific work his doctoral thesis under Max Born should be named first; in it, he calculated for the first time, in 1923, the optical rotatory power of a crystal. Sodium chlorate was chosen as a solid which owes its optical activity entirely to the crystal structure, since in solution it is non-active. The gyration vector had been introduced by Born in his general theory, but its actual determination for a given structure of such complexity as sodium chlorate presented difficulties of calculation which at the time appeared very formidable. Unfortunately Hermann went wrong on certain factors π, so that the numerical work had to be repeated later, but his work broke the ice.

Hermann's first salaried job after his graduation in Göttingen was in Hermann Mark's division of the Kaiser-Wilhelm-Institut für Faserstoffe in Dahlem. When this came to an end, Hermann came to me in Stuttgart in September 1925 as my assistant. I had the ambition of keeping the modest list of all known structures up to date which formed a twenty page appendix in my book of 1923, but soon became aware of the rapidly increasing number of yearly structure determinations which made it impossible to keep up with it by myself. So Hermann and I shared the burden of preparing the first volume of the *Strukturbericht*, his share of the work gradually increasing as my growing academic duties kept me otherwise engaged. For the next volumes Hermann took over the editorial responsibility.

I remember vividly how on one of the first days after Hermann's arrival in Stuttgart I showed him how, by forming $|F_{\mathbf{h}}|^2$, the series to which it is equal is the structure factor of a mass distribution containing the relative distances between the atoms—what is nowadays called the Patterson function or the 'folded' mass distribution of the crystal. The problem presented to Hermann was that of unfolding. Whereas my attempts at unfolding had been in general terms, Hermann at once became interested in the folding operation with symmetry elements present. Neither of us found a solution to the unfolding problem and therefore nothing of our work was published. But it had started Hermann on the introduction of 'Kennvektoren'—characteristic vectors for the symmetry elements—and 'Kennstellen'—characteristic components of these vectors—which he then made the foundation for his derivation of the 230 space groups and the main feature of his nomenclature which grew out of it (*Zeitschrift für Kristallographie* 1928, Vols. 68 and 69). The same derivation which, assuming three independent translations led to 230 space groups, gave 75 groups of a single translation (chain groups) and 80 groups of two translations (net groups), all of them in three-dimensional space. Except for slight changes, especially with a view of adapting the symbols to the use of a typewriter, Hermann's nomenclature, which came very close to one devised by Mauguin for his lectures, was adopted at the conference in Zürich in July 1930 where the plans for the *International Tables for the Determination of Crystal Structures* were worked out. (See also the papers by Mauguin and by Hermann in *Zs. Krist.* 1931, *76*, 542 and 559.)

Hermann's main interest at this time were the symmetry properties of matter. Thus he presented a new way of recognizing the influence of space-group symmetry on tensors of any order in *Zs. Krist.* 1934, *89*, 32–48 and tackled the systematic derivation of all possible symmetry cases for the arrangement of like molecules in a statistically symmetrical way in a paper written for a special issue of *Zs. Krist.* on liquid crystals (1931, *79*, 186–221). Including the amorphous and the periodical (crystalline) state, Hermann finds 20 types of arrangements with different statistical symmetries. This paper was used by Hermann as thesis for a Lecturership at the Technische Hochschule Stuttgart in 1931.

Hermann came from a protestant family from near Bremen in which matters of conscience were not taken easy. His father, captain of a freighter, asked his employers after years of service to relieve him of the responsibility and let him continue as a simple seaman. Of his brothers, two were in the church; one of his sisters became a very active philosopher and had to leave Germany under Hitler. Carl's wife Eva came from protestant theologians. Carl himself, after an early period of agnosticism (and of other 'isms') became devoutly religious, though not church-going, some time before the Second War. His wife and he joined the Friends and after the war became leading in the small group of such that had survived Hitler's enmity.

It was not astonishing that a man of such conviction and as fearless as Hermann had to give way to the mounting political interference in academic life two years after the Nazi had taken over. After abortive attempts at finding an academic position abroad Hermann accepted an offer from R. Brill to join his group of crystallographers at the I. G. Farbenindustrie laboratories in Oppau. Here Hermann became co-author of the well-known paper on the electron density distribution in rocksalt, diamond and other crystals (Brill, Grimm, Hermann, Peters, *Ann. d. Phys.* Lpz. 1939, *34*, 393). Although after having been called up for active duty at the outbreak of war, Hermann managed to convince the Commanding General personally that he would better be deferred since under no circumstances would he become a combatant; he and his wife were later imprisoned for having been found listening to the BBC radio news.* They were saved the death penalty only because it could be established that it was the news in English on which death penalty was not compulsory (Hitler did not count for much the 'ten thousand intellectuals' who knew English). They were separated for years from one another and their two adopted children, and found their adventurous way together—their home having been destroyed in the air raids on Mannheim—only after the liberation.

In 1946/47 Hermann was lecturer at the Technical University in Darmstadt until, in 1947, a special chair and institute for crystallography was created for him at the University of Marburg.

Hermann was one of the Germans attending the First Assembly of the International Union for Crystallography at Harvard in 1948 and he gave, on that occasion, a very brilliant and original paper on space groups in higher dimensions. It appears from the published part (*Acta Cryst.* 1949, *2*, 139–145) that Hermann had considerably more material on hand; the further progress hinged on the proof of some very simple assumptions, likely to be true and provisionally made. As it sometimes happens, this proof could not be found, and all the material Hermann had elaborated remains unpublished.

Hermann hardly ever felt the urge to apply his talents to actual

* This may have been a pretext only; the real but not legally punishable crime was that they had fed and hidden Jews in their plight.

structure determinations. The outstanding one among these is the analysis of the urea adducts. This work was done during his prison term in Halle from where he was, on the representation of the I. G. Farben, daily escorted by a guard to their Leuna works for 'work of national importance'. In these urea adducts long-chain molecules like paraffins are imbedded in cylindrical tubes formed by spiralling polyurea chains—an entirely unforeseen type of structure at the time.

Hermann's lectures on crystallography are said to have been very stimulating. They contained some of the material he could not bring himself to publish, and parts of his course were worked out by his students and checked by him. Perhaps, some day, they can be published, but they will now always remain a fragment.—Max Born, in a short obituary notice on his friend and pupil mentions his profound and expressive addresses at the Friends meetings which testified to his sincere search of truth, clarity of mind and moral fortitude. The longer the more, matters spiritual took precedence in Hermann's unfinished work and brought him nearer to the peace for which he always strove.

P. P. Ewald

Gösta Phragmén

1898-1944

Dr. Gösta Phragmén, who died on 21 August 1944, was born on 29 April 1898. He took his first degree in chemistry, physics and mathematics at the University of Stockholm in 1921. By this time he was already associated with the newly established Institute of Metallography where he acted as assistant first to Professor C. Benedicks and subsequently to Professor A. Westgren. In 1927 he succeeded Westgren as principal metallographer at the Institute, and when the Institute was reorganized in 1933 he was appointed head of its technical department. In 1934 he was awarded the degree of Licentiate of Philosophy in physics by the University of Stockholm. From 1939 he was assistant professor in metallography at the Royal Technical University of Stockholm, where from 1942 he lectured in physical metallurgy. When the Institute of Metallography was reorganized as a result of its expansion, he was appointed its head in 1943.

Even in his youth Phragmén showed unique qualities. In spite of remarkable personal modesty he had marked confidence in his knowledge and judgement and in consequence he was highly critical of the concept of authority. These characteristics were evidently inherited from his father, the distinguished mathematician Edvard Phragmén, and they were strengthened by an education free of all constraint. In addition to his critical scientific disposition, he soon showed a strong technical interest, and this probably influenced his choice of the new Institute in spite of all his prospects for a successful academic career. It is probable that his dislike of academic formality also contributed to this choice.

Phragmén's first efforts as an X-ray crystallographer were made in 1920, when G. Aminoff, then assistant professor at the University of Stockholm, and he investigated the structure of the mineral alloy osmiridium. The new techniques captured his interest, and when they were later used for alloy investigations at the Institute of Metallography, where Phragmén entered as a pupil, he very soon contributed to their development and application. His efforts in this field assured the success of the X-ray structure research carried out at the new Institute.

Phragmén used his considerable experimental skill to construct a reliable and simply operated apparatus for obtaining powder photographs. He constructed a rotation camera which was primitive but adequate for the first investigations of relatively simple crystal structures. During his experiments with different types of powder camera he also gradually improved the technique of powder photography, and finally he built a series of focussing cameras which, with their high resolution and very good line definition, proved particularly well suited for alloy investigations.

Studies of the crystal structures of the iron modifications and of steel, which were begun by Westgren in Professor Siegbahn's laboratory in Lund, were continued by Westgren and Phragmén together. They found that δ-iron is isostructural with α-iron, and they determined the crystal symmetry of cementite and also the unit-cell dimensions of martensite and austenite.

In 1923 Phragmén reported an X-ray investigation of the iron-silicon system which had led to the determination of the structures of FeSi and $FeSi_2$. In the course of further investigations of the same alloy system Phragmén found that silicon atoms substituting for iron atoms in the solid solution of silicon in α-iron tend to arrange themselves in a regular way when the silicon content exceeds about 7%. In this way a superstructure is formed which is fully established at a composition corresponding to the formula Fe_3Si. In 1926 Phragmén published a final report on his work on the iron-silicon system, a model combination of metallurgical microscopy and X-ray crystallography.

Together with Westgren, Phragmén studied the crystal structure of manganese. They found that this metal occurs in three different modifications. The X-ray data collected by Westgren and Phragmén, which were published in a paper in 1925, were later used by A. J. Bradley for a determination of the complicated crystal structures of α- and β-manganese.

As a preliminary to investigations of the technically important systems Fe—Cr—C, Fe—W—C and Fe—Mo—C, Phragmén and Westgren carried out X-ray analyses of chromium, tungsten, and molybdenum carbides; they also devoted a special study to the carbide occurring in high speed steel, which was later found to have the formula Fe_3W_3C. In cooperation with T. Negresco, a Rumanian guest at the Institute of Metallography, an investigation of the equilibria in the system Fe—Cr—C was undertaken, in which Phragmén took a particularly active part. It was no simple task to construct the equilibrium diagram for this complicated system from data provided by micro-structures and X-ray photographs, but in this very fact lay its appeal for Phragmén. This central part of the study was carried out mainly by him and he also wrote the section of the report dealing with equilibrium conditions in the system.

In parallel with these investigations connected with the physical metallurgy of steels, Phragmén also participated in a series of X-ray analyses of the systems Cu—Al, Cu—Zn, Ag—Zn, Au—Zn and Cu—Sn. In cooperation with Westgren he also published several papers in which the discovery of the structural analogies and the connection between valence electron concentration and structure type were reported.

Phragmén again showed his eminent skill as an experimenter in an investigation by X-ray diffraction techniques of the thermal expansion of invar, which he reported in 1931. For this purpose he constructed a camera, which recorded the X-ray reflections of a thin sheet of invar, which was heated electrically to about 200°C. He found that the reflections were so slightly displaced in relation to their positions at room temperature that the abnormally low thermal expansion of the alloy could not be ascribed to a partial phase transition. Thus, as Chevenard had already assumed, the practically constant volume of invar when heated depends on an expansion anomaly of the constituent γ-Fe-Ni-phase.

Phragmén was probably not by nature a good teacher, but his interest in teaching increased markedly as the years went by. One might perhaps say that he lacked a psychological interest in ignorance and stupidity. In his writing his presentation was concentrated and went straight to the point. Great concentration and a considerable amount of knowledge was needed in order to follow it. Phragmén sometimes showed an astonishing indifference to the publication of his own results. He probably thought that the whole thing was so simple that other people should be able to obtain the same results without difficulty. To strive for honours was unfamiliar and even repulsive to him.

One of Phragmén's earlier collaborators gives his impression of him as a leader in these words: 'Phragmén was to an unusually high degree a master of the theoretical basis of his field of research. This, when added to his requirement of consistency, order and clarity in reasoning and ideas, made all discussions with him extraordinarily instructive and interesting. Speculations without factual basis were aims of his quietly ironical scepticism. He was very much interested in the ideas of his assistants on aspects of the work, and stimulated their activities by demonstrating the gaps in their knowledge and the weakness in their reasoning. He permitted great freedom in the work, but in spite of this he was interested in every detail and one could always count on his assistance in all difficulties. When experimental problems occurred he always had simple and practical solutions to propose. With his considerable knowledge, his critical judgement and his mental balance he inspired a degree of confidence which made work with him very agreeable.'

Phragmén's experimental skill and interest in instrument construction has already been mentioned. Another of his collaborators testifies to these qualities, but adds that although Phragmén made exacting demands on the performance of an instrument, he was less concerned with its convenience in application.

Phragmén was an excellent adviser on scientific research. In this connection, the qualities which were particularly evident, in addition to his wide knowledge and critical outlook, were his never failing helpfulness and a lightning power of apprehension. One sometimes had the impression that he knew in advance what one intended to say. He was speculative within the limits imposed by known facts, but his imagination was always strictly controlled by observation. It is possible that Phragmén made his best contributions when he cooperated with a more optimistic and less critical colleague. It is quite natural that he turned out to be, as someone has said, 'an oracle, to whom one turned with all observations and ideas'.

With the years, Phragmén also became increasingly acknowledged abroad for his penetrating contributions to scientific discussion. In the field of theoretical metallurgy in particular, there are many proofs of his authoritative position. A list of 53 of his publications may be found in *Jernkontorets Annaler* 1944, *128*, 533–535.

In a relatively short working life, Gösta Phragmén was able to give a lustre to Swedish research in metallurgy and metallography through his scientific contributions. The level in this field of research was markedly raised under the influence of his example and criticism. His colleagues remember him with gratitude and admiration as an incomparable fellow-worker, and a straightforward and honest man.

A. Hultgren, B. Kalling, A. Westgren

Victor Moritz Goldschmidt

1888-1947

Victor Moritz Goldschmidt was born 1 January 1888 in Zürich. His father Heinrich J. Goldschmidt named his son after his teacher Victor Meyer. The Goldschmidt family came to Norway 1901 when Heinrich Goldschmidt took over a chair as Professor of Chemistry in Kristiania (Oslo).

Goldschmidt's first important contribution was within the field of geology and mineralogy. His two first larger works were his doctor thesis *Die Kontaktmetamorphose im Kristianiagebiet* and *Geologisch-petrographische Studien im Hochgebirge des südlichen Norwegens*.

Goldschmidt has been named the founder of modern geochemistry and crystal chemistry. A series of publications under the title *Geochemische Verteilungsgesetze der Elemente* is usually referred to as the start of geochemistry, the science that describes the distribution of the chemical elements in nature. The geochemistry has not only greatly inspired the field of mineralogy and geology but also theoretical chemistry and crystallography.—Goldschmidt's work on atom and ion radii has been of enormous importance for crystallography. His work in this area has no doubt inspired the introduction of the Pauling covalent, ionic, and van der Waals radii.

Goldschmidt took great interest in the technical application of his science; the utilization of olivine for industrial refractory goes back to him. He was for many years the head of the Norwegian Committee for Raw Material (Statens Råstoffkomité).

There has hardly ever been a person in the Norwegian university world who made such an early and rapid career as Goldschmidt. Withouth even taking the usual exams or degrees he got a post-doctoral fellowship from the University already at the age of 21 (1909). He obtained his Norwegian doctor's degree when he was 23 years old (1911). This is a degree that is usually obtained at an age of 30 to 40 years, even 50 years and more is not unusual. In 1912 Goldschmidt got the most distinguished Norwegian scientific award (Fridtjof Nansens belønning) for his work *Die Kontaktmetamorphose im Kristiania-gebiet.* The same year he was made Docent (Associate Professor) of Mineralogy and Petrography at the University of Oslo (at that time 'Det Kongelige Frederiks Universitet'). In 1914 he applied for a professorship in Stockholm. The selecting committee unanimously chose Goldschmidt for the chair. But before the Swedish king had made the final official approbation, the University in Kristiania was able to secure him a similar chair. This was quite an unusual procedure and speed for appointing a professor. Usually it will take at least two years to obtain a new chair at a Norwegian university and one or two years to have the professor appointed. In Goldschmidt's case it seems that all tradition of slowness was abolished, a fact that the University of Oslo shall always be grateful for. In 1929 Goldschmidt was called to the chair of mineralogy in Göttingen, but he returned to Oslo in 1935.

During the German occupation Goldschmidt was arrested but released by initiative of colleagues shortly before his planned deportation to a German concentration camp. He later fled to Sweden and went on to England.

After the war he returned to Oslo again where he died on 20 March 1947, only 59 years old.

A larger work *Geochemistry* was edited and published posthumously in England in 1954.

O. C. A. Bastiansen

Christen Johannes Finbak

1904-1954

Christen Johannes Finbak was born as the son of a farmer, on 4 June 1904 in Korgen, a small place in the northern part of Norway. He graduated from the University of Oslo where he afterwards taught for twelve years. From 1948 he was Professor of Theoretical Chemistry at the Technical University of Norway, Trondheim, until his death on 26 February 1954.

Finbak's earlier work was devoted to the study of the orientation disorder phenomena in crystals often referred to as 'rotation' of molecules and ions. It was from these studies that he was able to explain the low melting entropy and the large freezing point depression of a number of organic compounds.

This work naturally lead him to the study of the structure of liquids using monochromatic X-rays. Particularly his studies on aqueous solutions of salts, acids, and bases were of considerable importance.

Finbak's contribution to the method of electron diffraction in gases gave a great impetus to this subject. He was one of the independent inventors of the 'sector', and his early design of a sector electron diffraction camera contributed much to the modern development of the field. Finbak obtained the most distinguished Norwegian scientific reward (Fridtjof Nansens belønning) for this work.

Finbak was an unusually inspiring teacher. He was able to create a scientific atmosphere from which crystallography and molecular structure work in Scandinavia still benefit very much.

O. C. A. Bastiansen

Paul Knipping

1883-1935

Paul Knipping was born on 20 May 1883 in Neuwied on Rhine as the son of a medical practitioner. Already while attending secondary school he developed an outspoken interest in Physics. He studied in Heidelberg and Munich. Towards the end of his thesis work under Prof. Röntgen he interrupted the work in order to join W. Friedrich in carrying out the experimental proof of Prof. M. von Laue's theory of X-ray interference. Following his graduation he took up an appointment at the Siemens Laboratories in Berlin. After a short spell of military service at the outbreak of war, 1914, he joined the Kaiser-Wilhelm-Institut für Physikalische und Elektrochemie of F. Haber in Berlin-Dahlem where he was in charge of developing testing methods. He continued there after the war as assistant to James Franck and was co-author and author of papers on the ionization stages of Helium (Franck and Knipping), of Hydrogen (Franck, Knipping and Krüger) and of the Halogen Hydrides.

In 1923 he left the K.W.I. and, after one semester's work with Lenard in Heidelberg, settled at the Technical University of Darmstadt where he became lecturer in 1924. Here he soon opened a laboratory class in X-rays and, in the course of a few years and with the active support of the Technical University and industry, established a separate X-ray Institute which was inaugurated in 1933. He was made full professor in 1932. On 26 October 1935 he suffered a fatal accident on his motorbicycle.

P. P. Ewald

*Memorial Tablets**

Gregori Aminoff 1883-1947

Born 8 Feb. 1883 in Stockholm; died 11 Feb. 1947 in Stockholm.

1905 First academic degree, U. of Uppsala, after studying science in Stockholm.

1905 to about 1913 studied painting in Florence and Italy.

1913 Returned to science.

1918 Ph.D.; appointed Lecturer in Mineralogy and Crystallography U. of Stockholm. Thesis: *Calcite and Barytes from Längsbanshütten* (Sweden).

1923–47 Professor and Head of the Department of Mineralogy of the Museum of Natural History in Stockholm.

1930 Married Birgit Broomé, herself a crystallographer. *see Nature* (London) 1947, *159*, 597 (G. Hägg).

Dirk Coster 1889-1950

Born 5 Oct. 1889 in Amsterdam; died 12 Feb. 1950 in Groningen.

Studied in Leiden, Delft, Lund (with Siegbahn) and Copenhagen (with Bohr).

1922 Dr.-ing. Tech. University of Delft.
Thesis: *X-ray Spectra and the Atomic Theory of Bohr.*

1923 Assistant of H. A. Lorentz, Teyler Stichting in Haarlem.

1924–50 Prof. of Physics and Meteorology, U. of Groningen.

Bergen Davis 1869-1951

Born 31 March 1869 in White House, New Jersey; died 1951 in New York.

1896 B.Sc. Rutgers University.

1900 A.M. Columbia University (New York).

1901 Ph.D. Columbia University.

1901–02 Postgraduate work in Göttingen.

1902–03 Postgraduate work in Cambridge.

* The author (P.P.E.) is particularly aware of the incompleteness of this section and would be grateful for being sent additional data.

1903	Instructor	Columbia University, New York.
1910	Assistant Professor	
1914	Associate Professor	
1918	Professor of Physics	

Work on ionization, radiation, electron impact, physics of X-rays, X-ray spectroscopy with first two-crystal spectrometer.

Clinton J. Davisson 1881-1937

Born 22 October 1881 in Bloomington, Ill.; died 1 February 1958 in Charlottesville, Virginia.

Educated: Bloomington public schools; scholarship for U. of Chicago for proficiency in mathematics and physics.

1908 B.Sc. U. of Chicago.

1911 Ph.D. Princeton University.

1911–17 Instructor in Physics, Carnegie Institute of Technology in Pittsburgh.

1917 War-time employment with Western Electric Co., engineering department, which later became Bell Telephone Labs.

1919–29 Studied thermionics, thermal radiation and electron scattering (first with C. M. Kunsman, later with L. H. Germer).

1927 Electron diffraction by nickel crystal observed.

1928 Comstock Prize of National Academy of Sciences.

1931 Elliott Cresson Medal of Franklin Institute.

1935 Hughes Medal of Royal Society.

1937 Nobel Prize (shared with G. P. Thomson).

1930–37 Electron Optics, especially of very slow electrons.

1946 Retired from Bell Telephone Labs.

1947–49 Visiting Professor U. of Virginia, Charlottesville.

see Science 1945, *102*, 216.

Rosalind E. Franklin 1920-1958

Born 25 July 1920 in London; died 16 April 1958 in London.

1938–41 Cambridge Univ. (Chem., Phys., Miner., Math.)

1945 Ph.D. Cambridge. Thesis: *Physical Structure of Coal and related Materials*. Supervisor Dr. D. H. Bangham.

1942–46 Employed by Brit. Coal Utilisation Res. Ass.

1947–51 Worked at Lab. central des services chimiques de l'Etat in Paris; diffraction studies of coal and carbons.

1951–53 Introduced X-ray diffraction at King's College, London; work on DNA.

1953–58 Birkbeck College, London. Directed research on plant viruses by X-ray diffraction until her death from cancer. Papers in *Acta Cryst.* 1950, '51, '58 and *Nature* 1953 (*172*) and 1956 (*177*).

Hermanus Haga 1852-1936

Born 24 Jan. 1852 in Oldeboorn (Netherlands); died 11 Sept. 1936 in Zeist.

1871–76 Studied mathematics and physics at Leiden U.

1876 Ph.D. U. of Leiden. Thesis: *Absorption of radiant heat by water vapour.*

1886–1922 Professor of Physics, U. of Groningen.

1907 *Diffraction of X-rays by slit* (with C. H. Wind).

1914–16 *The symmetry of Laue diagrams* (with F. M. Jaeger).

Frans Maurits Jaeger 1877-1945

Born 11 May 1877 in 's-Gravenhage; died 2 March 1945 in Groningen.

1895–1900 Studied Chemistry in Leiden.

1900–02 Post-graduate work in Berlin.

1903 Ph.D. U. of Amsterdam. Thesis: *Crystalline and molecular symmetry of position-isomeric benzene derivatives.*

1904–07 Reader in Chemistry, U. of Amsterdam.

1909–42 Professor of Inorganic and Physical Chemistry, U. of Groningen.

1917 Book: *Lectures on the Principle of Symmetry and its Applications in all Natural Sciences.*

1919 *Historical Studies.*

1924 *Introduction to the Study of Crystals.*

1930 *Methods, Results, and Problems of Precise Measurement at High Temperatures* (George Fisher Baker Lectures).

Willem Hendrik Keesom 1876-1956

Born 21 June 1876 in Texel; died 4 March 1956 in Leiden.

1904 Ph.D. U. of Amsterdam. Thesis: *Isotherms of mixtures of oxygen and carbon dioxide.*

1900–09 Assistant, Physical Laboratory, Leiden.

1909–17 Conservator Physical Laboratory, Leiden.

1918–23 Prof. of Physics, Veterinary College, Utrecht.

1923–46 Prof. of Physics, U. of Leiden.
1924 X-ray structure analysis of solid CO_2, N_2O.
1926 Same for O_2, N_2, A and for ordinary and superconducting lead.
see Nature (London) 1956, *177*, 825 (Van Itterbeek).

Nicolaas Hendrik Kolkmeijer 1875-1950

Born 26 July 1875 in Amsterdam; died 15 June 1950 in Hilversum.
1915 Ph.D. U. of Amsterdam. Thesis: *Elimination of the concepts of axial system, length and time from the equations for the planetary movement.*
1915–40 Assistant, Physical Laboratory, U. of Amsterdam; Conservator, Veterinary College and Van 't Hoff Laboratory, Utrecht.
1920–21 Papers on time-space symmetry.
1928 Book: *Röntgenanalyse van Kristallen* (together with J. M. Bijvoet and A. Karssen).
1932 Co-author, *Internat. Tabellen zur Bestimmung von Kristallstrukturen.*

Walter Kossel 1888-1956

Born 4 January 1888 in Berlin; died 22 May 1956 in Tübingen.
1911 Ph.D. Heidelberg, in Physics (with Lenard).
1918–21 Privatdozent with Sommerfeld in Munich.
1921–31 Prof. Theoret. Physics, U. of Kiel.
1932–45 Prof. Theoret. Physics, T. H. Danzig.
1947–56 Prof. of Physics, U. of Tübingen.
see Naturwiss. 1957, *44*, 293.

Sir Kariamanikkam Srinivasa Krishnan 1898-1961

Born: 3 December 1898; died 14 June 1961 in New Delhi.
Educated: Christian Coll., Madras; U. Coll. of Science, Calcutta.
1923–28 Res. Asst. Ind. Assoc. for the Cultiv. of Science, Calcutta.
1929–33 Reader in Physics, U. of Dacca.
1933–42 Mahendral Sircar Prof. of Physics, Calcutta.
1942–47 Prof. of Physics and Head of Dept., U. of Allahabad.
1947–61 Director, National Physical Laboratory, New Delhi.

Alexander Müller 1889-1947

Born 29 April 1889 in Zürich; died 2 July 1947 in Essex (England).
1908 Leaves Industrieschule, Zürich.

1908–15 Studies Physics at the University of Zürich.
1915 Ph.D. Zürich. Thesis: *Messungen der thermischen Ausdehnung von kristallisiertem Quarz und von Gold zwischen 18° und 540°C.* (Supervisor Prof. A. Kleiner).
1915–18(?) University of Göttingen.
1918(?)–22 University of Cambridge (with G. I. Taylor).
1922 Research Assistant to Prof. W. H. Bragg, first at University College, London, then at Royal Institution.
1928 Assistant Director } Davy-Faraday Lab. at the R.I.
1946 Deputy Director }
1923 Fatty acids, paraffins.
1927–31 Rotating target X-ray generator.
1936–40 Physical properties of long-chain compounds.
see Nature (London) 1947, *160*, 323 (K. Lonsdale).

Paul Niggli 1888-1953

Born 26 January 1888 in Zofingen, Switzerland; died 13 January 1953 in Zürich.
1907–11 Student of geology and petrography at the Fed. Polyt. School.
1912 Ph.D. U. of Zürich. Thesis: *Die Chloritoidschiefer des nordöstlichen Gotthardmassives.*
1913 Post-graduate work at the Geophysical Laboratory of the Carnegie Institution in Washington.
1915 Assistant Prof. of Mineralogy, U. of Leipzig (with F. Rinne).
1918–20 Prof. of Mineralogy, U. of Tübingen.
1920–53 Prof. of Mineralogy at the University of Zürich and the Federal Polytechnic School.
1921–39 Editor of *Zeitschrift für Kristallographie*, Vol. 56–101.
1919 *Geometrische Kristallographie des Diskontinuums*
see Experientia (Basel) 1953, *9*,197–203 (F. Laves)
Neues Jahrbuch für Mineralogie, Monatshefte 1953, pg. 51–67 (H. O'Daniel, K. H. Scheumann, H. Schneiderhöhn)
Acta Cryst. 1953, *6*, 225 (P. P. Ewald)
Schweiz. Min. Petr. Mitt. 1953, *33*, 1–9 (F. de Quervain, and bibliography.)
Nature 1953, *171*, 675 (W. Campbell-Smith)
Zs.f.angew. Math. Phys. 1953, *4*, 415–418 (E. Brandenberger).

PART VI

Schools and Regional Development

CHAPTER 17

British and Commonwealth Schools of Crystallography

17.1. General Survey

by J. D. Bernal

The position of the British schools in the history of the development of our subject is necessarily quite a special one. Not only did Sir William and Sir Lawrence Bragg effectively start the study of crystalline structures by means of X-ray diffraction, but for many years their respective schools at the Royal Institution and in Manchester were the centres of world study in these fields. Naturally, important centres in other countries existed from the start and we have records of them in the succeeding chapters, but the primacy of the British schools was recognized, at the outset, by the large number of visits of young crystallographers, who were destined later to become the centres of schools of their own in other countries. Owing largely to the personal character of its founders the development of crystallography had, from the very outset, a peculiarly intimate and friendly character. All of those who worked at the Royal Institution or in Manchester carried away for the rest of their lives recollection of the atmosphere of active and exciting research which grew up around the Braggs, and the fact that they were father and son actually helped enormously to unify the whole subject.

To attempt adequately to deal in a few pages with the growing and diversifying field of structural crystallography over a whole of fifty years would be an impossible task. What we have chosen to do is to select the two principal schools, those of the Royal Institution and Manchester University, on which we have the detailed accounts of Professor Lonsdale and Professor James, each associated from the beginning with these schools and contributing greatly to them, and to add a necessarily more summary sketch by Professor Bernal of the other schools in Britain and the Commonwealth which in almost every case arose directly out of them.

To do this, the most arbitrary but, at the same time, necessary simplification is to attempt to divide the period into sections. It would be very convenient to take them as the first, second, third, fourth and fifth decade, but it did not come out quite like that because of the intervention of two World Wars. So we have first a short and intensely brilliant period beginning with the work of Sir William and Sir Lawrence Bragg on the simple salts, and corresponding work by Darwin and others on the theory and corrections of X-ray diffraction by crystals. This work was, effectively, broken off by the war. The second period may be reckoned from 1919 when Sir Lawrence Bragg went to Manchester or from 1923 when Sir William started work at the Royal Institution Davy-Faraday Laboratory. It may be considered to last until about 1929 when some of the first research workers from these laboratories set out and started schools of their own, notably in Cambridge and Leeds. This leads to the third period from 1929 to the beginning of the Second World War which had a large proliferation of schools, particularly in Leeds, Birmingham, Oxford and a further development in Cambridge and many other centres, a period which, effectively, came to an end with the transference of Sir Lawrence Bragg to the Cambridge Chair after Rutherford's death in 1937. Though the second war interrupted research, the period cannot in this case be passed over as such a blank in crystallography in Britain as was the first. Some very interesting work was done on structures of various explosive compounds, and the end of the war marked one of the great triumphs of crystallo-chemical chemical research, the elucidation of the structure of penicillin. The fourth and last period we may take as that which we are now in, although it might be possible to divide it into two with the line somewhere about 1957 when the influence of electronic machines was fully felt, and where such elaborate structures as vitamin B_{12} marked the high-water mark of the analysis of non-protein biological structures. However, here no attempt at this division will be made and the whole of the latest or modern period will be treated as one.

Another division which is, in a sense, imposed by the nature of the subject itself, is that between the different fields of study. There has been a continuous interaction between the *subjects* of crystalline materials studied on the one hand, and the *methods* used for interpreting them on the other. This is not an attempt to give a history of either of these aspects of research in Britain—for the world as a whole they are given elsewhere in this volume. Here they can only be alluded to in passing where the particularly important landmarks occur. But the division according to fields of study follow much more closely that of the different schools with which we are concerned here.

From the very outset there was an almost tacitly agreed separation between the work of Sir William and Sir Lawrence Bragg, that is between the Royal Institution and Manchester, corresponding to that between organic and inorganic chemistry. With the one important exception of crystalline forms of silica, Sir William's laboratory occupied itself with organic crystals and Sir Lawrence's with minerals and metals. In Britain and the Commonwealth the latter two fields of interest remained linked although, in fact, they corresponded to different methods of study, the metals in particular leading to refinements in powder diffraction techniques.

As further schools developed, and largely on account of whether their leader came from London or Manchester, very much the same specializations were carried over and when we speak of the schools we generally speak of bodies of research workers occupied in elaborating some particular field of work; this was often even more specialized such, for instance, as that of Astbury, starting in the Royal Institution and going on at Leeds, which was centered on the study of fibres and particularly fibrous proteins and nucleic acids.

When we look at the actual lines of development, we see very clearly that they depended on the possibilities available to the original founder, to get the necessary support and interest in his work. Those who were successful in achieving the professorial chair in a fairly large university were able to set up large schools which proliferated into many other places. Those, on the other hand, who occupied relatively subordinate positions in physics or chemistry departments, remained, for the most part, as isolated research workers or having one or two students at a time, and though the work they did was of the highest quality, it can hardly be said that they founded a school. This is brought out very clearly also by the way in which the transfer of an individual research worker from one university to another could result not only in the setting up of a new school in the second university, but often in the disappearance of crystallography altogether from the first. What we see, accordingly, is a fluctuating pattern lit up for a few years by the presence of a research director with drive for the time of his tenure there.

As it is a young subject, we crystallographers are still in the happy position of having with us many, indeed most, of the second generation of workers and one of the first, Sir Lawrence himself. The subject is still, in Britain, in a state of rapid growth and differentiation. We can only touch here on some aspects of the spread outside the field of fundamental science, that is, into industry and government service, although, in fact, in crystallography, industry contributed many elements to the fundamental study of crystals themselves, notably in the analysis of penicillin.

One conclusion is very evident, namely, that the development of this subject was a matter in which general or conscious planning had extraordinary little to do. Only in one or two cases, notably in Cambridge, did the University, itself, decide that it must have a crystallographic department, but in most cases, crystallography occurred almost unintentionally when a Chair of Physics or Chemistry happened to be awarded to a crystallographer as the most distinguished available candidate in a field which covered all branches of the subject. The non-establishment of chairs of crystallography in Great Britain has prevented the continuity which could so easily have been ensured in view of the availability of men of quite exceptional enterprise. There is no doubt that crystallography at several stages in its development in Britain was such an attractive subject that it automatically selected such people and the fact that a relatively unknown subject could acquire, in such a short time, no less than seven Fellows of the Royal Society, is some indication of it.

In a survey like this it is clearly impossible to be comprehensive. There are literally hundreds, the actual figure is around 820, of active crystallographers in Britain and the Commonwealth today. They cannot all, or even any large portion of them, be mentioned here by name as this would reduce this report to a mere catalogue. It is inevitable that the omissions may create the impression of invidious selection but all one can do is to use one's own judgment to pick out those who seem to have been able to contribute definitely new directions to the study of the subject.

The First Decade

In the first and glorious three years from 1912 to 1914, the study of the new-born subject, crystal structure and X-ray diffraction was necessarily limited to Cambridge and Leeds, the places where the Braggs were working, and to one other centre of vital importance, namely Manchester, in which the chair of physics was occupied by Rutherford then at his most creative time. This is the period culminating in the classic book *X-rays and Crystal Structure*, published in 1915.

The story of this period in Manchester is briefly told in Professor James' article. It must be emphasized that the work of Moseley and Darwin not only laid the foundations for the study of X-ray spectra and the principles of X-ray diffraction, but also included what is only now realized to be the important study of crystal imperfection, involving mosaic structure and primary and secondary extinction. In addition, and quite outside the fields of crystallography, the diffraction of X-rays by crystals furnished absolutely vital elements in the building up of the Rutherford-Bohr theory of the atom. It was in this period, too, particularly in Leeds and Cambridge, that the basic equipment for X-ray analysis was developed. The ionization spectrometer, with which much of the early work was done, was a product of earlier studies of nuclear radiations, an adaptation to the use of X-rays of Bragg's former study of the ionization produced by alpha particles. The photographic method also arose from the needs of X-ray spectroscopy as used by Moseley.

One other figure belongs to this early period and that is the veteran of crystallography, Professor Owen of Bangor, who in Richmond in 1913 started the study of metal structures in Britain which he subsequently carried out at Bangor. There was an even earlier influence, emanating from Edinburgh with Barkla, who had carried out the pioneer absorption measurements which had distinguished the K-, L- and M-levels of X-radiation. Unfortunately, his attachment to the non-existent J-radiation prevented him in his later days from making a serious contribution to the new diffraction crystallography. But some of his students started schools of their own, notably R. T. Dunbar at Cardiff.

The Second Decade

After the end of the First World War, that is, effectively in 1919 when Sir Lawrence Bragg took up his chair at Manchester to which he had been appointed in 1915, a new start was made and the corresponding transfer of Sir William, first to University College in 1919 and then to the Royal Institution in 1923, ensured the foundations by the beginning of the twenties of powerful new schools of structure analysis who set themselves to the essential task of working out the basic types of structure of solid substances, beginning very rationally with the simplest, the elements and simple salts. It is astonishing to think that, in fact, by putting together the pre-war and immediate post-war work, all the major types of structures known to us now had been studied in at least one example and essentially the right structures attributed to them. Most of the work in this decade is reported on here in the sections of Mrs. Lonsdale and Professor James because at that time the Royal Institution and Manchester schools effectively dominated structural studies in Britain. There were, however, important other elements which were just coming into play at this time. The old schools of crystallography of Oxford and Cambridge were still in vigorous life and it was owing to the inspiration of Dr. (and later Professor) Hutchinson at Cambridge and Professor Bowman and Dr. Barker at Oxford that the new methods were introduced. Dr. Wooster at Cambridge and Mr. Powell at Oxford were, in fact, the first X-ray crystallographers who had not been trained in the Manchester or Royal Institution schools. Another new school appeared at Bristol in the work of Dr. Piper on long-chain compounds which was to link up with those of Müller and Shearer at the Royal Institution and lead to the first effective break-through into the study of organic chemistry by X-ray methods.

The study of X-ray spectra was carried on with the original inspiration of Barkla by Professor Dunbar at Cardiff, followed by Professor Robinson. The corresponding study of absorption spectra with their implications on the theory of metals was particularly studied by Dr. Skinner at Bristol. However, it must be admitted that comparatively Britain has contributed little to the study of X-ray spectra.

The Third Decade

The years 1927–29 were to see the dispersal of the original schools and the start of the new ones. Bernal went to Cambridge, Astbury to Leeds, and Cox to Birmingham. The middle of the thirties was to see powerful schools of crystallography set up in Cambridge, Oxford, Leeds,

Birmingham, Liverpool and Bristol. Something can now be said about the individual character of these different schools.

Cambridge

The Cambridge school, in accordance with its tradition, for Bernal had been trained in classical crystallography by Professor Hutchinson, occupied itself with extensive studies in different fields of crystallography, both inorganic, including metals, and organic. From 1933 onwards the emphasis was on the organic side, corresponding to a division of the school between the Cavendish Laboratory on the one hand and the new Mineralogical Laboratory, which under W. A. Wooster devoted itself very largely to crystal physics and to the development of X-ray equipment and accurate intensity measurement. The study of metals in Cambridge, after a systematic start by Bernal, lapsed until it was again taken up by W. H. Taylor in the subsequent decade, but work went on very actively in the fields of inorganic compounds following the stimulus of the new Goldschmidt views of crystal chemistry and particularly in the study of water and hydroxyl compounds with Dr. Megaw. Here the link with the Cavendish was emphasized in the paper by Fowler and Bernal in 1933 on the structure of water and ionic compounds. This work was to form the structural basis for the understanding of hydroxides and hydrogen bonds.

At the same time in close connection with the Biochemical Laboratory of Professor Hopkins, work was started, first on amino acids and then on the sterols. There, owing to what was effectively a happy chance of being able to discover, by X-rays in the first place, the correct carbon skeleton of the sterols, Bernal was able to unify the structure of these important bodies which were then of particular interest in connection with vitamins and sex hormones. It was at this point that Miss D. Crowfoot (Mrs. Hodgkin) joined the laboratory and immediately became involved in both extensive and intensive structure work on the sterol compounds which was to lead later on to her great achievements in other organic fields. In 1934 the first successful photographs were taken of protein single-crystals due to a tactical break-through of examining them in the wet state. Miss Crowfoot continued her work with insulin in Oxford and the continuation of the sterol work was taken over by I. Fankuchen who had joined the laboratory from Manchester and originated from the United States. In 1935 Max Perutz came to the laboratory from Mark's laboratory in Vienna and started the studies of the haemoglobin systems which have now become classic. In 1936 another break-through was made in the examination of the structure of crystalline and paracrystalline viruses, in the first place of tobacco mosaic virus prepared by Bawden and Pirie. This led, in the first stages, to the use of very small angle scattering in order to elucidate the intermolecular structures of the liquid crystal aggregates formed in these viruses and incidentally to an understanding of long-range forces between colloidal molecules in solutes. The significance of the high-angle reflections which indicated something of the internal structure was, however, not worked out until much later, largely by Miss Franklin. Fankuchen also continued this work with Bernal after the latter's transfer in 1938 to the Chair of Physics at Birkbeck College, London, for the few months that elapsed before the beginning of the Second World War. During the same period Professor Ewald left Germany and joined the crystallographic laboratory where his influence was very large, incidentally, in setting up the discussion group known as the Space Group.

Oxford

Unlike the other schools mentioned, where the initiative had primarily come from physicists, in Oxford the impetus for crystal studies was that of chemical crystallography originating with Myers and with Barker who had been a friend of Fedorov. X-ray studies began with the appointment in 1929 of Mr. H. M. Powell as demonstrator of chemical crystallography. Powell's earlier work was largely with coordination complexes. The further developments of his work and his discovery and study of clathrate compounds followed naturally from this. His first student was Miss Crowfoot who worked with him on a thallium metal complex. She returned from Cambridge to the department in 1934 and went on there with her work on the analysis of complex organic compounds of the sterol type, particularly the structure of cholesterol iodide with Dr. Carlisle, while following her studies of insulin and lactoglobulin. Powell continued, now with A. F. Wells, on the structure of further complex compounds including the carbonyls and phosphorus pentachloride; the latter was proved to be an ionic compound in the solid state.

An entirely independent research section at Oxford was that of Metallurgy which was taken over by Hume-Rothery following on the work of the Swedish school and of Bradley. This led to the interpretation of the so-called alloy phase systems, particularly of the A and B group metals, which he has continued with his students ever since. The Hume-Rothery rules, which laid the foundation for the idea of electron compounds, were the first fruit of his work.

Birmingham

The beginning of modern crystallography at Birmingham came with the invitation in 1929 by Professor Howarth for Cox from the Royal Institution to set up a study of the constitution of the sugars. This problem, with the methods of the time, proved too difficult and Cox very wisely elected to apply an indirect approach and, with the help of Goodwin and Llewellyn, worked out the structure of pentaerythritol which was to turn out to have the most interesting physico-crystallographic properties. It was also an early example of three-dimensional Fourier analysis of an organic crystal structure. At the same time, he studied the structure of ascorbic acid—vitamin C—as well as glucosamine hydrochloride, the first optically active organic substance whose structure was determined by X-ray analysis alone using the ionic replacement method. Through this he established the stereochemistry of the pyranose rings in sugars. A number of inorganic coordination compounds were also studied. At this time Birmingham was one of the most fertile centres of X-ray analysis but most of this activity was switched to Leeds when Cox became the Professor of Inorganic Chemistry there in 1945.

Leeds

Already in the thirties, however, Leeds had become an important centre of structural studies with, in the first place, a bias towards those of biological origin and particularly of fibres. Very appropriately, Astbury had been invited to a lectureship in textile physics and immediately started his classical work on wool and other fibrous substances. Beginning in 1929, he established the fundamental character of the alpha-(coiled) and beta-(straight) configurations of wool and showed that they could be extended to cover most types of protein fibres, though collagen represented an exception. This new type of analysis brought X-ray crystallography for the first time in contact with the morphological and histological aspects of biology. Hitherto it had been limited to the biochemical aspects.

Another line of research opened in Leeds when in 1935 G. W. Brindley joined the Chemistry Department. There he developed accurate powder photography and developed the appropriate cameras, laying the foundation for future work on the structure of clay minerals which he was to continue in the next decade.

Liverpool

A small school existed at Liverpool consisting of Lipson and Beevers as mentioned in Professor James' article. Until they moved to Manchester in 1936 their work on the structure of salts might be considered as a part of that school.

Bristol

The crystallographic school at Bristol, under S. H. Piper, remained during this period an active but highly specialized one concerned, very largely, with the study of fatty acids and waxes of natural origin and their derivatives, ketones, secondary alcohols and other constituents.

Commonwealth Schools

The decade of the thirties also saw the beginning of X-ray work, generally inspired from the older centres, in India with Banerjee in Calcutta, and Krishnan at Bangalore, where a very interesting field of relations between the structure and the magnetic properties of crystals was first explored. It began then, too, in Canada with Barnes at Montreal and in South Africa with Professor James at Cape Town.

This completes a rapid survey of the major schools of crystallographic research in Britain and the Commonwealth in the decade before the war. It was a period of extremely happy activity over rapidly broadening fields. To sum up, the effect was to establish the approximate structures of most types of crystalline materials with a degree of accuracy and refinement which, though it would naturally not now be considered adequate, was quite sufficient in those pioneer days for establishing some of the major features of molecules, particularly in the organic field, and of the ways of linking them together. The concept of hydrogen bonding added to those of Van der Waals, and ionic linkage also appeared. At the same time the major structural types of the inorganic world, the fibrous and platy silicates, were worked out and a beginning made in the understanding of the rules of compound formation in the far more complex field of alloy structures. Effectively, this marked the creation of two new subjects, mineral chemistry and alloy chemistry, as rational disciplines, a task that had proved impossible and would probably have long continued to be so without crystal analysis.

17.2. Crystallography in Britain during and after World War II

by J. D. Bernal

Unlike the period of the First World War which marked a virtual cessation of structural crystallographic studies in Britain, the Second World War was one of considerable if limited activity closely linked with the needs of the war itself. It was marked by the greatest triumph of crystallographic technique that had yet occurred, namely, the elucidation, essentially by X-ray crystallographic methods, of the structure of penicillin. The story cannot be told here, but this achievement is a remarkable instance of the way in which research can be pushed forward if it is led by workers of genius backed by keen young collaborators. The molecule of penicillin was one of peculiar intractability by purely chemical means on account of its thiazolidine-β lactam ring system, so that knowledge of the molecular structure was essentially gained by X-ray analysis. Though the ultimate objective, a simple non-biological synthesis of penicillin, was not achieved until 1958, this is no reflection on the methods of analysis the results of which the final synthesis fully justified. The work is of interest from another point of view, as an example of fruitful co-operative research. Two groups, at Oxford and at the Northwich Division of the I.C.I., led by Dorothy Hodgkin and Bunn respectively working closely together on different varieties of crystals, were able to supplement each other's work. It was a magnificent start to a new era in crystallography.

These are not the only achievements of war-time crystallography. Some very interesting analyses of explosive substances were carried out under Cox at Leeds, leading to the understanding of the structures of nitrate groups and strong acids; and everywhere crystallographic methods were used for general auxiliary and identification purposes as well as in the study of metals and alloys. Everything was ready for a new burst of activity as soon as the war was over. However, there was one tragic loss, that of one of the founders of the subject. Sir William Bragg, who was still in his full mental vigour, and actively directing the Royal Institution laboratories died in January 1942. His last researches dealt with the pioneer field of the non-Bragg or diffuse reflections of X-rays. With him passed the first generation of X-ray crystallography, but fortunately his son, as inseparably linked with the original discovery, is still with us.

POST-WAR PERIOD, 1946–1962

After the end of the war, the expected new start of crystallographic research exceeded all expectations. War service released an augmented band of crystallographers including part of the second and now the beginning of the third generation of crystallographers, those taught by Bragg and those taught by his immediate pupils. All the old schools of crystallography renewed and multiplied their activities and new ones were founded in practically every university in England, Scotland and Wales.

During the war period and to an increasing extent after it, Britain had ceased to be the rather isolated centre of structural research and became part of an ever closer linked international exchange of persons and problems and materials. The formal side of this, the foundation of the International Union of Crystallography, is accounted for in another chapter, but here it can be said that world crystallography had grown, not only in extent, but in intimacy of cooperation during the whole period.

The characteristic of post-war work in crystallography in Britain has been the enormous increase, both in scope and in quality of the work, brought about by the new problems and the new methods. The much more critical understanding of the methods of diffraction analysis, though they contain few radically new principles, has made it possible to tackle crystals with molecules of enormous complexity up to the proteins and to a certain degree of far larger molecules like those of the viruses. At the same time it also has enabled much more precise information on structures to be obtained of crystals with relatively small molecules. These later developments would have been impossible without the increasing use of ever faster electronic machines, beginning effectively in 1957, and this latter period might even be called the first computer age of crystallography.

At the same time, the developments in theories of chemistry have given a much greater importance to the precise knowledge of structures, and the developments of theoretical chemistry in the hands of such pioneers as Coulson, Dunitz and Orgel have led to a new link between crystallography and organic as well as inorganic chemistry, including the intermediate field of organic metallic complexes.

A definite break into new ground came with the realization that the methods of diffraction could be applied with precision to structures without three-dimensional lattices, using in the place of the methods of Fourier analyses those of Bessel functions. This development was called for in the first place through the study of protein fibres but it was rapidly extended and deepened in the analysis of viruses and other irregular structures such as those of liquids. This was to lead to one of the greatest triumphs of crystallography in the biological field, the structures of nucleic acids, of which the pioneering work had been done long before by Astbury.

Cambridge

The transfer of Sir Lawrence Bragg from Manchester to Cambridge had occurred too shortly before the war to make a notable impact there until after the war but then crystallographic work began with redoubled vigour. The main strength of the Manchester school was soon effectively transferred to Cambridge where two closely linked groups of research workers grew up in the post-war years. One of these, representing a fusion between the Bragg and the Bernal schools there, developed the studies on organic crystals and proteins. On the other, the inorganic side, W. H. Taylor took over in 1945 the direction of the main laboratory and continued his work mainly on metal structures.

In the atmosphere of Cambridge, the metallographic side of the work developed in new directions. P. B. Hirsch, in particular, used micro X-ray beam methods to study dislocations and also imperfect crystallization in close connection with the electron microscope developments which were going on there under Coslett. Thus, the earlier Cambridge work of G. I. Taylor and Elam on metal deformation was blended with that of W. L. Bragg at Manchester. With Miss Megaw transferred from the Birkbeck team, they continued the work on silicate structures but particularly on the most intractable of the types of rock mineral structure, those of the felspars and, also, of hydrated calcium silicate compounds connected with cement, following up the work which was also being carried on by H. F. W. Taylor at Birkbeck. Miss Megaw has also, through her studies of the titanates, made notable contributions to the theories of ferroelectricity.

During the same period, the possibility of the use of electronic computing machines increased and the advantages of Cambridge were manifest in the brilliant work of Cochran in developing precise methods of crystal analysis applied in the first place to organic crystals but of perfectly general applicability. It was largely Cochran, in his very critical approach to crystal structure work, who raised the standard of over-all accuracy and in particular of bond length determinations by something like a factor of ten, but at the same time inevitably increased the amount of work that had to be done to determine the crystal structure properly. This, effectively, led to a division of the subject between rough analyses useful to the chemist from the point of view of determination of the main lines of structure, to the precise analysis now being required to check studies of theoretical chemistry.

Sir Lawrence Bragg, on moving to Cambridge, had taken over not only his own metal and silicate school, but the organic and biomolecular school which had been built up in Cambridge by Bernal. This, however, remained somewhat separate from the other group, partly for administrative reasons because it had acquired the support of the Medical Research Council; Perutz's researches on haemoglobins were continued on an ever expanding scale. Soon after the war, he was joined by Kendrew and a number of other fellow workers and between them they carried out the magnificent and ultimately successful attack on the structure of the haemoglobin and myoglobin crystals which were to result in the first strictly X-ray analysis of a protein structure. This, however, was not to be the only triumph of the school because, at the same time, one of the protein workers, Francis Crick together with Watson in the United States, put forward the hypothesis of the double spiral structure of nucleic acids, which was later to prove the clue to a fundamental understanding of biological structure and function, including the effective action of viruses, and a material explanation of genetic processes. The development itself is an example of international cooperation, the final proof by more careful X-ray analysis being left to another school, that of King's College in London with the work of Wilkins, Goodwin and Miss Franklin.

Another branch of crystallographic work at Cambridge was in the department of Mineralogy where Dr. W. A. Wooster continued to direct the only undergraduate courses on Crystallography given in Britain. His own researches now concentrated on the quantitative study of diffuse X-ray reflections, from which to determine elastic constants of crystals. This led, in conjunction with the work of Laval, to criticism of the long established theories of the relations of crystal elasticity to symmetry.

The London Schools

The Royal Institution. When in 1954 Professor Bragg retired from the Cavendish Chair at Cambridge, and moved to the Directorship of the Royal Institution which his father had held for many years, he carried with him in close cooperation with the Cambridge school some of the study of the proteins related to haemoglobin, and, as co-workers Dr. Green and Miss Scouloudi who had been attached to the Birkbeck school. Dr. Arndt at the new Royal Institution school has developed also very powerful X-ray tubes essential for the study of the most complicated proteins and viruses.

Birkbeck College. J. D. Bernal came to occupy the Physics Chair at Birkbeck College too shortly before the war for it to have had much effect at the time. The physical destruction of the college in the London raids resulted in a delay in setting up work again after the war. However, by 1947 a new school of crystallography had definitely been established in some ruined houses and was being gradually expanded in the years that followed. Postgraduate classes in Crystallography were started in Birkbeck in 1949 on a London intercollegiate basis. In research Birkbeck took over effectively part of the work of the Cambridge school with one important addition. Thanks to a grant from the Nuffield Foundation it was possible to set up a biomolecular unit concentrating largely on the structure of proteins and viruses.

Other organic structures were studied such as those of terpenes. More important, however, was the study of pyrimidene by Parry and of the nucleoside, cytosine, by the Norwegian research worker, Furberg, who was able to show that the planes of the pyrimidene molecules were arranged at right angles to the rings of the pentose sugars. This provided the essential clue for the idea of a helical structure of nucleic acid.

In the protein field, C. H. Carlisle concentrated mainly on the structure of the enzyme ribonuclease which was to prove a much more difficult problem for technical and structural reasons than was expected or was that of the haemoglobins. However, it seems to be approaching a definite conclusion.

The virus work hung fire for longer until it was taken up with great energy and success by the late Miss Franklin who was able to demonstrate, following the initial hypothesis of Watson, that the virus molecule possesses a helical structure, though not one corresponding to a single molecular chain, but rather to an aggregate of identical protein molecules, inside which are twined the molecules of ribosenucleic acid. The major success was the determination of the position of the ribosenucleic acid as a single helix in among the protein molecules forming the protective tube of the virus. In further work on the spherical viruses which was started just after the war with studies of bushy stunt and turnip yellow virus, A. Klug was able to prove that these, according to the hypothesis of Watson and Crick, consisted of polyhedral aggregates with the uncrystallographic symmetry of fivefold units. At this point the studies by X-rays became blended with those by the electron microscope.

The other field of research was initially of an industrial character but it led to very interesting scientific results. It was to determine the structure of cement and its hydration product, concrete. These studies were carried out largely by J. W. Jeffery, H. F. W. Taylor and their co-workers. They developed an ingenious combination of analysis of fine-grained products produced industrially, with structural studies of slow grown natural crystals of the same material. It turned out that the main hydration product of cement was an extremely rare hydrated silicate mineral called tobormorite whose structure was ultimately determined by Miss Megaw, one of the original members of the team who was later transfered to Cambridge. This work extended out to the discovery of a completely new series of silicate structures both unhydrated and hydrated, based on more complicated rhythms than the simple alternation which Sir Lawrence Bragg had found in the predominantly magnesium silicates of the pyroxene types, opening a new chapter in silicate crystal chemistry. The major work for this was that of Mrs. Dornberger, one of the original Birkbeck group, and later of her collaborators in Berlin, especially Liebau, and also of Belov and Mamedov in the Soviet Union.

It was about the same time that A. L. Mackay joined the Birkbeck group and opened up, with D. R. Das Gupta, a study on iron oxides and hydroxides using the new methods of electron microscopy and diffraction developed by Grudemo and Gard. In this way they elucidated a hitherto extremely confused chapter of inorganic chemistry. Through the magnetic properties of these compounds, this research was brought into connection with new studies of paleogeomagnetism which Professor Blackett was directing from Imperial College.

Professor Bernal, in the later years, has returned to his earlier interest in the structure of water and simple liquids by generalizing the model of a liquid to one formed by a random aggregate of spheres in contact. This he has been able to relate to the X-ray diffraction from regular to irregular structures.

King's College. Professor J. T. Randall, whose fame has been associated with the development of branches of physics, essential for the military successes in the war—luminescent materials and the magnetron valve—opened his new laboratory at King's College with a concentration on a very different field, that of biomolecular studies. Apart from the development of the electron microscope for these purposes, which does not come precisely into our field of concern, M. H. F. Wilkins began there with the study by X-rays of natural polymers with helical structure. One of these was the anomalous structural protein collagen, further studied by Miss Cowan and then by Ramachandran in Madras. Another was the vastly more important structure of the nucleic acids where Wilkins and his school, basing themselves on the Crick and Watson hypothesis, were able to verify the double helical structure and to produce the remarkable models which showed how this structure depended on the fit of a number of very closely packed groups of purines and sugars. The work of Randall and Wilkins definitely established biomolecular studies as a discipline of their own and, as a result, part of the new institute of biophysics attached to King's College is devoted entirely to this study.

*University College.** The revival of X-ray studies at University College came in 1946 when Kathleen Lonsdale, later Dame Kathleen Lonsdale, moved from her long held position at the Royal Institution to become Professor of Crystallography in the Department of Chemistry. This was particularly important from the point of view of teaching and research. In teaching for the first few years, she joined in an Intercollegiate Course which had been organized at Birkbeck College, and took part in the formation of the first generation of London students of crystallography. In her research she followed original lines, lying more in the development of crystal physics than crystal structure analysis. Some very interesting work was done by Dr. P. G. Owston on ice which much improved our knowledge of this really very complicated substance with special reference to the H positions. Professor Lonsdale's own chief contribution was to the study of the thermal vibrations of crystals. This was now becoming essential for structural analysis because refinement had reached such a stage that the movement of the atoms could no longer be neglected. However, she used this to study still further the peculiar anomalies in scattering in directions outside those predicted by Bragg's Law and which correspond to various irregularities, either intrinsic or thermally induced. She also continued her interest in the magnetic properties of crystals, linking them more quantitatively with the structure.

A separate school of work was on the subject of refinements in the structure of diamonds, now including artificial diamonds, carried out in conjunction with Miss Judith Grenville Wells (later Mrs. Milledge). The high standard of precision and of criticism in all this work puts it in the forefront of world crystal physical studies.

* See section 17.6 for the earlier work at University College, London.

Professor Lonsdale's school at University College has always had a particularly strong international character. Students from no less than twenty different countries, in all five continents of the world, studied there the methods she had developed and later several of them, having returned to their countries, were to found their own schools. Close association with Egypt and India were particularly fruitful.

Imperial College. Important crystal structural work started in Imperial College with the appointment in 1957 of A. R. Ubbelohde as Professor in the Department of Chemical Engineering and Technology. He continued there the work he had already started in the Royal Institution and added to it with the assistance of Dr. G. S. Parry. His particular interest lay in crystal transformations with temperature and the detailed study of phase changes, thermal expansion and molecular movement. This led to a further understanding of the nature of some thermodynamic transformations in solid systems which are not, as was previously thought, necessarily of a higher order but may depend on the linking and mutual strain of two forms above and below the transformation temperature. These studies necessitated the development of a number of X-ray cameras to be used at different temperature ranges, much extending the armoury of such instruments.

Oxford

The three main divisions of the Oxford school continued under the same direction that they had enjoyed before the war. Organic crystallography with particular reference to natural products was developed by Mrs. Hodgkin. Dr. Powell continued and extended his study of coordination and clathrates while Dr. Hume Rothery systematically extended his metal and alloy studies.

Mrs. Hodgkin and her team, in close cooperation with others in America, particularly with K. Trueblood and J. G. White, carried out successfully the determination of the structure of the most complex molecule known at the time. It was one of extreme biological interest, the anti-factor for pernicious anaemia, so-called vitamin B_{12}. This analysis has become a classic. This is partly on account of the interest of the structure itself, which is an extremely complex unit including both proteinoid and neucleotide elements with a central reduced porphyrinoid group containing cobalt, and linked in a most unexpected way. But the analysis was also of great significance on account of the elaborate and critically accurate methods applied. Great use was made of machines in America and in Britain (Manchester, Leeds and National Physical Laboratory). Ninety-three atoms not counting Hydrogens were placed one by one ending in 1959 with the positioning of all the twenty-two water molecules in the structure. The method of analysis so successfully used was one which gave great hope to other crystal analysts, yet there was hardly any structure short of the proteins that could not be tackled by strictly crystallographic methods, that is by methods which did not involve any chemical assumptions. The results of such analysis are already of value to chemists and are likely to be of increasing use to them in the future. It seems extremely unlikely that the structure of vitamin B_{12} could ever have been discovered by purely chemical methods. The structure of penicillin might have been, but was not actually done in time. But with vitamin B_{12} the difficulties of the analysis were proving too great with the older methods. It would be wrong, however, to oppose X-ray to traditional chemical analysis. The two must work closely together especially on the most difficult cases. For instance, in the process of X-ray analysis of B_{12} studies were made of a number of its chemical derivatives which assisted very much in arriving at the final structure.

In 1960 Dr. Hodgkin was chosen as special Wolfson Research Professor of the Royal Society.

Dr. H. M. Powell, now Reader in Chemical Crystallography, continued to devote himself more particularly to the structure of molecular compounds. A certain number of straightforward molecular compounds, addition compounds such as those of aromatic polynitro compounds and other aromatic substances, had been studied just before the war. The real break-through came with the study of clathrate compounds particularly those formed by quinols in which Powell and Palin were able to show the way in which the smaller in general non-polar molecules are caught in a kind of basket in which the meshes consist of hydrogen bonded molecules. This includes even such normally completely unreactive molecules as the atoms of the rare gases. Powell's chemical interests led him further into the discovery of a new type of clathrate compound where the optical activity resides in the structure and not in the optical activity of the resulting molecules. Clathrates formed of this type which are analogous to those of right or left quartz or of benzil, have naturally a way of separating the right- and left-handed forms of smaller molecular species and thus are, in a sense, defying Pasteur's principle that optical antipodes can only be separated by substances derived from living structures including, as in his classical separation of optically active crystals by hand of 1849, the living structure of Pasteur himself. In the case, however, tri-*o*-thymotide Powell was able to show that it is possible to separate a mixture of optical antipodes by methods involving neither biological products nor human intelligence, though the formation of the particular right- or left-handedness of the original clathrate crystal still depends on chance. Further work now going on is dealing with examples of ligands attached to particular inorganic ions. The laboratories of both Professor Powell and Mrs. Hodgkin were for a long time housed in the old Ruskin Natural History Museum in Oxford but since 1960 they have been moved to well-fitted and new laboratories in the chemical wing.

Dr. Hume Rothery and his school have continued their work on alloy systems of increasing range and complexity with the idea of arriving at a really quantitative alloy crystal chemistry. The later development of the studies has linked up the straightforward phase and structures analysis to the new considerations of dislocations, in particular to the immediate field of stacking faults and the study of the industrially enormously important martensitic transformations.

Leeds

In Leeds, after the war, the major lines of work of the vigorous Birmingham school previously described, were added to those already existing of Astbury's fibre structure work and of Brindley's studies on clays. The appointment of Dr. E. G. Cox to the Chair of Inorganic and Physical Chemistry in 1945, gave him opportunities to set up an even larger school than at Birmingham and the value of this has been proved by the work coming from it. Cox set himself essentially to study organic compounds of a relatively simple kind and to determine their structures with a very high level of accuracy. In this respect he and Robertson divided the field between them. While J. M. Robertson in Glasgow concentrated on aromatic compounds, Cox mostly concerned himself with aliphatic. With J. W. Jeffery he started the study of terpenes and other compounds related to isoprene. The structure of geranylamine hydrochloride is a landmark in structure analysis bringing out particularly the aspect of the variable heat motion of a chain attached at one end. The addition of D. W. J. Cruikshank to the team led to the introduction of rigour in the calculation which has helped very much in determining accurate bond lengths. Co-ordination compounds of stereochemical interest have been studied and further development of carbohydrates is under way.

This school has been particularly fertile in developing, not only physical methods of examination, but also machines of an analogue or digital type; particular use was made of low-temperature methods to reduce the effect of thermal vibration and it is largely due to the Leeds school that three-dimensional analysis in Britain became almost *de rigeur*.

G. W. Brindley's work was, as already indicated, mainly concerned with clays but in the years after the war he developed this method further and was one of the leaders of the world study of clay minerals and their various streaks of hydration.

After the war, Astbury's laboratory, whose interests were turning more and more biological, was transferred from the textile physics department to a newly created one of biomolecular structures. Here he widened the scope of his studies from X-rays to electron microscopy and infra-red spectroscopy and extended it further to cover other groups of compounds, especially a group of fibrous proteins which had hitherto not been adequately recognized, the so-called crossed β structures and also the highly orientated natural protein structures, including such unusual things as the egg-case of the prying mantis or the peduncle of the egg of the lace wing fly. It was clear that before his last illness he was on the way to a kind of structural natural history of proteins which he was admirably suited to pursue.

Glasgow

The rise of Glasgow to being one of the major schools of crystallography followed the transfer of Dr. J. M. Robertson to the Chair of Chemistry. His stay at Sheffield, 1939–41, had been too short to enable him to set up a viable school there. One of the major achievements of the school is in its teaching capacity, for it has taken in people from many different parts of the world and the students have gone out to found other research laboratories very widely, including one in Europe, Dunitz in Zurich. Robertson had already established his special competence in the study of aromatic substances and particularly condensed ring compounds. He continued new refinements of these to meet the needs of theoretical chemists but went further to observe a number of related compounds of essentially aromatic character or of terpenoid character such as limonin, cedrelone and calycanthine. These are magnificent examples of the use of X-ray analysis for solving problems which are extremely difficult for purely chemical methods and Robertson's knowledge of chemistry as well as of crystallography stood him in good stead. He also gave an account of rings of unusual numbers, particularly those of azunine and tropolone. It was largely due to Robertson's work that the idea of using X-ray crystal analysis has now spread widely in chemical circles and problems calling for them are very often sent to him or his students.

Manchester

In the University of Manchester physics department, the crystallographic work did not long survive the removal of W. L. Bragg to Cambridge where a number of his research workers joined him. However, one of them, H. Lipson was appointed Professor of Physics at the Manchester College of Science and Technology where he had been preceded by W. H. Taylor. He was soon able to set up there a small school of crystallography which was well able to carry on the Bragg tradition. He interested himself mainly in the methods of crystal analysis based on Fourier transforms and developed very beautiful optical representations which made the first stages of the analysis of many organic crystals almost an intuitive process. In this he was assisted by Dr. C. A. Taylor and others while Dr. M. M. Woolfson interested himself more in direct methods of analysis.

Bristol

The work of Piper on hydrocarbons, already mentioned, has been extended to artificial polymer hydrocarbons by Keller and has thus linked the work with that carried out in connection with the dislocation of crystal growth studies initiated by Professor Frank and the expression of his theory of spiral crystallization of substances. It is here, also, that the greatest concentration of work in connection with the physical properties, both electrical and magnetic, of ferro-electrics and ferro- and polar magnetics is being carried out. Resonance studies are fitted in with more precise knowledge of crystal structure of coordination complexes.

Cardiff

At Cardiff, under Prof. A. J. C. Wilson, another school of organic crystal chemistry has grown up, here concentrating on the alkaloids such as ephedrene to adharmine, and on a number of terpinoids such as longifolene. The research of Mr. Hine into amino acid derivatives led to the first determination of the stereoconfiguration of an optically active sulphur atom. Cardiff has thus become one of the leading schools of the refinement in organic crystal structure analysis. At the same time, much has been added to the theory, beginning with Wilson's enormously valuable method of determining the presence or absence of the centre of symmetry by the statistics of intensities, and a critical appreciation of the value of analysis of crystals both perfect and imperfect. Most recently, Professor Wilson has entered the field of precise X-ray pattern determination in connection with the A.S.T.M. index and has raised the standard of accuracy of powder patterns to make them a far more precise method of identification than they had been hitherto.

Edinburgh

The transfer to Edinburgh of Dr. C. A. Beevers, who had already put crystallographers in his debt by his contribution to the Beevers and Lipson strips, started a school of crystallography there remarkable for its attack on the really difficult organic structures, those of strychnine and sucrose sodium bromide. It was here that Cochran, already mentioned, first started his X-ray work. Now the emphasis is still on sugars and their derivatives and alkaloids. Work is also being done on inorganic salts, particularly on the phosphates and sulphates.

Further British University Schools

Crystallographic work in *Belfast* was initiated by Professor Ubbelohde before he transferred to London in 1957 and, as already indicated, dealt largely with the study of crystal transformations, isotope effects and thermal vibrations. Miss Woodward, one of the original Royal Institution research workers, has particularly studied the whole range of transformations of single crystals of potassium nitrate. Though not directly concerned with analysis, the stay of P. P. Ewald in Belfast strengthened the interest there in crystallography especially on the optical and theoretical side.

Since the departure of Cox and Llewellyn, from *Birmingham* crystallographic work has not been on a very large scale but of admirable quality. Dr. R. W. H. Small's work on the effects of hydrogen bonds in sugars and other compounds was done here.

In *Dundee*, a small but flourishing school under Dr. J. Iball has concentrated on a group of important hydrocarbons of aromatic character closely associated with carcinogenic properties beginning with kerosine and 3,4 benzpyrene. The emphasis is on accurate determinations capable of explaining properties by anomalies in bond distances.

At *Newcastle* (Durham University), Dr. H. P. Stadler has been working on problems of interest to the properties and transformations of coal substances, particularly high complex fused ring hydrocarbons and their derivatives. Work in the other parts of the University at Durham is being carried out by H. M. M. Shearer, especially on organometallic compounds.

Three new schools of crystallographic research deserve special mention. At *Keele*, North Staffordshire, the new University started in 1954 an X-ray crystallographic unit and Dr. S. C. Nyburg has been carrying out very interesting studies on inter-relations of organic structures which he has described in his book *Organic Crystal Chemistry*. At *Aberdeen*, H. F. W. Taylor, who transferred from Birkbeck in 1953, has continued work on silicates but made a welcome new approach by the use of a combination of electron microscopy and X-ray diffraction. Dr. Gard, who works with him in this field, has achieved the first analysis of X-ray structures by this method carried out in Britain. It was essentially a Russian and Swedish study hitherto (Pinsker, Vainshtein and Grudemo). A school of crystallography at *Nottingham* was set up in 1949 by Dr. Wallwork, formerly with Powell at Oxford. Its interests are in organic molecular complexes and anhydrous metal nitrates.

It can be seen that in practically every University in Great Britain, some research work in crystallography is being done and a certain amount of teaching is carried out. However, only in Cambridge for undergraduates and in Birkbeck College, London, for the M. Sc. degree, is the study entirely specialized. Elsewhere it is part of the general chemistry, physics or geology courses.

17.3. Post-war Commonwealth Development

by J. D. Bernal

Canada

The main centre for X-ray diffraction is in Ottawa at the National Research Council under Dr. W. H. Barnes (organic and inorganic structures) and Dr. W. B. Pearson (metals). Active research centres exist, however, at some of the Universities in the Departments of Geology and Chemistry. Of singular interest is the work done on neutron diffraction at the Chalk River reactor which, for many years, gave the highest neutron flux of all reactors. Considering the wealth of ores and wood, and the well developed metallurgical, pulp, and other industries, it is not astonishing that they have introduced X-ray analysis, even though on a very cautious scale.

South Africa

Largely under the influence of Professor R. W. James, an interesting school of X-ray analysis was started in the early forties in the University of Cape Town. He, himself, completed there the second volume of *The Crystalline State*, that dealing with the principles of X-ray optics and diffraction.

The principal interest of the Cape Town School in which Dr. and Mrs. Saunder have been most prominent, has been in the structure of aromatic molecules and molecular compounds using three-dimensional methods and extending the analyses to cover cases of molecular disorder. It was here, also, that Dr. Klug began his studies on the use of Fourier transforms in crystal analysis, a study that has stood him in good stead in his later work on viruses.

Australia and New Zealand

The record of Oceania in post-war crystallography has been a distinguished one. An important and autonomous school of crystal analysis has been formed in Melbourne in the Division of Chemical Physics of the Commonwealth Scientific and Industrial Research Organisation (C.S.I.R.O.) under Dr. A. L. G. Rees. Here X-ray analysis is being combined with the development of electron microscopy and electron diffraction. The principal interest of the group directed by Dr. J. M. Cowley is the development of experimental techniques and the theoretical basis of the subject. In particular, methods have been developed for the structure analysis of submicroscopic crystals from single-crystal patterns obtained by using electron microprobe techniques, and a new formulation for the theory of electron diffraction and microscopy has been evolved.

In the X-ray group Dr. A. McL. Mathieson with Dr. J. Fridrichsons has carried out many structure analyses of moderately large molecules (20 to 50 atoms in asymmetric unit) including plant alkaloids and peptides, incorporating developments of low temperature, heavy atom and generalized projection methods. Dr. B. Dawson has concentrated on the super-refinement of more simple structure and the accurate theoretical and experimental determination of atomic scattering factors and their modification by ionization and bonding.

This school has already made its mark internationally particularly in generating new ideas and methods.

In another section of C.S.I.R.O., that of Tribophysics, work is directed by Dr. W. Boas, formerly one of the Berlin Dahlem group which laid the foundation of structural metal physics. The interest of present work here is largely directed towards the study of crystal defects and plasticity as affecting metals and to surface phenomena of catalyst crystals studied by electron diffraction and electron microscopy.

There has been a very interesting beginning of organic structural studies in the University of West Australia under Professor Birkett Clews and Dr. Marsden which has now been concentrated in Government laboratories. In Sydney H. C. Freeman, who received his training at the California Institute of Technology, is building up a school for organic structure analysis at the University of Sydney. Interesting structural work and research on metal textures is being carried out at two departments of the University of New South Wales.

The school of crystallography in New Zealand was initiated by Professor F. J. Llewellyn, formerly at Birmingham, at the University of Auckland in 1948. It has been largely concerned with the structure of nitrogenous organic compounds. After his resignation in 1956 it has been carried on by Dr. D. Hall. In 1954 another school has been set up under the direction of Dr. B. R. Penfold at the University of Canterbury working on inorganic crystals.

India

At the Indian Association for the Cultivation of Science, Calcutta, Professor Banerjee's school has continued and extended its work in post-war years. The main interest has now turned to the study of thermal diffuse reflections from organic crystals, as studied by R. K. Sen, S. C. Chakraborty and R. C. Srivastava in relation to the elastic constants. Studies have also been carried out by G. B. Mitra on coals and B. K. Banerjee on glass.

At the Indian Institute of Science, Bangalore, under the direction of R. S. Krishnan, work on diffuse reflections of X-rays has also continued. With G. N. Ramachandran detailed studies on properties of diamond have been made including thermal expansion and relation of mosaicity and luminescence. Studies have been carried out on coordination compounds and on complex organic crystal analysis, where several new methods have been developed including low-temperature and anomalous dispersion techniques, especially in the work of G. Kartha and S. Ramaseshan.

One of the most important contributions of the Indian school has been the analysis of the structure of collagen as a three strand twined polypeptide, carried out by G. N. Ramachandran at the National Institute for Leather Research at Madras.

17.4. Research in Non-industrial Laboratories outside the Universities and the Royal Institution

by J. D. Bernal

Although there are no institutes where a large part of the work is devoted to X-ray crystallography, valuable work and new initiatives have come in the field of crystallography from several governmental and independent institutions, ancillary to their main objectives. These can be classified, in general, according to the nature of the work. Those at the National Physical Laboratory and the Atomic Energy Research Establishment at Harwell have been concerned very largely with problems of metals, while those at the British Museum of Natural History, the Building Research Station, the Agricultural Research Stations at Rothamsted and Aberdeen, and the Safety in Mines Research Establishment in Sheffield have been mainly concerned with studies in minerals, the last three particularly with clays. In the National Institute for Medical Research and to a certain extent at Harwell, organic substances have been the centre of interest.

National Physical Laboratory

The contribution of the National Physical Laboratory began at the very outset of structural crystallography with the work of Dr. (afterwards Professor) E. A. Owen, of the Physics Division who, in 1923 and 1924, started the study of metal structures by the X-ray powder methods. These were among the first studies in metals by X-rays to be carried out anywhere. Afterwards he was joined by Dr. G. D. Preston and together they developed further some of the most interesting studies of age hardening and precipitation processes in copper aluminium alloys, which were eventually to culminate in the demonstration of the existence of minute copper-rich regions in an aluminium matrix. These precipitations, the Guinier-Preston Zones, coincided with the onset of hardening. Preston was also a pioneer in the use of precision parameter measurements. In 1926, the Laboratory started a section on the application of X-ray methods to industrial research, some original members being H. M. M. Shearer, W. A. Wood and J. Thewlis.

Their main work was on phenomena concerned with the deformation of metals as well as the brittle fracture characteristics of iron. The value of this work was recognized in effect by the setting up of X-ray laboratories in many of the most important industrial firms so that the particular section was wound up after the Second World War and the whole of the X-ray work concentrated in the metallurgy division. Since the war the work has been continued on precipitation and binding processes but there has been also a condiderable development of X-ray techniques particularly in the direction of precision measurement.

Atomic Energy Research Establishment, Harwell

The work here is of two kinds, firstly that dealing with metals and other solids connected with the processes of atomic energy generation and secondly the use of neutrons for structural studies. The former part of the work concerns the properties of graphite, particularly very pure graphite, and studies, in particular by Kahn and Thewlis, on textural properties such as preferred orientation. Not only graphite itself but its reaction with other substances such as sodium, were examined and boron nitride was proved to have a different structure from graphite. Work has also been done on irradiation damage to crystal structure of these and simpler substances.

X-ray work has further dealt with the structure of compounds concerned with the trans-uranic and related fissile elements, such as the thorium halides, selenides and tellurides. Much work has been done on the properties of uranium and plutonium from the point of view of their metallurgical handling.

The important new contribution of Harwell to crystallographic studies has been the use of neutron diffraction which was started by G. E. Bacon and his co-workers in 1942. It is the only school of neutron crystallography in Britain. Bacon has concentrated on two kinds of problems for which neutron diffraction is especially suitable, the location of hydrogen atoms and the orientation of atomic magnets in compounds of transition elements. Starting with the ferro-electric KH_2PO_4, he was able to show that the hydrogen atoms in the hydrogen bond are not symmetrically placed and can be switched over to another oxygen atom by reversing the electric field. The bent nature of the hydrogen bond has, furthermore, been established in hydrated salts and acids. The location of C—H bonds has been used to study thermal motions of molecules in crystals.

In the magnetic field, studies of spinels have been used to elucidate ferromagnetic and anti-ferromagnetic states where neutrons with their own magnetic moment can detect differences inaccessible to X-rays. A beginning has also been made in the study of inelastic scattering of neutrons where exchange of energy takes place between the neutrons and the acoustic and magnetic-spin waves in the crystal.

Building Research Station

The main work of the Building Research Station, which began with that of Dr. G. E. Bessey in 1930, was the study of structures of aluminates and silicates of interest to cement chemistry. This work has been continued by Drs. H. G. Midgley, E. Aruja and M. H. Roberts in close conjunction with work on similar silicates carried out at Birkbeck College (pg. 389). One complete structure, that of dicalcium silicate was worked out and the cell dimensions and characteristics of a large number of the silicates and aluminates have been determined, such as chlorites and serpentines.

The most interesting development has been the use of X-rays in the study of silicate and aluminate materials at high temperatures, and in the preparation of phase diagrams with F. M. Lea, T. W. Parker, R. Nurse and J. H. Welch. Most recently, a new form of high-temperature single-crystal X-ray camera was developed capable of giving X-ray pictures at up to 1800°C. Much work was also done on the development of quantitative determination of phase by X-ray methods.

Macaulay Institute for Soil Research, Aberdeen

This has been a pioneer in the study of clay minerals which have formed part of the survey of soils for Scotland. In 1939 Dr. D. M. C. MacEwan started there his series of studies on the identification of clay minerals by the absorption of glycerol ethylene glycol which is the basis of all modern clay analyses.

Dr. G. F. Walker has studied the structure of vermiculite and other soil constituents including the materials derived from the weathering of rock minerals. X-ray methods have been devised for separately assessing the quantities of the different clay minerals in a sample, and this is used in conjunction with other methods such as differential thermal analysis.

The Macaulay Institute has become one of the leading world centres of X-ray studies of clay minerals and has established X-rays as one of the major tools in pedological research.

Rothamsted Experimental Station, Harpenden

Pioneer work on X-ray studies of clay structures were initiated by G. Nagelschmidt in 1934. His work dealt with the structure, lattice shrinkage and structural formulas of the montmorillonite group of minerals, the clay mineralogy of soils and sediments, and methods of investigation of clay fractions. From 1945 to 1958, D. M. C. MacEwan continued the work he had started at the Macaulay Institute, Aberdeen, on the absorption and complex formation between organic molecules, particularly alcohols, and the minerals montmorillonite and halloysite. MacEwan in collaboration with G. Brown also worked on the interpretation of diffraction patterns from clays consisting of interstratified layers of different kinds. This work has been followed up by their co-worker R. Greene-Kelly and O. Talibudeen. G. Brown located the exchangeable cations in the glycerol-montmorillonite complex by one-dimensional Fourier syntheses. More recently he has applied single crystal X-ray methods to the study of weathering products. Here he has shown that a large number of soil mineral 'crystals' are actually topotactic mixtures of interstratified decomposition compounds.

Safety in Mines Research Establishment

The X-ray group, under the direction of Dr. Nagelschmidt, Dr. R. L. Gordon and now Dr. C. Casswell, has been mainly concerned with studies of the minerals in coal mine dust causing pneumoconiosis. This has led to the development of quantitative methods of analysis. A special study has been made of the disturbed surface layers on ground quartz.

The Medical Research Council

The Medical Research Council has for many years supported crystallographic research through its various External Research Units. The work of these units, however, was mainly concerned with macromolecules and in 1951 a small crystallographic laboratory was established at the National Institute for Medical Research, Mill Hill, London, under the direction of Mrs. O. Kennard. The laboratory has been engaged in research into Chemistry, Biochemistry and Biophysics and with applied research like Biological Standardization. X-ray diffraction methods have been adopted as routine techniques in these fields for identification, molecular weight determination and physical characterization in a variety of problems ranging from the identification of the thyroid hormone, triiodothyronine, after its first isolation by R. Pitt-Rivers and J. Gross, to the diagnostic use of X-ray powder patterns in cystinosis. X-ray analytical work has been concentrated on organic compounds of moderate complexity, of which the structure of vitamin A acetate and the three-dimensional analysis of some steroidal sapogenins are examples, and to problems of bonding forces, as in the work on the structure of amidinium carboxylates. During the course of this work some new auxiliary techniques have been developed including a method of determining integrated intensities by the use of radioactive markers.

The laboratory has also been concerned with crystallographic documentation work and was responsible for the data on organic compounds in various compilations including the *Tables of Interatomic Distances in Molecules and Ions* and the projected second edition of *Crystal Data.*

17.5. Crystallography in British Industrial Laboratories

by C. W. Bunn

The great harvest of knowledge of crystal structure resulting from the fundamental discoveries of von Laue and W. H. and W. L. Bragg began to flow freely in the decade 1920–30, chiefly from the Royal Institution under the guidance of W. H. Bragg, and Manchester University where W. L. Bragg was professor; and it was natural that under this stimulus crystallographic methods, especially X-ray diffraction methods, were soon taken up in industrial and government laboratories and used either for direct practical applications or for studying the crystal structures of materials used in various technologies. The simplest of the practical applications was the use of powder photographs for identification, for it was evident from the work of Debye and Scherrer that the X-ray diffraction pattern of a powdered crystalline material is an unrivalled 'fingerprint' of identity; it was soon used to settle many questions of identity which had not yielded to chemical methods or the older crystallographic methods (optical or morphological) owing to the opacity or submicroscopic size of the crystals. But in addition to this empirical application, more sophisticated methods such as the study of crystal orientation in metal sheets and wires, the estimation of crystal size and distortion from broadened diffraction effects, and the study of atomic structure wherever it was felt that the knowledge was relevant, began to appear in non-academic laboratories.

One of the earliest organizations in the field was the General Electric Company, where F. S. Goucher at the Wembley Laboratories was studying the orientation of crystals in tungsten filaments by X-ray diffraction methods as early as 1923, and J. W. Ryde identified the crystals responsible for the scattering of light in certain opal glasses in 1926. H. P. Rooksby, who took part in some of this early work, has since then applied these methods to a great variety of problems. From 1930 onwards, with J. T. Randall, he made valuable contributions to our knowledge of the structure of glasses and liquids, and concluded that locally the atomic arrangement in these non-crystalline substances is similar to that found in crystals of the same substances. Other applications for which Rooksby was responsible, either alone or with others, range from the identification and quantitative analysis of refractory materials and thermionic cathode coatings to studies of the structures of ferromagnetic and antiferromagnetic crystals and luminescent materials, and metallurgical problems of lattice distortion and crystal orientation. This activity at the G.E.C. continues unabated.

In Imperial Chemical Industries, the first X-ray crystallography was done in the laboratories of Nobel Division at Stevenson in Ayrshire by F. D. Miles, who had worked for a time in Sir Wililam Bragg's group at the Royal Institution. Before 1930 he had followed the course of the nitration of cellulose fibres and studied the structure of cellulose trinitrate; later, he reported the crystallography of lead azide and other sensitive materials used as detonators. He also studied habit modification by dissolved impurities—a subject taken up again much later (1947–9) by J. Whetstone, with practical results in the control of the caking of ammonium nitrate. At Alkali Division at Northwich in Cheshire, C. W. Bunn, who had worked there on crystal growth problems and petrological methods of identification of inorganic substances since 1927, started using X-ray diffraction methods in 1933, first of all for identification but later for structure determination. Long-standing problems of inorganic chemistry which were cleared up by the use of X-ray powder photographs included the question of the constitution of the chlorinated lime product known as 'bleaching powder', the identity of the variously coloured precipitated iron oxide pigments, and the constitution of boiler scales, cements and plasters. The discovery of polythene there in 1933 led to an X-ray determination of its structure, and this started a series of similar investigations (with

E. V. Games and A. Turner-Jones) on other crystalline polymers such as rubber and related substances, and polyamides and polyesters. Meanwhile, other crystallographic work included further studies (with H. Emmett) of crystal growth from solution (the role of layer formation, surface structure and concentration gradients), the development (by H. S. Peiser) of the method of estimation of the degree of crystallinity in polymers which has since become one of the main technological applications of X-ray diffraction in the polymer field, contributions to methods of interpretation of X-ray diffraction patterns, and the determination of the crystal structure of sodium benzylpenicillin (Bunn and Turner-Jones) which together with the work on the potassium and rubidium salts at Oxford (Hodgkin and Rogers-Low), settled the chemical constitution of that substance. The work on polymers was continued later (from 1946 on) at Plastics Division at Welwyn Garden City in Hertfordshire by Bunn, R. de P. Daubeny, D. R. Holmes and A. Turner-Jones; X-ray methods are used for characterizing new polymers, estimating crystallinity, studying crystal orientation in films and fibres, and for structure determination ('terylene', polyvinyl alcohol, nylon 6, polyisobutene, polytetrafluoroethylene) as far as the pressure of practical affairs permits. Similar work is now done also at Fibres Division at Harrogate in Yorkshire, and at British Nylon Spinners at Pontypool.

During and after the war of 1939–45, X-ray crystallographic work was started in several other I.C.I. laboratories. At Metals Division in Birmingham, T.Ll. Richards studied crystal orientation in rolled metal sheets. At Dyestuffs Division in Manchester, A. F. Wells, who in earlier years solved several structures of inorganic and metal-organic compounds in various University laboratories, found time amidst the pressure of practical problems to develop a systematic treatment of network structures (those held together by localized directed bonds, as opposed to 'packing' structures where local directional effects are less important), while C. J. Brown and others concentrated on the structures of organic compounds much used in the Dyestuffs industry, such as aniline hydrochloride, p-aminophenol and acetanilide. At the Akers Laboratories at Welwyn in Hertfordshire, P. G. Owston and others, by solving several metal-organic structures, have played an important part in the development of the chemical studies of such substances by J. Chatt's group. At General Chemicals Division at Widnes in Lancashire, and Billingham Division at Stockton-on-Tees in Co. Durham, both X-ray diffraction and X-ray spectrographic methods are used for identification and quantitative analysis—as indeed they are in many chemical laboratories in other organizations.

The development of the use of X-ray diffraction methods in the metallurgical, electrical and other industries in the North of England in the 'thirties owed a good deal to the encouragement of W. L. Bragg when he was Professor in Manchester, and to the existence of the group in his laboratory working on metallurgical problems under A. J. Bradley. A. H. Jay, who first worked on powder-camera design with Bradley, later applied powder methods to a variety of problems involving metals, alloys and refractory materials in the United Steel Companies laboratory at Stocksbridge. Others who made similar contributions were C. Sykes at Metropolitan-Vickers in Manchester, H. J. Goldschmidt at William Jessop and Sons Limited in Sheffield, and J. A. Darbyshire at Ferranti Limited in Manchester. Since that time the methods have been taken up in many laboratories, too numerous to mention individually. Most of this work is never published: it uses established methods and knowledge in order to fulfil its function in practical affairs, and makes a valuable contribution without necessarily revealing anything fundamentally new. Moreover, background studies of crystal structure can only be undertaken when the pressure of practical affairs is not too great; notable contributions of this sort were made by E. J. W. Whittaker (Ferodo Limited, Chapel-en-le-Frith, Cheshire) in his studies of the structure of chrysotile and the theory of diffraction by cylindrical lattices, and by V. Vand (Unilever Limited, Port Sunlight, Cheshire), who determined the structures of several organic chain compounds, discovered in a potassium soap a new type in which the chains are crossed, and contributed to the methods of interpretation of diffraction patterns.

Natural textile fibres like cotton, wool and silk, which were shown, very early in the history of X-ray diffraction, to contain oriented crystalline arrangements of molecules, offer good opportunities for both structure studies and technological applications. Much of the early work on the structures was done in academic laboratories—by Mark and Meyer and others on the continent and by Astbury in this country. X-ray methods were introduced at the Shirley Institute at Manchester in the early 'thirties by Dr. Pelton, who had worked under W. H. Bragg at the Royal Institution. They were at first used for technological problems, such as in the study of crystal orientation in cotton and its relation to the physical properties, but later on, more fundamental structural studies of cellulose and its derivatives and silk fibroin were undertaken by J. O. Warwicker and others. The British Rayon Research Association at Manchester also made contributions in this field. In the laboratories of Courtauld's Limited, the group led by C. H. Bamford at Maidenhead, together with L. Brown at Coventry, have made very valuable contributions to our knowledge of the structure of the synthetic polypeptides, while C. Robinson discovered a very remarkable new type of liquid crystal structure in solutions of one of these substances.

The use of crystallographic methods in industry continues to grow. The extent and value of the work are not to be judged by the volume of published work, large though this is, for the majority of the work done is unpublished. Although a substantial amount of fundamental structure work has been done in industrial and government laboratories, and although new facts and phenomena are sometimes discovered and reported, the justification of the use of these methods in industry lies in the direct technological applications, such as identification and analysis, and the correlation of crystal size, orientation and texture with the properties of materials. Scientific discoveries and advances in fundamental understanding can and do come out of industrial research, but they are not usually regarded as primary objectives.

17.6. Early Work at University College, London, 1915–1923

by Dame Kathleen Lonsdale

W. H. Bragg was appointed Quain Professor of Physics in 1915, but did not take up his duties fully until the end of World War I. Then he began to gather around him a group of young physicists which included:

I. Blackhurst, who studied thermal vibration effect on intensity, using the ionization spectrometer with diamond, graphite, aluminium and designed a Hg high-vacuum pump; and G. Shearer and A. Muller, who respectively developed home-made hot-wire and self-rectifying gas tubes for photographic studies of long-chain compounds. For this purpose they used foot-operated fore-vacuum pumps, home-made high-vacuum pumps and induction coils, and Wehnelt interrupters—constructed out of aluminium hot-water bottles.

W. T. Astbury joined the team in 1921 and used the ionization spectrometer, complete with Mo Coolidge tube (air-cooled), lead box, gold-leaf electroscope, Ruhmkorff coil with Hg interrupter, and metronome (to time the rate at which the crystal was rotated through a reflecting position). He was fascinated by the problem of the optical activity of tartaric acid. His training in chemistry enabled him to understand better, probably, than any of the other workers the potentialities of the X-ray method of structure analysis in respect of the light it might throw on both physical and chemical problems of the solid state. His personality and enthusiasm led to the liveliest discussions in the laboratory and over joint meals, about everybody's research studies, crystallographic problems in general, politics, religion and almost everything under the sun.

Other workers in the UCL Physics department forty years ago included R. E. Gibbs (who studied the structure of quartz, with W.H.B.), W. G. Plummer (preliminary investigation of C_6Cl_6 and C_6Br_6, an isomorphous pair, by the photographic technique) and K. Yardley, who began the study of succinic acid and simultaneously the application of space-group theory to structure analysis. Thomas Lonsdale, to whom she was married five years later, also worked in the laboratory, but on the elastic properties of metal wires, under Professor Porter, whose main interests were in the phenomena of radiation and convection. (One of his more original lectures was entitled 'Why the Daddy-long-legs doesn't wear Stockings'!)

During this period W. H. Bragg himself worked mainly on the structures of anthracene, naphthalene and naphthalene derivatives. He noted that the difference in length of the *c* axis of anthracene and naphthalene, whose *a*, *b* axes and β angles were very similar, was about equal to the diameter of a six-membered carbon ring in the structure of diamond (which had been fully worked out) or in graphite (which had not, but the unit-cell dimensions and the fact that it was a layer-structure were known). From this he deduced the general dimensions and positions of the anthracene and naphthalene molecules, although he wrongly assumed the benzene nuclei to be 'puckered'; and the assumed orientation of the molecular normals in the (001) planes was not correct.

It is interesting, however, to note that he initiated the studies of *isostructural* series both of aromatic and of aliphatic compounds (long-chain paraffins, fatty acids, esters, alcohols, ketones, etc.) and was aware of the value of the *isomorphous + heavy atom* technique. Indeed in a paper on the structures of NaCl, KCl, KBr and KI (*Proc. Roy. Soc.* A*89*, 468, 1913) W. L. Bragg had written:

'By noticing what differences were caused in the photograph by the substitution of heavier for lighter atoms in the crystal, a definite arrangement was decided on as that of the diffracting points of the crystalline grating.'

It was this that led to the examination of C_6Cl_6 and C_6Br_6 by W. G. Plummer and later to a study of the substituted ethanes by K. Yardley. Neither study was successful, however, at that time!

When W. H. Bragg moved to the Royal Institution in June 1923 he took most of this team with him, and for a time Professor Porter was head of the department. The Quain chair was held from 1928 onwards by E. N. da C. Andrade, whose main interest was in the mechanical properties of solids, especially metals, and in epitaxial growth, and who used X-ray techniques as an auxiliary tool.

17.7. Crystallography at the Royal Institution

by Dame Kathleen Lonsdale

While Sir William Bragg was Quain Professor of Physics at University College, London, he gave a Friday Evening Royal Institution Discourse (19 May 1922) on 'The Structure of Organic Molecules'. In it he discussed naphthalene, anthracene, α- and β-naphthol and acenaphthene, all substances that he was then engaged in studying by means of X-ray diffraction, and he compared them with diamond and graphite, (although at that time it was not known that the graphite layers were plane) and showed that in all of them six-membered carbon rings were present. He had already invited to work with him at University College a group of young people, most of whom he took with him to the Royal Institution when, in June 1923, perhaps as the result of this lecture and of the course of Christmas lectures he gave in 1919, he went there to succeed Sir James Dewar as Director of the Davy-Faraday Laboratories and Fullerian Professor of Chemistry.

The Royal Institution, founded in 1799 by Benjamin Thompson, Count Rumford, had housed a succession of famous men such as Thomas Young, Humphrey Davy, Michael Faraday, John Tyndall and James Dewar, and the lectures delivered within its walls had covered almost every new discovery in science as well as a wide range of other subjects. Almost a century later (in 1896) the Davy-Faraday Laboratory, formerly a private house with many rooms, both large and small, had been added, by the munificence of Dr. Ludwig Mond, to provide wider opportunities for research to men *and women* of any race or nationality. The historic laboratories in the Royal Institution had not been open to women; the early 19th century records plainly suggest that they would be expected to be only a nuisance there. On the other hand they had always been cordially invited to attend the Royal Institution Lectures; it would, as Thomas Young gracefully put it, be an alternative to their 'insipid consumption of superfluous time'. And their subscriptions were needed to maintain the Institution in the style to which it was accustomed.

W. H. Bragg included three women among the twelve research workers whom he had gathered around him by the autumn of 1923.* He found the Davy-Faraday Laboratory almost moribund. He made it not only into a lively international school of research but also into a centre to which famous men of science gravitated naturally when they were in London. As one of the Davy-Faraday research workers wrote to me recently: 'the triple appeal of laboratory, library and lectures was an inspiration. My main impressions were of the happy family atmosphere with formality in the background; the casual way world figures appeared at tea-break; the loose organization...; the dearth of mathematical texts marking the emphasis on experimental science.'

In spite of his mathematical background Sir William Bragg was indeed an experimentalist. He did encourage his students to study mathematical crystallography, but it was in order to apply it, not as an end in itself. He was keenly interested in the discussions that went on in the laboratory concerning the extent to which the molecular symmetry was used in the building of crystal symmetry. The results of these discussions had been partly summarized in what were called 'Shearer's Rules', which expressed the empirical fact that up to that time no structure had been found in which the 'asymmetric unit' contained more than one molecule, although it was sometimes a submultiple of the molecule. Sir William was not so keen on the idea of tabulating all the symmetry properties of the 230 space groups and their implications in terms of diffraction theory. He felt that this savoured of mathematical perfectionism and that it was simpler and more realistic to examine each case as it occurred in the course of research. (In later days he never became wholly reconciled to the use of reciprocal space; and preferred a more complicated but, as he felt, more realistic picture of what was actually happening to the X-ray

* W. T. Astbury, J. D. Bernal, R. E. Gibbs, L. C. Jackson, Miss I. E. Knaggs, Miss G. Mocatta, A. Muller, W. G. Plummer, G. Shearer, C. H. Weiss, J. F. Wood and Miss K. Yardley. In all, 12 of the 70 workers admitted before 1940 were women.

beam within the crystal.) But he allowed himself to be convinced that the Astbury-Yardley Tables were worth publishing and then he convinced the Royal Society that they were worth the considerable expense involved in having the 230 diagrams professionally re-drawn for publication. It was one of the few *Philosophical Transactions* publications that had to be reprinted.

Those of us who worked with W. H. Bragg in the Davy-Faraday Laboratory got the impression that we were allowed to choose and develop our own research themes entirely independently. Sir William certainly never dictated; and he expected his team (later classified in the D.F. records as 'Research Assistants' who were in receipt of an annually renewed salary and 'Research Workers' who were of independent means or in receipt of supporting grants) to have original ideas and to develop them independently. But, looking back, it is possible to see that in fact he directed the research by means of silken reins that were hardly felt but were very effective.

To begin with, it was understood that we would choose some problem connected with organic structures. Apart perhaps from quartz, which had interested Sir William since 1914, the inorganic world was left to the Manchester school under his son, W. L. Bragg. Then he guided the general trend of research by injecting, from time to time, a new worker having a different background and outlook. Miss Knaggs, for example, had worked with Professors Pope and Hutchinson at Cambridge and brought with her the ideas of the importance of valency and spatial considerations in determining structures and a wide knowledge of mineralogical and optical methods. J. M. Robertson was first and foremost a chemist, whereas the earlier workers all had a background of training in physics; Miss Woodward was a mathematician; and A. R. Ubbelohde, who has described himself as the Benjamin of the family, was interested primarily in the thermodynamical problems of crystals. But the principal way in which Sir William guided us without our really being aware of it was by asking us to help him with the preparation of lectures. He had a habit of taking subjects of research in the D. F. Laboratory, thinking about them, looking up related papers, talking about them to visitors and then lecturing about them so clearly that the research worker engaged on the problem became aware of all sorts of possibilities that he had somehow overlooked before. Early in 1926 Sir William gave an afternoon course of R.I. lectures on 'The Imperfect Crystallization of Common Things', which was repeated with additions in the autumn. He asked W. T. Astbury to assist him in the preparation of this lecture by taking X-ray photographs of natural fibres, such as were being taken at the Kaiser-Wilhelm-Institut für Faserstoffchemie. This Astbury did with such thoroughness that he became interested in the field and when an opening occurred in Leeds for an X-ray physicist to study textile fibres, W. H. Bragg persuaded Astbury to go. He needed a good deal of persuasion; none of the workers in the D.F. Laboratory ever wanted to go. The salaries there were not particularly good, but the atmosphere was so pleasant that the idea of staying on indefinitely was most alluring. But if Sir William thought that one of his people was ready to take more responsibility and if a suitable job presented itself, he pretty well pushed them out.

It really was astonishing that he should have had such vigour, for although we affectionately called him 'the old man' none of us really thought of him as old. He was still publishing vigorously when in his 80th year, and seemed to have the mind of a young man, able to take a keen interest in such a new phenomenon as the diffuse scattering of X-rays due to thermal vibrations.

When I look back at the early days at the Davy-Faraday Laboratory disconnected memories come to the surface. We had plenty of time for discussion. It was not possible to sit *all* day long with one eye glued to a microscope taking readings of the movements of a gold leaf, although quite a lot of our time was spent this way. From time to time our Hg interrupters had to be thoroughly cleaned out, a filthy job; but we could do it and talk simultaneously. Several of us brought sandwiches and lunched together in a room on the premises and then played table tennis in the basement afterwards. Within a year or two we could put on international tournaments: France was represented by C. H. Weiss (study of alloys) and later by M. Mathieu and M. J. H. Ponte (scattering factor of the carbon atom); Holland by W. G. Burgers (study of i-erythritol and other crystals too complicated to solve then); the Soviet Union by Boris Orelkin (preliminary study of 1,3,5-triphenylbenzene). W. T. Astbury (who was universally known as 'Bill' and who insisted on calling everyone else—including me—'Bill' also) was the life and soul of these table tennis sessions and introduced various hazards, such as matchboxes at strategic points on the table, to make them more exciting. Most of us were pretty good.

Tea-time at 4 p.m. was something not to be missed. To begin with, W. H. Bragg was nearly always there and there were generally Bourbon biscuits too. And all sorts of interesting visitors turned up. Some of them were Friday Evening lecturers come to prepare the experiments for their discourses. It might be Sir Ernest (afterwards

Lord) Rutherford, about to talk on the 'Life History of an α-particle from Radium' or on the 'Nucleus of the Atom'; J. H. Jeans on 'The Origin of the Solar System'; H. E. Armstrong on the 'Scientific work of Sir James Dewar'; G. Elliott Smith on the 'Human Brain' or Lord Rayleigh on the 'Glow of Phosphorus' or Charles Darwin on 'Recent Developments in Magnetism'; Sir William Pope on 'Faraday as a Chemist' or Sir J. J. Thomson on 'Radiation from Electric Discharges'. Or sometimes there were rich business men from whom Sir William was busy extracting money for the D.F. Laboratory. One I particularly remember was one of Sir William's own past students who was quite disturbed that Sir William's talents were being wasted in such a dead-end, poorly-rewarded job when he might be making top money, as he himself had done.

From time to time there were open days, or 'conversaziones', at the Royal Institution (as there still are), and these involved the construction of illustrative charts and models which we had to be prepared to discuss with experts and laymen alike. (One old lady asked Sir William why his naphthalene model didn't smell like mothballs?) Or Sir William would himself bring an eminent visitor round the laboratory. One visit that has remained in my memory was that of Sir Alfred Yarrow, who had endowed some attractive research fellowships. He propounded a theory that brilliant men inherited their intellect on the *maternal* side and asked Sir William what he knew about his mother's people? Sir William, looking slightly embarrassed, said all he knew was that they had something to do with the Church. Sir Alfred went on, rather inconsistently, to deplore the fact that young women scientists were apt to leave their professions in order to get married and was taken aback when I asked where his intelligent mothers would come from if only those with no professions were allowed to marry?

The Royal Institution library was well-stocked with books (although not on the mathematical side, as has already been remarked) and especially with periodicals. It was thrilling occasionally to open a very early back number of, say, the *Phil. Trans. Roy. Soc.*, and find that Faraday had made some comment in the margin, and even more thrilling to meet an aged member of the R.I. (Mr. William Stone) who remembered, as a small boy, sitting next to Faraday and talking to him, in the gallery of the R.I. Lecture Theatre, during some Christmas Lectures. Much of the early historical apparatus, both research and demonstration, was also kept in the Royal Institution, and occasionally one could find, in bottles that had not been disturbed for many years, simply enormous single crystals that had grown very very slowly by sublimation on to the walls. When work was begun on the magnetic anisotropy of crystals these were exceedingly useful, because the use of large crystals minimized the effect of the suspension.

It would be impossible to mention apparatus and crystals without speaking of Mr. Jenkinson and Mr. Smith. Mr. Jenkinson we called 'Jenk' in our irreverent moments when he was not there but never when he was, whereas 'Smithy' was so-called to his face and liked it. Mr. Jenkinson had come from University College with Sir William and was a superb instrument maker. He made the ionization spectrometers and gold-leaf electroscopes (but we put on our own gold leaf if we could, although he would show us how to, the first time). He made the ionization chambers (but we filled them) and had an assistant (who for many years now has been head of his own crystallographic instrument workshop) who helped to make the huge lead-covered box that housed both the Coolidge tube and the electric fan used to cool it. What exotic radiations we used: Mo, Rh, Pd, as a rule! Even with the fan, the anticathode would get white-hot after a short time of running. Our one fear was lest we should become so mesmerized with taking readings of the movements of the gold-leaf to the sound of the metronome by our side that we would forget to look down the collimator to see the colour of the massive target before it *sagged*. While the tube cooled off, we recorded our measurements and interpreted them. We ran the Coolidge tube with an induction coil and the aforesaid Hg interrupter, with a condenser in parallel. In the *secondary* circuit there were a milliameter and a spark-gap. The latter was set so that we got both a visible and an audible signal if the voltage rose above the 60 KV which was the normal running condition. These stood, with a battery of accumulators that gave us the voltage for our ionization chamber, on a small insulated table just at the side of the lead box. By stretching out our hand we could just touch it. We had to remember *not* to. I am not the only worker whose hair has stood permanently on end, more or less, ever since.

Smithy was the laboratory steward, but he was much more than that. He was skilful with his hands and could make the pyrex high-vacuum pumps, originally designed by I. Backhurst at University College, better than anyone else. If we could not find the leaks in the gas tubes used for photographic work (that gradually replaced the ionization method), Smithy would help us. Later he took entire charge of the maintenance and running of the 5 kW tube and could design all kinds of auxiliary equipment for special purposes; and when Drs. Müller and

Clay were both absent through illness he ran and repaired the 50 kW equipment also. The 2 metre diameter spectrometer and 50 kW tube used in combination gave really good resolution, but the outfit was not foolproof enough to become standard equipment.

Mr. Green, the lecture assistant in the Royal Institution, was also very helpful with ideas if asked, but although workers in the D.F. Laboratory were permitted, by grace, to use the R.I. library, they were not expected to make themselves too free of R.I. facilities unless they became Members of the Royal Institution, which not all of them could afford to do. It paid off very well if one could, but in those days the 5 guinea entrance fee and 5 guinea annual subscription seemed a terrible lot of money. Sir William was sometimes called in to soothe old gentlemen snoozing in the upper library who complained that the young D.F. workers had disturbed them by *walking through*. He was excellent at soothing. I doubt, however, whether even he could have soothed the indignant passer-by who brought in from Albemarle Street the pieces of a steel file that had been hurled out of a third-floor window by the irritable worker in whose hands it had broken.

Mention has been made of the international character of the Davy-Faraday research school. Apart from those already mentioned, there were Miss N.C.B. Allen (Australia), A. L. Patterson (Canada), C. C. Murdock (U.S.A.), M. Prasad (India), W. H. Barnes (Canada), W. P. Jesse (U.S.A.), Miss T. C. Marwick (New Zealand), D. O. Sproule (Canada), Miss B. Karlik (Austria), A. A. Lebedeff (U.S.S.R.), F. Halle (Germany), N. Japolsky (U.S.S.R.), K. Banerjee (India), Miss L. W. Pickett (U.S.A.) and E. Pohland (Germany), who had come, in that order, before the end of 1932.

In 1933–5 several refugee scientists from Nazi Germany found a welcome in the D. F. laboratories: R. Eisenschitz, G. Nagelschmidt, A. Schallamach, A. Lowenbein. Then in 1938–9 came W. Boas (Germany and Australia), J. J. de Lange (Holland), L. O. Brockway (U.S.A.) and J. Monvoisin (France).

Most of these have since become heads of departments or of institutions, in various parts of the world.

The same is true, of course, of the British workers, although by now some of them have retired or died. Five * of the twenty-four pre-1930 vintage have become Fellows of the Royal Society, and all these established flourishing crystallographic schools in Universities. Others went into Government service or industry; or obtained senior academic

* W. T. Astbury, J. D. Bernal, E. G. Cox, K. Yardley (Mrs. Lonsdale), J. M. Robertson.

posts. The second or even third generation of crystallographers are now making their own marks on the pages of scientific history.

W. H. Bragg did not regard it as any part of his duty to *train* or *teach* research workers. The Ph. D. student who expected to be spoon-fed with pre-digested pap would have had short shrift from him. Unless a worker had some interesting results to show him, or some promising problem to discuss, he simply took no notice of him and in due course he disappeared quietly from the scene. He treated all his people as responsible colleagues and gave them the encouragement they needed; he found money and facilities for them. But he expected them to build their own apparatus out of bits and pieces, or to superintend its making in the workshops;—and above all he did not expect, except in the indirect ways mentioned earlier, to produce research problems for them or to have to tell them what to do next. He did insist on seeing the manuscripts of papers before they were sent for publication, and if he thought them worthy of it, he would communicate them to the Royal Society. An almost complete record of the research work done can be found in the *Proc. Roy. Soc.* and the *J. Chem. Soc.* (London) or for the later R.I. period (from 1937) in the *Research Abstracts* in the *Proc. Roy. Institution.*

It is only possible to give some of the highlights here. Early attempts on the crystal structure analysis of aliphatic and aromatic compounds were largely unsuccessful, apart from those of the series of long-chain compounds. These were prepared by W. B. Saville (1923–34), J. W. H. Oldham (1936–) and later by Miss H. Gilchrist (1927–37); and studied by A. Müller and G. Shearer. Shearer's greatest triumph was in the correct identification, by X-ray measurements of spacing and intensity only, of the values of n and m, first in a single long-chain ketone $CH_3(CH_2)_nCO(CH_2)_mCH_3$, and then in a mixture of two such ketones. These were prepared for him by Professor (later Sir) Robert Robertson, who as a chemist was greatly impressed by this proof of the power of the new method. Müller, who later became Assistant Director, designed a successful 'Spinning Target X-ray Generator' (water-cooled) as early as 1929. He was particularly interested, not merely in the structures of the odd- and even-numbered chain compounds and of the 'state of rotation' that set in a little below the melting point for some of them, but in their related physical properties (lattice energy, dielectric polarization, torsional flexibility, etc.).

W. H. Bragg had long been puzzled (as Faraday was before him) by the hardening of metals produced by cold-working. In February 1924 he gave an R.I. Discourse on the research work going on in the

D.F. Laboratory in which he referred to X-ray studies by Müller of Au, Ag, Cu and Al leaf or foil. Some work on metal structure continued in a minor way until 1951, when under Professor Andrade it became a major interest in the laboratory.

In 1924 J. D. Bernal (independently of Hassel and Mark) successfully proved the planarity of the layers in graphite, but his crowning achievements were the production of charts for the interpretation of X-ray single-crystal rotation photographs and the design of a universal X-ray photogoniometer. W. T. Astbury, in addition to attempting the structure analyses of tartaric acid and of Al and Ga acetylacetones, produced an ingenious integrating photometer for the photographic method. E. G. Cox and W. F. B. Shaw worked out correction factors for photographic measurement of intensities, and Cox also made the earliest determination of the structure of benzene. W. H. Barnes studied the structure of ice from 0°C to −183°C. Later Mrs. Lonsdale was able to take 'Laue' photographs of the diffuse scattering from ice grown on the D.F. Laboratory roof, simply by opening the windows of the laboratory on a wintry day and thus making the whole room into a refrigerator. It was not possible, however, to give an exposure of longer than about half an hour, because by then the X-ray beam had bored a neat hole right through the ice plate. Hailstones collected from the windowsill were also studied.

J. M. Robertson, who spent altogether some twelve years in the D.F. Laboratory, carried out a series of brilliant investigations of the crystal structures of aromatic compounds, beginning with naphthalene, anthracene, resorcinol, durene and benzophenone; and going on, partly with the later collaboration of Ida Woodward, to oxalic acid dihydrate, the phthalocyanines and the dibenzyl series, including stilbene, tolane, trans- and cis-azobenzene. He was particularly interested in the development both of special apparatus and numerical and mechanical computing techniques; and made great advances in the use of isomorphous and isostructural series and of the heavy-atom methods. Together with A. R. Ubbelohde he studied the effects of isotopic replacement (H by D) and of transitions such as $\alpha \rightarrow \beta$ resorcinol, and he carried out most valuable work on the relationship between bond character and interatomic distance, on the basis of his Fourier analyses.

A. R. Ubbelohde, in addition to his work in collaboration with J. M. Robertson, made fundamental investigations on the thermodynamic properties of the metallic state, on the mechanisms of combustion of hydrocarbons, on melting and on various irreversible processes.

Sir William's interest in fibre structure has already been mentioned. In 1933 he extended this to include 'crystals of the living body' and in a lecture which included references to the work of W. T. Astbury (Leeds) on silk, wool, nerve and muscle, and to that of J. D. Bernal (Cambridge) on the aminoacids, vitamins etc., he emphasized especially that it was now becoming possible to correlate the magnetic, electric, optical and thermal properties of crystals with their structure and, conversely, to use such properties to assist in the determination of unknown structures.

From 1932 onwards measurements of the diamagnetic anisotropy of aromatic and aliphatic crystals were made first by Mrs. M. E. Boyland, then by Mrs. K. Lonsdale with assistance from C. H. Carlisle and in parallel with structure determinations. Some horrible risks were taken in ignorance. For example, Miss Knaggs published a structure of cyanuric triazide, for which magnetic measurements were made by Mrs. Lonsdale. Professor K. S. Krishnan, then in Calcutta, decided to repeat some of these measurements; but the crystals, which are a somewhat erratic explosive, detonated overnight and wrecked his laboratory. The D.F. Laboratory evidently had a better guardian angel.

In 1933 W. H. Bragg became absorbed in the problem of so-called 'liquid crystals', following a general discussion at the Faraday Society in April of that year. The lecture in which he gave an account of the optical effects shown by smectic, nematic and cholesteric classes was one of the few in which he showed that his fundamental mathematical training could still stand him in good stead. His later interest in clays derived partly from the Christmas lectures he had given on 'Old Trades and New Knowledge', partly from his son's work on the silicates and partly from the current investigations being made in the D.F. Laboratory by G. Nagelschmidt.

Shortly before Sir William's death in 1942 there began in the laboratories and elsewhere the studies of diffuse scattering by the thermal waves in crystals and of the anomalous scattering in type I diamonds which interested him so much that he arranged a Royal Society Discussion on the subject. These researches were continued during the subsequent years when first Sir Henry Dale and then Professor E. Rideal was Director of the D.F. Laboratory. At the same time Miss Woodward and A. R. Ubbelohde were studying the sub-crystalline changes in structure of Rochelle salt and potassium dihydrogen phosphate in their ferroelectric regions, and studies of texture and extinction were being made by means of Laue and divergent-beam photographs. In 1950 L. R. G. Treloar began his studies of polymers and D. P. Riley those of DNA and other globular proteins, while the coming of Professor Andrade as Director in 1951 brought R. King to study metals and U. W. Arndt to develop Geiger and proportional counters for intensity measurements.

Since 1954, when Sir Lawrence came from the Cavendish Laboratory to take the position formerly held by his father, the laboratories have developed very much along these three lines: the study of metal structures (illustrated by the delightful Nye-Bragg bubble models); the building of equipment for the automatic recording of large quantities of crystal data; and with the help of A. C. T. North, D. W. Green, D. C. Phillips and Miss Helen Scouloudi, and in collaboration with the Cambridge Medical Research Council Unit, the successful attack on the sperm whale and seal myoglobin structures.

Less than 40 years from naphthalene and anthracene to the structures of complicated protein crystals: what a pity that Sir William did not live to see the latter! One feels that even at 100 years old he would still have been thrilled at this crowning achievement of the science that he helped to found.

17.8. Early Work on Crystal Structure at Manchester

by R. W. James

Manchester's connection with the diffraction of X-rays by crystals began very shortly after the discovery of the effect at Munich, when H. G. J. Moseley, working in Rutherford's laboratory, and following up W. H. Bragg's discovery that the characteristic X-ray spectra from the elements were line spectra, used a crystal as an optical grating to establish his famous relation between the characteristic frequencies and the atomic numbers of the elements. At the same time, C. G. Darwin, then Lecturer in Mathematical Physics in the laboratory, extended W. L. Bragg's treatment of diffraction by crystals, taking into account multiple reflections from plane to plane within the crystal, which Bragg had neglected. He showed that a crystal with small absorption, consisting of a perfectly ordered array of planes, should reflect the radiation totally over a very small range of angles, proportional to the amplitude scattered by a single crystal unit, and that the middle of this range occurred at an angle rather greater than that given by the simple Bragg law. He showed too that while total reflection was taking place the rays could penetrate only to a very small depth in the crystal, an effect known as primary extinction; and he obtained an expression for the refractive index of the crystal for X-rays *less* than unity by a few parts in one hundred thousand.

On Darwin's theory the integrated intensity as the angle of incidence varied through the reflecting range ought to be proportional to the first power of the amplitude scattered by a single crystal unit, but if multiple reflections were neglected it should be proportional to the square of this quantity. Moseley and Darwin made some measurements with rocksalt and white radiation to try to test this point, and found results that seemed rather to support the simpler theory. This work was published in 1913 and 1914 in two remarkable papers that laid the foundations of the quantitative measurement of the intensities of X-ray spectra, and drew attention at the very outset to what has always been the main difficulty in such work, allowance for the state of perfection of the crystal. Because of the war, the implications of Darwin's work were not at once appreciated, and meanwhile Ewald had handled the same problem in a more fundamental manner in his dynamical theory; but his papers too were overshadowed by the war, and did not become well known in England until some years after their publication.

In 1919 W. L. Bragg succeeded Rutherford in the Manchester chair, and one of his first objects was to put reflection of X-rays by crystals on a proper quantitative basis. With this end in view he, with R. W. James and C. H. Bosanquet, began a series of measurements of the absolute intensity of reflection of X-rays from rocksalt, a crystal whose structure was definitely known, with no uncertain parameters. In this way it was hoped to test the applicability of the reflection formulas. It was realized, moreover, that in view of the relation of the wavelength of the radiation to the dimensions of the atoms the amplitude scattered by an atom would not be proportional at all angles to the number of electrons it contained, but would decrease with increasing angle of scattering, merely as a consequence of the increasing phase differences between the contributions scattered by the different parts of the atom. The measurement of the so-called *f*-factor, the ratio of the amplitude scattered by an atom to that scattered by a single classical electron under the same conditions, was one of the aims of these experiments; for, in the first place, from its angular variations it was hoped to get direct optical evidence of the distribution of the electrons in the atoms, and secondly, it was clear that information about the way in which the scattering powers of different atoms depended on the angle of scattering would be essential if any but the simplest crystals were to be analysed.

The apparatus available for this work was by modern standards primitive. An ionization spectrometer was used, one of the original instruments constructed in W. H. Bragg's laboratory at Leeds, which was to have another good fifteen years of useful life at Manchester. The source of radiation was at first a gas tube, excited by an induction coil with a Wehnelt interrupter; and conditions were often so unsteady that it was impossible to obtain readings of any real value. A little later, when the gas tube was replaced by a Coolidge tube, and the induction coil by a more suitable transformer, reproducible results of fair accuracy were obtained. The integrated reflection was measured by a method first suggested by W. H. Bragg, in which the crystal was rotated through the reflecting range with a uniform angular velocity, the total ionization produced in the chamber during the rotation being taken as a measure of the intensity. The lead-screw of the crystal table was fitted with a capstan-head with four spokes, and this was turned by the index finger of the observer, one flick of the capstan to each beat of a metronome, a simple device that proved surprisingly adequate. For absolute measurements the radiation had to be made monochromatic by reflection from a crystal, for which purpose resorcinol was used in these experiments, and then reflected a second time from the rocksalt crystal. After two reflections the intensity was small, and only strong spectra could be measured in this way.

The rocksalt crystals used in these experiments were found to reflect very nearly according to the simple formula that neglected multiple scattering, and the more intense spectra were as much as twenty or thirty times stronger than the Darwin perfect-crystal formula indicated. It was noticed, however, that these strong intensities varied considerably from specimen to specimen. A freshly cleaved face might give an abnormally low reflection, but this could perhaps be increased eight or ten-fold by grinding the face on fairly coarse emery paper, and it was found too that as a result of this treatment the intensities from different specimens tended to approach the same limit.

It appeared that, if crystals were so imperfect that exact regularity did not persist over a large enough region for the Ewald-Darwin dynamical field to be set up within it, the formula that neglected multiple reflection still applied. Ewald suggested the name *mosaic* for crystals of this kind, and the experiments suggested that the Manchester rocksalt crystals approached this type closely. But even so, the strong spectra were weaker than they should have been, and this was ascribed to the shielding of regions deeper in the crystals by nearly parallel regions above them, which reflected away radiation that would otherwise have reached them. This shielding effect was reduced by grinding, which presumably reduced the degree of parallelism of the different regions, but it could not be entirely removed. Darwin called this effect *secondary extinction*, although it is of course of quite a different nature from primary extinction in perfect crystals. It proved possible to estimate the enhanced linear absorption due to this effect, and to some extent to correct for it; and ultimately a set of absolute *f*-curves for sodium and chlorine were obtained which were in fair agreement with what was to be expected from what was known at the time of the electron distribution in these atoms. Bragg, James and Bosanquet published these results in three papers that appeared in 1921 and 1922, and their implication in terms of crystal perfection was discussed by Darwin soon afterwards.

The choice of the very imperfect crystal rocksalt for these experiments was fortunate, for it led to fairly unambiguous results; but it was soon clear that not all crystals behaved in the same way. W. H. Bragg, for example, observed the reflections from a diamond to be nearly proportional to the structure factors, and not to their square as the mosaic theory required. There was considerable lack of understanding among many practical workers of the implications of the dynamical theory, which was not easy to read; and there was a corresponding ignorance on the part of the theorists as to how much crystals did actually reflect, as distinct from what, on certain assumptions as to their nature, they ought to reflect. Ewald, who had kept in touch with the work at Manchester and elsewhere, realizing this, organized in September 1925 an informal conference at Holzhausen in Bavaria, consisting of about a dozen members, and in a week's discussion a great advance in the general understanding of this rather difficult subject was made. It became clear that the perfect crystal is rather rare, that most crystals are neither perfect nor mosaic, but something between the two, and that the most reliable test of perfection or imperfection is probably the intensity with which they reflect X-rays. The problem, if not solved, had become defined. It seems proper to mention this conference in discussing the Manchester work, for the Manchester measurements had a good deal to do with its being held, and the Holzhausen discussion certainly had a great influence on later work there. It may be interesting to mention the members of the conference. They were P. P. Ewald, M. von Laue, W. L. Bragg,

P. Debye, C. G. Darwin, L. Brillouin, H. Mark, K. Herzfeld, I. Waller, H. Ott, A. D. Fokker, R. W. G. Wyckoff and R. W. James.

In 1925, D. R. Hartree, afterwards Professor of Applied Mathematics at Manchester, but then still at Cambridge, calculated average electron distributions for sodium and chlorine atoms based on the Bohr orbit model of the atom. The *f*-curves calculated from these distributions fell away less uniformly and less rapidly than the measured curves, and the effect was one not to be explained by an imperfect knowledge of the temperature factor. It was in fact mainly due to the concentration of charge density at certain radii produced by averaging the circular orbits of the Bohr model.

In the same paper Hartree estimated the dimensions of the atomic orbits for a number of ions, and *f*-curves based on these were used by James and W. A. Wood in a determination of the structure of barytes, published in 1925. There is little doubt that the obvious need of crystallographers for reliable information about *f*-factors had considerable effect in directing Hartree's attention to the calculation of atomic electron distributions, and so led him to devise the method of the self-consistent field.

The effect of the thermal vibrations of the atoms in reducing the intensities of X-ray spectra had been pointed out by Debye, and demonstrated experimentally in 1914 by W. H. Bragg. James continued the experimental work on rocksalt by measuring the intensities of a number of spectra from room temperature to 600°C in 1925, and in 1926 and 1927, with Miss E. M. Firth (Mrs. W. Taylor), extended the measurements to the temperature of liquid air. By this time Schrödinger's theory had been developed, and Hartree had calculated the atomic wave-functions of sodium and chlorine by the method of the self-consistent field. Wentzel had shown that for the usual X-ray wave-lengths the scattering of radiation from an atom could be obtained by treating the Schrödinger charge distribution as a classical charge distribution, and Waller had extended Debye's theory of the temperature effect. A detailed comparison with theory was therefore possible. In papers published in 1927, which had their origin in a visit to Manchester of Ivar Waller of Uppsala, he with Hartree and James showed that there was good absolute agreement between the measured and calculated *f*-curves, provided that in allowing for the temperature correction the existence of zero-point energy was assumed; and these results were confirmed by James, Brindley and Wood with potassium chloride, and with aluminium.

Concurrently with the quantitative work a considerable amount of structure determination had been going on in the laboratory, and in due course two main lines of investigation developed, those on the structure of the silicates and the structure of the alloys. Bragg insisted that structure determination ought to be considered primarily as a physical problem, and not merely as a geometrical one, and that, to make progress, relevant information of all kinds should be sought and used. A structure determination was in those days usually a matter of trial and error, and success was likely to depend on the skill and physical insight displayed in guessing an initial structure. That Bragg himself possessed this particular skill in an unusual degree was a great factor for success in the work. Ideas of what was a suitable crystal for a structure determination were still largely governed by whether it could be obtained in specimens large enough for use on the spectrometer, and whether it had a high symmetry. It was customary to grind faces on the crystals if they did not occur naturally. Another technique in which a slice of crystal was mounted on the spectrometer in such a way that it could be rotated in its own plane was often used, and in this way, by reflecting through the slice from planes perpendicular to its surface, the intensities of spectra round a zone axis could be compared with considerable accuracy. Photographic methods came into use comparatively slowly, and were mostly limited to oscillation photographs, first with flat photographic plates, and afterwards with cylindrical cameras.

Bragg laid stress on the idea that an atom in a crystal had a characteristic size, and that in deciding on likely structures packing had to be taken into account. He encouraged his pupils to make models, and to see how best the available material would fit into the available space. A structure ought to look sensible, to be so to speak a good engineering job. In 1920 he showed that in many crystals interatomic distances obeyed a simple additive law, and these results were reinterpreted in 1923 by Wasastjerna in terms of atomic sizes, and checked by atomic refractivities. To the same period belong papers by Bragg on the refractive indices of calcite and aragonite, in which the double refraction is calculated by taking into account the varying interaction in different directions of the atomic dipoles produced by the optical electric fields.

The realization that the refractivity in such compounds was due principally to the relatively large and easily polarizable oxygen atoms was of influence in the early development of the work on silicates; for it was seen that many of the simpler silicates could be regarded very nearly as an array of close-packed oxygen atoms, with the relatively small metallic ions tucked in the crannies between them, the crystal as a whole having a refractive index not very different from that of a close-packed array of oxygen. Bragg saw that the ruby, Al_2O_3, not of course a silicate, could be regarded in this way, and the same idea was of importance in some of the earlier silicate analyses. An interesting example was cyanite, which although triclinic, was seen to be essentially, so far as packing was concerned, a close-packed array of oxygen.

From 1926 onwards a long series of papers by Bragg himself, by his own Manchester pupils, and by many workers from laboratories abroad, issued from the laboratory. It was found that the silicon atoms always lay at the centres of tetrahedra formed by four oxygen atoms. In the orthosilicates these were independent groups, but sometimes by sharing oxygen atoms they might form rings, or endless chains, or ribbons in which two parallel chains shared oxygen atoms, as in the fibrous minerals like asbestos; or they might form sheets of linked hexagonal rings, as in the flaky minerals like mica or talc; or they might form cage-like structures in three dimensions, as in the felspars and zeolites. These extended groups are in effect extended negative ions, and valency requirements must be fulfilled when they build into a structure by including suitable positive ions. The extreme variability in composition of the silicates is accounted for by the fact that in these groups a certain number of silicon atoms may be replaced by aluminium, which alters the effective valency of the group, and allows a corresponding variation in the number and nature of the metallic ions in the structure. Bragg showed as a result of this work that the chemistry of the silicates was a chemistry of the solid state, intelligible only in terms of the three-dimensional structures. Quite precise valency requirements have none the less to be obeyed, and this was well understood and used by the Manchester workers sometime before the full development of the idea by Pauling.

These structures were at the time among the most complex that had been attempted, and a paper by W. L. Bragg and J. West, who took a large share in the silicate analyses and in the training of those who worked on it, entitled 'A Technique for the X-ray Examination of Structures with Many Parameters', was published in the *Zeitschrift für Kristallographie* in 1928, and summarized the methods then in use in the laboratory.

A very important development in 1929 was the introduction by Bragg of the method of two-dimensional Fourier synthesis, which was first applied, as an illustration, to the analysis of one of the silicates, diopside, the structure of which had been obtained by other methods. The importance of the method was that it allowed all available measurements to be used in the determination of the structure, and it rapidly became and has remained, a standard method of crystal analysis. The work on silicates lasted from ten to twelve years at Manchester. A very active worker was W. H. Taylor, now Head of the Crystallography Department of the Cavendish Laboratory, who made the silicates and zeolites his special field and has continued to work on them.

The use of powdered-crystal methods in the laboratory was developed by A. J. Bradley and J. C. M. Brentano. The latter was interested in its development as a method of obtaining quantitative measurements of intensities, unaffected by secondary extinction. Bradley was responsible for developing the Debye-Scherrer method and for applying it to a long series of determinations of alloy structures. He began work with a small and rather primitive powder camera and a Shearer gas tube, and with this he determined the structure of the hexagonal crystal lithium potassium sulphate, making perhaps the first attempt to allow for the decrease of scattering power of oxygen with increasing angle of scattering. The structures of arsenic, selenium and tellurium followed in 1925, and this led on to work on the structure of alloys that occupied Bradley and his fellow workers for the next twelve or fourteen years. He greatly improved the powder technique, and with A. H. Jay in 1932, made it into a precision method for the determination of lattice spacings. The work on alloys was helped by Messrs. Metropolitan Vickers of Trafford Park, who installed a vacuum induction furnace for the preparation of the alloys, and a number of members of their research department, among them Dr. C. Sykes, worked from time to time in the laboratory.

Notable achievements were the determination of the structure of γ-brass, with 52 atoms to the unit cell, and the recognition of the relation of the γ-phase to the simpler β-phase in such alloys; and of the structure of α-manganese, with 58 atoms to the unit cell, by Bradley and Thewlis. These, and the later determinations of the structures of phosphotungstic acid by Bradley and Keggin, and of phosphomolybdic acid by Bradley and Illingworth, were notable examples of what can be done by the powder method in analysing a complicated cubic structure. The iron-aluminium superlattices were investigated by Bradley and Jay, the Heusler alloys by Bradley and Rodgers, the nickel-aluminium system by Bradley and A. Taylor, and the chromium-aluminium alloys by Bradley and Lu. Later important work was that on the ternary alloys and the relation of the lattice structure to the magnetic properties. Professor Bragg took a deep interest in this work on alloys, and it led to a much better understanding of the nature of an alloy and the significance of the different phases, and in this connection there were many helpful discussions with W. Hume-Rothery, who visited the laboratory from time to time. Bradley too spent some time in Sweden in Westgren's laboratory where similar work was in progress.

In 1934 Bragg and E. J. Williams discussed theoretically the effect of thermal movement on the atomic arrangement in alloys, and problems of annealing and kindred subjects, in three papers, one of which formed the subject of a Bakerian Lecture to the Royal Society.

Two Liverpool students, C. A. Beevers and H. S. Lipson, had been working at Liverpool on the structures of the sulphates of beryllium, copper, nickel, and cadmium, and also of the alums, all crystals with water of crystallization, the elucidation of which was an important piece of work, and had been in touch with the Manchester laboratory. In 1935 Beevers moved to Manchester, and continued his work on magnesium sulphate, and Lipson followed in 1936. During 1935 and 1936 they had worked out their method of summing Fourier series, and the wellknown Lipson-Beevers strips were prepared in the Manchester laboratory. Lipson was in due course to succeed W. H. Taylor as head of the Physics Department at the Manchester College of Technology, where he has become well known for his work on optical transforms and their application to structure determinations; and Beevers migrated to Edinburgh, where he has been the centre of an active school of X-ray work.

A feature of the Manchester school during these years was the large number of visiting research students who came to work in the laboratory from all over the world. Among them may be mentioned J. M. Bijvoet, F. W. H. Zachariasen, L. Pauling, B. E. Warren, I. Waller, J. A. Santos, H. Brasseur, F. Machatschki, Sc. Naray-Szabo, H. Strunz, I. Fankuchen, E. Onorato and O. R. Trautz.

Whenever a structure had been determined in the laboratory a model was, if possible, made of it, and various types were fashionable from time to time. In the earlier days, when crystals were still relatively simple, packed spheres often made of dental wax were used. The silicate models were usually constructed by cutting thin glass tubing into lengths equal to the interatomic distances, stringing them together with thin wire to form the structure, and showing the positions of the atoms, without regard to size, by balls of coloured wax at the junctions of the glass tubing. These models made an impressive show on days when exhibitions of the work of the University were held. They were the trophies of the school, and the memory of them must linger in the minds of all who were privileged to work under Bragg's leadership during those early years.

CHAPTER 18

The Development of X-ray Diffraction in U.S.A.

18.1. The Years before 1940

by Ralph W. G. Wyckoff

The investigation of crystal structure in the United States belongs to two very different periods, a pre- and a post-Second World War epoch. Before this war, and its urgent needs for applied crystallographic information, crystals were investigated for their structure in only a few places, and with one or two notable exceptions this scattered work was carried out with too limited financial support to allow long-range programmes of research. Since the war and the general appreciation of the usefulness of structural information it brought, this situation has drastically improved. We now have vigorous and effective programmes being carried out in many of our better universities as well as in large industrial laboratories; and crystallographic knowledge is being widely accepted as an essential ingredient of modern natural science. Much of this rapid, recent expansion has roots in our earlier, restricted activities and it is with these roots that this preliminary discussion deals.

The concern with crystal structure has never had in America the kind of unity which it has had in Great Britain as a natural outgrowth of W. H. Bragg's early preoccupation with the nature of X-rays. Here the first people to use the experiments of von Laue and the Braggs were interested in specific applications. American physicists dealing with X-rays were not attracted by the possibilities for a new knowledge of the solid state of matter which these X-ray experiments had opened up, and neither were our mineralogists.

To the best of my knowledge, the first American investigations of structure were carried out independently of one another during the period of the First World War. One began at the Massachusetts Institute of Technology, one in the Research Laboratory of the General Electric Company at Schenectady and the third at Cornell. The first two were stimulated by direct contact with the Braggs, the third had a different origin.

In the personal reminiscences of C. L. Burdick (Part VII) we read how, after spending some time in the laboratory of W. H. Bragg, he returned to carry through an analysis of chalcopyrite with the help of J. H. Ellis. This was when A. A. Noyes, at whose instigation this X-ray work had been undertaken, was retiring from the Massachusetts Institute to build up a chemistry department in the newly organized California Institute of Technology. Though neither Burdick or Ellis continued to work with X-rays, Ellis remained for some years at the California Institute where he retained an interest in the further development of the subject. The first studies of structure at Pasadena centered around Dickinson who came with Noyes from Cambridge. He carried through a number of structures, partly alone and partly with students. Most of these men did not continue in crystal structure but one who did was Pauling who took over when Dickinson's research interests turned to other matters and whose problems have since determined the pattern of X-ray research at the California Institute.

In his reminiscences (Part VII) Hull recounts how a visit of W. H. Bragg to Schenectady led him to examine the structure of metallic iron and thus through the lack of single crystals to invent the powder method, independently of Debye and Scherrer. He also describes how, having used this method to deduce the structures of many of the commoner metals, he abandoned crystal structure as being a field too remote from the fundamental objectives of the General Electric Laboratory. Davey had, however, been working with him at Schenectady and had there obtained the structures of a number of chemically simple crystals using Hull's powder method. He went to Pennsylvania State College where for the next thirty years he devoted himself mainly to analytical uses of X-ray diffraction, establishing as part of this effort the A.S.T.M. powder diffraction file.

While Nishikawa was a visitor in the department of physics at Cornell, I began under him in 1917 a thesis on the structures of sodium nitrate and cesium dichloroiodide. We had no X-ray spectrometer and our data were drawn from single-crystal photographs, both spectral and Laue, and were interpreted with the aid of space-group theory. Before coming to the United States Nishikawa had used these methods to establish several structures. He had learned of space-group theory from a Japanese professor who twenty-five years earlier had been working in Germany when Schoenflies was developing this theory. Communication with German-speaking Europe was at a low ebb immediately after as well as during the war and we had been using space-group results for several years before becoming aware, through

Niggli's *Geometrische Kristallographie des Diskontinuums*, of a rebirth of European interest in the theory.

On getting my degree I went to the Geophysical Laboratory and thus it came about that of the three original studies of structure in America only one had a continuing existence in a university. In this way the California Institute assumed from the outset a dominant position both as a center of crystal structure research and as a source of trained personel, able, when opportunities later arose, to initiate new centers in other institutions; and the fact that this activity was in the chemistry department is one reason why so much of subsequent crystal structure in this country has been under the sponsorship of chemistry.

Though none served as the basis for a developing school of crystal structure, several physics departments of American Universities were in the 1910's and 1920's actively studying X-rays, and their work has contributed in important respects to the ultimate growth of our subject. Thus Bergen Davis and his students at Columbia were engaged for many years on fundamental problems of X-ray production and its quantitative measurement. Duane's school of X-ray physics at Harvard touched closely on problems of crystal structure. In the early 1920's he became interested in the use of Fourier methods to study the electron distribution in atoms and Havighurst's work with him was a pioneering application of these methods to the determination of electron distributions in the atoms of crystals. Allison is another of Duane's students. And so is G. L. Clark who went to Illinois to become the chief American exponent of applied X-rays. The other American school of that time whose work has strongly influenced the course of crystal analysis was that of A. H. Compton at Washington University and at Chicago. It requires only a glance at his book with Allison to appreciate how much of the quantitative measurement that underlies all modern structure analysis is based on his investigations and those of his school.

Early in the first half of the 1920's there was little extension of crystal structure work in this country. Nevertheless it was during this period that Mc Keehan worked on metallic systems at the Bell Telephone Laboratories and that Davey and Clark began to apply X-ray diffraction to practical problems.

A second stage in the development of crystal analysis in the United States began during the later 1920's and early 1930's with the gradual appearance of more opportunities for research. People to start this new work came from both the California Institute of Technology and from among those who had been gaining experience abroad, mainly with the Braggs. Thus Warren, returning from work with W. L. Bragg, started the investigations in the physics department of the Massachusetts Institute of Technology which he has been continuing ever since. Patterson came via McGill in Canada from a stay in W. H. Bragg's laboratory at the Royal Institution. Also during this period Jette was working on metals at Columbia after being with Phragmén in Sweden. During the 1930's Zachariasen came to the physics department at Chicago from Goldschmidt's laboratory after a stay with W. L. Bragg.

Several workers trained at the California Institute of Technology established laboratories for further X-ray work during this period. Among the more senior of these is Hendricks whose laboratory in the Department of Agriculture dealt for years with the structures of mica and related minerals as well as with problems of electron diffraction. Another is Huggins who has recently retired after many years at the Eastman Kodak Laboratories. More recent people from Pasadena who set up laboratories in the years before the second war are Hoard who went to Cornell, Brockway in Michigan and Hultgren in Harvard, Harker now in Buffalo.

During this period there also sprang up in several universities groups which took their origin less directly from any of the older schools here or abroad. One of these, initiated by F. C. Blake (an earlier student of Duane), was active for several years at Ohio State; Havighurst had worked with him before going to Harvard to study with Duane, and Klug also started with him. Another programme of this period, due to Gruner in Minnesota, seems to have been the first to have been started in the mineralogical department of an American university; for a decade beginning in the later 1920's he studied the clays and other minerals. Ramsdell at Michigan and Pabst at California began working at this time and so did M. J. Buerger, whose work in the mineralogy department at Massachusetts Institute has, however, always reached beyond a preoccupation with minerals.

18.2. From the Beginning of World War II to 1961

by Elizabeth A. Wood

Although publication dwindled almost to nothing during the 1940's because of the war, there was a great deal of crystallographic research going on which was to come to light in a flood of post-war publication.

Neutron Diffraction

Because of nuclear research during the war, neutron beams from atomic piles became available. The phenomenon of the diffraction of a neutron beam by a crystal, predicted by Elsasser in 1936, had been demonstrated in the same year by Mitchell and Powers of Columbia University and by Halban and Preiswerk (reported in Comptes rendus) using a radium-beryllium neutron source.

With the stronger beams from atomic piles, monochromatization became feasible and, with it, quantitative neutron diffraction work on crystals. The horizons opened up by the existence of a set of atomic scattering amplitudes entirely different from those for X-rays (even including negative scattering amplitudes) were quickly recognized by Shull, Wollan, et al. of Oak Ridge National Laboratory and pointed out in such papers as the discussion of NaH and NaD by Shull, Morton and Davidson in 1948.

In addition, the fact that the neutron has a magnetic moment made possible the determination of the arrangement of magnetic dipoles in magnetic materials. Shull and Smart's pioneer work in 1949 at Oak Ridge on the magnetic structure of MnO has been followed by extensive structure work by Henri Levy, W. C. Koehler and Michael Wilkinson at Oak Ridge and also by Lester Corliss and Julius Hastings of Brookhaven National Laboratories.

5f Series of Elements

Another beneficial result of the nuclear research effort was the attention paid by W. H. Zachariasen of the University of Chicago to the 5f series of elements and their compounds. From his research on these substances, there resulted a series of papers in *Acta Crystallographica* extending over 11 years.

Computing Machines

Another important contribution to crystallography traceable to advances made during the war years was the development of powerful electronic computing machines and all their associated automation of data recording, storage and transfer.

In this country one of the first to make advantageous use of this development was R. Pepinsky who in 1947 produced XRAC, the X-ray analogue computer which showed a contour map of the projected electron density on an oscilloscope screen when the Fourier coefficients were set on a system of potentiometer dials. More recently 'programs' have been written which will instruct the various commercially available digital computing machines to do a large assortment of crystallographic chores.

David Sayre in 1945 devised the first least-squares program to be available as a package.

Punched cards came into use in the forties. Effective use of these was made in 1950 by David P. Shoemaker, Jerry Donohue, Verner Schomaker and Robert B. Corey in their three-dimensional refinement of the L-threonine structure.

S. C. Abrahams and Emanual Grison enlisted the aid of the M.I.T. computer *Whirlwind* in the solution of the structure of cesium hexasulfide in 1953 while they were in the Laboratory for Insulation Research at the Massachusetts Institute of Technology.

The widespread sharing of programs that has contributed to the rapid increase in the use of high-speed computers for crystallographic work in the United States is due partly to the personal generosity of individuals who have devised programs and partly to the far-sightedness of the companies who produce the computing machines.

Automatic Data Collection

Automation of computation has inevitably led to automation of data collection. A machine was devised by W. L. Bond and T. S. Benedict in 1955 which will seek rapidly for a reflection and, finding it, record its azimuth and elevation relative to the crystal axes and its integrated intensity.

Subsequently X-ray diffractometers with automatic crystal setting and data recording have been designed by Diamant, Drenck and

Pepinsky of Pennsylvania State University, by J. Ladell of Philips Laboratories and by S. C. Abrahams of the Bell Telephone Laboratories.

Soon such intensity data will be fed directly into the calculating machine and the next contact of the crystallographer with the problem may be his examination of the three-dimensional Patterson diagram!

In the field of neutron diffraction automation is especially important. Because of the long time required to scan a reflection, it is desirable that the diffractometer operate without interruption and without requiring the presence of an operator.

H. A. Levy and S. W. Peterson at Oak Ridge National Laboratory were using an automatic single crystal neutron diffractometer throughout the late 1950's. Langdon, Frazer and Pepinsky described the design of a new type of neutron diffractometer in 1956, and in 1959 Langdon and Frazer published design details of an instrument they had built and tested in which probably the most important new feature was the use of digital angular settings. This instrument has now been modified for tape input and output and made more flexible. Meanwhile, E. Prince and S. C. Abrahams have been using a single crystal automatic neutron diffractometer with paper tape input and output which was designed by them.

Phase Determination

The availability of high-speed computing facilities has made possible the investigation of structures that might not have yielded to the efforts of one individual in a lifetime of old style computation. Partly as a result of the tackling of more difficult structures urgent attempts were made to solve the problem of determining the phases of the various scattered rays which have to be known for a Fourier synthesis of the electron density.

The attempts to use such information as the fact that the electron density is nowhere negative had their first great success in the work of David Harker and John Kasper of the General Electric Research Laboratories in 1948. They applied inequality theorems well known to mathematicians. Although their method has limitations, its publication stimulated thinking on the subject. Subsequent papers pointing out the application of inequalities were published by David M. Sayre and W. H. Zachariasen in 1952, the latter introducing the use of equations that are only statistically true. H. Hauptmann and J. Karle of the Naval Research Laboratory, using the statistical approach and working with joint probabilities concerning several intensities, feel that in this approach lies the true solution to the phase problem. Several structures have been solved with the assistance of the Hauptmann and Karle method but it has not yet come into widespread use.

Patterson Function

The limitations inherent in the Patterson method of vector representation of the data, published in 1934 by A. L. Patterson then at Bryn Mawr College, have not prevented its widespread, almost universal use. Computer programs have made possible the rapid construction of three-dimensional Patterson diagrams. This, along with systematic procedures for extracting the maximum amount of information from the vector representation, will result in even greater use of the Patterson function. This subject has been developed over a period of time by M. J. Buerger of Massachusetts Institute of Technology in a series of papers and in 1959 in a book called *Vector Space*.

It was Dr. Dorothy Wrinch of Smith College who suggested in 1939 techniques which were a powerful aid in deducing the atom positions from a knowledge of the Fourier Transforms of special groupings and their Patterson representations.

'Least Squares' Analysis

In speaking of analytical techniques, the important contribution o E. W. Hughes of the California Institute of Technology should be mentioned. He was the first to introduce, in 1941, the use of the 'least squares' fitting of the data as used by statisticians. In 1946 Hughes and Lipscomb further pursued this attack in applying error theory to least squares analysis.

One of the most powerful computer programs today is a three-dimensional least-squares program devised by W. R. Busing and H. A. Levy of Oak Ridge National Laboratory.

Structure Analyses

The increasing complexity of the organic structures which have succumbed to solution has been mentioned. A plot of their cell content versus year of successful solution anywhere in the world has been published by Jerry Donohue of the University of Southern California. It has a slope of 30 atoms per decade, but every time a new high speed computer comes out, this slope increases. Jerry Donohue belongs to the impressive group of structure analysts that has fanned out in all directions from the California Institute of Technology at Pasadena where the dynamic teacher, Linus Pauling, has inspired the crystallographic research in the Chemistry Department. Beyond this department, Pauling's influence has been widely felt through his book *The Nature of the Chemical Bond* which has had an important effect in stimulating thought on this subject throughout the field of crystallography. Some of his students, such as J. L. Hoard at Cornell, R. E. Rundle at Iowa, W. N. Lipscomb at Minnesota and later at Harvard, and David Shoemaker at Massachusetts Institute of Technology, have established similar fountainheads elsewhere in the country. J. L. Hoard and his students have specialized in complex inorganic structures, including boron and its compounds. R. E. Rundle's group has worked on many inorganic and metallic structures.

Proteins and Related Structures

The concept of the complex spiral structure in proteins introduced by Linus Pauling was an important step toward the understanding of protein structures and resulted in intensified work on this subject at the California Institute of Technology.

During the 1950's David Harker and a small group working with him at Brooklyn Polytechnic Institute concentrated on the protein ribonuclease and related structures. Murray Vernon King of this group developed new techniques for producing, 'staining' and maintaining single crystals of these difficult substances for structure analysis. David Harker and T. C. Furnas Jr., developed the use of the 'Eulerian cradle' for taking complete diffraction data from a single crystal without remounting it, working always in the equatorial zone, a technique of special importance for work on the large-celled organic substances.

Low-Temperature Work

In the middle forties I. Fankuchen and his colleagues developed an 'in-place' method of growing single crystals in a capillary while it was at the center of the X-ray camera by using a regulated stream of cold air together with a heater for remelting undesirable crystals (Kaufman, Mark and Fankuchen, 1947; Kaufman and Fankuchen, 1949; Post, Schwartz and Fankuchen, 1951). W. N. Lipscomb and his school have done pioneering research in low-temperature diffraction work. Some of the early techniques in this field were reported in 1949 by Abrahams, Collin, Lipscomb and Reed. Since 1950, when Collin and Lipscomb published their paper on hydrazine at low temperatures, much work has been done at temperatures down to 78°K, not only at Brooklyn and Minnesota, but at an increasing number of other centers of research.

The work comprises not only determinations of crystal structures of substances liquid at room temperature (Fankuchen, et al., v.s.) and phase-transition studies, but it also holds promise of very accurate structure determinations on substances in which thermal motion presents difficulties at room temperature, that is to say, most substances. Indeed the difficult problems associated with the boron-hydrogen groups whose complexity was pointed out in the decaborane paper in 1950 by Kasper, Lucht and Harker have received much light from the low-temperature structure work done on boron hydrides by Lipscomb and his students. Very low temperatures, in the liquid helium range, have been used by Charles S. Barrett in his studies of metal structures.

Both at Oak Ridge and at Brookhaven single crystal and powder neutron diffraction work has been done at temperatures as low as 1.4°K.

Ferroelectricity and Ferromagnetism

R. Pepinsky, Y. Okaya and others at Pennsylvania State University have without a doubt determined the structure of more ferroelectric and piezoelectric crystals than any other single group. It is not the least of his contributions that Pepinsky has brought to this country many students from all over the world whose interaction with our crystallographers has resulted in mutual enrichment.

Among the other laboratories especially interested in ferroelectric and ferromagnetic crystals has been the Laboratory for Insulation Research at the Massachusetts Institute of Technology where A. von Hippel has directed a series of young research people, many of them visitors from abroad, and the Bell Telephone Laboratories. Of special interest in crystal chemistry and ferrimagnetism as well as for practical applications is the work of S. Geller at Bell on a wide range of crystals with the garnet structure in which the site preference of various ions has been demonstrated.

Electron Diffraction

It was at the Bell Telephone Laboratories in 1927 that C. J. Davisson and L. H. Germer showed that a beam of electrons may be diffracted by a crystal just as an X-ray beam is diffracted, thus demonstrating the wave nature of physical particles and at the same time making available a tool especially valuable for the study of thin surface films and isolated molecules. Germer and his colleagues K. H. Storks and A. H. White proceeded to apply the new tool to the study of the structure of forms of carbon, of soaps and polymers. The folding of polymer chains was suggested in 1937 by Storks to explain the electron diffraction effects he had observed.

The use of electron diffraction for the study of interatomic distances in gas molecules has been the primary interest of Lawrence O. Brockway of the University of Michigan who, particularly in the forties, showed the power of electron diffraction in giving us information about complex organic molecules. His student and collaborator, Isabella Karle, whose husband's work on the phase problem is mentioned elsewhere in this history, has carried on the investigation of molecular structure at the Naval Research Laboratory. Verner Schomaker and his students at the California Institute of Technology, and S. H. Bauer and his students at Cornell have made major contributions to the investigation of gas molecules by electron diffraction.

Because of their shallow penetration, diffracted electrons give us information about the atom layers near the surface of a solid specimen which is not available from the diffraction pattern from the more deeply penetrating X-rays. The extensive, careful research by E. A. Gulbransen of Westinghouse Electric Corporation on thin films of oxides and other tarnishes on metal surfaces has lead to a better understanding of corrosion processes and so has the work of Lorenzo Sturkey of the Dow Chemical Company.

At the University of Virginia, A. T. Gwathmey and K. R. Lawless have used electron diffraction to study the epitaxial relations between surface coatings and substrate metals.

R. D. Heidenreich of the Bell Telephone Laboratories who worked with the electron microscope in its early years of development was a pioneer in 1946 in the use of transmission electron diffraction for the study of thin metal films.

All of these workers and many others who came into the field of electron diffraction from electron microscopy are now using the very powerful selected-area methods which combine electron microscopy and electron diffraction to give detailed structure and compositional information about thin films and surfaces.

The use of very slow electrons, accelerated by a few tens of volts, was early suggested by Germer for the investigation of surface structures and was for many years pursued persistently by F. E. Farnsworth of Brown University who detected the weak diffracted beams with a Faraday cage. Recent investigations using a postdiffraction accelerating technique first suggested by W. Ehrenberg in 1934 are yielding information about the structure of surfaces that are completely free of any foreign atoms and also about quantitatively controlled deposits on these surfaces made possible by present day high-vacuum techniques. L. H. Germer, collaborator in the initial electron diffraction experiment in 1927, is one of the most active investigators in this field today.

Liquids

Short-range order which has been investigated by means of electron diffraction in gases has also been studied in liquids and glasses by means of X-ray diffraction. B. E. Warren of Massachusetts Institute of Technology did some of the early work in this field, perhaps led into it by his pioneer work on the silicates, some of which was in collaboration with Sir Lawrence Bragg. Newell Gingrich, working initially with Warren, has continued his X-ray diffraction studies of liquids at the University of Missouri.

Metals

Recently, B. E. Warren with his many students has been making most careful quantitative analysis of diffraction effects other than sharp Bragg reflections, in particular as they relate to information about lack of perfection in metal structures.

C. S. Barrett of the Institute for the Study of Metals of the University of Chicago has, perhaps more than any other one person, brought X-ray diffraction into the field of metallurgy with his studies of deformation and transformation of metals over a wide temperature range, his investigations of preferred orientation and especially through his book *Structure of Metals* written primarily for the metallurgist.

Because of the widespread practical applications of metals, many of the workers of metals crystallography have emphasized the applied aspect of their work. However, the theoretical approach by Linus Pauling toward giving a unified explanation of known metal structures has stimulated structural research in metals.

Norman C. Baenziger and his students at the State University of Iowa have determined the structures of a large number of intermetallic compounds and alloys with special attention to those involving uranium and thorium.

Teaching

A teacher of many crystallographers at Massachusetts Institute of Technology, M. J. Buerger is perhaps more gratefully regarded by crystallographers for his invention in 1944 of the precession camera than for any of his many other contributions, although the equi-inclination Weissenberg technique introduced by him has also saved hours of distress for many crystallographers. He is one of the editors of the *Zeitschrift für Kristallographie*. A conscientious teacher, he has produced in the last ten years four text books of crystallography and a fifth as co-author. These will be listed below.

X-ray crystallography seems indeed to have reached the text book stage in its fourth decade. The sudden spate of text books has included the following from the United States:

BOOKS

1. Azaroff, L. V. and Buerger, M. J. (Mass. Inst. of Technology): *The Powder Method in X-ray Crystallography*, McGraw-Hill, N.Y., 1958.
2. Barrett, C. S. (Carnegie Inst. of Technology): *Structure of Metals*, McGraw-Hill, N.Y., 1943.
3. Buerger, M. J.: *X-ray Crystallography*, Wiley, N.Y., 1942.
4. Buerger, M. J.: *Elementary Crystallography*, Wiley, N.Y., 1956.
5. Buerger, M. J.: *Vector Space*, Wiley, N.Y., 1959.
6. Buerger, M. J.: *Crystal-Structure Analysis*, Wiley, N.Y., 1960.
7. Clark, G. L. (Univ. of Illinois): *Applied X-rays*, McGraw-Hill, 4th edition, 1955.
8. Compton, Arthur H. and Allison, Samuel K. (Univ. of Chicago): *X-rays in Theory and Experiment*, Van Nostrand, N.Y., 1935.
9. Cullity, B. D. (Univ. of Notre Dame): *Elements of X-ray Diffraction*, Addison-Wesley, Reading, Mass., 1956.
10. Klug, H. P. and Alexander, L. E. (Mellon Institute): *X-ray Diffraction Procedures for Polycrystalline and Amorphous Materials*, Wiley, N.Y., 1954.
11. McLachlan, Dan, Jr. (Stanford Research Inst.): *X-ray Crystal Structure*, McGraw-Hill, N.Y., 1957.
12. Pauling, Linus: *The Nature of the Chemical Bond*, Cornell University Press, 3rd edition, 1960.
13. Zachariasen, W. H. (University of Chicago): *Theory of X-ray Diffraction in Crystals*, Wiley, N.Y., 1945.

Of the books that cannot properly be called text books, two deserve special attention because they are on the desk of every practicing crystallographer. They are R. W. G. Wyckoff's *Crystal Structures* and Donnay and Nowacki's *Crystal Data*. The staggering amount of work that the authors have put into these two books is vastly exceeded by the amount of work they have saved their readers.

A somewhat similar statement could be made concerning the work of the editors of the *Structure Reports* and the *International Tables for X-ray Crystallography* throughout the world. In the United States, C. S. Barrett of the Institute for the Study of Metals and Norman C. Baenziger of Iowa State University have been metals editors for the *Structure Reports* and John S. Kasper of the General Electric Laboratories was one of the two editors of Volume II of the *International Tables*.

The teaching of crystallography in American Colleges may take place in any one of several different departments and some of our best teachers manage to have one foot in each of two departments. In this category we have J. D. H. Donnay, professor of crystallography and mineralogy in the chemistry and geology departments of the Johns Hopkins University, and George A. Jeffrey, professor of chemistry and physics at the University of Pittsburgh.

Rose C. L. Mooney Slater was a pioneer in the teaching of crystallography at a women's college, Sophie Newcomb, Tulane University, and in addition has made important contributions to our knowledge of inorganic crystal structures. A. L. Patterson, who formerly trained crystallographers in the women's college of Bryn Mawr, is continuing to train young men and women at the Institute for Cancer Research in Philadelphia.

Dan McLachlan at the University of Utah, the Stanford Research Institute and now at Denver, has taught with that special sort of imagination and ingenuity that has characterized all of his work.

The teacher whose students are probably distributed in the most heterogeneous assortment of organizations is I. Fankuchen of Brooklyn Polytechnic Institute. This is partly because, in addition to the regular academic courses, he teaches an intensive short summer course for anyone who wants to learn about crystallography. He has taught enthusiastic doctors, dentists, newspaper science writers, high school students, and professors of physical chemistry in medical schools, among others.

Small-Angle Scattering

In 1938, Bernal, Fankuchen and Riley, working in University College, London, described X-ray diffraction measurements of spacings as large as 394 Å by use of techniques involving long exposure of a plate 40 centimeters away from the specimen, the technique later known as 'small-angle scattering'. In this country the technique has been refined by R. S. Bear and O. E. A. Bolduan at M.I.T. as well as by W. W. Beeman and his colleagues at the University of Wisconsin in studies of collagen fibers. I. Fankuchen and his students at Brooklyn Polytechnic Institute have applied small-angle scattering techniques not only to organic substances such as the polyamides, but also to the mineral fibers of chrysotile asbestos.

Minerals

Returning again to M. J. Buerger's laboratory at the Massachusetts Institute of Technology, we see it as a source of mineralogical crystallographers. It was in this laboratory that the structure of tourmaline was determined by Gabrielle Donnay and M. J. Buerger. Gabrielle Donnay has continued in mineralogical crystallography at the Geophysical Laboratory of the Carnegie Institution in Washington. She and her husband, J. D. H. Donnay of the Johns Hopkins University, have enriched the mineralogical crystallographic literature with many papers, both singly and jointly, and the data compilation literature with the book *Crystal Data.* In Washington, too, is Howard T. Evans Jr., also a former student of M. J. Buerger's and now associated with a very active mineral structure group at the U. S. Geological Survey under the leadership of Charles L. Christ. Evans, with the cooperation of M. E. Mrose, has accomplished what is probably the most complete structural analysis of the associated minerals in an ore deposit ever made. Working on the vanadium and uranium minerals of the Colorado Plateau ore deposits, they were able to achieve chemical analysis of the various species by means of structure analysis and by doing so to confirm the genetic role played by weathering that had been postulated by the petrologists.

At the University of California at Berkeley, A. Pabst and his students have studied the structures of thorium silicates and uranothorite.

The wide variety of fields in which X-ray diffraction is now being applied as a tool was emphasized by G. L. Clark of the University of Illinois who in the fourth edition of his book *Applied X-rays*, published in 1957, lists more than a hundred types of industries using powder diffraction in routine industrial analysis. He describes the X-ray diffraction control of the orientation of enormous numbers of quartz crystal oscillator plates as 'the largest scale single-crystal achievement in industry'. The magnitude of the achievement had to do not so much with the number of plates produced as with the development of techniques whereby totally unskilled workers could determine the orientation of a plate to $\pm$ 10′ by X-ray diffraction.

Powder Data Card File

A tool which has greatly aided the use of X-ray diffraction methods in industry is the powder data file jointly sponsored by the American Society for Testing Materials, the British Institute of Physics, and the American Crystallographic Association. It was through the hard work and personal sacrifice of Wheeler P. Davey of the Pennsylvania State University that this file came into existence and persisted through the difficult early years. Now, under the editorship of J. V. Smith, working with special assistant editors for each different crystal field and receiving contributions from large numbers of crystallographers, its usefulness is constantly increasing and receiving wide appreciation.

* * *

Several hundred nonacademic organizations are now using X-ray diffraction equipment. Among these are aircraft companies, hospitals, oil companies, paint manufacturers, museums, steel companies, police departments, glass manufacturers, rubber companies and even gas companies.

One of the most difficult powder diffraction experiments ever to be performed will be attempted in 1963 by William Parrish of the Philips Laboratories. Drawing on his extensive experience in instrument design, he has designed an X-ray diffractometer to be landed on the moon. With the cooperation of Clifford Frondel of Harvard University, the analysis of the data which this instrument will send back to earth may tell us more about the composition of the moon than man has ever known before.

The story of crystallography in the United States would not be complete without an acknowledgment of the important part that visitors from all over the world have played in enriching our crystallographic development. Many have been with us briefly, as André Guinier, Rudolph Brill, John Nye, Victorio Luzzati, Emmanuel Grison and Andrew Lang. Many others have stayed, making their home in the United States, so, for instance Peter Debye, Paul Ewald, Kasimir Fajans, Herman Mark, and William Zachariasen.

Every crystallographer who reads this historical essay will discover grave omissions that have been made. It would be quite impossible to tell the whole story of the extensive spread of diffraction work in the United States.

With our colleagues throughout the world we look forward with confidence to the great achievements that the next fifty years will bring.

CHAPTER 19

The New Crystallography in France

by J. Wyart

19.1. The Period before August 1914

At the time of Laue's discovery, research in crystallography was carried on in France principally in two laboratories, those of Georges Friedel and of Frédéric Wallerant, at the School of Mines in St. Étienne and at the Sorbonne in Paris, respectively. Jacques Curie, it is true, also investigated crystals in his laboratory and, together with his brother Pierre, discovered piezoelectricity which soon found extensive application in the measurement of radioactivity; but he had few students and formed no school. Among the research of Georges Friedel that on twinning is universally known and accepted, as are his studies of face development in relation to the lattice underlying the crystal structure. Frédéric Wallerant's best known work was on polymorphism and on crystalline texture. In 1912 both Friedel and Wallerant were deeply interested in the study of liquid crystals, a form of aggregation of matter only recently discovered by the physicist in Karlsruhe, O. Lehmann. Friedel's co-worker was François Grandjean, Wallerant's Charles Mauguin. Laue, Friedrich and Knipping's publication immediately drew their attention, and Laue's remark that the diagrams did not disclose the hemihedral symmetry of zincblende prompted Georges Friedel to clarify, 2 June 1913, the connection between the symmetries of the crystal and the diffraction pattern. If the passage of X-rays, like that of light, implies a centre of symmetry, i.e. if nothing distinguishes propagation in a direction AB from that along BA, then X-ray diffraction cannot reveal the lack of centrosymmetry in a crystal, and a right-hand quartz produces the same pattern as a left-hand one. This remark is well-known among crystallographers as Friedel's Law, and its more thorough proof has given rise to a fair number of theoretical papers throughout the years. Though usually true, Friedel's Law does not hold generally, even for non-absorbing crystals. But this does not impair the usefulness of the 11 classes of 'Laue symmetry' which Friedel enumerated.

Neither Friedel nor Wallerant had equipment for working with X-rays at their disposal. So it came about that in the hectic 1913/14 pre-war period Maurice de Broglie was the first in France to obtain X-ray diffraction. He was working in the laboratory of Paul Langevin at the Collège de France on the ionization of gases, so that he was well acquainted with the required technique. In his first communication to the Academy of Sciences (31 March 1913) de Broglie reports on results he obtained by the Laue-Friedrich-Knipping method with cubic crystals (ZnS, CaF_2, NaCl, magnetite). He studied the influence of temperature by cooling the crystals to liquid nitrogen temperature, and of magnetic fields (up to 10^4 gauss) in the case of magnetite. He finds that cooling neither increases the number of diffracted spots, nor sharpens them; that the magnetic field at right angles to the primary beam (which has the direction of a ternary symmetry axis) does not destroy the crystal symmetry; and incidentally, that the secondary rays which produce the diffracted spots, are not deflected by the magnetic field, as they would be were they electron beams.

In the following three months de Broglie continued using the Laue method and photographic recording; striations appearing in the diffracted spots were at that time under discussion both by de Broglie and others (Barkla, Hupka), because it was not clear whether they were to be explained as further diffraction fringes or as produced by irregular growth of the crystal.

De Broglie's first fundamental contribution came in his two notes of November and December 1913 introducing the rotating crystal method; this soon became one of the most useful methods for X-ray spectroscopy as well as crystal structure analysis. The first note announces a provisional camera in which the crystal was mounted on the drum of a recording barometer which gave it a rotation of 2° per hour. Evidently the clockwork was not running smoothly enough, for a number of lines appeared which had not been registered by W. H. Bragg or Moseley and Darwin. A double film technique was used for distinguishing by absorption between coincident first and higher order lines. In the second note successful double-sided registrations of the platinum and tungsten spectra are recorded, obtained on a variety of crystals. On rocksalt complete spectra were registered inside 15 minutes. An interesting remark in this paper is that the effect of temperature on the diffracted intensity depends on the order and not on the angle of diffraction.

This first paper on X-ray spectra by de Broglie opened up a long series of spectroscopic papers. Together with F. A. Lindemann, the later Lord Cherwell, he introduced the method of secondary excitation of the emission spectra of substances from which targets inside the X-ray tube could not be formed, by irradiating them with harder X-rays. He could hereby complete the systematic exploration of X-ray spectra throughout the periodic system, except for the elements of order below 30. On this occasion the absorption edges of the silver and bromine in the photographic emulsion were also observed for the first time (May 1914). Whereas the atomic weight of Te is smaller than that of I, the X-ray spectra show that Te has the higher atomic number.

Among the other papers by M. de Broglie focussing devices by reflection on bent mica flakes and diffraction obtained with thin metallic sheets, a precursor of the powder method, should be mentioned. This most fertile activity came to an abrupt end with the outbreak of war in August 1914 which put a stop to all research in this field in France for five long years.

19.2. The Period 1918–1950

M. de Broglie's Laboratory

When scientists went back to the laboratories after the end of the First World War, Maurice de Broglie was the first to organize X-ray research. He did this by transforming his private mansion in central Paris, not far from the Étoile, into a makeshift laboratory where he, together with a brilliant group of young collaborators ran their X-ray tubes in rooms still panelled with oak or hung with gobelins. The main line of de Broglie's own work remained X-ray spectroscopy which at that time was one of the chief sources of evidence for the correctness of the rapidly developing ideas of atomic structure.

Alexandre Dauvillier was M. de Broglie's first helper in establishing the new laboratory. During the war he had been in the army medical corps operating a motorized radiological field unit. He stayed with de Broglie for twelve years, from 1919 onward, and extended the technique and knowledge of X-ray spectra to the long-wave region where tube, grating, and plate had to be kept in the same vacuum in order to eliminate unwanted absorption. A. H. Compton wrote of his work in the *J. Optical Society of America* 1928: 'Dauvillier was making rapid strides, working from the soft X-ray side of the gap (between light and

X-rays). First, using a grating of palmitic acid, he found the Kα-line of carbon of wave-length 45 Å. Then, using for a grating a crystal of the lead salt of melissic acid, with the remarkable grating spacing of 87.5 Å, he measured a spectral line of thorium as long as 121 Å, leaving only a small fraction of the interval to the shortest ultraviolet lines. The credit for filling-in the greater part of the original gap must thus be given to Dauvillier.'

Soon after Dauvillier, Jean Thibaud, Jean-Jacques Trillat and Louis Leprince-Ringuet joined de Broglie.

As far as his X-ray work is concerned, Jean Thibaud is mainly known for his measurement of X-ray wave-lengths by means of a ruled grating in 1927, a method which originated with A. H. Compton and R. L. Doan in 1925. From his measurements Thibaud expressed X-ray wave-lengths directly in terms of the standard meter and thereby related Siegbahn's X-unit, which was based on crystal diffraction, to the Angström unit which is 10^{-10} of a meter. This work, and subsequent measurements of even greater precision, led to the revision of one of the most fundamental constants in physics and chemistry, Avogadro's number.

In the first period of the eight years that J. J. Trillat worked in de Broglie's laboratory, he investigated long-chain organic compounds by means of X-ray diffraction. These studies, which paralleled those of A. Müller of the Royal Institution in London, were of twofold interest: paraffins and fatty acids were among the very limited group of organic compounds for which the full structure could be determined with the means of discussion then available; besides, they were suitable crystals for the spectroscopy of very soft X-rays. Trillat studied their orientation on glass or metallic supports and invented the 'method of the tangent drop'. In this method a drop of the melt is allowed to solidify in air, whereby the surface layers form a structure of equidistant curved sheets which reflect locally according to Bragg's law. The diagram obtained from a tangentially incident X-ray can be compared to a wide-angle diagram with ordinary crystals and gave Trillat and his co-worker Dupré La Tour information about the structure of the surface layer.

When in 1927 Davisson and Germer in U.S.A., G. P. Thomson in England, and little later Maurice Ponte in France confirmed Louis de Broglie's idea of the wave nature of electrons by diffraction experiments it was natural that Trillat, being in close contact with the latter, should take up electron diffraction. He became the inventor of many ingenious methods and applied the new technique to the study of the oxidation of metals and alloys, the cementation of iron and the kinetics of other chemical reactions, especially of the surface type.

Jean Perrin's Laboratory

Jean Perrin, Nobel laureate and professor of physical chemistry at the Sorbonne, was one of the most spirited and stimulating teachers of Science in France. His 'Laboratory of Physical Chemistry' might well have been named better so as to indicate its main field of research more precisely, namely the structure of matter. Typical of the fundamental simplicity of Perrin's work are his determination of Avogadro's number from the microscopic observation of the density gradient of emulsified droplets, and his observations of the colours of patches of uniform thickness in films of fatty acid soaps from which he deduced the lengths of the soap molecules before the days of their X-ray determination. Similar observations of the uniformly coloured patches on thin mica flakes between crossed nicols led his co-worker René Marcellin (killed in September 1914) to a correct value of the cell length of the mica structure.

X-ray spectroscopic work began in this laboratory when Miss Yvette Cauchois constructed, in 1931, the first of a series of spectrographs using a bent crystal for intensifying weak spectral lines and increasing the resolving power by focussing the rays diffracted by the crystal. From that date onwards there was a steady flow of X-spectroscopic results, often under the common authorship of her and H. Hulubei (later professor of physics and rector of the University in Bucharest, Rumania), which made this laboratory rank in this field second only to M. Siegbahn's laboratory in Stockholm.

Her own work, and that of her pupils Manescu, Despujols, Barrère and others, covers the details of emission and absorption spectra of the K and L series of a very large number of elements, including some, like krypton and neon, in the gaseous state. Working with a wide source, obtained by secondary excitation of the samples outside the X-ray tube, she overcame the difficulties of low intensity and high absorption. The results, collected in 1947 in a book by Cauchois and Hulubei, are about to be published in a revised and enlarged second edition. The details of both emission and absorption spectra contain a great deal of information about the electronic states of the atoms in the crystal medium, but to find a simple interpretation for them is still a major problem of solid state physics. Professor Cauchois' work included ultra-soft X-rays and γ-rays besides the more usual X-ray range.

Georges Friedel's Laboratory

When the University of Strasbourg became French at the end of the First World War, Georges Friedel was appointed in 1919 to the chair of mineralogy. Being particularly interested in clarifying the nature of liquid crystals, he asked his son, Edmond Friedel, to investigate the hypothetical stratified structure of the smectic phases with X-rays. Since there was no suitable X-ray equipment in Strasbourg, Edmond Friedel performed his experiments in the laboratory of Maurice de Broglie. He confirmed the stratified structure and followed ethyl azoxybenzoate and ethyl azoxycinnamate through the temperature range from crystalline to mesomorphous to liquid state. In the temperature interval of smectic ethyl para-azoxy-benzoate he was able to measure the sheet thickness of 19.9 Å. He also found for sodium oleate a sheet thickness of 43.5 Å, thus refining the value which Perrin and Wells had obtained by their optical methods.

When X-ray equipment became available in Strasbourg, Louis Royer was in a position to check some of the basic data on which his theory of the mutual orientation of different kinds of crystals is based, a phenomenon he called 'epitaxis'. Raymon Hocart similarly used the equipment for the study of the twinning which is a standard feature in some minerals, such as boleite, pseudoboleite, cumengite, boracite; he and his group of students also investigated epitaxis.

The Laboratoire de Minéralogie et Cristallographie in Paris and its Spread

In 1919 Frédéric Wallerant succeeded to get Charles Mauguin appointed as Associate Professor and helped him to set up his X-ray equipment. Mauguin became the first in France to be interested in the analysis of individual crystal structures rather than in general physical problems associated with the crystalline state. After a first checking of the known quartz and calcite structures (1921 and 1925), Mauguin determined independently the structure of graphite (1925) (A. W. Hull in 1917 had found it already, and in 1924 other independent determinations by Hassel and Mark, and by J. D. Bernal appeared). He found unknown structure types in investigating cinnabar (1923) and calomel (1925). After that he devoted several years of study to the various kinds of micas, without, however, solving their structures. In his teaching, Mauguin developed a simplified presentation of space-group theory, including a symbolism which was later combined with that of Carl Hermann to form the Hermann-Mauguin notation now universally used. As early as 1924 his book *La structure des cristaux* appeared which presented an introduction and survey of the methods and results of X-ray diffraction backed by Mauguin's own experience.

Around Mauguin grew a school of crystallographers of the new type. Jean Wyart, Stanislas Goldsztaub, Jean Laval, and Pierre Chatelain were the first of these.

Jean Wyart, in 1926, determined the structure of a basic zinc acetate which he had synthetized. Later (1929) he studied several zeolites and more particularly 'chabazite', working out the position of atoms and following with X-rays the changes of their positions when the percentage of zeolitic water varies, or water is replaced by mercury or ammonia, or when the nature of the alkaline or alkaline-earth ions is changed by diffusion. He worked out the atomic structures of paratoluidine and leucite and studied several cases of polymorphism. When Charles Mauguin succeeded in 1933 to Frédéric Wallerant, Jean Wyart became his Associate Professor. With his students, he used X-rays chiefly for determining the atomic structure of minerals, and as a mean of identification of finely crystallized products formed during the hydrothermal synthesis of silicates. He remained permanently with Charles Mauguin, and succeeded him in 1948, when Mauguin retired.

St. Goldsztaub, while he was working with Mauguin, studied the iron oxides and determined the atomic structure of lepidocrocite. After the Second World War, he took over the former laboratory of Georges Friedel at the University of Strasbourg, and became engaged mainly in electronic optics and crystal growth.

Jean Laval investigated the diffuse scattering of X-rays. Using monochromatic radiation, he measured accurately with an ionization spectrometer the intensities of the incident and the scattered beams. He showed that diffuse scattering was essentially a temperature effect, and he established experimentally the laws for different crystals, in particular for sylvine. As these results were not in good agreement with the theories of Debye, Waller and Born, he worked out a quantitative theory which is generally adopted nowadays and which is based on Born's work on harmonic vibrations in solids. The diffuse scattering of X-rays is interpreted as selective reflection from the elastic waveplanes. From this theory and his X-rays measurements, he was able to obtain the coefficients of elasticity and the speeds of the acoustic waves in crystals such as sylvine. These remarkable results have since been widely confirmed, in France first by Philippe Olmer for aluminium; for iron Hubert Curien was able to determine only from X-ray data the atomic oscillation spectrum in iron crystals, the binding forces between an atom and 26 of its neighbours, and the specific heat curve.

Further confirmations of Laval's theory were found by D. Cribier regarding fluorine and P. Mériel regarding sodium chloride. Theoretical and experimental research of Laval and his students on the Compton effect in crystals should also be mentioned.

Since 1950 Laval occupies the chair of Theoretical Physics at the Collège de France and his research, mainly on the fundamentals of crystal elasticity, has moved there.

Pierre Chatelain, while working at the laboratory of Mauguin, took up once more the optical study of liquid crystals. To begin with, he rarely used X-rays, but during recent years in his laboratory at Montpellier, he and his co-worker J. Falgueirettes obtained important results on the structure of these substances by means of X-ray diffraction.

A large number of scientists have since studied at the Laboratory of Mineralogy and Crystallography at the Sorbonne, amongst the earliest and the best known, are C. Kurylenko, known for his studies on X-ray absorption; J. Rose for improvements of the X-ray spectrometer and methods of determining crystal structure; J. Barraud for his research on X-ray optics; Robert Gay and his group for his determinations of the atomic structures of organic compounds; M. M. Herpin, Rimsky, for their determination of crystal structures. André Guinier, though he never worked in this laboratory, is also one of Charles Mauguin's students. His thesis on the small-angle scattering methods made him known in the world of crystallographers, and the curved monochromator which he introduced, and his focussing powder cameras are used in every laboratory. These methods allowed him to study small-angle scattering in much greater detail and he clearly recognized what a powerful tool this was for the study of the first stage of a transformation of an alloy; the slight disturbance of the initial lattice does not change appreciably the diffraction pattern, but has a striking effect on diffuse scattering. He has studied binary alloys like Al—Cu, Al—Zn, Cu—Be, in which the scattering power of the two atoms is widely different and which are obtainable as single crystals. In these alloys he could prove the segregation of solute atoms. Among his early students, Fournet has applied the small-angle scattering to hemoglobin and given a theory of X-ray scattering in an order-disorder assemblage and in liquids; Devaux and Brusset have studied the hole size in various charcoals. Castaing is very well known for the perfecting of his 'Sonde' which permits him to determine, by the X-ray emission spectrum, the chemical composition of microscopic inclusions of the order of one micron. Guinier's work was done at the laboratory of the Conservatoire des Arts et Métiers from 1940 to 1960. It has now moved to the new research centre of the Faculty of Science of Paris University at Orsay (S. & O.).

Urbain's Laboratory and its Spread

The first *chemical* laboratory in France to obtain X-ray equipment was the laboratory of Georges Urbain at the Sorbonne. Here Delaunay obtained his X-ray tube at the same time as Charles Mauguin, and he soon had as assistant, later as successor, Marcel Mathieu. The latter had already worked in 1925 and 1926 at the Royal Institution of London, in Sir William Bragg's laboratory. He applied X-ray analytical methods to several chemical problems, to what he called the topo-chemical reactions, such as the gelatinization of nitro-cellulose. He studied solid catalysts and found that catalytic activity is related to the existence of holes in the solid phase. In some cases he was able to estimate the shape and size of these holes. His laboratory rapidly became very active and was attended by numerous scientists, a number of which have become renowned crystallographers. One of the first of these was Jacques Mering whose work deals with X-ray scattering by highly imperfect crystals, such as charcoal and clay. For several years the lamented Miss Rosalind Franklin worked with him.

E. Grison specialized in the determination of atomic structures and V. Luzzati was initiated by him to X-ray techniques. After several years spent in Mathieu's laboratory, the latter, well known for his researches on the phase problem and for his determinations of the atomic structures of nitric acids, organized an X-ray laboratory for biology at Strasbourg.

With Georges Urbain, G. Champetier also used X-rays to study the properties of macromolecules, for instance the mechanism of the nitration of cotton by nitric acid vapour.

Among the first students of Mathieu was Miss Cécile Stora. She remained in charge of his laboratory when he left thbrnonee So with his whole group to get installed in larger laboratories, first at the Institut de Recherches Appliquées, and later at the Office National d'Études et de Recherches Aeronautiques. The structures which Miss Stora studied by means of X-rays are those of organic dyes, derivatives of triphenylmethane and of Indigo (Mrs. von Eller-Pandraud). G. von Eller, besides for publishing interesting memoirs on the phase problem, is well known for his much used photosummation machine for Fourier synthesis, which was first constructed in Stora's Sorbonne laboratory.

Other Laboratories

Another chemical laboratory using X-rays is that of Professor Chaudron in Vitry. He and his first students Benard, Faivre, Lacombe, Michel, nowadays directors of important laboratories, have perfected the methods of precise lattice parameter measurement in particular by the back-reflection method. Much of their work deals with very pure metals, alloys, and the mechanisms of oxidation and of corrosion as linked to crystal defects.

During the Second World War, Mathieu's associate Mering, having taken refuge at the University of Grenoble, set up there an X-ray laboratory and was joined by F. Bertaut. The latter remained in charge of this active laboratory after the end of the war, when Mering returned to Paris. Bertaut and his coworkers have determined the structures of mineral compounds, more particularly of iron and rare earth garnets. Bertaut also published important memoirs of a theoretical character on the direct determination of atomic structures from X-ray data alone.

Needless to say, X-ray diffraction methods have penetrated, slowly but irresistibly, into many more governmental and industrial laboratories in France. Among the first, mention should be made of Mrs. Adrienne Weill's X-ray section of the Navy Research Laboratory, of the National Telecommunications Laboratory, and the National Nuclear Research Laboratories. Important industrial X-ray laboratories are to be found in the steel and mining industries, the tire manufacture and in photographic works. There is room for expansion here, and it may be hoped that as more students are now getting a training in crystallographic methods, so will the use industry makes of X-ray diffraction continue to rise.

CHAPTER 20

Germany

by E. E. Hellner and P. P. Ewald

When M. von Laue left Munich in the fall of 1912 to go to Zürich, W. Friedrich remained as experimental assistant to Sommerfeld at the Institute for Theoretical Physics. P. Knipping obtained his doctoral degree soon after the publication of the Laue-Friedrich-Knipping paper and took a job in the Siemens Laboratory in Berlin for development work on Coolidge type X-ray tubes. He later became assistant to J. Franck in the Kaiser-Wilhelm-Institute for Physical and Electrochemistry in Berlin-Dahlem during the war, and after that settled in Darmstadt as Lecturer at the Technical University; in one of the following years, however, he had a fatal collision with a truck while riding his motorbicycle.

Friedrich left Munich in 1913 to become the X-ray specialist in Prof. Krönig's Gynaecological Section of the Freiburg University Clinic, then—as today—famous for its advanced methods of operating and of radium and X-ray treatment. Friedrich was soon fully absorbed by the medical problems and eventually became Head of a special medical X-ray Institute attached to Berlin's most renowned hospital, the Charité.

The Munich equipment remained in the hands of Ewald, but was soon diverted from its original purpose by being transferred (including Ewald) to an emergency hospital which was set up in a Munich elementary school at the outbreak of war. The initial experience in medical use led, a year later, to the appointment of Ewald (and similarly to that of Glocker and other physicists) to the newly created position of Field X-ray Mechanic of the Army. While Glocker, Regener, Spiess and others saw heavy fighting and had plenty of work on the frontier with France, Ewald was assigned to the northern part of the Russian front and when he arrived by the fall of 1915, fighting had practically ceased. The tranquillity, together with the isolation of the field hospital enabled Ewald to carry through his dynamical theory of X-ray diffraction which he then used as thesis (Habilitationsschrift) for becoming a lecturer (Privatdozent) in Physics at the University in Munich in 1917.

Compared with the progress made by W. H. and W. L. Bragg by August 1914, the progress made in Germany was not spectacular. The Munich group worked exclusively with 'white' X-rays and Laue diagrams. The photographs could be explained in detail, once the crystal structure was known, but they proved little suited for finding the structure. On Friedrich's excellent Laue photos of pyrites (FeS_2) and hauerite (MnS_2) Ewald developed a method of determining a parameter with great accuracy, but this could only be achieved while checking the type of structure found by W. L. Bragg. Much later work, by Parker and Whitehouse in Manchester (1932) confirmed by Fourier methods the accuracy of the parameter determination.

In Röntgen's Institute his senior lecturer, E. Wagner, took up diffraction work. Wagner was an experimentalist who had great understanding for physical argument and practically none for mathematical proof. He would not be convinced by any formula that a diffracted ray in a Laue diagram contains only a single wave-length or perhaps also an overtone or two. So he asked his student, R. Glocker, to analyse one of the rays by reflecting it on a second, parallel crystal. When the second crystal gave a diagram with only a few spots, this finally convinced Wagner. Later (1920) Wagner did important work on X-ray spectra; in particular he showed the influence of the silver bromine absorption edges on the photographically recorded spectral intensity. In the course of this work he found that spectra, obtained from a slowly rotating crystal, were traversed not only by the dark lines of the intensive characteristic radiation, but also by faint white lines ('Aufhellungslinien'), less dark than the continuous background. These were either parallel to the dark lines or inclined at definite angles. The faint lines, it was soon shown, occur whenever the Bragg condition is fulfilled for more than one atomic net plane, so that energy is deviated into a further direction. O. Berg, working in the laboratories of the Siemens concern in Berlin, made a very thorough study of these 'Aufhellungslinien' in 1926.

During the war, 1914–18, experimental research came to a near standstill, except in Göttingen where P. Debye, of Dutch nationality, and a Swiss postgraduate worker, P. Scherrer, were not affected by conscription. They discovered in 1916 the method of powder diagrams while trying to obtain evidence of the electron clouds surrounding the atoms in lithium fluoride—but this story is better read in Part VII in

Scherrer's Personal Reminiscences. The enthusiasm which this method provoked is well shown by the remark of one of the leading crystallographers, A. Johnsen, then professor in Kiel, later in Berlin. When he explained the method to the writer, he ended up by saying in all seriousness—to his listener's horror—: 'We have now got rid of all troubles of finding crystals good enough for structure analysis, we simply go to the collection of minerals and grind up whatever crystal we find, good or bad.' Two interesting papers in *Physikalische Zeitschrift* 1917 and 1918 on the indexing of powder diagrams appeared, by C. Runge the first, and the later by A. Johnsen and O. Toeplitz.

At that time the nature of the brightly coloured colloidal metals was still unknown and a matter of speculation by Szigmondy, the Professor of Colloid Chemistry in Göttingen. Scherrer, employing his new means of investigation, showed that colloidal gold gave essentially the same diagram as gold filings, but with much broadened lines. The broadening he interpreted as due to very small particle size and developed a formula by which the particle size could be determined from the broadening. This opened up a field of research which was to lead later to many important applications, especially in relation to catalysts (R. Brill).

Also during the war, H. Seemann, then in Würzburg, began an extensive series of studies of the geometry of image formation by the diffracted rays. This led him to the construction of new types of cameras, both for the conventional methods of obtaining photographic diagrams, and for wide-angle and 'complete' diagrams. He later manufactured in his own laboratory, and still does so, demountable X-ray tubes and cameras of many types and great workmanship. The wide-angle method gives no easily interpreted intensity data and is therefore not so suitable for structure analysis as for the precision determination of lattice constants.

After the war many changes took place. E. Wagner became professor in Würzburg and, after Röntgen's death in 1923, organized a small museum in memory of Röntgen's discovery in the very house where it was made. This museum, later looked after by Wagner's successor H. Ott, miraculously survived the fierce bombing which destroyed large parts of Würzburg in 1945 in one of the last bombing attacks of the Second World War; it has now been partly transferred to the museum in Röntgen's birthplace Lennep.

R. Glocker, after his discharge as Field X-ray Mechanic, settled in 1919 in his native Stuttgart and began building up an X-ray laboratory attached to the Technical University and financed largely by contributions, in kind or in money, of industry. In spite of the post-war financial crisis this became in a few years the best equipped and best staffed X-ray laboratory. Often the factories of X-ray equipment sent there, for a test period, the latest improved high tension and stabilizing machines, and its choice of X-ray tubes of a variety of shapes, form of focus, and target material roused the envy of visitors, especially from U.S.A., where the production of tubes was standardized to meet medical needs, and the importing of foreign makes required special permits. (Besides, the C. H. F. Müller tubes were half the price of the American ones.) An excellent workshop with first-rate instrument makers formed an important section of the institute. Apart from routine and testing work necessary at times for its financing, the work of the institute was largely governed by Glocker's own interests. These were:

1) Metal structures and textures, in particular hardening by annealing, and the effect of impurities on recrystallization.
2) From about 1935 onwards, the X-ray determination of stress in built-up parts, for instance in the girder of a bridge in situ. Portable X-ray tubes and backreflection cameras were constructed and tested in the laboratory.
3) The study of the photographic and biological action of X-rays, and later of penetrating electron rays.
4) Dosimetry and radiation protection. As a long-time member of the Association of Radiologist's Commission on this subject, Glocker and his co-workers made valuable contributions to the definition of the practical units like the röntgen and rad, and to the correct method of their measurement.

Among the many long-time co-workers of Glocker are U. Dehlinger (Stuttgart), K. Schäfer (now at I. G. Farbenindustrie, Ludwigshafen), R. Berthold (now manufacturing X-ray equipment in Bietigheim), H. Kiessig, W. Frohnmayer, L. Graf, R. Glauner, E. Oswald. In fact, this institute, besides those in Dahlem, was the main seed-bed for X-ray crystallographers. Structure analysis was only incidental and remained on a primitive level; the emphasis lay on problems of an engineering type which require a physicist's training for their solution. When a Kaiser-Wilhelm-Institute for Metals was created in Stuttgart in 1932, it was built next to Glocker's Institute which then became an independent section of the K.W.I. for Metals, and this is its present state. The close association works out well to the advantage of both parties. Under the direction of Köster and with the aid of Scheil, Nowotny (now in Vienna), Schubert, etc., a great number of intermetallic systems were investigated, their crystal structures and other physical properties determined. Glocker, supported by Richter, Hendus and others, was in addition interested in the structure of liquid metals and alloys. Dehlinger, Ganzhorn and Bader developed a theory of bond structure of transition elements in respect to their ferromagnetic properties. In the section of Grube, Kubatschewski (now in England) and Weibke did research on the physico-chemical properties of metals and alloys; Kochendörfer pointed out a theory of fracture and elastic constants of crystals in relation to vacancy energy and surface energy. Seeger in his section has published several papers about the electron theory causes of defects in metals and the interpretation of small-angle scattering of X-rays in plastically deformed metals. A recently opened section for 'Sondermetalle' is going to investigate systems of alloys of uranium, zirconium, etc.—Glocker's textbook on the Testing of Materials with X-rays, now in its fourth edition, and Dehlingers book *Theoretische Metallkunde*, 1955, should be mentioned as a valuable outcome of the Institute's work.

In 1930 an Institute for Theoretical Physics was established at the Technical University in Stuttgart, headed by P. P. Ewald, and modelled after the Munich example in that it provided for modest experimental work, mainly on crystals. The experimental assistant was M. Ruhemann (now in Manchester), who had just graduated under F. Simon in Berlin. His construction of a simple low-temperature camera for the growing of nitrogen crystals in a capillary tube at the centre of the camera was a very original contribution which may be well worth resuming and developing further. After two years Ruhemann was succeeded by M. Renninger (now at Marburg) who constructed a double-crystal spectrometer for the investigation of the intriguing 222 reflex of diamond which was first noticed by Sir William Bragg and which, since it cannot occur for point atoms, gives an indication of the actual shape of the carbon atom in diamond. In the course of these experiments Renninger discovered that the intensity reflected under the Bragg angle from an octahedral face of the crystal varied, if this face is turned in itself, that is, by changing the plane or azimuth of incidence without changing the Bragg angle. The normally very weak 222 reflexion becomes ten times stronger under certain azimuths. It was soon shown that this happened when by simultaneous reflection on other planes energy was being chanelled into the direction of the direct 222 reflection; this phenomenon, named 'Umweganregung' (simultaneous reflection), is thus a counterpart of the faint lines which Wagner found crossing the spectra.

The theoretical work at this institute comprised the collection of known structures (*Strukturbericht*) by Ewald and C. Hermann, the co-editing of *Zeitschrift für Kristallographie*, the planning, and later the editing of the *International Tables for the Determination of Crystal Structures* (Hermann), review articles on X-ray Diffraction in *Handbuch der Physik* and on Solid State in Müller-Pouillet, and C. Hermann's papers on Structure Theory which paved the way for the replacement of the Schoenflies space-group symbols by the more systematic modern ones. H. Hönl developed the theory of the atomic factor including dispersion, which, very much later, gave the theoretical foundation of Bijvoet's elegant method of distinguishing the atomic arrangements in d- and l-crystals of optically active compounds

Before Ewald left Germany in 1937 Hermann had already joined R. Brill at the I. G. Farben Laboratory in Oppau and Renninger soon followed; the Institute passed to the hands of U. Dehlinger and was closely linked to the K.W.I. für Metallforschung. It was fully destroyed by a direct hit during the bombings of Stuttgart.

One of the most brilliant physicist-crystallographers in Germany was Walter Kossel who, after his graduation with Lenard, came to Munich at the end of the First World War, joining Sommerfeld's group. He, as well as the Munich professor of physical chemistry, K. Fajans, discussed the early examples of crystal structures from the point o- view of chemical and electrical bonding forces, charges and polari. zabilities of the atoms according to the early stages of the Bohr theoryf In 1921 Kossel became professor in Kiel, and later the head of the physics laboratory at the Technical University in Danzig. It was here that he developed the 'crystal source diagrams', diffraction effects not of plane X-waves, but of the spherical waves emitted by atoms within the crystal on excitation by electron bombardment or by the incidence of harder X-rays. One of M. v. Laue's most elegant papers discusses these diagrams in terms of the dynamical theory of X-ray diffraction. Among Kossel's pupils and co-workers in Danzig are Möllenstedt, Borrmann and Voges. After the Second World War Kossel became professor of physics in Tübingen, where he died in 1956.

A very important centre of X-ray diffraction developed in Berlin-Dahlem (see the Reminiscences of H. Mark and M. Polanyi in Part VII). From 1920 onward the Kaiser-Wilhelm-Institut für Faserstoffchemie under R. O. Herzog was foremost in the application of diffraction methods to the study of fibres and other high-polymers. Herzog himself, and W. Jahnke, who had, while working with Scherrer in Göttingen, acquired the art of constructing reliable dismountable X-ray tubes, began the work and they were soon joined by M. Polanyi. Following the lead given by Nishikawa and Ono as early as 1913, diagrams were obtained of fibres of cellulose, silk fibroin and other materials which lent themselves to being partially orientated by stress. The repeat distances along the fibre axes could be determined, and Polanyi, Mark and Weissenberg were stimulated to using rotation diagrams also in single-crystal structure determinations.

Weissenberg developed the idea of spreading a single layer line to a picture in two dimensions in the goniometer, and he constructed the first such instrument. Meanwhile the analogy between the diffraction pictures of fibres and of work-deformed metals prompted Polanyi to investigate the deformation of single metal crystals. He and H. Mark discovered the slip properties of single-crystal tin, and soon after that E. Schmid found the dynamical condition of slip, namely the law of the critical shear stress component along the slip direction in a slip plane.

When Herzog's institute was discontinued in 1926, its buildings and installations were taken over by W. Eitel for a new K.W.I. for Silicate Chemistry. Mark, who had brought into being a vigorous school for crystal structure analysis, left Dahlem in 1927 and created a prominent industrial X-ray laboratory in the Ludwigshafen works of the I. G. Farbenindustries. This firm had already one X-ray laboratory at its Oppau works, headed by R. Brill.

After F. Haber's dismissal (see Laue autobiography, pg. 298) the Kaiser-Wilhelm-Institut for Physical and Electro-chemistry in Dahlem came under the direction of Thiessen. Here M. v. Ardenne developed a universal electron microscope, which was used mainly for research on high polymers. Later on, the name of this institute was changed into 'Fritz-Haber-Institut' and Max von Laue was its director until February 1959. In his section on X-ray optics G. Borrmann studied the effect, named after him, of anomalous transparency of a crystal under the conditions of Bragg reflection, adding later to his observations those on the effects of deformation and temperature gradients on the transparency of the crystal and the path of the energy flux in it. The latter was the subject of important theoretical studies of von Laue. A field electron microscope was developed by Müller at the Technical University in Berlin (now in Pennsylvania State University). His work is being continued in the Haber Institute by Drechsler. In the section of Molière, the interference-refraction of electron beams, especially in MgO crystals, was one of the important problems.

Besides the structure determination of claudelite, complex cobalt compounds and some organic substances, Stranski's section continues the calculation of surface energy of homopolar crystals and studies the influence of different conditions of crystallization and nucleation on the face combinations and other physico-chemical properties.

Hosemann and collaborators were interested in several fields: the calculation of the electron-density distributions in crystals with the help of convolution operations, the analysis of small-angle scattering by means of the folding operation (Faltungsintegral), and its application to high polymer and colloidal substances, paracrystalline structure, etc. His work with S. N. Bagchi on the theory of Fourier transforms and unfolding procedures has recently been summarized in a book *Direct Analysis of Diffraction by Matter* (North-Holland Publ. Co., 1962).

In 1959 the Fritz-Haber-Institut came under the direction of R. Brill, who returned from the U.S.A., where he had gone in 1946. Among many other topics he is interested in the influence of bonding electrons in diamond on X-ray intensities (anisotropic atomic scattering factor). A new section for the determination of crystal structures has been created in the institute, including a group for neutron diffraction.

In the K.W.I. für Silikatforschung under W. Eitel, Schusterius, Radczewski and O'Daniel were interested in the application of the electron microscope to mineralogy, especially to clay minerals and phases in cement. Dietzel's interest belonged to the investigation of glasses and the different industrial applications. Eitel, (now in Toledo, U.S.A.), wrote his important book *Physikalische Chemie der Silikate*, which appeared after the war in a new edition in the U.S.A. (1954). During the Second World War the institute was partly transferred to the Rhön and from here Dietzel created a new institute in Würzburg. Since 1952 Jagodzinski (who since 1959 is also professor in Karlsruhe) formed a special section for the investigation of crystal structure where he continued his research of order and disorder problems, of chrysotile asbestos and antigorite (the latter problem was continued in Darmstadt by Kunze). Flörke and Saalfeld too belonged to this institute at that time, working on the modification of SiO_2, respectively $Al_2O_3.nH_2O$. (Flörke is since 1959 in Zürich, Saalfeld since 1960 in Saarbrücken.)

Among the other Kaiser-Wilhelm-Institutes, the Düsseldorf K.W.I. for Iron Research should be mentioned because of the intensive study of iron alloys phase diagrams and deformation properties of iron and steel there undertaken since 1922 by F. Wever.

The main industrial laboratories for X-ray work, besides those of the

I. G. Farben, were those of the electrical industry, especially the Siemens and the Osram works. Metals were also investigated in the Materialprüfamt (Materials Research Station) in Gross-Lichterfelde, close to Dahlem. G. Sachs worked there for some years before he joined the Metallgesellschaft in Frankfurt/Main.

In the Universities, crystallography was coupled to mineralogy, and some of the Departments of Mineralogy used X-ray methods, for instance R. Gross in Greifswald. In Leipzig F. Rinne and E. Schiebold developed the teaching aspect, contributed to the development of methods and instruments (the Schiebold-Sauter goniometer) and had a fair number of advanced students, among them E. Onorato (structure of gypsum, 1927).

Niggli with his broad and excellent survey of mineralogy recognized very early the importance of mathematical crystallography for the study of morphological, physical and chemical properties in crystals. Between 1915 and 1918 he was in Leipzig in Rinne's Institute. A year later he wrote in Tübingen *Geometrische Kristallographie des Diskontinuums*. In his famous textbook *Lehrbuch der Mineralogie und Kristallchemie*, which appeared in three editions, he pointed out the importance of crystal chemistry to mineralogy, geochemistry, etc. Besides his *Grundlagen der Stereochemie* (1945) he published in the following years (1949/50) a complete condensed symbolism of space groups which later formed the main subject of the book by P. Terpstra *Introduction to the Space Groups* (J. B. Wolters, Groningen 1955).

In 1929 V. M. Goldschmidt in Oslo accepted the chair of mineralogy and crystallography at the University of Göttingen; here he continued his fundamental work on the radii of elements and ions by the investigation of crystal structures. His results are published in the fundamental series *Geochemische Verteilungsgesetze der Elemente* by the Norwegian Academy of Sciences, Oslo. At that time, Ernst, Laves and Witte were his assistants. Early in 1935 V. M. Goldschmidt had to leave Germany for political reasons. The investigations of crystal structures were mainly continued by these three assistants who became lecturers later on. Laves together with Witte and Wallbaum did important work on intermetallic compounds. In addition, he developed his nomenclature of structure types, which is used in many textbooks.

Machatschki, who had been working with V. M. Goldschmidt in Oslo and also with W. L. Bragg in Manchester, Professor of Mineralogy at the University in Tübingen, where he continued his work on structure analysis of silicates and oxides, the discussion of the chemical formulas of silicates with respect to their atomic structures and the relationship between silicate structures. After a three years' period in Munich, he became director of the Mineralogical Institute in Vienna in 1944.

After the Second World War, Laves began to create a centre for structure analysis in Marburg/Lahn, by expanding his Mineralogical Institute with the assistance of Jagodzinski and Hellner. In 1947 he succeeded in negotiating with the Government the creation of a second part of the centre, a new Institute for Crystal Structure Research which was headed by C. Hermann.

Hermann's work was mainly theoretical, applying group theory to mathematical crystallography in three and more dimensions and devising a nomenclature for lattice complexes. In connection with Ludwigshafen, in particular with H.-U. Lenné, he also did interesting structural work on urea adducts. At the beginning of the war, Hermann and Renninger had both been working under R. Brill at the Oppau Laboratory of I. G. Farben. It was here that the well-known investigation was undertaken by Brill, Grimm, Hermann and Peters (*Ann. d. Phys.*, Lpz., 1939) to learn as much as possible of the electron distribution in simple crystals by accurate intensity measurements and critical discussion of the Fourier syntheses. Renninger made additional intensity measurements at 20°K on rocksalt, but owing to the war, publication was delayed. Now, having joined Hermann at the new institute in Marburg, Renninger constructed a three crystal spectrograph with which he measured the reflection curve of a near-perfect crystal with greater accuracy than ever before.

At the end of 1948 Laves left Marburg for Chicago, and H. Winkler succeeded him. He discovered the crystal structure of eukryptite and of some alkali-fluoroaluminates in connection with the structural discussion of thermal polymorphism. E. Hellner started his research on sulfide structures in Marburg; after a short interlude in Chicago, he is now continuing them, since 1959, in Kiel.

O'Daniel became director of the Mineralogical Institute in Frankfurt in 1947 where he continued to investigate problems of cement and other silicates which he had begun at the K.W.I. für Silikatforschung. Together with Th. Hahn he created a new school for the X-ray investigation of analogues of silicate structures, such as the fluoroberyllates. The programming of electronic computers and the correction of measured X-ray intensities were other fields of interest.

As a student of A. Johnsen in Berlin, G. Menzer succeeded 1926 in the structure determination of garnet. At the end of the Second World

War—after seven years in the K.W.I. of Physics—he left Berlin and became after a two years' stay in Tübingen Professor of Crystallography in Munich at the University, where he continued his structural work together with Dachs and others. Like that of C. Hermann, his interest belongs to homometry, nomenclature of space groups and lattice complexes. As a possible example for homometry the structure of bixbyite was reexamined. Besides X-ray work he applied neutron diffraction to the location of hydrogen atoms in crystal structures.

In Göttingen a new center for X-ray analysis grew up, when Zemann became director in 1952. A large number of structures of silicates, phosphates and sulfates have been determined in his school.

Apart from the application of X-ray analysis, much of interesting work is being done in mineralogical institutes in such classical fields of crystallography as crystal growth and solubility, twinning, epitaxy, etc., for instance by Neuhaus in Darmstadt, later Bonn; Spangenberg in Breslau, Kiel and Tübingen; Seifert in Halle and Münster; Kleber in Bonn and since 1953 in Berlin, etc.

A number of institutes of inorganic chemistry took up the determination of crystal structures. Thilo, since 1938 in Berlin except for a short period in Graz (1943–46), was interested in the chemical and structural properties of silicates and their analogues, the phosphates and arsenates. After the Second World War an Institute for Crystal Structure Research was founded with the aid of the Deutsche Akademie der Wissenschaften, in (East-) Berlin, headed by Mrs. K. Boll-Dornberger. Besides structure determination she is especially interested in disordered structures and their interpretation by 'order-disorder' or OD-groupoïds.

Starting with the investigation of alcali alloys in liquid ammonia, Zintl (1933) created a school for inorganic and physical chemistry at the Technical University in Darmstadt. He was especially interested in the crystal chemistry of alloys of those elements which form anions in crystal structures (Zintl-Grenze). A lot of new structure types were found. Zintl died in 1941, at the age of only 43. Brill became his successor up to the end of the war. After the war Witte succeeded Brill and together with Wölfel he reexamined the Fourier synthesis of NaCl, LiF, Al etc. by measuring the intensities even more accurately.

Another inorganic school was founded by Klemm, first in Danzig and after the Second World War in Kiel and later in Münster; he correlated magnetic properties and crystal structures. A book (*Magne-*

tochemie, 1936) gave first results in this field and a lot of inspiring suggestions.

A centre for structure investigation of organic and biochemical substances was founded in 1959 in the Max-Planck-Institut für Eiweiss und Lederforschung headed by W. Hoppe, who is also lecturer in the physico-chemical institute of the Technical University in Munich. The interpretation of diffuse thermal-wave scattering outside the directions of Bragg scattering, and the determination of crystal structures by the convolution molecule method are two of his favourite subjects. A goniometer for automatic recording of X-ray intensities of single crystals is under development. In addition schools were created by Bauer in Freiburg (1944), by Juza in Heidelberg (1942) (since 1952 in Kiel); and by Krebs in Bonn.

The important theoretical development represented by Max Born's books *Dynamik der Kristallgitter* (1915) and *Atomtheorie des festen Zustands* (1923) opened up the understanding of many of the physical properties of the solid state, first based on the principles of classical dynamics, and later on quantum and wave mechanics. A new field of physics developed rapidly and concurrently with, if not always closely related to, the increasing knowledge of crystal structures. In the work of Mott and his school in Bristol a second great step forward was taken in the years before World War II. A. Smekal in Halle had stressed the difference between physical properties which are explainable by the model of a perfectly periodic crystal, as used by Born, and the 'structure-sensitive properties' (by this he meant properties determined by imperfections, impurities, or domain structure of crystals) for which Born's theory is evidently inadequate. Among the latter are mechanical strength, diffusion, fluorescence and phosphorescence, often dielectric properties and surface tension, etc. The new ideas applied by Mott centered around the energyband picture of the crystal as a wave-mechanical system; besides, for the explanation of the mechanical properties 'dislocations' of various types in the crystal structure became essential. During and after the Second World War the structure sensitive properties of Smekal's classification underwent intensive study, especially in U.S.A., because of their technical importance (phosphors, rectifiers, transistors, ferromagnetic and ferroelectric materials). But also the German industrial laboratories, especially those of the Siemens concern, now near Erlangen, contributed their share to these advances.

The various facets of solid state physics and the New Crystallography were repeatedly summarized in the well-known German Hand- and Yearbooks.

Before ending this report the support of the entire field of research by the 'Notgemeinschaft der Deutschen Wissenschaft', now renamed 'Deutsche Forschungsgemeinschaft', which began more than four decades ago, should be gratefully acknowledged. This support included in 1956 the first electronic computers, and in 1960 also an IBM 704, available for basic research.

In recognition of the need to promote basic research in Germany the 'Wissenschaftsrat', a council created by the President of the German Federal Republic, recommended in 1960 an expansion in German universities, especially with respect to the number of staff and assistants, but also including facilities and housing. It may be expected that this improvement will prove beneficial also to the field of crystal structure research.

CHAPTER 21

The Netherlands

by J. M. Bijvoet

Immediately upon the news of von Laue's discovery there was great interest in the Netherlands in its application to a number of fields.

In Groningen there were Jaeger and Haga; the former had a life-long intense interest in Nature's symmetry, and in the crystal world in particular—witness his *Principles of Symmetry and its Application in all Natural Sciences* (1916)—; the other had carried out, searching for the nature of X-rays, the well known diffraction experiments of X-rays through a slit. It is therefore not surprising to see them working together in 1913 on the symmetry of Laue diagrams. Jaeger tested experimentally the *Laue symmetry* for nearly all crystal classes in a series of careful investigations. Nowadays one probably feels such a systematic investigation to be somewhat superfluous. For that reason it is interesting to notice that Jaeger for a short time thought to have to conclude that von Laue's theory was exact for isotropic crystals but that something was lacking in applying it to others. It soon appeared that the small deviations could be explained by wrong adjustments of the specimens. Of these first investigations we mention that as early as 1913 Jaeger developed an apparatus for X-ray exposures at high temperature. Jaeger—Terpstra—Perdok—Hartman and their pupils form a *Groningen chemico-crystallographical* school. For each school we shall mention its typical fields of study and structure determinations, even if here and further on we shall have to make some rather arbitrary choices. Here we find the structure determinations of the *ultramarines* with their vagrant atoms; of $NaSb(OH)_6$, whose structure is in agreement with a suggestion of Pauling; attack on the formulation $Na_2H_2Sb_2O_7$ 5 aq., which in this customary form brotherly unites the elements pyro and aqua! This school has always kept up the closest ties with the classical geometrical crystallography. Its contribution to *The Barker Index of Crystals*, and a theory of the relation between external form and internal structure of crystals are examples of it.

From the first year after the discovery of X-ray diffraction date the theoretical papers of Lorentz and Debye, which have already been dealt with in this memorial volume (Ch. 5). In 1916 Lorentz delivered three lectures on X-rays and crystal structure, which decided Professor Keesom to start on X-ray work. Already then Keesom, who later solidified helium, had in mind to extend the investigation to low temperatures. At that time he was professor at the Veterinary College in Utrecht—Van 't Hoff too started his career there—and he and his co-worker Dr. Kolkmeyer tackled the first structure determination in the Netherlands. The initial investigations were largely connected with the studies on allotropy of the Utrecht professor of chemistry Ernst Cohen. It is indeed remarkable how great a part was played by allotropy in the first planning of the Dutch X-ray work. There were two professors of chemistry who had strong personal, and sharply opposite views regarding this phenomenon. Cohen at Utrecht wanted to call every substance physically impure, a *mixture* of different modifications. Smits, in Amsterdam, wanted to assume for every crystal the character of a *solid solution*, a dynamical equilibrium between different kinds of molecules. Both were eager to see their views confirmed with the aid of X-ray analysis—less willing to modify them according to its results!—. Close co-operation arose between the X-ray workers of Utrecht (Kolkmeyer) and of Amsterdam (Karssen and Bijvoet, who passed a short apprenticeship in Utrecht). From this *Utrecht-Amsterdam school*, orientated mainly chemically, came many X-ray workers. This resulted in a large penetration of the Dutch Universities by X-ray analysts, of whom a considerable number later on as professors, mostly in physical chemistry, remained faithful to their first love.

Karssen, who went over to the application of X-ray diffraction to histology, died early. Kolkmeyer had a great liking for time-space symmetry considerations, which he developed extensively after an investigation of bond orbitals in diamond, in which he considered the electrons to move with mutual phase relations.

In Amsterdam, as mentioned already, the phenomenon of *allotropy* predominated in the early days. Studies were made of the crystal structures of red and yellow HgJ_2, the transition of which appeared to be brought about by the shift of the Hg-ion from a hole surrounded by 4 atoms to one with (2 + 4) atoms (molecule-formation); disorder was observed in the sequence of the layers in the $CdBr_2$-structure, in the random distribution of the Hg-ions in $Hg(NH_3)_2Cl_2$—which makes the X-ray diagram of this compound practically identical to that of the element Ag!—and in rotation-transitions ($NaNO_3$, NaCN).

When changing over to organic structures, after Professor Robertson's famous phtalocyanine synthesis, the isomorphous replacement method was successively extended in Utrecht to the more general cases: campher, strychnine, tyrosine. In the last case 'anomalous' scattering was called in for the *determination of phases or signs*, which settled the old problem of absolute configuration of optically active compounds. Structure determinations of particular chemical interest are those of gamhexane and muscarine, one of great accuracy that of NH_4H-tartrate where the standard deviations of the coordinates amount to a few thousands of an Ångström.

Professor MacGillavry investigated in Amsterdam, among other things, the cause of the well known *alternation in properties* of aliphatic dicarboxylic acids, which was shown to be due to the packing of the carboxylic groups. While this leaves the 'even' molecules planar, it induces a twist in the molecules with odd number of carbon atoms. In cooperation with Prof. W. G. Burgers the classical example of *habit modification*, octahedral growth of rocksalt induced by urea, was explained by temporary adsorption of a nucleus of the compound urea-$NaCl \cdot H_2O$ on the rocksalt octahedral plane. The nitronium ion NO_2^+, supposed *nitrating agent* for aryl compounds, was shown to exist in the solid state of nitrogenpentoxide ($N_2O_5 = NO_2^+ \cdot NO_3^-$) and in some combinations of nitric acid with sulphuric acids. Oxygen compounds of S^{VI} and P^V reveal the same structural principles as the silicates: tetrahedra sharing corners form chains, plates, rings and interlocking spiral ramps. Studies on vitamin-A related compounds reveal the influence of *steric hindrance* on a conjugated bond system. Disorder and twinning were studied in several cases. Out of a seminar grew an interpretation of peculiar *diffraction fringes in electron diffraction*.

It is characteristic for the Laboratory of Professor MacGillavry that many foreigners work there: there were guests from the U.S.A., Canada, Uruguay, Italy and Spain. Needless to say that workers from all Dutch laboratories go regularly to England and U.S.A.

The laboratory of Professor Wiebenga, the successor of Professor Jaeger, at Groningen is at present the only centre in the Netherlands where *protein structures* are investigated (papaine, mol. weight 21 000). From this laboratory comes an *integrating Weissenberg camera* (Wiebenga and Smits) which is widely used. Examples of structure investigation are the determination and interpretation of *interhalogens and polyhalogenides*—compounds which of old had been the stumbling block for the classical valence theory—and structure determinations of the *phosphor sulphides*, which with P_4S_5 and P_4S_7 led to a chemically unexpected result. X-ray intensities of crystals are measured accurately at low temperature in the hope that the distribution of the valency electrons in molecules can be determined.

Dr. Jellinek (muscarine; investigation and systematization of structures related to the NiAs type) has recently found an intriguing effect in the alternating bond length in the sandwich structure of chrome benzene. When writing this survey (June 1961) this effect was still theoretically unexplained.

Let us go back to former times. Keesom, returning to Leiden as the successor of Kamerlingh Onnes together with his pupils—among whom Professor De Smedt from Louvain—carried out X-ray investigations at *extremely low temperatures* including structure determinations of the inert gases. These were experiments that could be performed only in very few places in the world. Röntgenographically the transition to the supra-conducting state in Pb, the difference between the crystals of hydrogen and parahydrogen, between He I and He II were studied without obtaining a revealing answer. After Keesom had resigned, the X-ray work in the Kamerlingh Onnes laboratory was not continued.

In Groningen Professor Coster, who already had attained international fame at M. Siegbahn's laboratory in Lund, and his pupil Prins concentrated in the first place on *X-ray spectroscopy*. Among their results is the remarkable connection found between the fine structure of the absorption edges and the crystal structure of the absorber. Soon they also started diffraction work. Among other things an absorption correction was introduced in the dynamical theory and measurements were made of the effect of anomalous scattering in the reflections from the tetrahedral plane of zincblende in which Friedel's law proved to be violated. The underlying phase shift, known for long in optics, refound in X-ray diffraction, later had to be discovered again for electron waves. Zernike and Prins gave the theory of liquid-diffraction (radial distribution curve); preceded by Ehrenfest's derivation of the diffraction effect in a bi-atomic gas, Keesom and De Smedt, and independently Debye, had pointed out the part played by the intermolecular distances in the liquid diffraction. The basis for this theory was already laid by Ornstein and Zernike in their study of the light scattering by density-fluctuations of a gas at the critical point.

After Coster's death and Professor Prins's move to the Technical University at Delft, X-ray research in the physical laboratory at Groningen came to an end.

As a matter of course Philips Laboratory at Eindhoven took up the

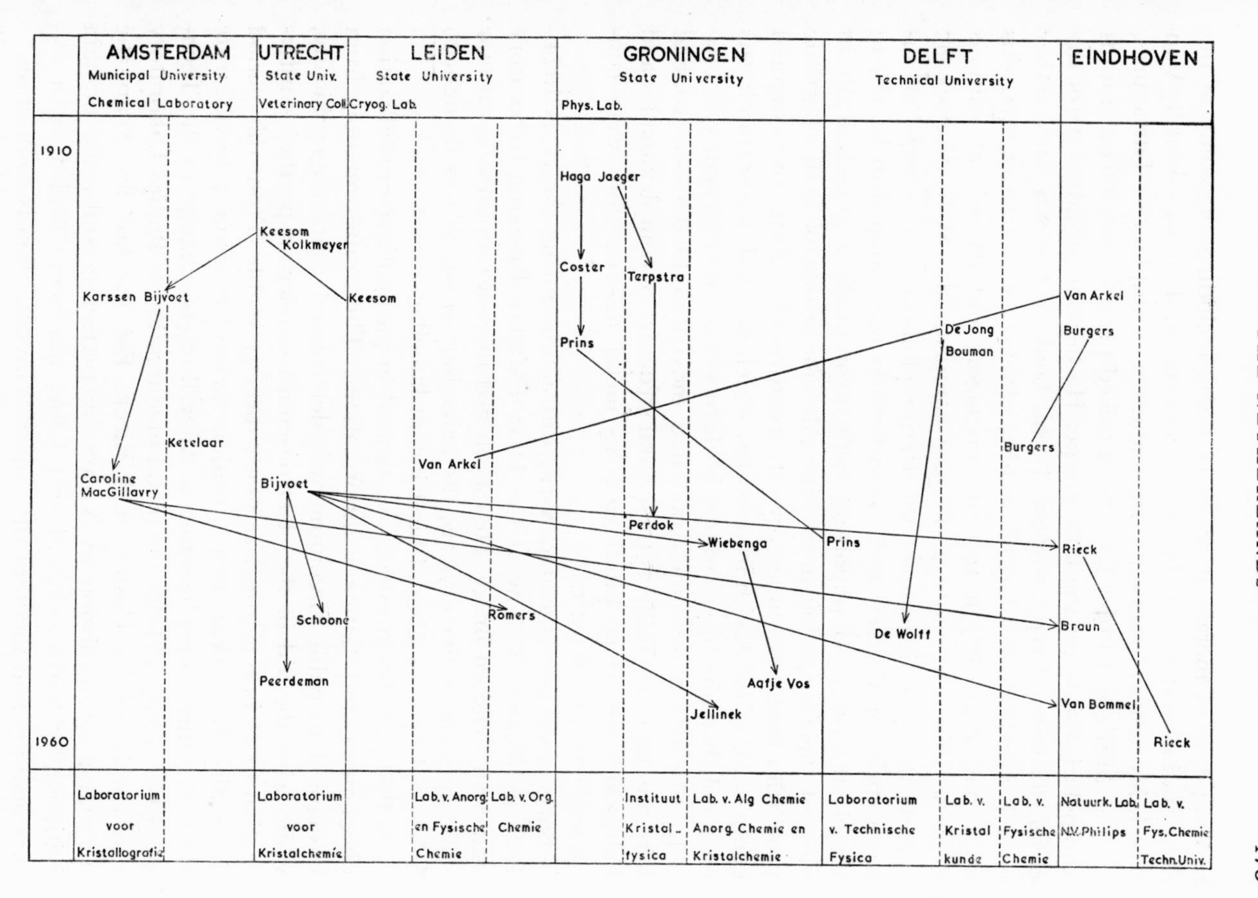
AMSTERDAM
Municipal University
Chemical Laboratory
UTRECHT
State Univ.
Veterinary Coll.
LEIDEN
State University
Cryog. Lab.
GRONINGEN
State University
Phys. Lab.
DELFT
Technical University
EINDHOVEN
1910
1960
Karssen
Bijvoet
Keesom
Kolkmeyer
Ketelaar
Caroline MacGillavry
Peerdeman
Schoone
Romers
Van Arkel
Haga
Jaeger
Coster
Prins
Terpstra
Perdok
Wiebenga
Jellinek
Aafje Vos
De Jong
Bouman
De Wolff
Burgers
Rieck
Braun
Van Bommel
Laboratorium voor Kristallografie
Laboratorium voor Kristalchemie
Lab. v. Anorg. en Fysische Chemie
Lab. v. Org. Chemie
Instituut Kristal-fysica
Lab. v. Alg. Chemie Anorg. Chemie en Kristalchemie
Laboratorium v. Technische Fysica
Lab. v. Kristalkunde
Lab. v. Fysische Chemie
Natuurk. Lab. N.V. Philips
Lab. v. Fys. Chemie Techn. Univ.

investigation of metals at an early stage, including the study of the deformation and recrystallization phenomena. This was done by Van Arkel. Moreover, by determining a number of simple crystal structures, such as MgF_2 and PbJ_2, Van Arkel traced the influence of ion radius and polarizability on the lattice type. He became an advocate of the electrostatic bond type, witness also his book, at first written together with J. H. de Boer, *Chemische binding als electrostatisch verschijnsel*; this book has appeared in several languages, and its several editions describe this field more and more comprisingly. With Verwey he started the investigation of the ferrites, followed by a series of investigations of Verwey and co-workers on the cation distribution in spinel structures, so important both theoretically and technically in view of their magnetic properties. Also for reason of their magnetic properties Braun examined crystal structures of a group of hexagonal compounds of BaO with iron oxides, which are built up of thin layers of spinel structure alternating with layers of another structure.

In the meantime Van Arkel had moved to Leiden—he studied here among other subjects the crystal structure of complex oxides and halogenides—and in Eindhoven his metal investigations were continued by W. G. Burgers.

Burgers was introduced to X-ray diffraction during the years 1924–1927 as Ramsay Fellow at the Davy-Faraday Research Laboratory under Sir William Bragg. He established there a normal structure for some optically active crystals, the behaviour of which was thought to be caused by a packing of birefringent lamellae.

At Philips Laboratory he investigated in particular recrystallization phenomena and lattice transformations. The transformation from hexagonal to cubic zirconium was determined röntgenographically, while with the aid of the electron-emission-microscope the transformation of α- into γ-iron was made directly visible. From this period dates also an optical demonstration method developed to illustrate electron diffraction phenomena. After Burgers' change to the Technical University at Delft the metal investigation at Philips Laboratory was continued by Custers and Rieck. For the last few years the anomalous transmission of X-rays by perfect crystals, the domain initiated by Borrmann and von Laue, has been studied both experimentally and theoretically, and this investigation is continuing.

At the Technical University Professor Burgers established the *Delft metallographical school* (Tiedema, Verbraak). The idea was developed that in many cases the origin of preferred orientation in the recrystallization of metals is due to oriented nucleation by polygonization of lattice-domains deformed in a certain way. In the case of copper also the possibility of a martensitic process of nucleation in the recrystallization was suggested. Continuing Cohen's work the transformation from white to gray tin was again studied and compact crystals of the gray modification were successfully prepared by alloying with mercury.

At the Delft Technical University the earliest contributions came from Dr. De Jong, formerly assistant of the late mineralogist Professor Grutterink. De Jong's thesis (1928) described the crystal structure of three sulphide minerals; he also wrote one of the first papers on a focussing powder method (1927). In the middle 30's a close cooperation arose with the theoretician Dr. Bouman. Their joint invention, the retigraph, was the forerunner of Buerger's precession camera and is still being used and further developed for its proper merits. Bouman's later work covers a great variety of aspects of diffraction theory. Some of it was done jointly with younger physicists at the Delft University, e.g. Professor De Wolff who developed a version of the Guinier camera allowing simultaneous exposure of 4 specimens.

Professor Prins left Groningen for Wageningen and later on Delft, where he continued some X-ray investigations, especially of liquids and vitrification. He cleared up the interesting allotropy of sulphur and its melt, which, in the heyday of the phase theory, had intrigued a number of Dutch investigators.

Finally some places of diffraction work should be mentioned which have not (yet) grown into schools.

Professor Goedkoop who specialized in neutron diffraction at Kjeller, is now director of our pile at Petten. Dr. Romers at Leiden is doing more particularly electron diffraction work. As regards industry, besides Philips, diffraction is being used in the laboratories of the Royal Shell Co., initiated by Ir. De Lange who had worked with Professor Robertson in Glasgow; at the State Coal Mines (Dr. Westrik); and at the Royal Dutch Explosive Manufactories (Dr. Trommel). A long series of investigations has been devoted to cellulose at the A.K.U.-Arnhem (Algemene Kunstzijde Unie) (Dr. P. H. Hermans and co-workers). We may regard J. R. Katz as the precursor in our country of this work on fibre structures. In 1925 he made the spectacular discovery that rubber, when stretched, becomes crystalline. Katz was a quite different type of X-ray investigator: a psychiatrist of enormous literary culture, but with an extremely narrow basis of mathematics. He travelled much and published a good deal on the diffraction of liquids and starch, and on the physico-chemical background of why bread grows stale.

The Dutch X-ray workers are united in the F.O.M.R.E., Stichting voor Fundamenteel Onderzoek der Materie met Röntgen- en Elektronenstralen (Foundation for Fundamental Research of Matter by X-rays and Electron Rays). This foundation was set up after the war, mainly for the purpose of obtaining more easily money from the Government so as to supplement equipment, which had become quite antiquated. The most important activity nowadays is the yearly scientific meeting. The board of this body also functions as National Committee for the Union of Crystallography. The foundation employs a praeparator, A. Kreuger, who, having grown up with X-ray analysis in Amsterdam from its beginnings, now places his great skill at the disposal of all departments.

After a preliminary period in which incidental use was made of I.B.M. accounting machines in crystallographic computations, the computer age was ushered in in 1958 by the installation of a medium-size electronic computer, the ZEBRA (Zeer Eenvoudig Binair Reken Apparaat, very simple binary calculation apparatus), at the Universities of Delft, Groningen and Utrecht, and some other type computers at industrial centres. A very close cooperation between the Universities resulted in a well equiped arsenal of interchangeable programs, preventing any essential duplication. Less intensive but still fruitful contacts are kept with ZEBRA users abroad (Cardiff, Pretoria). Very large computations, such as ultimate refinements including anisotropic temperature factors, are usually carried out abroad (Manchester, Leeds).

This survey may be concluded with a very brief summing up of the present centres of X-ray investigation. These are at the Universities of Amsterdam, Utrecht and Groningen, where structure determinations are the main subject, the metallographical school of Burgers at the Technical University at Delft and the Philips Laboratory at Eindhoven.

The author wishes to thank Dr. G. A. Bootsma for his help in assembling this survey.

CHAPTER 22

Scandinavia

by G. Hägg *

Denmark

Denmark was the last Scandinavian country to take up extensive X-ray analysis of crystals. Pioneer work in the twenties by the Dutch physicist J. R. Katz on high polymers, notably his discovery of the 'crystallization' of rubber on stretching, was not continued, nor was the work by the mineralogist H. Clausen (a student of Goldschmidt and of Aminoff) on fluoride minerals. Almost all subsequent research has been done by chemists. In pure chemistry, powder method phase identifications on calcifications, minerals, ammonia catalysts and drugs in forensic medicine have been published. Only one worker was doing crystal structure work at the end of World War II, but lately the subject has been gaining momentum, though much is at a preliminary stage. At the Royal Veterinary and Agricultural College, A. Tovborg Jensen (a student of W. L. Bragg and of Hägg) has done structure work on salt hydrates, and C. Knakkergaard Møller has investigated double halides. At the Royal Danish School of Pharmacy, Bodil Jerslev (a student of Hägg and of Dorothy Hodgkin) has studied oximes. Four chairs in chemistry are now occupied by X-ray crystallographers and an expansion of research in the field may be anticipated.

Finland

The application of X-ray diffraction techniques to structure analysis was started in Finland in 1925 by Professor S. A. Wasastjerna. His first studies in this field dealt with the structure of dolomite and in 1925 included work on the atomic configuration of the sulphate group. Of his later work, the research begun in 1939 into the structure of the

* The reviews on Denmark, Finland and Norway are based on material kindly supplied by Professor A. Tovborg Jensen, Professor M. Kantola and Dr. H. Viervoll respectively. In addition I gratefully acknowledge the assistance of Professor A. Westgren in acquiring data for the review on Sweden.

solid solutions of alkali halogenides, is worthy of note. In this investigation Wasastjerna carried out precision measurements of the diffracted intensities and proposed a new theory for the structures of the solid solutions which was related to his earlier calculations of ionic radii. In papers dating from 1945 he discussed the phenomena of thermal vibrations and lattice deformation.

Wasastjerna's pupils have continued his work in the fields mentioned above. Professor E. Laurila has measured the atomic form factors of the noble gases, Professor P. E. Tahvonen studied rotation of the nitrate group and Professor U. Korhonen developed Fourier methods and investigated the deformation of electron clouds by X-ray intensity measurements. Professor M. Kantola used X-ray diffraction for the study of the solid solutions of the alkali halogenides, paying attention to their miscibility, thermal expansion and formation by diffusion in the solid phase. Professor V. Hovi has developed the solid solution theory proposed by Wasastjerna and examined the overlapping of the electron clouds in ionic crystals.

Norway

In Norway, X-ray crystallographic investigations were initiated by Professor L. Vegard at the Institute of Physics of the University of Oslo. Vegard's first paper on this subject appeared in 1915, and in the following years he described the crystal structures of several metallic elements, oxides such as rutile and anatase, ammonium halides, zircon, nitrogen and carbon monoxide; he also proposed the well-known Vegard's Law for the atomic distances in solid solutions.

V. M. Goldschmidt's contributions in the twenties and thirties are well known, and are noted elsewhere in this volume.* F. W. H. Zachariasen's work will not be listed here, although his first X-ray analyses were carried out in Norway.

About 1925 Professor O. Hassel began using X-ray crystallographic methods at the Institute of Chemistry of the University of Oslo, and this has since been the centre for X-ray structure research in Norway. Hassel's first work was connected with the determination of the structures of graphite, boron nitride and several other inorganic compounds. One of his main interests has been the structure of organic ring compounds, particularly cyclohexane and its derivatives. Of the several problems of special interest which he has investigated in recent

* Unfortunately not, as planned, in an In Memoriam in Part V (Editor).

years are addition structures (donor-acceptor systems), mainly of halogens and halogen compounds.

One of Hassel's pupils, S. Furberg, has recently carried out a series of investigations on biologically important substances.

In ca. 1935, in cooperation with Hassel, and later independently, Chr. Finbak started investigations on the X-ray diffraction of rotating groups. As their diffraction patterns possess the same character as patterns from liquids and amorphous materials, it was natural that Finbak turned to studies of liquids. In this field, he solved a number of methodical problems. Since his death in 1954, this work has been continued by N. Norman among others. At the Central Institute for Industrial Research in Oslo, Norman has investigated fibres and high polymers, for example cellulose.

At the end of the thirties, Hassel and Finbak took up electron diffraction as a useful tool for the investigation of the structures of molecules of gases. They were joined by H. Viervoll and O. Bastiansen who developed the method to one of very high accuracy. Bastiansen has also started work on crystal diffraction at the Norwegian Institute of Technology in Trondheim. At the same Institute, H. Sørum has investigated certain minerals. After Goldschmidt, mineral structures have been further studied at the Institute of Mineralogy of the University of Oslo by T. Barth, I. Oftedal and H. Neumann. At the Institute of Chemistry, H. Haraldsen, and later F. Grønvold, studied the chalkogenides of the transition elements, and at the same Institute O. Foss started the investigation of inorganic sulphur compounds, which he then continued at the University of Bergen.

Dutch-Norwegian cooperation at the nuclear reactor at Kjeller has introduced neutron diffraction into Norway as a tool for the study of crystal structures. Organic structure problems, related to Hassel's work, have been started here; J. Goedkoop and T. Riste have also begun to examine magnetic structures.

Sweden

The first major application of X-ray diffraction in Sweden occurred in 1914 when M. Siegbahn made his first experiments in X-ray spectroscopy. The great number of fundamental studies which was then initiated is recorded in Part IV, chapter 16. In the following account, only those aspects of the development of X-ray diffraction which bear on structure analysis will be reviewed.

In 1918 G. Aminoff, who was then lecturer in mineralogy at the University of Stockholm, started structure determinations of minerals by means of the Laue method. At first he studied several simple structures, and among them found the first representatives of the C6 (brucite and pyrochroite) and B8 (nickel arsenide) types. In addition he soon began very extensive studies of the minerals from Långban, and this also included much X-ray work. A pupil of Aminoff's, N. Alsén, investigated a series of sulphides, selenides and arsenides mainly of the B8 type (1925).

Independently of Aminoff, A. Westgren, working in 1919 at the metallographic laboratory of the Swedish Ball Bearing Company (SKF) in Gothenburg, established contact with M. Siegbahn (at that time in Lund) to inquire whether the structures of the iron modifications might be determined by means of powder diffraction methods. Westgren was invited to Lund in order to attempt this and the first experiments were carried out in collaboration with A. E. Lindh. The work was continued in the autumn of 1920 with a high-temperature camera, and it was then found that the α-β transition was not accompanied by any structural change, whereas γ-iron showed the same cubic face-centered structure as the austenitic steels at room temperature.

Laue photographs of felspars were taken in Lund in 1921 by A. Hadding. From 1923 onwards, S. Holgersson, also at Lund, determined lattice dimensions of sulphides and spinels as well as of some metallic elements and alloys by powder methods.

In 1921 Westgren obtained an appointment at the Institute of Metal Research in Stockholm and he immediately began to cooperate with G. Phragmén, who at that time had made his first attempts in X-ray crystallography under the guidance of Aminoff. Westgren also succeeded in obtaining funds for enlarging the X-ray diffraction equipment of Stockholm University with cameras, for powder as well as rotating crystal methods. The new equipment was to a great extent built by Phragmén.

The studies of the iron modifications were now continued and the structure of δ-iron was found. The unit-cell dimensions of cementite were determined. Phragmén investigated the system iron-silicon and in 1923 determined the structures of FeSi and $FeSi_2$. Together with E. Jette from U.S.A., Westgren and Phragmén studied the system copper-aluminium, in which a phase with the γ-brass structure was found for the first time (1924).

The structure analogies in the systems copper-zinc, silver-zinc and gold-zinc were found by Westgren and Phragmén in 1925. At about the same time they investigated the modifications of manganese, and studied the carbide systems of chromium, molybdenum and tungsten. They also succeeded in characterizing the so-called high-speed steel carbide (1927).

In connection with a study of the copper-tin system, Westgren and Phragmén found the structure analogies between the three systems copper-zinc, copper-aluminium and copper-tin (April 1926). At a discussion meeting in the Institute of Metals in London one month earlier, Hume-Rothery had called attention to the similarities between the β-phases of these systems observed under the microscope, and had suggested that these similarities were connected to the fact that the ratio of the number of valency electrons to the number of atoms in all three phases is 3 : 2. Through the studies of Westgren and Phragmén on these and several other binary systems, this rule of Hume-Rothery's obtained more solid support.

In the autumn of 1926 A. J. Bradley came from Manchester to Stockholm in order to take part in the X-ray work on alloys. He was particularly interested in the structure of α-manganese and succeeded in determining this for an element unusually complex structure. Using diffraction data obtained in Stockholm for γ-brass and for analogous phases in several other systems he was also able to determine the structures of the γ-phases in the copper-zinc and copper-aluminium systems. The characteristic formulas of these phases derived from the structures, together with the characteristic formula found by Westgren and Phragmén for the γ-copper-tin phase, showed that a valency electron to atom ratio of 21 : 13 is the deciding factor for phases of this type to occur.

In the meantime, Aminoff had been appointed Professor and head of the Department of Mineralogy of the Museum of Natural History in Stockholm (1923). After he had obtained adequate laboratory equipment at the Museum, he was able to resume his research work in the new surroundings. The Långban minerals occupied much of his time, but he also devoted himself to general problems, for example the mechanism of the evaporation and dissolution of crystals. His theoretical study of the contact zone between the two individuals of a twinned crystal is of great value. Aminoff also made use of electron diffraction, for example in his study of the oxidation of zinc sulphide crystals. In the last-mentioned studies, as in much of his other work, Aminoff was assisted by his wife, Birgit Broomé-Aminoff. Aminoff died in 1947 after illness had greatly reduced his capacity for work over a period of about two years.

In 1925 X-ray diffraction was also started in the Physics Department of the Royal Institute of Technology in Stockholm by Professor G. Borelius together with C. H. Johansson and J. O. Linde. Their main investigations concerned alloys between metals forming continuous ranges of solid solution, and they made the important discovery of superstructures (1925–28).

In 1927 Westgren was appointed Professor in General and Inorganic Chemistry at the University of Stockholm. The close collaboration with Phragmén at the Institute of Metal Research was maintained, however, and the use of the X-ray diffraction equipment was shared by the two institutes. The equipment was also modernized and extended. Pupils in increasing numbers began to come to Westgren's laboratory, and the field of action was widened.

Westgren's first pupils were H. Arnfelt (the sytems iron-molybdenum and iron-tungsten, stacking disorder in layer lattices) and G. Hägg (systems of iron with elements of the fifth group, nitrides, carbides, borides, and hydrides of transition elements, solid interstitial solutions, spinels, tungsten bronzes, molybdenum and tungsten oxides). Then came, amongst many others, E. Öhman (manganese modifications, the system iron-manganese, martensite studies), W. Ekman (systems of zinc, cadmium and aluminium with transition elements of the eighth group, in which it was found that a number of structures of electron compounds could be explained by the Hume-Rothery type of rule if the transition elements were regarded as having zero valency), C. Brosset (alkali tungsten(III)-chlorides, fluoroaluminates, Mo(II) complexes, structures in liquid phases), A. Ölander (alloy phases for which electrochemical measurements of disorder were also made) and H. Perlitz (different alloy phases). There were also L. G. Sillén (bismuth oxides, numerous oxide halogenides, molybdates and tungstates), and A. Byström (minerals, manganese and lead oxides). Westgren himself took an active part in much of this work and solved other problems, e.g. complete structure determinations of several binary and ternary metallic carbides.

Hägg left Stockholm in 1937 for a chair in General and Inorganic Chemistry at the University of Uppsala. Among his collaborators in Uppsala the following may be mentioned: A. Magnéli (molybdenum and tungsten oxides, tungsten bronzes), I. Lindqvist (polyanions, coordination compounds, bond problems), R. Kiessling and B. Aronsson (metallic phases of transition elements with non-metals), E. von Sydow and S. Abrahamsson (long-chain organic compounds), and I. Olovsson (low-temperature work on ammines of ammonium halides and other substances with hydrogen bonds).

In 1943 Westgren was appointed Secretary of the Academy of Sciences, thus leaving his chair at the Stockholm University, and in 1944 Phragmén died. X-ray work was carried on at the University mainly through Sillén, Brosset, and Byström. B. Aurivillius (bismuth compounds, mixed oxides with layer lattices) should also be mentioned. Sillén moved his work to the Royal Institute of Technology when in 1950 he obtained the chair in Inorganic Chemistry there. His work has continued with investigations on uranium and thorium compounds as well as on solid phases connected with his extensive studies of hydrolysis. He has also been joined by G. Lundgren (oxo and hydroxo salts of tetravalent heavy metals). In 1953 Brosset obtained the corresponding chair at Chalmer's Institute of Technology in Gothenburg and was also appointed director of the Institute of Silicate Research. He has started structure analysis of glass with X-ray diffraction methods and has also cooperated with N.-G. Vannerberg (peroxides) and J. Krogh-Moe (borates).

Byström died in 1952. In the following year Magnéli went to the Stockholm University as associate professor. Since then he has organized a fairly extensive programme of research on oxides along the lines he had been following in Uppsala.

Structure determinations on minerals are being done by F. Wickman, Aminoff's succesor at the Museum of Natural History in Stockholm. Important applications of X-ray diffraction to structure problems in medicine are being carried out by E. Stenhagen (long-chain compounds), previously in Uppsala and now in Gothenburg, as well as by A. Engström and D. Carlström (structures in bone tissue) in Stockholm.

CHAPTER 23

Japan

by I. Nitta

As described in Part V, the history of fifty years of X-ray diffraction in Japan was inaugurated by T. Terada, who found in 1913, immediately after Laue's discovery and also quite independently of W. L. Bragg, the law of X-ray reflection based on his original visual as well as photographic observation of the movement of the Laue spots as the crystal was being turned.[1] In his diffraction experiments using an X-ray bulb of the Müller-Uri type operated with a Toepler influence machine, he examined single crystals of rocksalt, fluorite, quartz, mica, gypsum, borax, tourmaline, epidote, penninite, cane sugar, etc. For the visual observation with a fluorescent screen he used a very wide beam of X-rays collimated by a diaphragm with a circular aperture 5–10 mm in diameter. Terada continued his diffraction studies for about one year, investigating the effect of the bending of rocksalt [2] and analyzing to some extent the crystal structure of alum.[3]

The second name appearing in this history is that of S. Nishikawa, of whom we have spoken in detail in Part V and whose rôle was of primary importance to the progress of diffraction studies in Japan. As already mentioned, Nishikawa published in 1913 his first paper, with S. Ono, on the study of diffraction photographs of fibrous substances such as asbestos, silk and *asa*(*Cannabis sativa*—a kind of hemp), of lamellar substances like talc and mica, and of granular substances such as marble, finely pulverized rocksalt, quartz, etc., using continuous X-rays.[4] Moreover, Nishikawa studied diffraction patterns of rolled sheets of metals such as copper, iron, zinc, etc., and the effect of annealing them. Such a study is nothing other than that of the diffraction characteristics of polycrystalline textures, which has later found many applications in the fields of metallography, polymer science and other sciences. His earliest introduction of the theory of space groups as a general and logical means for crystal analysis appeared in his paper in 1915 on the crystal structure of some crystals of the spinel group and magnetite.[5] In this way began the first page of the history of X-ray diffraction in Japan.

Now, in order to describe the development, it may be convenient first to name the schools or the regional research groups in more or less the chronological order of their origin. These include the Nishikawa school, the Honda school, the Kyoto school, the Ito school in Tokyo, the Osaka school, the Hiroshima school and the Nagoya school. Besides these, there may be given names of individuals who were known to be active in the early period. They are G. Asahara, S. Kôzu, A. Ono and others.

We begin briefly with the Nishikawa school in Tokyo, as it has already been described in Part V in some detail. In the early twenties, within the so-called Nishikawa Laboratory in the Institute of Physical and Chemical Research, Nishikawa was engaged in analysing some orthorhombic crystals by means of an ionization chamber spectrometer, and his early students, Y. Sakisaka, I. Nitta, Z. Ooe and S. Shimura carried out structure analyses of crystals of inorganic compounds, minerals and compounds of metallographic importance. A little later, Nishikawa, Sakisaka and I. Sumoto, using a double crystal spectrometer as well as Laue photographs, investigated the effects of various physical treatments of crystals on the reflection intensities from the point of view of extinction and crystal imperfection. Thus they investigated the effect of surface grinding,[6] that of thermal strain or inhomogeneous temperature distribution,[7] and that of piezoelectric vibration.[8] Such investigations were later extended by his students, E. Fukushima, Y. Kakiuchi, S. Miyake, S. Yoshida and others. Fukushima studied the effect of inhomogeneous elastic deformations under an external force.[9] Kakiuchi examined that of long impressed strong electric fields.[10] Miyake observed an anomalous change in the intensity of reflection from Rochelle salt on passing the Curie point.[11] Yoshida found that the relative intensities of X-ray spectral lines changed with the degree of imperfection of the crystal used for spectrometry.[12] T. Muto made a theoretical calculation of the intensity of reflection from an alloy with a disordered structure.[13]

Since Kikuchi's experiments of electron diffraction by crystals carried out in the Nishikawa Laboratory in 1928, most of Nishikawa's students in the Institute of Physical and Chemical Research and in the Department of Physics of the University of Tokyo turned to this field. Thus K. Matsukawa, M. Miwa, T. Muto, T. Yamaguti, K. Shinohara,

S. Nakagawa, S. Miyake, Ryozi Uyeda and others played important parts in the development of wide studies on electron diffraction by crystals. Shinohara noticed that, in order to elucidate the observed Kikuchi-lines, -envelopes and -bands, one should start from the dynamical theory put forward by Bethe.[14] Besides Kikuchi, Shinohara and Nakagawa, T. Yamaguti also determined precisely the inner potentials of a series of crystals by the rotating crystal method with a knife-edge.[15] The interesting complexities of electron diffraction by crystals led Miyake and Uyeda to study very keenly and thoroughly the dynamical theory developed by Bethe, Harding, Laue and others. Their effort had a very favourable influence upon the later development of younger students, and there was gradually formed a strong electron diffraction group. Of the younger students graduating before 1945, the names of S. Takagi, K. Kimoto, G. Honjo, H. Yoshioka, K. Kohra, N. Kato and Y. Kainuma will be given. To the experimental development too, contributions of Miyake and Uyeda have been made in various respects. The instrumentation for the study of electron diffraction has been greatly improved by them before and after the War. In this connection it may be added that the manufacture of electron microscopes in Japan has greatly benefited from the co-operation of electron diffraction scientists possessing long and valuable experiences. As for theoretical development, there have been published important papers on such topics as the problem of simultaneous reflection, including the violation of Friedel's law (Miyake, Uyeda and Kohra); an anomalous phenomenon found by Kikuchi and Nakagawa (Miyake, Mieko Takagi and Kohra); determination of phase angles (Miyake and K. Kambe); dynamical theory for a finite polyhedral crystal (Uyeda and Kato), explanation of Kikuchi patterns (Kainuma); theory of absorption (Yoshioka); effect of thermal vibration (S. Takagi), etc.

The Nishikawa school, which had started as an important center of X-ray diffraction, changed its character, as described above, to become an active center of electron diffraction. In the meantime the school branched in many lines. Thus in 1933 Kikuchi and Nitta went to Osaka University, with the former beginning nuclear research and the latter continuing X-ray diffraction work. Around 1940 Miyake entered the Kabayashi Institute of Physical Research, Tokyo, and there he carried out X-ray and electron diffraction studies. In 1948 he became a professor of physics at the Department of Physics, Tokyo Institute of Technology, and led a group of scientists in both fields of X-ray and electron diffraction. Honjo, Mrs. Mieko Takagi (formerly

Miss Mieko Kubo), S. Hoshino and others were the members of the group. Uyeda remained in the University of Tokyo up to 1942, when he became a professor of physics at the Department of Physics, Faculty of Science, Nagoya University. There he formed the Nagoya school which is very active in the field of electron diffraction. Among the scientists of the Nagoya school the names of Kimoto, Yoshioka, Kato and Kainuma have already been given above. Speaking of the Nagoya school, it is to be added that a colleague of Uyeda, Y. Morino (1908–) began a series of electron diffraction studies of gas molecules with M. Kimura and others. Morino became later a professor of chemistry at the University of Tokyo. Coming back once again to Miyake, he became very recently a research professor at the newly established Institute for Solid State Physics, which is attached to the University of Tokyo, and is working there with his colleagues, Y. Saito, S. Hosoya, S. Hoshino and others.

In the Tokyo region, G. Asahara (1891–) was active, in the early period under review, as the leader of the Asahara Laboratory, Chemistry Division, Institute of Physical and Chemical Research. As mentioned in Part V, Nishikawa and Asahara began the X-ray study of metals at Cornell University in 1920. On returning to his Institute in Tokyo, Asahara soon published his X-ray studies of graphite and amorphous carbon [16] and of thallium.[17] Of his research group, H. Nakamura examined by means of X-rays the structure of electrolytic brass,[18] and T. Sasahara studied the solid solution system KCl—KBr [19] and also the structure of α-thallium.[20] Tokunosuké Watanabé (1904–) determined the crystal structure of northupite, brominated northupite and tychite.[21] By the time he published this paper, Asahara had retired from the Institute and had been succeeded by H. Shiba. Besides the two groups of Nishikawa and Asahara, M. Majima, S. Togino and K. Yamaguchi in the Engineering Division of this Institute were also active in carrying out X-ray studies of metals. S. Yamaguchi made independently a series of electron diffraction studies of metals and chemical reactions on their surfaces.

Next to the above groups in Tokyo, it will be appropriate to turn to the Sendai region, where the Tôhoku University is located. The application of the X-ray method was first attempted by S. Kôzu (1880–1955), a professor of petrology of the Faculty of Science, Tôhoku University. In collaboration with Y. Endö, a physicist, he began in 1921 his notable X-ray study of the felspar group, especially adularia and moonstone.[22] Later Kôzu and K. Takané made structure determination of cancrinite, bronzite, vesuvianite, diaspore, enargite, etc.[23] In the Research Institute for Iron, Steel and Other Metals, founded by K. Honda (1870–1954) and attached to Tôhoku University, the X-ray method was introduced by M. Yamada. The first two X-ray papers by Yamada published in 1923 were a note on the reflection of X-rays from a fluorite crystal [24] and on the occlusion of hydrogen in palladium.[25] As is well known, Honda led in his Institute a large number of scientists in the fields of metallurgy, metallography and physics of magnetism for a long period of time. He was also one of the sponsors of the *Zeitschrift für Kristallographie*. Honda's students who were engaged in X-ray metallographic studies are, besides M. Yamada and Y. Endö mentioned above, A. Ôsawa, S. Sekito, S. Ôya, K. Endo, T. Sutoki, Z. Nishiyama and many others. They made use of the X-ray diffraction method for the identification or confirmation of a definite phase, for the determination of equilibrium phase diagrams, the study of occlusion of gases in metals, solid solution formation, phase transformation, etc. E. Matsuyama improved the high-temperature camera, and I. Edamoto constructed an X-ray tube with oscillating target. I. Obinata, who had made an X-ray study of the β-phase of Cu-Al alloy at the Ryojun College of Engineering and then worked with E. Schmid and G. Wassermann at the Kaiser-Wilhelm-Institut für Metallforschung on the solid solubility in the Pb-Sn system, the plastic deformation of metal single crystals, etc., joined the Honda school. Shiro Ogawa (1912–), who is known with M. Hirabayashi, D. Watanabe and others for his studies of antiphase domains of some alloys, is leading the research group of X-ray and electron diffraction in the Institute.

In Kyushu University X-ray work dates back to 1922, when A. Ono (1882–), at the Department of Mechanical Engineering, College of Engineering, attempted an X-ray examination of the inner structure of strained metals such as copper, aluminium, and α-iron from the standpoint of material testing. The results of this investigation were reported in a series of papers from 1922 to 1930.[26] In his third report in 1925 he noted that the findings of G. I. Taylor and C. F. Elam (1925) and of M. Polanyi and E. Schmid (1925) about the slip resistance of crystals stood in conformity with his view concerning the cause of strain-hardening.

Kyoto University did not stand behind in the introduction of the X-ray method to various fields of scientific studies. This was mainly by virtue of M. Ishino and U. Yoshida in the Department of Physics,

College of Science. Thus in 1925 S. Tanaka (1895–) and T. Fujiwara (1897–), students of Yoshida, published, respectively, papers on the X-ray study of the polycrystalline texture of rolled platinum sheet[27] and of aluminium and copper wires.[28] In 1927 Yoshida proposed some experimental devices which facilitate the determination of the orientation of crystal axes.[29] Yoshida introduced many students to the X-ray investigation of metallurgical and various other problems. Such students are K. Tanaka, J. Tsutsumi, S. Shimadzu, G. Okuno, K. Hutino, M. Kabata, S. Nagata and others. H. Hirata, G. Shinoda, and C. Matano of Kyoto University are also known for their X-ray studies, mainly of metallographical and metallurgical problems, during the decade around 1930. I. Sakurada (1904–) of the College of Engineering, Kyoto University, once a student of Hess at the Kaiser-Wilhelm-Institut für Chemie, Berlin-Dahlem, began a series of X-ray investigations of natural and synthetic high polymers with Hutino. K. Tanaka (1904–) became the professor of physics succeeding Yoshida and the leader of the X-ray and electron diffraction group at the Kyoto College of Science. A book on X-ray crystallography written by Yoshida and Tanaka appeared in the thirties and was of great help to young students who wanted to advance in this field. E. Suito, who has for many years been engaged in electron microscopic studies of fine powder systems, has extended his study into electron diffraction at the Institute for Chemical Research attached to the University.

Now we pass on to the Ito school in Tokyo. T. Ito (1898–) finished in 1923 his student course of geology at the University of Tokyo, and went to Kyoto University for further study in petrology. After a short time he was called back to the University of Tokyo, and in 1925 went to Zurich to work with P. Niggli at the Eidgenössische Technische Hochschule. There he formed his thorough background of structural crystallography, and studied especially the topological structure analysis of the Niggli school. His paper on the diamond lattice complex in the orthorhombic system appeared in 1928.[30] The next year he visited W. L. Bragg in Manchester and learned the methods of X-ray crystal analysis. His X-ray papers, with J. West, on the crystal structure of hemimorphite [31] and of bertrandite [31] were published in 1932. When he came back to Tokyo, he became professor of mineralogy, University of Tokyo, and since then has been very active in the field of X-ray crystallography. Influenced by the study of Kôzu mentioned above, he has long been interested in problems such as crystal structures of rock-forming minerals, especially the felspar group, mode of twinning, theoretical extension of space groups, etc. In 1950 he published a book entitled *X-ray Studies on Polymorphism* (Maruzen, Tokyo), which covers the work done in his school during the last War and was not published elsewhere. In 1949 Ito devised a new general method of lattice determination based on the Debye-Scherrer pattern. This is a development of the old idea of C. Runge (1917). In Ito's method use is made of the method of lattice reduction devised by Delaunay (1933), which Ito had noticed quite early. His students active in the X-ray analysis of minerals and organic compounds are R. Sadanaga, Y. Takéuchi, N. Morimoto, Y. Iitaka, K. Doi and others. Sadanaga succeeded Ito after his retirement. M. Nakahira, who was once with Nishikawa and then with G. W. Brindley, later became a lecturer in the Department of Mineralogy of the University. Nakahira and T. Sudo, of the Tokyo University of Education, are known for their X-ray investigations of clay minerals.

In the Osaka region, Osaka University was established in the early thirties. In the Department of Applied Physics and Precision Machinery, Faculty of Engineering, S. Tanaka, G. Shinoda, K. Kojima and S. Nagata, all from the Kyoto school, have been engaged in applications of X-ray methods to metallographic and other problems and in the instrumentation for X-ray and electron diffraction studies. In the Department of Chemistry, Faculty of Science, I. Nitta and T. Watanabé, both from the Institute of Physical and Chemical Research already mentioned, undertook the project of crystal analysis of organic compounds. In 1937 two-dimensional Fourier syntheses of electron density distribution in tetragonal pentaerythritol were carried out by them for the first time in Japan. They became interested in hydrogen-bonded structures of mainly organic crystals and also in orientational and rotational disorder in molecular crystals; the latter are related to the phase of the so-called plastic crystals of J. Timmermans (1938) and form a significant approach to the physics and chemistry of the liquid state. The X-ray and other physico-chemical studies of such problems have been made by them and their students such as R. Kiryama, S. Seki, K. Sakurai, T. Oda, I. Taguchi, K. Osaki, Y. Saito, M. Kakudo, S. Hirokawa, R. Shiono, Y. Okaya, Y. Tomiie, M. Atoji, and many others. In 1952 Okaya and Nitta published a paper containing an elementary derivation of linear inequalities for phase determination. Sakurai devised a graphical method applicable to the Harker-Kasper inequalities. Recently

Taguchi, S. Naya and Oda developed a theory of inequalities in a general manner by use of matrix theory. Y. Saito and K. Nakatsu determined absolute configurations of some complex salts. T. Matsubara of the Department of Physics made some improvements in the theory of X-ray diffuse scattering by using matrix calculations, and together with Oda applied these to some actual cases of plastic crystals. It is to be added that Ryuzo Ueda, who is now a professor of the Department of Applied Physics, Waseda University, Tokyo, was before with Nitta and Watanabé. Dating back to 1935, Y. Go, once with O. Kratky in the Kaiser-Wilhelm-Institut für Faserstoff-forschung and then with K. H. Meyer at the University of Geneva, returned to Japan and joined the staff of the Department of Chemistry of Osaka University bringing the technique of the Weissenberg goniometer. He established a laboratory of polymer science and began X-ray and electron diffraction studies of polymers with S. Nagata and J. Kakinoki. After the War the latter became a professor of physics of Osaka City University, and there he has led a group in X-ray and electron diffraction studies. Recently Kakinoki and Y. Komura developed the theoretical calculation of intensities from irregular layer lattices by use of a matrix method. The Institute of Industrial and Scientific Research attached to Osaka University was opened in 1939. There Z. Nishiyama, already mentioned as a student of Honda, carried out X-ray studies of martensite, of age-hardening of alloys and of the nickel oxide structure. He has trained Y. Shimomura, S. Nagashima and others. K. Kojima, from the Kyoto school, has been engaged in the determination of internal stress in metallic materials by means of X-ray diffraction, along with S. Karashima.

In Hiroshima University, until very recently T. Fujiwara, (1897–) as a professor of physics, Faculty of Science, was active in research and teaching of metal physics using X-ray methods. As already mentioned he is a student of U. Yoshida of Kyoto University. He is known for his studies of divergent beam X-ray photographs and for growing single metal crystals. Incidentally, T. Imura, now at the Institute of Solid State Physics, University of Tokyo, developed the study of divergent beam photographs in the Department of Metallurgy, University of Osaka Prefecture. S. Yoshida is one of the best known among the students of Fujiwara and has followed the same line as his teacher. In the same Department H. Tazaki is also known for his structure analysis of boric acid and other inorganic compounds.

References

1. T. Terada, *Proc. Math. Phys. Soc. Tokyo, 7,* 60 (1913); *Nature, 91,* 135 (1913).
2. T. Terada, *Proc. Math. Phys. Soc. Tokyo, 7,* 290 (1914).
3. T. Terada, *Proc. Math. Phys. Soc. Tokyo, 7,* 292 (1914).
4. S. Nishikawa and S. Ono, *Proc. Math. Phys. Soc. Tokyo, 7,* 131 (1913).
5. S. Nishikawa, *Proc. Math. Phys. Soc. Tokyo, 8,* 199 (1915).
6. Y. Sakisaka, *Jap. J. Phys., 4,* 171 (1927); *Proc. Phys. Math. Soc. Japan, 12,* 189 (1930).
7. Y. Sakisaka and I. Sumoto, *Proc. Phys. Math. Soc. Japan, 13,* 211 (1931).
8. S. Nishikawa, Y. Sakisaka and I. Sumoto, *Phys. Rev., 38,* 1078 (1931).
9. E. Fukushima, *Bull. Inst. Phys. Chem. Research, Tokyo, 14,* 1105, 1199 (1935); *15,* 1 (1936).
10. Y. Kakiuchi, *Proc. Phys. Math. Soc. Japan, 23,* 637 (1941).
11. S. Miyake, *Proc. Phys. Math. Soc. Japan, 23,* 377, 810 (1941); *J. Phys. Soc. Japan, 2,* 98 (1947).
12. S. Yoshida, *Sci. Papers Inst. Phys. Chem. Research, Tokyo, 38,* 263 (1941).
13. T. Muto, *Sci. Papers Inst. Phys. Chem. Research, Tokyo, 31,* 153 (1937).
14. H. Bethe, *Ann. Physik* (Lpz.), *87,* 55 (1928).
15. T. Yamaguti, *Proc. Phys. Math. Soc. Japan, 12,* 203 (1930); *14,* 1, 57 (1932); *16,* 95 (1934); *17,* 58 (1935); *18,* 372 (1936); *21,* 375 (1939).
16. G. Asahara, *Sci. Papers Inst. Phys. Chem. Research, Tokyo, 1,* 23 (1922); *Jap. J. Chem., 1,* 35 (1922).
17. G. Asahara, *Sci. Papers Inst. Phys. Chem. Research, Tokyo, 2,* 273 (1925).
18. H. Nakamura, *Sci. Papers Inst. Phys. Chem. Research, Tokyo, 2,* 287 (1925).
19. T. Sasahara, *Sci. Papers Inst. Phys. Chem. Research, Tokyo, 2,* 277 (1925).
20. T. Sasahara, *Sci. Papers Inst. Phys. Chem. Research, Tokyo, 5,* 82 (1926).
21. H. Shiba and T. Watanabé, *Compt. rend., 193,* 1421 (1931); T. Watanabé, *Sci. Papers Inst. Phys. Chem. Research, Tokyo, 21,* 40 (1933).
22. S. Kôzu and Y. Endö, *Sci. Reports Tôhoku Imp. Univ.,* (III) *1,* 1 (1921); S. Kôzu and K. Wada, *J. Geol. Soc. Tokyo, 30,* 342 (1923).
23. S. Kôzu, *Jap. J. Geol., 9,* Abstr. 1–2 (cancrinite); K. Takané, *Proc. Imp. Acad. Japan, 8,* 308 (1932) (bronzite); S. Kôzu and K. Takané, *ibid., 9,* 56, 105 (1933) (cancrinite); K. Takané, *ibid., 9,* 9 (1933) (vesuvianite); *ibid., 9,* 113 (1933) (diaspore); *ibid., 9,* 524 (1933) (enargite).
24. M. Yamada, *Sci. Reports Tôhoku Imp. Univ., 11,* 447 (1923).
25. M. Yamada, *Phil. Mag., 45,* 241 (1923).
26. A. Ono, *Mem. Coll. Eng. Kyushu Imp. Univ., 2,* 241, 261 (1922); *3,* 195, 267, 287, (1925); *Proc. 2nd. Intern. Congr. Appl. Mechanics,* Zurich (1926); *Proc. 3rd. Intern. Congr. Appl. Mechanics,* Stockholm (1930); *Proc. Japan Acad., 26,* (9) 14 (1950).
27. S. Tanaka, *Mem. Coll. Sci. Kyoto Imp. Univ., 8A,* 319 (1925).
28. T. Fujiwara, *Mem. Coll. Sci. Kyoto imp. Univ., 8A,* 339 (1925).
29. U. Yoshida, *Jap. J. Phys., 4,* 133 (1927).
30. T. Ito, *Z. Krist., 67,* 341 (1928).
31. T. Ito and J. West, *Z. Krist., 83,* 1 (hemimorphite), 384 (bertrandite) (1932).

CHAPTER 24

Schools of X-ray Structural Analysis in the Soviet Union

by A. V. Shubnikov

The discovery of X-ray diffraction coincided with the centennial celebration of the eviction of Napoleon from Russia. The Czarist government attempted to use this day for boosting the patriotic spirit of the people, but without much success. In 1911 a mood of opposition prevailed among students and professors which was brought about by mass discharges of students and professors from universities, the closing of a series of departments at universities, and other measures taken by the Minister of Public Education, L. A. Kasso. Among the creative workers in the field of crystallography who left the Moscow University were V. I. Vernadskii, A. E. Fersman, Ia. V. Samoilov, and G. V. Wulf. At that time I was finishing my studies at the Moscow (State) University, working under the guidance of G. V. Wulf on a paper on the symmetry of $K_2Cr_2O_7$ crystals, and at the same time acting as the unofficial assistant of my teacher at the Peoples' City University. The further research and teaching activities of G. V. Wulf continued at this Peoples' University, organized with great difficulties from private means in Moscow in 1908. A year before the described incidents in Moscow the well-known Russian crystallographer E. S. Fedorov, the Director of the Mining Institute in Petersburg named for Empress Catherine II, was discharged from his post (approval not granted by Minister Timashev after a second election by the Scientific Council of the Institute). This was the general picture of the political circumstances which befell Russian crystallographers in the year of Laue's magnificent discovery.

Despite the extremely unfavourable circumstances for the flourishing of the Sciences which prevailed in Czarist Russia at the beginning of the 20th century, crystallography in this country was developed to a rather high level. This was evidenced by the fact that at the time of Laue's discovery, a series of original textbooks had been written for the teaching of crystallography at higher schools. Research in the field of crystallography underwent intensive development and was concentrated around two schools: the Petersburg school, headed by Fedorov at the Mining Institute, and the Moscow School, headed by Wulf at the Peoples' University. Their many similarities notwithstanding, these two schools essentially differed from each other in the purpose, the method, and the role of crystallography in the development of Natural Sciences.

In his renowned courses of crystallography, Fedorov treated crystallography as the 'base of all sciences of inorganic nature'. He placed theoretical crystallography 'on the same level with the most precise of the existing sciences'. Fedorov was of the opinion that theoretical crystallography could be built up, 'without the risk of even the smallest conflict with experimental data', from the 'initial experimental state of crystallography' and the 'immutable fact', which was that 'all particles of a crystalline substance are identical and arranged in parallel positions', that these particles—'crystalline molecules'—taken as a whole 'completely fill space', and that the 'parallelohedra should be considered as portions of space, belonging to separate crystalline molecules'. It is interesting to note that Fedorov remained faithful to this 'basic law' even after the completion of the space-group derivation (1900) which clearly indicated that a non-parallel arrangement of identical particles was possible. It should be added that Fedorov did not follow up his conclusion because he assumed that crystals belonging to asymmorphic groups could not exist in nature. His *Brief Course of Crystallography*, published ten years after the derivation of the space groups, does not even mention these groups.

My teacher Wulf held different views of crystallography. In contrast to Fedorov who over-estimated crystallography, considering it as the 'base of all sciences of inorganic nature', Wulf under-estimated crystallography, which, in his opinion was simply a 'chapter in physics', 'did not deserve to be called a separate science'. Combining forces with physicists (Voigt), Wulf identified crystallography with the 'study of a solid as a certain *medium*'. In this Wulf definitely set himself apart from mineralogists who, until recently, considered a crystal as an '*individual of inorganic nature*'. Wulf was disturbed because the idea of crystallography as a part of physics did not find unanimous acceptance; for this reason, our university does not include crystallography in its physics course.

Von Laue's discovery left very deep impressions on the two schools of crystallography in Russia and brought with it different reactions from the leaders of these schools.

When the news of the discovery of X-ray diffraction reached Wulf, he immediately expressed his desire to work in this field. Studying von Laue's equations to the very end, Wulf derived from them the relationship

$$\frac{\lambda}{2} = \frac{\Delta\varepsilon}{m},$$

which is of identical meaning with Bragg's well-known formula

$$n\lambda = 2d \sin\theta$$

The paper was published in *Phys. Zs.* 1913, *14*, 217. All of Wulf's further scientific work was largely determined by von Laue's discovery. In 1916 Wulf translated into Russian the book by W. H. and W. L. Bragg—*X-rays and Crystal Structure.*

Fedorov's reaction to Laue's discovery was quite different. Evaluating correctly the 'change brought about in crystallography by the application of X-rays to the study of crystals', Fedorov could not overlook the fact that new experimental data of the structures of diamond, rocksalt, and other crystals refuted the 'unalterable fact' (the unfailingly parallel distribution of identical particles in a crystal), which is contained in the 'basic law'. Compelled to acknowledge that 'under no circumstances can crystals be considered as simple space lattices of particles', Fedorov nevertheless attempted in various ways to change this law, but without success. Fedorov passed away in 1919, in Leningrad.

Fedorov's students at the Leningrad Mining Institute attempted to organize experimental research in the field of X-ray structural analysis. However, the difficult post-war days seriously impeded these attempts. Somewhat more favourable conditions existed in Moscow, where Wulf was in charge of studies by means of X-ray diffraction. It must be emphasized, however, that this work in Leningrad as well as in Moscow developed extremely slowly. It need be said only that until the death of Wulf (in 1925), the crystal structure of only one substance ($NaClO_3$) had been studied in our country, by Wulf himself. The next twenty years could be characterized also as a preparatory period for serious experimental research in the field of X-ray structural analysis.

The research in this field underwent an intensive development only after World War II and was localized in certain research centres in Moscow: the Institute of Crystallography of the Academy of Sciences,

USSR, under the supervision of N. V. Belov; the L. Ia. Karpov Physico-Chemical Institute, under the supervision of G. S. Zhdanov; the Institute of General and Inorganic Chemistry of the Academy of Sciences, USSR, and the Moscow University, under the supervision of G. B. Bokij; and the Institute of Organic Chemistry of the Academy of Sciences, USSR, under the supervision of A. I. Kitaigorodskii.

Belov devoted his studies to the structure of silicates. In collaboration with his numerous students Belov succeeded in deciphering a whole series of very complex silicate structures (ilvaite, epidote, zoisite, cuspidine, xonotlite, wollastonite, gadolinite, seidoserite, lovenite, lovoserite, epididymite, rhodonite, and hillebrandite). 'Direct' methods of X-ray structural analysis were, and still are, developing parallel to the experimental research in Belov's school. The complex methods of structural calculations are now carried out by machines. Belov is now rightfully accepted as the leading figure in the field of structural crystallography in our country and an outstanding expert of space groups.

Zhdanov is known here as the author of the first textbook on X-ray analysis. At the beginning of his independent scientific activities, Zhdanov had a somewhat limited choice in the selection of substances for his studies which had to be in the field of interest of the Karpov Institute. This resulted in studies of the crystal structures of a series of inorganic compounds (carbides, cyanides, rhodanides, borides, oxides), and organic substances (nitro and halogen derivatives of naphthalene and benzene, organometallic substances, and organic dyes). At the present time Zhdanov is interested in crystals with special physical properties (ferroelectrics, piezoelectrics, superconductors, and others). This research is being conducted at the Department of Solid State Physics of the Moscow University.

Bokij chose complex compounds as the object of his crystal chemical investigations. A new group of complex compounds with multiple bonds (within the molecule) was discovered and investigated recently. The crystal chemical theory of daltonides and bertolides was formulated. The Department of Crystallography and Crystal Chemistry was organized by Bokij under the Geological Faculty of the Moscow University. His scientific work has been conducted primarily at the Institute of General and Inorganic Chemistry of the Academy of Sciences, USSR. Bokij is the author of the book *Crystal Chemistry* (1960), and editor of the *Journal of Structural Chemistry*, in the founding of which he has been instrumental. Among Bokij's numerous students, M. A. Porai-Koshits deserves special mention.

Organic crystal chemistry is the speciality of A. I. Kitaigorodskii. The underlying idea of this science, according to the author, is the close-packing of nonspherical particles—molecules—, i.e. arrangements of molecules such that the 'protrusions' of one molecule fit into the 'depressions' of the other. This idea is developed in detail by A. I. Kitaigorodskii in his book, *Organic Crystal Chemistry* (1955), and is substantiated in studies by the author and his students and by thorough investigations of the structures of organic crystals according to data available in literature. Kitaigorodskii is known here as the author of basic textbooks on X-ray structural analysis.

CHAPTER 25

The World-wide Spread of X-ray Diffraction Methods

by P. P. Ewald

In the foregoing chapters the development of centres and schools of X-ray diffraction has been described in some detail for the leading countries. Early work has, however, not been restricted to these countries, and the spread of the application of these methods to new countries and to new laboratories took place in the twenty years between the wars and continues today at an even increased rate. This is a natural development for a method of such general importance for problems in many sciences and in technology, and the present chapter can not and need not go into the often repetitive detail of the development in all countries.

In collecting material for this chapter, the author sent out a questionnaire form to the sub-editors of the 1960 edition of the *World Directory of Crystallographers*. The 19 forms brought back 10 answers, most of them with very complete and interesting information, and the author wishes to acknowledge his great obligation to those who assisted him in collecting the material. It was essential in compiling the short passages following below; at the same time it should be made clear that the author in many cases has added material from his own knowledge or recollection, and mistakes which may have crept in through him should not be blamed on his informants.

We follow the alphabetical order.

Argentina

Probably the first Argentine scientist to get in contact with an X-ray diffraction laboratory was Dr. Horacio Damianovich. He travelled to Paris in 1929 taking with him some platinum electrodes that had been working for many hours in helium-filled discharge tubes. Prof. Trillat took some Debye-Scherrer patterns, but the results regarding a possible Pt-He interaction were not conclusive (Damianovich and Trillat, 1929).

In 1933 Prof. E. E. Galloni of the Natural Sciences Faculty in Buenos Aires worked for five months with a fellowship grant with Prof. J. Palacios in Madrid (space group of gypsum), but it was not until 1938 that a powder camera could be installed in the Institute of Experimental Medicine (Director Dr. A. H. Roffo) and the structure of platinum oxide studied (Galloni and Roffo, 1941).

In 1941 the Science Faculty of the University in Buenos Aires purchased a General Electric diffraction unit including powder and rotating crystal cameras. Work on metallic oxides and mineral identification was done with this unit up to 1952, and a similar unit purchased at about the same time for the Physics Institute of the University of La Plata was in use there.

The Argentine Atomic Energy Commission in Buenos Aires uses powder and Laue methods in its metallurgical Department and electron as well as X-ray diffraction in its diffraction laboratory which is directed by E. E. Galloni. This Commission has recently acquired Weissenberg, small-angle, and microcameras from the Navy Medical Institution.

Several more installations can be found at the Solid State Physics Institute at San Carlos de Bariloche (founded in 1954) and in academic physics and geology departments.

A National Committee of Crystallography was founded in 1959. A rapid increase in research is to be expected thanks to scholarships which enable students to study in well-known laboratories, and to visiting lecturers. The first meeting of the Ibero-american Association was held in Corboda Province in 1960; twenty papers were read on that occasion.

Austria

The first serious attempt at establishing structure analysis seems to have been made by Prof. Fr. Raaz of the Department of Mineralogy of the University in Vienna. Raaz had been studying the subject with Rinne and Schiebold in Leipzig and used Seemann equipment in 1929. Before that, from 1927 onwards, F. Halla and M. C. Neuburger used the powder method for studying metal oxides, alloys, and the lattice constants of metals at the Department for Physical and Inorganic Chemistry of the Technical University in Vienna. A considerable amount of work on light and X-ray scattering was done at the First Chemical Institute of the University of Vienna while H. Mark had the chair there (1932–38) in connection with the nature of liquids and of high-polymers. The mineralogical department, now under Prof. F.

Machatschki, and the chemical department of the university are still the centres of Austrian structure analysis. O. Kratky in Graz together with G. Porod devoted much work to the technique and interpretation of small-angle scattering by colloids and high-polymers. The main industrial laboratory using X-ray diffraction is that of the Plansee-werke in Reuthe which specialize in powder metallurgy.

Chile

The application of X-ray diffraction dates entirely from after the Second World War. In the initial period, 1951–53, such work was begun in the physics departments of the universities in Concepción with English, and in Santiago with German (Siemens) equipment. Further X-ray facilities are to be found at the Technical University at Valparaiso and the State Geological Institute in Santiago. At the university in Santiago N. Joel applies X-ray diffraction to the physical problems of thermal vibrations as well as to structure analysis, and E. Grünbaum is paralleling this with electron diffraction. At all three universities the work is being done in the physics departments. The number of scientists trained in these methods is still small—about 12 in 1960—but as the methods now penetrate into the teaching curricula, a rapid advance seems likely.

Czechoslovakia

A first independent start with X-ray analysis was made at the Charles University in Prague in 1923 in a joint effort between the theoretical physicist K. Teige and the mineralogist B. Ježek. In 1923–25 F.Ulrich received a training under V. M. Goldschmidt in Oslo. A year or two earlier V. Dolejšek had gone to Lund in order to study X-ray spectroscopy with Manne Siegbahn. Two important papers resulted from this in 1922: the first measurement of the lines of the N-series; and, together with Siegbahn, the improvement of the accuracy of line measurement and its application to the K-spectra of elements from 17 Cl to 30 Zn. On his return to Prague, Dolejšek established a school of X-ray spectroscopy of great precision.

In the period between the wars, G. Hüttig did interesting work as Professor of Chemistry at the Charles University on 'vagrant constituents' in crystals using X-rays in conjunction with other methods. More recently, active schools of X-ray crystallography are to be found in Prague (Praha) and Brno in academic laboratories and in the industrial laboratory of the Skoda works.

Israel

At the Weizmann Institute in Rehovoth X-ray crystal analysis was from its beginning (1948) in the hands of the chemists. G. M. J. Schmidt in particular studied the effects of 'molecular overcrowding' on the structures of organic crystals. J. Gillis at the same institute is interested in the mathematical and computational problems of crystal structure analysis.

In the Physics Department of the Hebrew University in Jerusalem (Prof. E. Alexander) problems of X-ray optics and spectroscopy are being studied. Further laboratories using X-ray diffraction are those of the Atomic Energy Commission (governmental) and of Tadir, Ltd. (industrial).

Italy

In spite of the great tradition in mineralogy and crystallography which in Italy goes back to the 17th century, the first X-ray work came from the *chemical* institute of the Polytechnic in Milan and was directed to problems of analytical and physical chemistry. A very viable school, which still persists, was created here in 1924 by G. Bruni, G. R. Levi and G. Natta with many co-workers. Their investigation of oxides, hydroxides, fluorides, carbonates, etc. was based on powder diagrams, the technique of which G. R. Levi had been studying with Debye and Scherrer in Zürich. In the course of time, single-crystal methods were adopted, and now the use of Weissenberg and precession cameras, optical and machine summation leads to single-crystal work of high quality and chemical significance.

Among the mineralogists, E. Onorato was the first to enter the field (1927) by his determination of the space group of gypsum from Laue diagrams during his studies with Rinne-Schiebold in Leipzig. In the same laboratory V. Montoro received his training; he later devoted most of his work to analyses of metallurgical interest.

The lack of equipment which was largely responsible for the relatively small amount of work coming from Italy in the period between the wars was overcome in the last decade; the number of well equipped laboratories has doubled and a large group of young and well trained X-ray crystallographers is now at work in academic, industrial and governmental laboratories on chemical as well as physical and mineralogical problems. Fluorescence analysis and electron diffraction are also being used.

Spain

Within a year after Laue's discovery F. Pardillo gave a report on it and on the first work of the Braggs to the Royal Society of Natural History (*Bol.* 1913, *13*, 336), and two years later (1915) Blas Cabrera wrote a similar report in the *Anales Soc. Esp. Fis. y Quim. 13*, 7. It took, however, another ten years before the new experimental technique was actually introduced in Spain, first at the Physics Department of the University in Madrid, but soon after, and on a larger scale at the Instituto Nacional de Física y Química which had meanwhile been established with the help of the International Education Board (Rockefeller) in Madrid. Here J. Palacios assembled a group of keen young workers to whom he lectured and for whom he invited guests from abroad. A Weissenberg camera was built in the workshop of the institute from the drawings which Hengstenberg had provided, and he and Wierl, both from H. Mark's laboratory in Ludwigshafen, initiated electron diffraction which was taken up by L. Brú. J. Garrido, O. Foz, L. Rivoir and R. Salvin worked together with Palacios on structure determinations of inorganic and organic crystals and on the perfection of the Fourier methods of analysis. The Civil War brought all this to a near standstill from 1935 onwards. The same is true for the work which F. Pardillo had inaugurated quite independently and without external help in the mineralogy department of the university in Barcelona.

A revival took place after the Second World War, and there were considerable shifts of location of the scientists. L. Brú who had no X-ray instrumentation as long as he was professor of physics at La Laguna in the Canaries, was appointed to the physics chair at the University of Sevilla in 1949, and there, gathering some of his former collaborators around him, he began work on structure problems. In 1956 he came in the same capacity to the University of Madrid and is continuing both X-ray and electron diffraction there with an increasing number of co-workers.

L. Rivoir heads the X-ray department in the converted Instituto Nacional de Física y Química, now Instituto de Física 'Alonso de Santa Cruz'.

The most active among the Spanish crystallographers is J. L. Amoros who after graduating at the university in Barcelona in 1942 worked there in the Instituto Lucas Mallada of the Higher Research Council (CSIC) before becoming the professor of crystallography in Madrid. Together with his numerous co-workers he is investigating structures as well as physical properties of crystals and the methods of their analysis. Interesting and useful publications have been issued by this group which provide the new generations of students with texts to study.

It should be mentioned that also J. Garrido and J. Orland published in 1946 a book *Los rayos X y la estructura fina de los cristales* which forms a good introduction to the subject and contains some methods developed by the authors.

A Crystallographic Society was formed in 1950 with some 35 members; in 1960 membership had risen to 60. In 1960 an Ibero-American Association for Crystallography was founded with the intent of tying together crystallographic research in Argentina, Chile, Uruguay and Spain.

Switzerland

The development of diffraction methods in this country was shaped mainly by the prominent men in Zürich: P. Debye, P. Scherrer, and P. Niggli. They came to Zürich in 1920, the former two to the Physics Department, the latter to that of Mineralogy and Petrography of the E.T.H. In neither of these places did much interest evolve in crystal structure determination as such, although Niggli analysed tenorite, and under Scherrer the structures of ferroelectric crystals like Rochelle salt were investigated. In Niggli's institute the emphasis lay on the general laws of crystalline architecture in a rather abstract, geometrical and classifying sense; E. Brandenberger's important analysis of the laws of extinctions, and H. Heesch's extension of the theory of space groups to include anti-symmetry elements should, however, be mentioned.

Crystal structure analysis in the accepted sense was used on a large scale from 1930 onwards by the physico-chemist W. Feitknecht in Bern, initially in conjunction with W. Lotmar. Systematic chemical surveys of series of compounds, such as the basic salts of bivalent metals, double hydroxides, and hydroxy salts were carried through mainly using powder diagrams, and led to extensive series of papers; Lotmar later turned to high-polymer and protein diffraction. In the department of mineralogy of the same university, Bern, W. Nowacki has built up an active and well equipped centre of structure analysis. A great variety of crystals, ranging from minerals to sterines and other organic crystals of high molecular symmetry have been determined here. The principles of crystalline structure, and the statistics of the distribution of structures among the space groups have drawn the particular interest of Nowacki (for the latter see Donnay and Nowacki, *Crystal Data, Memoir 60*, Geological Soc. of America, 1954).

In the years 1932–38 (approximately) a very active school of X-ray optics flourished in Geneva under the physicist J. Weigle. In a number of papers Weigle and his co-workers tested the validity of the dynamical theory of X-ray diffraction, and also laid the foundations for the understanding of the moiré patterns which are very revealing in electron diffraction and microscopy. This school came to an end when Weigle turned his interest to genetics and settled in U.S.A.

A re-activation has taken place in the mineralogical institute in Zürich since the appointment of F. Laves as professor of crystallography; an interesting program of research into the relations of structural and physical properties is developing on a broad front. A. Niggli has further extended the mathematical theory of space groups.

Needless to say that in a highly industrialized country like Switzerland X-ray diffraction is being used in the chemical, metallurgical, electrical industries and in the special Laboratoire suisse des Recherches horlogères in Neuchâtel.

Yugoslavia

The first paper to be published was a powder diffraction study for the identification of bauxites by M. Karšulin, A. Tomič and A. Lahodny and dates from 1949. But the introduction of single-crystal methods and of Beevers-Lipson strips is due to D. Grdenić (1950) who received his training with A. I. Kitaigorodskii at the Institute for Organic Chemistry in Moscow. The centres of research are the physical, physico-chemical and mineralogical institutes at the universities in Zagreb and Belgrade, the Institute for Nuclear Science in Belgrade and the industrial institute for light metals in Zagreb. In Prof. Grdenić' chemical institute 'Ruder Bošcović' of the University in Zagreb inorganic and organo-metallic compounds are being investigated, while in the physics institute small angle scattering is applied to colloid systems.

* * *

The above discussion represents a fairly good sampling of the growth of diffraction methods in countries with a long tradition in crystallography and mineralogy, and in those where research is only recent. Among the countries not mentioned above are Belgium and India; Poland, Pakistan, Korea; and the Chinese Peoples Republic.

Of the last, too little is known, except for the fact that there are first rate crystallographers (for instance S. H. Yü, see Reminiscences of Lipson and Wilson); this is counteracted by the fact that in the planned programme of education and research of this country X-ray diffraction has not yet reached a high priority. Once this stage is reached, an outburst of structural work may be expected.

In Korea and Pakistan there are well trained research workers and useful equipment, but the entire educational work has first to be carried out before a group of research people can be brought together —and experience shows that isolation of a single, or only a few scientists usually squashes research. In Poland there are X-ray laboratories at the six universities and some more at state or academy instituions, but the output seems not to be high.

India has been included in the survey of the British and Commonwealth Schools (Chapter 17), and the same holds for Canada, Australia, New Zealand and South Africa. Some additional information will be found for India and South Africa in the Personal Reminiscences of K. Banerjee and R. W. James, respectively.

In Belgium, X-ray diffraction goes back to H. Brasseur, who in 1930 worked on malachite and azurite first in Manchester, and then in Pasadena. Back in Liège at the mineralogical laboratory, he and his co-workers investigated the structures of double cyanides, like the hydrated barium-platino-cyanide. Apart from this laboratory, X-ray methods seem not to have aroused much active interest in Belgium until in the 1950's. Then J. Toussaint in Liège, and later H. van Meersche in Louvain, a student of Bijvoet's, took up organic structure work, and H. J. Lambot, who had studied with Guinier, made careful studies at the Université Libre in Brussels of the precipitation mechanism in the ageing of duralumin. X-ray methods were also introduced in several industrial laboratories.

* * *

There are a number of other countries where X-ray diffraction methods are being taught and used, though in a rather isolated way, such as Greece, Egypt, Bulgaria, Rumania, Brazil and other South American countries. Experience shows that the development of the subject is usually a cooperative one. As long as the chemists, both in academic and industrial laboratories, are not actively interested, the group of crystallographers remains small, and from the point of view of the student the special training required does not hold promise of a well paid and interesting employment. Once this first stage of indifference is overcome, it takes not long to prepare a good seed-bed for the subject to develop and take root. The sending abroad of suitably prepared students to well-established teaching laboratories, or the invitation of workers from such laboratories for a visit of a few months, has proved to be the most efficient means for getting the subject established in good form and without avoidable experimentation.

It is interesting to note the important effect which the problems of agriculture have had on the spread of these methods in the less industrialized countries. 'Clay minerals' is usually one of the first subjects tackled after the establishment of the first X-ray diffraction machines.

PART VII

Personal Reminiscences

Some Personal Reminiscences

E. N. DA C. ANDRADE

After being at school at St. Dunstan's College I went, as a youth of seventeen, to University College, London, where I studied physics under F. T. Trouton. His name is known to all physical chemists from Trouton's Rule, giving a relation between latent heat, molecular weight and boiling point, which he discovered when a student as the result of an afternoon's manipulation of physical constants. He carried out, as a collaborator with that genius G. F. FitzGerald, fundamental experiments on electromagnetic waves just after their discovery by Heinrich Hertz, and was subsequently much occupied with experimental attempts to establish the ether drift. The negative results of these experiments helped to pave the way for the first theory of relativity. Trouton was a charming, sympathetic and humorous Irishman, with a great love for physics. When, at the age of nineteen, I took my B.Sc. with first-class honours in physics, he, with great kindness, encouraged me to do research and I carried out my first work on the creep of metals in his laboratory, work which established the $t^{1/3}$ law.

In 1910 I was awarded an 1851 Exhibition scholarship and went with it to Heidelberg, as I wanted to work in a German university and was attracted by the name of Philipp Lenard, Professor of Physics there, who had been awarded the Nobel Prize in 1905 for his fundamental work on the electron. I had hoped to carry out research on electron vacuum physics, but Lenard decided that I should work on the electrical properties of flames, in which he was particularly interested at the time. Accordingly I settled down to this subject and duly obtained my Ph. D. (summa cum laude) at the end of 1911, when Lenard, who had apparently, from his conduct at the time and after, taken a liking for me, offered me a post in the laboratory. This I declined, probably wisely. Lenard was an experimental physicist of the highest distinction, undoubtedly a genius, and he had treated me with great kindness and wrote me many friendly and flattering letters after I left Heidelberg. He was, however, a difficult man to work under, who failed to awaken the independence and sympathy of those working beneath his rule, as is shown by the fact that, with all his qualities and in spite of the fact that some brilliant Germans worked in his laboratory, he did not found a great school of physics. As Carl Ramsauer, who was closely associated with him, wrote 'Als Doktorvater hatte Lenard keine glückliche Hand'.

At Heidelberg I made some close friends with whom I remained on intimate terms until, alas, they all eventually died. Prominent among them were Wilhelm Hausser, Walther Kossel and Carl Ramsauer. The two first named took their doctorates in the same academic year as I took mine, Ramsauer was some years older and was already a Privatdozent when I came to know him. Hausser carried out fundamental work on the production of *Lichterythem* (sunburn) by ultraviolet light. After the First World War he became director of research at the Siemens and Halske works in Berlin, where he not only promoted the application of electron physics to practical ends, but played a great part in the discovery of rhenium and masurium,* for which X-ray crystal spectroscopy was essential. With typical modesty he withheld his name from the discovery. He died in 1933. Walther Kossel achieved fame for his work in explaining the excitation and absorption of the different series of X-ray lines, for making clear the relation between X-ray and optical spectra and for other important work in physics. He died in 1956. Carl Ramsauer, whose name is perpetuated in the Ramsauer effect and who carried out other important work on the passage of electrons through matter and on the ionization produced by ultraviolet light, became director of the great research institute of the Allgemeine Elektrizitätsgesellschaft. He died in 1955. All three were men of outstanding personality. From Hausser and Ramsauer I learnt much of German student lore, including the legends of Bonifacius Kiesewetter. From Ramsauer I learnt to admire and revel in Goethe's *Faust*, in which I have since continued to take delight. I hope that present day research students find such company. We spent many merry evenings together and worked hard in the laboratory.

After quitting Heidelberg I went to the Cavendish Laboratory at Cambridge, directed by the world-famous J. J. Thomson, as what was called an 'advanced student', a man who was allowed to work in a research laboratory but had no status whatever. There went up with

* Masurium was the name assigned to the element of atomic number 43 when its discovery was claimed. This element is now called technetium.

me at the same time my friend S. E. Sheppard, the famous photochemist, who was already a doctor of the Sorbonne. Needless to say, our degrees were not recognized and we were neither of us called doctor. I remember saying to Sheppard that the University of Heidelberg, having been founded in 1385, later than the oldest Cambridge colleges, was perhaps unworthy of recognition, but I thought that the Sorbonne, founded in 1256 and famous before Cambridge was heard of, might be accepted as a centre of learning. Having no recognized degrees we both had to ask permission of a tutor, unheard of outside Cambridge, if we wanted to be out after 10 o'clock at night, as we occasionally did. No doubt it was with this kind of thing in mind, although his status was higher than ours, that the physicist H. R. Robinson later remarked that he was not a Cambridge man, but had been an out-patient there for a year or two.

Finding that one had to work for a few years at Cambridge before one had the status that entitled one to apparatus, I returned to University College, London, while Sheppard at the same time went to the Kodak research laboratories in the U.S.A., where he made capital discoveries in photochemistry. I continued my work on the creep of metals and accidentally made the first single crystal metal wires. Then an opportunity arose for me to go to Manchester to work with Rutherford as John Harling research fellow. I embraced it with enthusiasm and accordingly started in Rutherford's laboratory in the autumn of 1913.

The analysis of the wave-lengths of X-rays by means of crystals had been initiated in the previous year by von Laue and the Braggs, and Rutherford chose me to collaborate with him in determining the wave-lengths of the gamma rays of radium. For this we used crystals of rocksalt, in the first place a rather dirty little slab of which the surfaces, although cleavage surfaces, were not very plane. With this we obtained lines in positions not to be anticipated with a perfect crystal, as everyone can well understand today. As was stated in our first paper (*Phil. Mag. 27*, 854, 1914) 'All our photographs showed similar peculiarities, but the outside lines which appear are very variable for different angles of the crystal. This behaviour of rocksalt led us to make many fruitless experiments to obtain a more perfect crystal, but all the crystals of rocksalt we have examined showed similar imperfections, though in varying degree.' The attempts to obtain more perfect crystals included a descent which I made into a salt mine in Cheshire. I well remember being lowered in a kind of great bucket into hacked-out corridors and halls with glistening walls of salt. I collected many pieces of salt with, apparently, beautiful smooth surfaces, and similar pieces in surface pools, but, as can well be understood today, they did not give any better results than our original little slab—in fact, if my recollection is correct, not such good results. C. G. Darwin, who was working in the laboratory at the time, wrote a classical paper on the theory of X-ray reflection in which the effect of imperfections of crystals was for the first time considered.

In this work with Rutherford the first divergent-beam transmission photographs, showing deficiency lines, were observed by passing the gamma rays through a slab of rocksalt.

I have been asked recently if, when we decided to try this method, we anticipated the dark and light lines which appeared. My recollection of those distant days is that we certainly expected the dark lines caused by the direct reflection and that it was to obtain these that the experiment was planned, but that we had overlooked the fact that this reflection must cause deficiency lines. As soon as we saw the white lines, however, we naturally recognized their origin.

Working in Rutherford's laboratory in 1913 was H. G. J. Moseley, who was one of the first to obtain results of fundamental importance with the newly-born method of X-ray analysis. With reflection by a crystal of potassium ferrocyanide he measured the wave-lengths of the K- and L-series of the elements and established the basic significance of the atomic number. Everybody knows this, but not everybody knows how definite Moseley was on the point, so I quote words that he used. 'We have here a proof that there is in the atom a fundamental quantity, which increases by regular steps as we pass from one element to the next. This quantity can only be the charge on the central positive nucleus, of the existence of which we already have definite proof.'

Moseley was an outstanding character, who would undoubtedly have accomplished further great things had he not been killed in the First World War. Others working in the laboratory when I went there were Ernest Marsden, Charles Darwin, Walter Makower, H. R. Robinson, E. J. Evans, D. C. H. Florance, Stanislaw Loria and Bohdan de Szyszkowski. Niels Bohr was not working there at the time, but was a visitor. Life was good fun—and hard work—in Rutherford's laboratory in those days. I occasionally did a bit of boxing with the undergraduate champion of my weight. I had won the (London) University College and Hospital light weight championship in 1910.

In the summer of 1914 I went to pass a few weeks in Sweden at the invitation of Svante Arrhenius, a great man of most kindly, amusing but incisive character. I had every intention of continuing to work in

Rutherford's laboratory, where my Fellowship was open to me. One of the things on which I intended to work was the investigation of the structures of my single-crystal metal wires by the new X-ray method. The outbreak of war, which took place while I was in Sweden, put a stop to all such projects. On my return to England I became a very junior artillery officer, spent two and a half years at the front in France and the rest of the war on trivial 'scientific' tasks supposed to be of significance for the conduct of the war.

At the end of the war, having recently married, I had to find a post which would enable me to keep a family and became Professor of Physics at the Artillery College, a staff college for artillery officers destined for technical posts, which later took the name of the Military College of Science. Here there was little opportunity for research. I wrote a book on *The Structure of he Atom*, which I dedicated to Rutherford. This came out in 1923 and was widely read at the time: a third much revised edition appeared in 1927. In this period I also published a volume of poems. In 1923 I visited Germany and had the great pleasure of staying with my old friend Wilhelm Hausser; we had long talks together of our old Heidelberg days. This was the period of the collapse of the value of the German mark and it was a strange experience to see the expense of posting a letter to England go up ten times, in German marks, in a day or two. Luckily I had English money with me, for which I could get anything I liked. Hausser and I met again at the Physikertag in Danzig in 1925. Being in Danzig at that time was another queer experience. I last saw Hausser when I visited Heidelberg in 1931. He foresaw already the rise to power of Hitler and told me what a disaster it would be. At the time I did not understand, as he did, the seriousness of the situation.

In 1928 I was appointed Professor of Physics at University College, London, the post carrying the title of Quain Professor of Physics in the University of London. The laboratories were very poorly equipped and funds were scanty, so that there was no question of experimental research requiring elaborate apparatus. I and my students worked on problems in sound, on the viscosity of liquids and on the mechanical properties of metals, devoting much attention to the preparation and behaviour of single crystals of metals. In sound I was able to investigate and explain the mechanism of the sensitive flame, the behaviour of small particles under the influence of air vibrations and to deal with other classical problems. In the field of viscosity I put forward the formula for the variation of the viscosity of a liquid with temperature which is generally accepted and with my collaborators measured, by special methods, the viscosity of certain molten metals, as being the simplest liquids. With Dr. Cyril Dodd I established the influence of an electric field on the viscosity of liquids, about which there had been considerable confusion. But these and other such matters have nothing to do with the X-ray diffraction in crystals.

In our work on metals, however, particularly that on metal single crystals, we naturally made considerable use of X-ray diffraction. The break-up of asterism which occurs with strained sodium and potassium single crystals was discovered and discussed by Dr. Tsien and myself. In the work of Dr. C. Henderson and myself on single crystals of face-centred cubic metals, in the course of which 'easy glide', a term now generally accepted, was established, much use was made of back reflected X-rays, asterism being again much in question. In fact it is natural that in all our investigations of the deformation of metals, whether single crystals, as in the work carried out with Dr. Aboav on copper single crystals, or polycrystalline metals, as in the work which I carried out with Dr. Jolliffe on the flow of metals under simple shear, the X-ray method is indispensable in studying the behaviour of the glide planes and the distortion of the crystals, or crystal grains, in general. I am glad to say that my collaboration with Dr. Jolliffe on the problems of metal flow continues.

It is strange to look back over fifty years to the days when von Laue made his great discovery and the Braggs, Moseley and Darwin carried out their fundamental work. In those days the physics research laboratory was directed by men in close contact with all their research students, men who gave all their time and thought to the problems of their choice. The students were mostly enthusiasts, for there was practically no career except the academic one open to them. Money was scarce: Rutherford, H. R. Robinson has written, ran his laboratory on about £400 a year, the equivalent of which, even allowing amply for the change in the value of money, can easily be spent on one student, finding out nothing in particular, in these days. Everything has changed—has, it need hardly be said, immensely improved. Possibly, however, to work in a research laboratory under one of the great leaders of the time, was a more exciting adventure, a gayer and more satisfying life, than to labour seven hours a day in one of the splendidly equipped and highly organized discovery factories of today.

Development of X-ray Crystallography Research in India

K. BANERJEE

X-ray crystallography began in India with the successful determination of the structures of naphthalene and anthracene crystals in 1929 by the method of trial and error. The results of these determinations were published in 1930. This beginning however, has an interesting background. C. V. Raman and his students at the Indian Association for the Cultivation of Science in Calcutta were greatly interested in the study of optical anisotropy of molecules from scattering of light. He and Ramanathan extended the optical theory of scattering to X-ray diffraction by liquids and attempts were started in this laboratory by Sogani and others to compare the theory with experimental results. It became clear from these experiments that the shapes of the molecules and the intermolecular forces had profound influence on the diffraction patterns. I found at about the same time that phase changes in the liquid alloys of sodium and potassium could be studied by X-ray diffraction in a way similar to that possible in the case of solid alloys. All these studies led to the idea that the shapes of molecules and the intermolecular forces correspond to some degree in the solid and the liquid states.

C. V. Raman and K. S. Krishnan published at this time their highly interesting work on the magnetic birefringence of liquids which led to a method of measuring also the magnetic anisotropies of molecules. The experimental investigations that followed made it clear that the benzene ring has very strong magnetic as well as optical anisotropy. The direction of the optical anisotropy can be easily deduced theoretically from the idea that the anisotropy originates from the anisotropic polarisation field and that therefore the refrangibility is greater when the electric vector is in the plane of the molecule than when normal to it. From experiments with benzene and other aromatic liquids it was found that the diamagnetic susceptibility is greater numerically for the magnetic field normal to the benzene ring than parallel to it. This is exactly what one should expect theoretically according to quantum mechanical ideas. At about the same time the attempt of W. L. Bragg to explain the optical birefringence of some crystals from their structures by calculating the polarisation fields was of great interest to us, and I tried to see whether stress-optical effects in alkali halides and amorphous solids could be explained by this method. The attempts proved unsuccessful and it was concluded that there should be changes in the deformations of the atoms due to their displacements and change of crystalline field in the strained state. These changes alter their polarizabilities, and thus may be the cause of the discrepancies.

Since it was clear that knowledge of the magnetic and optical anisotropies of crystals may give a clue to the orientation of the molecules, and as the benzene ring was found to have very marked anisotropy, crystals of the aromatic compounds naphthalene and anthracene were the first to be tried for finding the molecular orientation. Bhagavantam measured the magnetic anisotropies of these two crystals and found that the benzene planes should lie nearer to the *bc* plane than the *ac* plane. This is contrary to the structure given by W. H. Bragg in which he puts the molecular plane along *ac*. The optical anisotropy also corroborated the results of the magnetic measurements. It was therefore felt necessary to re-examine the structure on the basis of X-ray diffraction.

The task of cutting this new ground was taken up by me as crystals fascinate me. The reason for this fascination to my mind is that when I started my research career under Prof. Raman the first problem that he set me was the study of an interesting optical phenomenon in amethyst quartz, namely a phase alternation diffraction pattern produced by twinning in this crystal. The amethyst quartz crystal is a lovely jewel and the pattern seen under the polarising microscope is extremely fascinating.

Trial and error method was therefore applied to find out the structure and it was found that not only the benzene rings are plane but also the rings in both naphthalene and anthracene crystals are coplanar and the molecular planes are inclined not only to the *ac* planes but also to the c-axis.

While these results were in course of publication in the *Indian Journal of Physics*. J. M. Robertson published a paper on the structures of naphthalene and anthracene in the *Proceedings of the Royal Society*. This was based on the structure given previously by W. H. Bragg but obtained by using much more extensive and accurate measurements than I could think of obtaining at that time in India. So I sat down to compare the relative intensities calculated on the basis of these two structures and compare them with those measured by Robertson and found that the results were clearly in favour of the structure obtained by me and I sent a note to *Nature* stating these results. I was highly pleased to find this note published with an addendum from Robertson that Bragg and he had also arrived at the same conclusion later and that he fully agreed with my results.

Krishnan and his students calculated the actual anisotropy of the molecules of naphthalene and anthracene from the orientations determined by the X-ray method, and utilized the results for developing a method of finding the molecular orientations in crystals of a large number of aromatic substances.

A Calcutta University Fellowship for foreign travel enabled me to proceed to London and work at the Royal Institution of Great Britain under Sir William Bragg. The contact with the brilliant group of X-ray crystallographers there assembled was of great benefit to me. J. M. Robertson was engaged in carrying out the structure determinations of aromatic crystals including naphthalene and anthracene by the two-dimensional Fourier synthesis method. In spite of the knowledge of the rough structure already determined in these cases, the labour involved in determining the signs of the terms seemed to be depressing. I also took up the determination of the structure of paradinitrobenzene of which the preliminary structure was not known. The attempt brought me face to face with the labour involved in carrying out a successful Fourier synthesis by fixing the signs by trial. The ambiguity of phases in a Fourier synthesis, as I began to ponder, arises from the fact that theoretically any structure whether physically significant or not, obtained by synthesis with any combination of phases, would give the same distribution of diffracted intensity as is given by the actual crystal. But all except the correct combination can be eliminated because of our knowledge of the chemical nature of the unit cell as containing a definite number of atoms of roughly spherical shape. If we could utilize this knowledge in a mathematical form which enables us to find a priori a consistent set of phases for the structure factors, the problem would be solved. If we regard the electron distributions in the individual atoms to be known from the wave-mechanical pictures of the isolated atoms, the atomic coordinates are the only unknown parameters. The known quantities are the absolute values of the structure factors whose number far exceeds the number of unknown parameters in the cases that interested crystallographers. The problem did not seem to be an impossible one.

About that time a paper by Ott on the determination of the atomic coordinates in graphite attracted my attention. He set up equations whose roots are the coordinates of the atoms in the unit cell. When the equation is expressed as a polynomial equated to zero, the coefficients of the terms of the equations were shown to be obtainable from the structure factors. The method, however, succeeded in locating the coordinates of the atomic centres in the case of graphite only, because of the high order of symmetry and the small number of unknown parameters involved. In such cases there is no difficulty in locating the atomic centres even without recourse to this method. But this method showed the possibility of setting up equations giving the relationships between the structure factors and the coefficients of terms in the equation whose roots are the atomic coordinates. The number of these coefficients is limited by the number of unknown parameters while the number of equations depends on the number of structure factors measured, which is ordinarily very much larger than the number of unknown coefficients. Thus one can eliminate these coefficients and obtain simple relations between structure factors only. The numerical values of the structure factors being known, their signs can now be found out since these relationships are to be satisfied.

I continued in this line of thinking after returning to India and rejoining the Indian Association for the Cultivation of Science and found out relations that would give the signs of the structure factors in carrying out a Fourier synthesis. The results were published in *Proc. Roy. Soc.* in 1933. After a short stay at the Indian Association for the Cultivation of Science, I joined Dacca University as Reader in Physics and had to devote myself mainly to teaching and to giving some research training to new graduates. So work was restricted mainly to rather routine type of determinations of space groups and to measurements of magnetic anisotropies from which to obtain some preliminary ideas about the structures. Only in the case of anthraquinone a Fourier synthesis was carried out by Dr. S. N. Sen under my direction. Some problems of chemical interest such as an X-ray diffraction study of jute fibres, isomorphism of fluoberyllates and sulphates, and structure of nickel catalysts were carried out. Interest was also taken in the study of isodiffusion lines of diffuse reflections.

In 1943, I returned to the Indian Association for the Cultivation of Science as the Mahendra Lal Sircar Professor and in 1952 I joined the Allahabad University as the Professor of Physics. During these periods, I had the opportunities of building up a research school in X-ray crystallography in India. In structure analysis, J. Dhar made a fairly accurate determination of the structure of diphenyl by the trial and error method quite early. Dr. B. S. Basak determined the structure of phenanthrene by Fourier projection methods, Dr. B. V. R. Murthy refined the structure of anthraquinone of which the preliminary structure was determined by Sen by two-dimensional Fourier synthesis. The refinement was carried out by a number of successive difference syntheses, the atomic positions being finally fixed by utilizing three-dimensional data. In this determination an innovation was made regarding the evaluation of the anisotropic temperature factor by obtaining it from electron distribution plotted without this correction. This automatically takes into account the effect of overlapping of electron densities of adjacent atoms. Very interesting results were found regarding bond lengths in this compound.

S. N. Srivastava has carried out the Fourier synthesis of anthrone. A very remarkable phenomenon has been observed in this crystal. The chemical formula with one oxygen on one side and two hydrogen atoms on the other side is not regularly repeated in the same order, but oxygens and hydrogens are arranged in a random way on either side of the molecule, although the carbon structure maintains complete regularity as is expected in a crystal. Srivastava is continuing his studies on this statistical structure. Dr. S. C. Chakraborty and U. C. Sinha have carried out the analysis of benzalazine by two-dimensional Fourier synthesis and refined the structure by difference synthesis. Dr. B. S. Basak and M. G. Basak have carried out the analysis of 1, 2-cyclopentenophenanthrene by two-dimensional synthesis and refinement by difference synthesis. The atomic positions were finally fixed by utilizing three-dimensional data.

A very interesting type of sharp extra reflections was observed in the case of phloroglucine dihydrate crystal. These reflections were studied in very great detail in order to find out the type of lattice defect producing them. Another interesting piece of work that was carried out during this period is the development of a new method of studying monochromatized low-angle scattering in which the direct beam is avoided by placing the monochromator crystal after the scatterer.

Dr. R. K. Sen developed the photographic method of studying the intensities of diffuse reflections. He was thus able to extend the applicability of the X-ray diffraction method of determining elastic constants to crystals of symmetry lower than are possible by the Geiger counter method and also where large single crystals are not available. The corrections that are necessary for this method have been systematized and the accuracy of the method has been tested against the counter method as well as by means of a crystal whose elastic constants are known. Thus he and Dr. S. C. Chakravorty determined all the elastic constants of the rhombohedral class in which benzil crystallizes.

Dr. S. C. Chakravorty and Dr. R. C. Srivastava have determined also the elastic constants of pentaerithritol which belongs to a tetragonal class. An attempt has also been made to distinguish between the classical and the atomistic theories of elasticities in these cases but the differences are too small even for the photographic method. S. K. Joshi in Allahabad is engaged in studying the elastic spectrum from the distribution of the intensities of diffuse reflections and trying to obtain in this way the vibration spectra of the crystal lattice.

Structure of glasses of various kinds have been studied by Dr. B. K. Banerjee. He has made extensive studies of the nature of alkali halides and colouring materials dissolved in the glasses. The mechanism of devitrification and the natures of glass surfaces and glass fibres were also studied. Dr. G. B. Mitra evolved a method of classifying coal by X-ray diffraction. He also systematized the method of testing the efficiency of coal-washing by X-ray diffraction of the samples. Mineral contents and particle sizes of the minerals of several coal samples have been studied by him by means of X-ray diffraction. At Kharagpur he and his students are studying the thermal expansion of crystals and the relations between lattice defects and particle sizes. He has also developed a micro-wave analog for the calculation of structure factors for use in crystal structure analysis.

After independence, there has been a great impetus to scientific research in India and a very large number of active research centres have developed. In X-ray crystallography also a number of other centres that have grown up are showing great promise.

I wish to make particular mention of a developing school in Madras under G. Ramachandran at the Physics Department of the Madras University, and another at Bangalore at the Indian Institute of Sciences under R. S. Krishnan and S. Ramaseshan.

G. Ramachandran and his students have been engaged in finding the structure of collagen and chitin by a combination of optical and X-ray methods. Ramaseshan, with a number of collaborators, has been carrying on an extensive series of investigations in structure analysis both organic and inorganic. The inorganic series covers potassium permanganate, cupric complexes with varying amounts of water of crystallization, and barium complexes. The organic series includes 3-3′ dibromobenzophenone, 4 bromo-diaminobenzophenone, 4 bromo-dicarboxylbenzophenone and nitrobromofluorene. R. S. Krishnan has been engaged in designing new types of analogue computers useful for crystallographic work.

Personal Reminiscences

N. V. Belov

As an alumnus of the then famous Polytechnic Institute of St. Petersburg (Leningrad) I enjoyed the personal influence of the best Russian professors of the first 15 years of this century, among them Prof. A. Joffé. During the years of the Revolution I happened to live in a small provincial town without industry, and once was head of a 'Sovnarhoze'.

Then I became an ordinary analytical chemist in a Leningrad factory. The well-known geochemist and mineralogist A. E. Fersman selected me to become one of the chief contributors for his magazine *Priroda* ('Nature'). I was for him a person who could write something about anything in all fields of pure Science. These were the times of V. M. Goldschmidt's mighty intrusion into the mineral sciences with ionic radii, coordination numbers, etc. Nobody in Russia was able to get beyond the very elements of Goldschmidt's geochemistry and crystal chemistry, and so Fersman decided that here was a job for me. The first result was a translation of Hassel's *Crystal Chemistry*; the Russian version was twice the size of the original and contained ten times as many figures. After that, I compiled *Fundamental Ideas of Geochemistry* with translations of papers of authors such as Bragg, Schiebold, Machatschki, W. H. Taylor (felspars). I became an authority for explaining to the geologists the inner mechanism of their minerals.

After some adventures in mineral technology, I entered quite by chance a new Laboratory (some years later it became an Institute) of Crystallography, created by Shubnikov, and here I was soon made Head of the X-ray Department, without my having an adequate knowledge of, or practice in X-rays (by the principle of 'better you than anyone not belonging to our clan'). I was then 45. The competition with the old-timers in X-rays was at the time not too hard, because it was the era of Patterson syntheses, of new Fourier methods (Beevers and Lipson strips), etc., and it was an easy matter to get ahead of the classical X-ray people of Glocker's type. In Russia most of these were metal physicists and convinced that crystallography is just a small part of metal physics.

Also quite accidentally I became Professor of Crystallography and

X-ray Analysis in the Faculty of Physics of the Gorki University. *Docendo discebam* and acquired the knowledge which I was still lacking, being primarily a chemist and mineralogist.

In my teaching I adopted a method of introducing space groups which was generally accepted in Russia and is partly presented in my booklet on *Space Groups*. We believe that our students should play with space groups in a similar manner as they play, a year earlier, with wooden models of crystal shapes which are traditional in crystallographic collections. For introducing the students to the more complicated structures, a theory of closest packing was developed and explained in a book which is well known in Russia. About 15 of my original figures may now, after 12 years, be seen in Azaroff's *Introduction to Solid State Physics*, two of them adorning the inner and outer covers of the volume. Gorki, where I am spending no more than 50–60 days a year, has become for me the source of my best pupils. They readily come to me with the quite realistic hope of being advanced to Moscow; in contrast to geologists, they usually have a good training in general physics and mathematics. Their diploma theses all contain contributions connected with my name, mainly in the field of space-group theory of different dimension and kinds: ordinary, black-white, and colour groups. Their Ph. D. theses deal with new direct methods of solving crystal structures: inequalities, statistical approach, superposition methods, and contain many original ideas immediately applied to the analyses of new structures.

As a mineralogist (my colleagues from Geology made me a member of the Academy) I always chose as thesis subjects for my students significant silicate structures. When the number of solved structures exceeded twenty, I tried to invent a system covering all of them, and this is the 'Second Chapter of the Crystal Chemistry of Silicates' which I am at present preaching to the world. Most of these silicates are the Zr, Ti, Nb, Ta, Be, and Li silicates which are so important for the mineral technology; others are calcium hydrosilicates which are no less important for the cement industry. Both parts have been shown to be governed by the same set of crystal-chemical laws. Only this year I was exceedingly happy and felt myself rewarded by the discovery by one of my pupils (she is a mineralogist) that there exists a dimorphous silicate which forms two types hitherto regarded as exclusive in one and the same substance: epididymite, $NaBeSi_3O_8H$, is a chain silicate, while eudidymite, having the same formula, is a layer or net silicate (Phyllosilicate).

My Time at the Royal Institution 1923-27

J. D. Bernal

I would like to add some of my personal recollections to supplement to some extent those of Kathleen Lonsdale.

These early years at the Royal Institution were certainly the most exciting and the most formative of my scientific life. We who had the privilege to be among the first of those who worked there were quite exceptionally lucky because we were at a point when a new field, the arrangement of atoms in crystals, was just being worked out. No one will ever have precisely that kind of excitement again now that the main types of crystal structures are known. Just because we did not know what to expect, every discovery opened up new possibilities. For instance, Sir William's and Sir Lawrence's work at the outset of the analysis of crystals like rocksalt seemed to have destroyed the idea of the chemical molecule. Sir William's later work on the organic compounds restored it.

There were really two experimental schools of work in the Davy-Faraday Laboratory, those of the old hands following the Bragg method—Kathleen Lonsdale herself and Astbury, for instance—based on the use of that very exact tool, the ionization spectrometer, and those who were developing for the first time the new method of X-ray photographic analysis, incapable of giving anything like the same accuracy of intensity estimations but able to cover a far larger number of reflections.

I myself remember being given a few days of instruction on Kathleen Lonsdale's spectrometer and deciding that I was not made for it: to spend a whole day for only two accurate reflections was quite beyond my patience. So I was put onto a very different apparatus, not something that I just had to sit down to, but something which had to be made from the very beginning. In my small room I was given a few pieces of brass made up by Jenkinson, some miners' lamp glasses, a little aluminium foil for the window, plenty of sealing wax to stick everything together. I was given some glass tubing and a little mercury to make a diffusion pump, some copper and iron wires for the transformer and as an essential ingredient an aluminium hot-water-bottle and a small piece of platinum to make the interrupter.

I had, of course, the invaluable guidance and assistance of Shearer, the designer of the efficient gas tube, and of Müller, who had the general direction of the younger arrivals. He was most helpful but also extremely abusive if one did things wrong, which I was continually doing. I remember his unprintable remarks when the Winchester bottle in which we kept the back vacuum broke with a loud explosion, or, rather, implosion.

It was after three months when I had not been able to get even the trace of an X-ray out of the apparatus, after endlessly burning myself and breaking things, that I decided that experimental physics was not for me and I went up to Sir William with the plea to be allowed to go back to the *theory* of crystallography. It was the first time I had seen him since I had been taken on. I had done a paper on the 230 space groups which had given me my job. But when he said 'You don't think I *read* your paper?' this, somehow or other, cheered me up enough to go back and finish the work and in the end to get out the structure of graphite and have it printed before the end of the year.

To do that I had to make my own cylindrical camera which I did in the most amateur way out of a piece of brass tubing which I had cut with a hack-saw, bored a hole in it, stuck in with sealing wax a smaller piece of brass tubing with two bits of lead with pin holes through them for the aperture. The film was held together in place with bicycle clips and I used an old alarm-clock and a nail to mount and turn the crystal. It worked and remained as a prototype of all existing cylindrical cameras as is shown by the fact that the diameters of the cameras have remained essentially the same ever since.

The hazards of work at the Royal Institution can be illustrated by what happened at the crucial stage of my graphite investigation when I had with great trouble managed to secure a few flakes of graphite from the Museum of Natural History, very much begrudged by the Curator, Dr. Spencer. I made a selection under the microscope and put them out on a nice piece of paper by the window and then went down to eat my sandwich lunch with the pingpong players. When I came back the graphite crystals had completely disappeared! I could not make out what had happened and finally I tackled the charwoman and asked her whether she had seen them. She said 'Them smuts?' (there were quite a lot which used to come through the window) 'I swep' 'em up.' So I had to start all over again. Mrs. Dyke had an extreme and quite unnecessary interest in laboratory cleanliness and I was not entirely upset when very shortly after that she tried to clean the X-ray apparatus while it was going. The terminals always collected an enormous amount of dust and she could not resist it but, she said! 'It jumped out at me, like.' She was lucky, I suppose, to have survived!

Working a Shearer gas tube, with its hand adjustable leak, was more art than science. Some people, like Gibbs, could get their sets to work for days on end without any trouble but I had to sit over mine all the time. It had a disconcerting habit of suddenly going hard or soft. At longer intervals the interrupters would develop leaks which would lead to their filling up with explosive gas and going off with a loud bang. In the end we found the noise was intolerable, so we put them up on the roof which of course made it impossible to keep an eye on them; you just had to wait for your bottle to 'pop off' and then go for an expedition among the chimney pots to replace it.

What was especially good about the laboratory was that it worked as a group of young people who were quite unashamedly keen at discovering this new world of molecular and atomic pattern. We did not shut ourselves up but wandered in and out of each other's rooms seeing how the work was going, talking all the time, not only at tea time, and I had the pleasure of doing vicariously a great deal of research. I always felt I was saved from the worst trouble, that of actually writing out the papers.

An element of laboratory life which Mrs. Lonsdale has not mentioned, were the colloquia which were instituted, I seem to remember, about 1925, where we were permitted to go up into the sacred regions of 'the flat' and, in Sir William's study, sit down on easy chairs and discuss the work of one of our number with continual comments and interruptions, particularly from Astbury.

The characters, scientific and otherwise, of the earliest workers were themselves markedly different. Kathleen Lonsdale, despite her unobtrusive ways, had such an underlying strength of character that she became from the outset the presiding genius of the place; her opinions were sought for and her judgment was always respected. Some of the research workers, like Gibbs, were very quiet and unassertive. His admirably executed work on the structures of the various forms of tridymite, only stable in narrow ranges of temperature, has never been improved on to this day. In contrast, Astbury was slap-dash and imaginative with unflagging enthusiasm which kept us going. At a time when there was no possibility of working out the structure of compounds much more complicated than naphthalene, he willingly attacked the totally unknown field of proteins where even chemical analysis was absent. Crazy experiments were the rule. I remember very well taking an X-ray photograph of the leg of a live frog stimulated by an induction motor to see whether the contraction made any difference to the X-ray pattern. The frog was released later and seemed to be none the worse, but the patterns, of course, were quite undistinguishable, just typical protein patterns, though really we would not have recognized it then.

Research workers came and went. They were accommodated wherever an inch could be found for them in the old building. I remember Patterson settling down as best he could between the glass cases that enshrined the historic apparatus.

After graphite, I did very little more experimental work at the Royal Institution. I was more interested in developing methods of analysis and this led to preparing charts both for flat and cylindrical cameras which involved a lot of calculation. I remember asking Sir William whether I could have £20 for the hire of an adding machine but as this was apparently beyond the capacity of the Institution, I had to do the whole thing by hand and eye using seven figure logarithm tables. It took me about two months: I suppose it would be equivalent now to about a minute machine computing time.

Almost the last experiment I did at the Institution concerned the structure of alloys and here I remember well working in Faraday's own cellar in the basement, still festooned with flasks full of Dewar's rare gases, but an admirable low-temperature place where one could crystallize things slowly. I was trying, by a flotation method, to measure to several points of decimals the density of a very small single crystal of a delta bronze. It involved a great number of weighings and measurings and on the very last weighing on which all the rest depended, the little crystal slipped out of my fingers and disappeared. An eighteenth century cellar is not the best place to look for a crystal of less than half a millimetre in size and I never found it again.

In spite of all these difficulties I do not think that any of those who were then in their twenties would ever have wished to work anywhere else. As Mrs. Lonsdale has said, we had to be kicked out!

Reminiscences

J. M. Bijvoet

When asking for an autobiography of a kind Prof. Ewald wisely added a few directives for its contents.

'*What made you a chemist?*' No doubt the excellent training in chemistry I received at the secondary school, where the understanding was emphasized and not the learning by heart. In the beginning of this century, before the theory of Bohr and the data of the X-ray analysis, and before the recent astounding development of chemistry, the subject matter treated at the time offered hardly any high lights. But the well planned way in which it was built up, and how, with the aid of balance and logic, one arrived at the atomic composition of matter—this deeply impressed us. And now that we have witnessed the spectacular growth of X-ray analysis and the no less amazing progress in the determination of chemical conformation by entirely different means, our admiration for the direct optical Fourier method can only increase that for its surprisingly complicated older partner.

At the time of my final examination at the secondary school one had to take a tough supplementary examination in Latin and Greek in the Netherlands for the admission to the study of the natural sciences at the University, and this took another year or two! Not until 1917 was there a change. Nevertheless it is with pleasure that I recall this time, since it provided the scarce chance of leaving me ample time for music.

'*Does the physicist or the chemist prevail in you?*' I studied chemistry in Amsterdam, which study in so far as it concerned the chemical part gave me but little satisfaction owing to a lot of elementary analytical chemistry in the first few years—this state of affairs is still often found here—and later on much phase theory, a continuation of the famous work of Bakhuis Roozeboom. Only lately has far more choice been provided at the Netherlands Universities of drawing up one's own programme of study (especially after the B.A. examination). Fortunately physics made up for a good many failing expectations. As a second-year student I prepared my first publication, together with the professor of experimental physics, Dr. Sissingh, a man of unparalleled enthusiasm in lecturing, whose memory I cherish most gratefully. I always remember these months with pleasure. One day when measuring the angles of a nearly equilateral glass prism with the goniometer, on reflecting from a lateral face more than one image appeared. The assistant was called in—it was a horribly cold day—and he pronounced: it must be the cold! Solving even such a childish little problem—the light beam is reflected directly but also leaves after circulating the prism—will fascinate the beginning student. I mention this occurence to bear witness to experimental work in the first years of study of a more romantic nature than usual, a subject that of late has come up for discussion now and than in *J. Chem. Education.*

The First World War I spent in military service; it gave me the opportunity—it was shortly after my B.A. examination—to study quietly thermodynamics and statistical mechanics. My later liking for teaching these subjects dates from that time. I even made the attempt —though without success—to invite attention for the 2nd Law of Thermodynamics already at the secondary school, as it moves head and heart more strongly than many other topics learned.

As a physico-chemist I have later always felt somewhat envious of theoretical physicists. In this volume I should like to make particular mention of how much I admired Ewald's classic articles on light propagation in crystals when I was re-reading them recently. I also remember vividly a lecture on this subject delivered by him in 1924 in Leiden.

'*What made you a crystallographer?*' The excitement aroused by Bragg's NaCl model at the chemical laboratory in Amsterdam. I arrived there to continue my studies having been demobilized. My tutor Smits was against the model: 'In the representation of the solid substance given by the Bragg model the considerations are perfectly ignored which have led to the firm conviction that the atoms in the molecule are bound by forces which are characterized by their localized nature and by their definite number.... so that any representation which leaves out of account these circumstances so exceedingly important from a chemical point of view, must be erroneous.'

The happy consequence which this view brought about was that the decision was taken to start X-ray investigation in Amsterdam; in this endeavour A. Karssen, the friend I lost too early, joined me. For a couple of months we studied at the Laboratory of the Veterinary College at Utrecht, where Keesom and Kolkmeyer had started this research shortly before. The latter was to remain my fatherly friend all his life. Our first investigation was concerned with the crystal structure of $NaClO_3$. I well remember Keesom's virtuosity in motivating these investigations for the Ministry of Agriculture—the Veterinary College came under its competence.—In our publication one reads: 'For biological science every deepening of our insight into the nature of the chemical bindings of the element carbon, so important for the organic world, will be of great value.... sodium carbonate and sodium-hydrocarbonate first came into consideration because of their importance for animal life. Considering, however, that we could expect to meet with great difficulties in these investigations.... we first investigated some substances with analogous(!) structure. We chose sodium-chlorate and sodium-carbonate.'

The apparatus was made up from an old induction coil and a self-constructed tube on a Gaede pump. One of the most important occupations when experimenting was fanning away again and again the spark discharge which jumped between a disk and point when the tube became too hard. In toy-shops we hunted for the little balls for our atom models. At that time the 'playing' with these models tickled everybody to laughter. Instead of the *International Tables* there were only the space-group tables of Niggli and, a few years later, those of Wyckoff. Then Wyckoff was a help and a name to me, afterwards he became a warm-hearted friend.

For our doctor's degree Karssen and I had determined the crystal structures of LiH and $NaClO_3$, and we drew lots who was to take which for a doctorate thesis. The investigation of LiH was a first, bold attempt—on the ground of Bohr's orbitals—to localize the valency electrons. Shortly afterwards we spent a brief but extremely instructive period at the laboratory of Sir Lawrence Bragg in Manchester. This allows me to count Sir Lawrence Bragg among my personal tutors, a priviledge not all crystallographers enjoy, even though we all learnt from him. Dr. James had the daily supervision—I recall his saying: 'Don't think too highly of our physics', a pronouncement that is surely easily disproved by his Volume II of *The Crystalline State*.

I retain a painful memory of a visit Sir William Bragg paid to the Amsterdam laboratory. So persistent were they in pouring out the hackneyed arguments against the results of the X-ray analysis that Sir William in despair, raised his arms to heaven. I need not mention that in later years Sir Lawrence's lectures at our Universities were the summits of röntgenographical life. Some ten years ago we were predicted that the solution of the structure of haemoglobin 'was just around the corner'. Many were in doubt. We now know: wrongly so.

In Amsterdam I became lecturer in crystallography and thermodynamics. And there I passed through a time of struggle for the raison d'être of X-ray analysis in crystallography. A large geological institute was being built and crystallography was to get a floor. The professor in geology had the overall directorship. But he first wanted us to make a study-tour through Germany to ascertain whether X-ray analysis was of real importance for mineralogy (ten years after Bragg's silicate structure and Goldschmidt's Geochemische Verteilungsgesetze!). When this had been convincingly demonstrated to him the intended report of our findings was not drawn up. Shortly afterwards I was informed that chemists were not permitted to work in the new institute so as to prevent it from becoming top heavy. Of course I did not give up my pupils, though this set-back resulted in my being without any official subsidy for some years. Of this fight mainly two points remained with me: I had the support of my friend Nieuwenkamp, then my assistant, now my colleague in petrography in Utrecht, who never lost his sense of humor. And furthermore, the interest of the physicists, who did see the importance of X-ray analysis. They made the training in elementary crystallography obligatory for students of physics; this was in contravention of our official examination statutes which still require crystallography for chemists but not for—solid state!—physicists. This Amsterdam regulation is still being continued there, and there alone, also under my excellent successor Caroline MacGillavry.

In 1939 I changed over—at first not without heartache—from crystallography in Amsterdam to physical chemistry at Utrecht, as successor of Prof. Ernst Cohen, well known for his investigations into allotropy. During the war Cohen, being a Jew, was killed, though it was he who, after the First World War, had been the great advocate for admitting the Germans again into the international organizations.

My teaching task—physical chemistry, excepting only electrochemistry and colloid chemistry—was a heavy one, but it always pleased me provided I was not too troubled by imperfections in such an extensive teaching job. I once told a Swedish colleague that I lectured ten hours, he answered: I give five. A little later it appeared that with him it meant five hours yearly, whereas mine were given weekly! By dividing and doubling chairs this has much improved the last few years and at present my job is shared by my colleague Peerdeman. Another pleasant change was that my Röntgen department was separated from the physico-chemical laboratory ten years ago, and put up—odourless—in a gentleman's house. Students call this the Crystal Palace. Living as I do in my laboratory I could observe a co-worker—Dr. van Bommel—who for some years slept from say 4 a.m. till 10 a.m. and for the rest measured and calculated. This tenseness, when eagerly awaiting results so directly and strikingly visible, is quite characteristic for the Röntgenanalyst. Exciting also was our first record with anomalous scattering, which required a continuously watched exposure of some hundreds of hours. It had to be successful in view of an intended journey, and was daily threatened with failure because of the improvised Zr-tube and the freakishness of a pump that had been hurriedly put back into use. Twenty-four hours before the time of departure half of the Weissenberg (deflection 0/180°) was developed and revealed nothing, the next day the further exposed other half (0/−180°) showed the effect looked for.

Often one forgets—or one is anxiously silent—about that which lies uppermost in our heart: Very much I owe to my co-workers, to their friendship too. A close friendship with Paul Rosbaud dates from the German version of my X-ray textbook; his flying visits are a delight to me and my family.

With the publications of the Union I had for many years a pleasant contact as co-editor of *Structure Reports* and here—beside the lot of work—I especially think of Wilson's occasional, welcome appearance in Utrecht when on his way to his publisher. My most remarkable recollection from my time as President of the Union was—beside that of the cheerful meeting in Paris—the reversal of financial distress into prosperity by increasing the subscription price of *Acta*. I worried about this daring solution. Fortunately my worries were lived down.

In a public lecture given in 1925 at Leiden Dr. van der Veen ended, after describing Von Laue's, and the Braggs' discoveries: 'There we have the jousting ground, where the natural sciences may compete and where chemistry and physics unite.' This jousting ground has indeed shown exciting and unexpectedly grand victories, where these pioneers did not confine themselves to formulating the rules of the game, but themselves remained the first players.

Personal background behind it all. I admire Mozart, crystals (let him who visits Paris take delight in Professor Wyart's wealth of beautiful crystals) and cats. My wife was amongst my student audience in Amsterdam; after having lost sight of her for five years, I asked her to marry me.

Personal Reminiscences

W. L. Bragg

I lived for the first eighteen years of my life in Adelaide, South Australia. My father had come to Adelaide in 1886 to succeed Sir Horace Lamb as Professor of Physics and Mathematics at Adelaide University; three years later he married my mother, who was the daughter of Sir Charles Todd, Postmaster General and Government Astronomer of South Australia. My parents sent me to St. Peter's College, a fine Adelaide school, and afterwards to Adelaide University. School education was then much less specialized than it is now. I had instruction in eight subjects to an equal standard, in fact in all the subjects which were taught at the school with the exception of German and Physics—and I have always regretted that I did not learn German. I did not have an entirely happy time at school, though no fault of the school. I was no good at games, not for physical reasons but because I had not the right temperament, lacking the drive and self-assurance necessary to a good games player. Being rather advanced in my school work I was a very young member of the sixth form, while at the same time making such a poor showing in games that I could only be put to play in the 'sets' with boys in much lower forms. Schoolboys only accept the normal boy into their full fellowship and, although they regarded me with kindly tolerance as a strange freak, I was very much cut off from my fellows, and driven into finding solitary interests of my own. One was the collecting of shells along the coasts and on the reefs in the neighbourhood of Adelaide. I had quite a fine collection for a youngster and amongst other rarities I discovered a new species of cuttlefish; I still feel proud that 'Sepia Braggi' has continued to be accepted by conchologists as a distinct species.

Chemistry was well taught at my school, and the chemistry master, who took a kindly interest in me, used to let me help him to set up the class experiments. It was this, I think, which gave me my first interest in the methods of science. My brother and I, also, were always making things in a shed in the garden which we used as a workshop. My father's University mechanic used to give us bits of scrap and apparatus which were being discarded and we rigged up crude motors, morse sounders, a telephone system and so forth. Much of the Post Office stores was kept at the Observatory and I am afraid we sometimes raided them for some coveted piece of insulated wire or zinc or chemicals. I did not get any formal instruction in Physics till I went at the age of fifteen to Adelaide University, and then it was only as a subsidiary course to my main subject of Honours Mathematics in which I took my degree at the age of eighteen. It was at that time, however, that my father started his research career. Up to the age of 43 he had devoted himself entirely to teaching and administration and had been one of the stalwarts who almost recreated the University and made it into a fine institution. He was one of the first in Australia to experiment with a Röntgen tube and I think I was one of the first patients to be X-rayed, my elbow having been smashed in an accident at the age of six. He and my grandfather put up the first Marconi set in South Australia, transmitting messages from the Observatory to the coast five miles away, and he also installed a seismograph at the Observatory. As a boy I saw a good deal of these ventures. When I was at the University my father began his investigations of the ranges of the α-rays and used to talk to me about his results, so I lived in an inspiring scientific atmosphere.

My father was appointed to the Physics chair at Leeds University in 1908 and the family came to England. I was entered at Trinity College, Cambridge, for a course in Mathematics and at the end of my first year was awarded a Major Scholarship in that subject. I was taken ill with pneumonia just before the examination and had to take it in bed with quite a high temperature. Our beloved Master, Butler, read all the essay papers himself and particularly commented on mine, which he said showed great dash and fire—I think the temperature helped very much towards the happy outcome. In fact, I was headed towards being a mathematician till my father urged me, in my second year, to change to Physics for my final examinations.

After getting my degree at Cambridge, I started research under J. J. Thomson at the Cavendish. He set me a problem of measuring the mobility of ions in gases. The great reputation of J. J. Thomson's school had attracted many young researchers to Cambridge, and the facilities in the way of apparatus were quite unequal to the strain put upon them. I remember, for instance, that there was only one foot bellows between the forty of us for our glass-blowing which we had to carry out for ourselves, and it was very hard to get hold of it. I managed to sneak it once from the room of a young lady researcher when she was temporarily absent, and passing her room somewhat later I saw her bowed over her desk in floods of tears. I did not give the foot pump back. Then again, when I first got a reflection of X-rays from a mica sheet, I kept the induction coil running so hard in my enthusiasm that the platinum contact on the hammer-break burnt out. Lincoln, our head mechanic, was very angry and told me I must wait two months before he could get me another one because it had cost ten shillings. I have still not got used to the cheerful and airy way in which young research students nowadays ask for some item of equipment costing many thousand pounds.

My first researches into the analyses of crystals by X-rays were made at Cambridge in 1912. My father had aroused my interest in Laue's work; it was the help which I got from Professor Pope, the Professor of Chemistry, and the inspiring influence of C. T. R. Wilson, which led to my analysis of sodium chloride and potassium chloride by the method of the Laue photograph. Pope and Barlow had developed a valency-volume theory of crystal structure, and when my first studies of Laue's diffraction patterns led me to postulate that zinc sulphide was based on a face-centred cubic lattice, Pope saw in it a justification of his theory and urged me to experiment with sodium chloride and potassium chloride crystals which he got for me from Steeg and Reuter in Germany. These experiments were made in the later part of 1912 and early in 1913. Simultaneously, my father seized on the conception of the reflection of X-rays by crystal planes to design his X-ray spectrometer, and discovered the X-ray spectra. I did not start to work with my father till I came home for the long vacation in the summer of 1913. He was still principally interested in X-ray spectra, but it was immediately clear that the spectrometer provided a far more powerful method of crystal analysis. After showing its power in the analysis of the diamond structure (a research in which I played a very minor part) he continued to establish the relations between the X-ray spectra and the absorption bands, and I concentrated on the interpretation of crystal structure. We had a thrilling time together in an intense exploitation of the new fields of research which were opening up, brought to an end by the start of World War I in 1914.

I volunteered for service when the war started. We were allocated to regiments in a very haphazard way and of all things I found myself thrown into a horse artillery battery. The other officers were all hunting men, and I was very much a fish out of water, though it was an interesting experience. We trained in England for a year, but just before the battery went abroad for active service I was called to the War Office and told that I was to go to France to test a new invention called Sound Ranging for the British forces. The French army had started this method of locating the enemy guns by sound and I was first sent to the Vosges to learn its methods. A small British section was then established on Kemmel Hill just south of Ypres and we set up our base of microphones and recording headquarters. At first the method had a very poor success, though we tried to gloss this over in our reports to headquarters. The trouble was that the loud crack made by a shell travelling faster than sound muffled the fainter report of the gun we were trying to locate, when recorded by the carbon microphone we were using. Then one of the men in my section, Corporal Tucker, had the brilliant idea of using a device which later came to be known as a 'hot-wire microphone'. The report of the gun, though it produces little effect on the ear, creates a powerful pressure wave. We bored a hole in the side of an ammunition box, and stretched a fine platinum wire across it. A current heated the wire a dull red, and the report of the gun created a rush of air through the hole, cooling the wire and reducing its electrical resistance. It worked like a charm. I shall never forget the thrill when we first saw the recording apparatus in our dugout paying almost no attention to the 'shell-wave' and other incidental noises, but giving a powerful response when the real report of the gun reached it. Sound Ranging then developed rapidly and we established sections to cover the whole Western Front. It could only work in conditions of calm or easterly wind, but it gave valuable counter-battery information because the shell-burst as well as the gun report was recorded, and so we were able to establish not only the gun position but also the calibre and the target on which the battery was registering.

It was in 1915, when we were setting up our first Sound Ranging base in Belgium, that I heard of the award of the Nobel Prize to my father and myself. I remember that the genial Curé in whose house we were billeted brought up a bottle of wine from his cellar to celebrate the occasion.

When the war ended, I had a brief time in Cambridge as a lecturer, and then went to Manchester as successor to Rutherford in the Physics professorship. It was a difficult time. Not only was I inexperienced, but also the war years had made us forget all our knowledge of physics and the classes, being largely composed of men returning from war service, were by no means easy to handle. My main helper in getting the laboratory into full activity again was R. W. James. He had been the physicist in the Shackleton expedition which was on the way to the South Polar regions when war broke out. The expedition lost its ship, which was crushed in the ice, but managed to make its way to the barren Elephant Island whence Shackleton set out on his famous boat journey to South Georgia to get help. James played a crucial role in the escape of the party. They of course had no radio in those days and were quite uncertain as to their longitude, but had saved a nautical almanac; James studied this and managed to determine the time by observation of stellar occultations. This information gave them their position and enabled the boats to reach Elephant Island. Soon after his return to England, James joined our Sound Ranging section in Belgium and we worked together for the greater part of the war. When we were demobilized he joined me at Manchester and together we built up a research school.

At first we concentrated on the quantitative measurement of the efficiency of X-ray diffraction, using the formulas which C. G. Darwin had developed before the war. This led to our examining more complex crystal structures, in which the values of a number of paramaters had to be determined in order to fix the positions of the atoms. Some of our more ambitious crystal analyses were at first viewed with deep suspicion by our colleagues in other laboratories, but I think that one of the chief contributions of the 'Manchester School' was the demonstration that these quantitative measurements enabled one to tackle such structures with confidence. I can only mention a few of the men who worked with us at Manchester. Hartree came to the University as Professor of Theoretical Physics and did his famous work on atomic scattering factors. Bradley started as a chemist, but then joined our group. He developed the powder method as a means of determining the structure of metals and alloys. Bradley was unique in his management of the powder method, a brilliant researcher. The great theoretical physicists Waller, Bethe, Peierls were amongst those who came to the Manchester laboratory. That volatile genius, E. J. Williams, whose early death was such a profound loss to science, was our Reader for a number of years. I remember that when Bradley and Sykes had got some interesting results with iron aluminium alloys, and I had given an account at a colloquium in which in a vague way I ascribed their findings to an order-disorder transformation, Williams came to me next morning with practically the whole thermodynamic treatment of the problem which he had worked out on the preceding evening.

Warren and Zachariasen played a great part in the elucidation of the silicate structures.

I first met Alice Hopkinson in Cambridge just after the War, and two years later we became engaged and were married in December 1921. My wife comes from a famous Manchester family of engineers, though her father, Albert Hopkinson, broke from the family tradition by becoming a doctor. Our elder boy, Stephen, is an engineer and chief scientist to Rolls Royce in Derby. Our second boy, David, is an artist. Our elder daughter, Margaret, married Mark Heath in the Foreign Office, and our younger daughter, Patience, is married to David Thomson, the son of my closest friend, Sir George Thomson, the Master of Corpus Christi College in Cambridge. My wife was Mayor of Cambridge in 1945, is a Justice of the Peace, and has served on many public bodies, including the Royal Commission on Marriage and Divorce. She is at present Chairman of the National Marriage Guidance Council.

I left Manchester to become Director of the National Physical Laboratory in 1938, and at about the same time James went to South Africa to become Physics Professor at the Cape University. My directorship of the National Physical Laboratory only lasted for a year, because soon after my going there Rutherford died and I was invited to succeed him at the Cavendish Laboratory in Cambridge. I remained based on Cambridge during the Second World War, though for eight months I was abroad in Canada as Scientific Liaison Officer between that country and Great Britain. I acted as scientific adviser to the Admiralty on underwater detection of submarines, and was also connected with the development of Sound Ranging during the war.

The reorganization of the Cavendish Laboratory after the war was not easy. Cockcroft, my fellow Professor of Experimental Physics, was not released by the Government when the war ended, and so we had not got his help, which would have been invaluable, and at the same time could make no alternative appointment. He was finally appointed head of Harwell. We had suffered a great loss by the death of R. H. Fowler, who directed the Theoretical Physics, and several of the most important members of the staff did not come back till some time after the war ended. Large sections of the main building were still occupied by Government departments which had been installed there during the war. Eventually Hartree succeeded Fowler in Theoretical Physics, and Frisch came to direct the nuclear research. W. H. Taylor was appointed Reader and built up a flourishing school for X-ray analysis. He has been a very active participant in all our international gatherings. Perutz was working in the Cavendish on protein structure and was later joined by Kendrew, and this created the starting point for the Medical Research Council's unit for Molecular Biology which has produced such spectacular results and become world-famous. I of course had a particular interest in the protein work. For a long time we only had the most meagre success, just sufficient to encourage us to continue. The Crick and Watson solution of the DNA structure was a notable triumph coming from the research group, but the analysis of the protein diffraction patterns was a formidable problem, eventually solved by Perutz' discovery that heavy atoms could be attached to the molecule at specific points without any alteration of the crystal framework. The successful analysis of molecular structures containing many thousands of atoms is an event which has given me the very greatest pleasure, after a lifetime of research devoted to X-ray analysis.

When I came to the Royal Institution in 1954, I was faced with the problem of starting research in its Davy-Faraday Laboratory practically afresh. All costs had risen so much that the income of the laboratory was insufficient to meet even the basic running expenses, and it was drawing heavily on its research fund which was almost depleted. The Davy-Faraday is unlike a university laboratory in that there is no steady stream of young people wanting to do research after taking their degrees, so we had to find both the men and the money before we could start work. I could not have done it but for the help of Perutz and Kendrew. We agreed to pool resources between the Cambridge unit and the D.F. Laboratory, and make a joint attack on protein molecules. Perutz and Kendrew were formally appointed as Readers in the Royal Institution, and the Medical Research Council appointed me as a general adviser on the research. We got going by transferring a living part, as it were, of the Cambridge research to Albemarle Street. The Rockefeller Foundation, the National Institute of Health, the Medical Research Council, and a number of industrial firms generously financed the research. The D.F. Laboratory has participated in the analysis of myoglobin and of haemoglobin, its particular contribution, by Philips, Arndt, Green, North and others, being the automatic collection of data and programmes for the electronic computer.

The Royal Institution receives no government support. It is maintained by its endowments, the subscriptions of its members, and donations by industrial and other bodies interested in its activities. It is primarily a centre for the diffusion of scientific knowledge, in particular by popular lectures to a non-scientific audience. Founded by Rumford in 1799, it has great traditions and a famous history. Davy, Faraday, Tyndall, Dewar and W. H. Bragg lived and worked in it, and it was the home of many of the outstanding advances in science in the nineteenth century. It also has a priceless asset in its building, which provides assembly rooms and a world-famous lecture room right in the centre of London. But when I came here seven years ago, a good deal of thought had to be put into its function in the modern world. The Friday Evening Discourses and the Christmas Lectures, which had been held for a century and a half, were as successful as ever, and it was a much sought after centre for meetings by other scientific bodies. However, other activities had ceased to be effective in the changed conditions of modern times. It used to be a main centre in the country for courses of lectures dealing with the latest advances in science, but this function has now been taken over by the Universities and by societies of many kinds. The afternoon lectures were still being held seven years ago, but their usefulness had almost ceased. At the same time, finance was a major problem; one had to reconsider the function of the Royal Institution both to take full advantage of its unique advantages and to form a basis for an appeal for support. The Christmas Lectures had always been so successful that we conceived the idea of lectures of the same kind all through the year to boys and girls in Greater London, and at the same time of making the Royal Institution a place where science teachers could hear about the latest developments in their subject. Schools lectures are held twice a week, when the young people are shown experiments illustrating scientific principles. These lectures are run like the standard plays of a repertory theatre. Each is repeated a number of times so that a large number of boys and girls can attend it, and a store of the most effective demonstration apparatus is kept up for the repeat of the programme every three years, when the school population has changed. We arrange occasions when the science teachers can meet representatives of famous research schools in the country and hear about their work, and we maintain a large collection of demonstration apparatus for the teachers to study. It has proved to be a very interesting and rewarding experiment which is much appreciated by teachers and by their pupils, and I have taken a deep interest in this work. Although some 20000 young people come each year to the Royal Institution to see the experimental demonstrations, they only represent a small fraction of the potential young scientists. The schemes on a far larger scale initiated by the British Association and by other bodies must meet requirements all over the country. The Royal Institution, however, with its traditions and its setting is an ideal place for maintaining the standards of lecturing and demonstration and for exploring new techniques. It is a school for the art of lecturing, the influence of which extends far beyond the area which it mainly serves.

Personal Reminiscences

J. C. M. Brentano

I was born on 27 June 1888 and brought up in Florence, Italy, where I attended high school and also took one year of university courses.

From there I went to the University of Munich, and after the preliminary mathematics courses, entered Röntgen's laboratory. Röntgen belonged to the classical school of experimental physicists. It was the time of plausible theories or models and Röntgen was fond of such interpretations. He was an experimentalist, not in the sense of a technician or gadgeteer, but one who stressed accuracy and insisted that everything observed should be accounted for. It was not fortuitous that he became the discoverer of X-rays. Many others, in pursuit of their research, would have disregarded the patches which appeared on the photographic plates, or would have attributed them to some chemical or other indefinite action. For him, nothing could pass by without his establishing a definite reason, and an experiment was deviated from its course or halted until this was found. In his later years he was engaged in the study of physical phenomena of crystals, which he considered to be better defined objects than other materials. He was retiring, self-effacing, almost shy, creating the impression of being unapproachable. Actually he was kind, and keen on encouraging the work of his students. Röntgen declined to enrich himself from his discovery, and sought to stop others from doing so; in fact, no major patents were taken on the field of X-rays until the advent of the Coolidge tube. When in the early days of X-rays a tube maker submitted a new model to Röntgen's endorsement he would enquire on its price and say that it was too expensive, and that it should be possible to make it for much less. X-ray tubes cost then 7 to 8 Marks.

The trend of the laboratory contrasted with that of the great experimental schools of Rubens, Lenard and Rutherford, where the whole institute is set to work on one field and the beginner is immediately provided with all the accumulated knowledge and given a place on the common front, which puts him in the position of making his own contributions. In Röntgen's institute each worker pursued a distinct line, he had to study the literature from scratch, and was left to his own devices. To many this was discouraging. I rather enjoyed it to be left to find my own way. As a preliminary to a thesis I was to examine the residual charges of dielectrics after they had been exposed in succession to electric fields in opposite directions. Röntgen appeared at intervals of four to six weeks, when he would spend a whole afternoon looking for any possible sources of error. He was satisfied when I could show that I had foreseen them, and was pleased when I had definite results. Actually my results disproved a theory he was developing. Not only did he get them published, but he offered his help to improve the wording of my article. When I ended with the customary acknowledgement of thanks, which in this case was well justified, he insisted that it should be cut out. After this first step, I enjoyed full freedom of using the facilities of the laboratory without having to submit to detailed scrutiny.

The chair of theoretical physics was then held by Sommerfeld, who was a brilliant lecturer as well as an inspiring leader of research. He ran a colloquium in which theoretical and experimental developments were equally reported. He took great interest in the work of his students and proposed problems covering a wide range of topics. Among his students at the time were men like Epstein, Ewald, Landé.

When I was in the midst of my studies, a third physicist appeared, Max von Laue, who established himself as a Privatdozent. He came from Berlin as a former pupil of Planck. Laue settled in the same boarding house in which I was staying. With his fine features, the open expression of his blue eyes, he won the affection of those he met. He was unconcerned of external social conventions and was somewhat absent-minded in the face of gossipy conversation, but always ready to take up the discussion of any topic of physics. He took a strong stand for whatever he considered just, and held high the integrity of science and of the scientist. He was very fond of nature; mountains and lakes held a great attraction for him and so also works of art. He was particularly fond of classical music, loved especially Beethoven. Soon after coming to Munich he built a boathouse with a living-room standing on posts in the way of a prehistoric dwelling on Lake Starnberg, where he owned a sailboat. Sundays I was often invited there and so I first became his pupil not in physics, but in sailing. While in Munich he married, and his wife became a most gracious hostess to the friends he used to invite.

At one time, when a group of students, including myself, wanted to take his course in thermodynamics, we sat waiting in the lecture room at the appointed time, but Laue did not appear. This repeated itself three or four times, when we learned from the Dean that Laue had come to give his lecture, but at a different hour, and finding the room empty had gone home. After the Dean had synchronized the hour we were given a course, not brilliant in delivery, but most inspiring in its logical development, and in the presentation of new aspects like the entropy of coherent radiation. Such an error in timing could have happened to him at any time, but in this case there was a special reason; Laue was pursuing with Friedrich and Knipping the experiments on the diffraction of X-rays which have been fully described in Part II of this book.

When the first diffraction patterns were observed Laue sent out copies to his friends, asking: 'what is this?' Soon after, he presented a complete mathematical treatment of the phenomena.

In 1914, when I obtained my Ph. D. degree, Laue made me his assistant. I was then to join him in Zurich, where he held an appointment for theoretical physics at the University. Before I arrived, however, Laue had accepted a chair of theoretical physics at the newly-founded University of Frankfurt and I became his assistant there. The 1914 World War had broken out and my initial task was to assemble equipment for X-ray diffraction. No instruments could be bought and I had to scan secondhand metal dealers; one shop in the house 'Zum goldenen Kopf', which had been the home of my ancestors and now had become a tool and metal store, played a great part in providing supplies. Other useful ties were established. Friedrich Dessauer, the director of Veiva, then one of the leading manufaturers of X-ray equipment, came from a family closely befriended with mine. For a mechanic I found an excellent craftsman beyond military age, who after his daytime work came for nightwork to the University. My hours at the University extended thus into the night, when one could hear from the street the marching songs of the young soldiers who were led for embarcation in troop trains for the front. After some time the mechanic collapsed because of overwork and had to stop. It was not the atmosphere for indulging in X-ray diffraction, particularly for an alien, so in March 1915 I left for Zurich to join my ageing father. The impact Switzerland had on me was bewildering; in Germany I thought I had kept aloof of war bias and had distinguished between true news and war propaganda. Suddenly confronted with news from all parts of the world, I became aware of how much I had been the victim of the psychological experience that repetition makes truth.

I was at once taken on by Auguste Piccard, then in charge of the physics department of the Polytechnical Institute, while Weiss, the head of the department was engaged in war-work in France. Piccard had then already conceived the idea of the high altitude balloon and of the bathysphere with the ingenious oil compensation to counteract the effect of pressure, and I had to promise as unwritten part of my contract that I would be willing to take part in an ascension of the first. This did not come off. Piccard had come to physics after having studied mechanical engineering, and this reflected on his choice of problems and on the methods of attack. He was responsible for a great part of the instrumentation in Weiss's laboratory and was very imaginative in devising original and unconventional methods. The Weiss laboratory was concerned with the measurement of magnetic susceptibilities and Piccard reduced the uncertainty in the measurement of magnetic field strengths with a paramagnetic liquid, by introducing small particles held in suspension and observed with a microscope as indicator for any small displacement. I adopted this method with the object of measuring the X-ray yield as a function of wave-length and tube potential by observing the increase of pressure when the X-rays are absorbed in a closed vessel. The apparatus had the required sensitivity. The work was interrupted because I was called upon to replace Professor Schweitzer, who held the second chair in physics and was struck by illness.

Another problem, somewhat related to the first, was the design of an X-ray monochromator giving the highest possible yield. It implied developing the geometry for the diffraction of a diverging beam of X-rays of given wave-length and of large aperture so that, emerging from a source point, it would be diffracted by a given spacing to another point. This means distinguishing between the geometrical locus of the surface from which such diffraction takes place, which is a toroid, and the orientation of the diffracting lattice planes, which in general are inclined to it like the tiles of a roof. Later we referred to this geometry as para-focusing. The general aspects were presented in *Arch. Sc. Phys. et Nat.* (4) *44*, 66 (1917); the application to a powder goniometer in *Arch. Sc. Phys. et Nat.* (5) *1*, 550 (1919).

Using gypsum we constructed high-intensity monochromators. When high resolving power rather than high intensity is required the

Cauchois ground and bent crystal is the elegant solution of the para-focusing condition for one plane.

Edgar Meyer held the chair of physics at the University in Zurich. He ran a colloquium which we used to attend. Einstein had come to enjoy the freedom of Switzerland for part of the war. He gave a fascinating course in special theory of relativity and attended the colloquium, where, to the general surprise, he took a lively interest in the discussion of experimental papers. I became particularly good friends with the mathematician George Polya. He would open any discussion with a sentence reminding of scholastic dialectics; 'Also die Frage ist die—'followed by the statement of the premises made. He is now in the U.S. actively engaged in raising the level of high school mathematics.

From Zurich, after the war, I visited von Laue who had not taken root in Frankfurt and had returned to Berlin. He opened the door in person, a revolver in one hand and a dagger in the other. It was the time when the new German Republic was weak and terror bands of opposing parties had a free hand. He was relieved to find that the ringer of the doorbell was a harmless individual.

With the death of Professor Schweitzer and the termination of my duties of replacing him I went to England in 1920, where young W. L. Bragg was establishing an X-ray centre at Manchester University. I became a member of the staff, and found myself suddenly transplanted in quite new surroundings. Bragg had come to Manchester shortly before, and the memories of Rutherford's activities were still very much in evidence. They were passed on to the newcomers by W. Kay, the steward of the department, who used to assist Rutherford during the summer months in exploratory experiments to set out the research topics for the work to be proposed to students in the coming term. Rutherford was more interested in qualitative results than in exact quantitative measurements. He had no time to wait for the building of elaborate instruments and so the department had no proper mechanics shop, and what tools there were had been misused for purposes other than those for which they were intended, but there was a tradition how to set up experiments quickly and effectively. Outside firms, a mechanics shop, a glassblower and a tinsmith were under contract to give precedence to university work when required and these were called in whenever the resources of the department were insufficient. Of these, the tinsmith was most in demand. In England the trend of a department is determined by its head, so the delicate optical instruments assembled by Sir Arthur

Schuster, Rutherford's predecessor, were short of reading telescopes and eye pieces, which Rutherford had used for scintillation counting. With Bragg, the aspect of the department changed again and a further change occurred when he was succeeded by P. M. S. Blackett, when the activities were concentrated in one single large room which happened to have a light roof so as not to interfere unduly with cosmic rays.

When I came to Manchester, Bragg, James and Bosanquet, in consultation with Darwin, were engaged in the problem of primary extinction, a problem which had also been covered by Ewald's dynamic theory of diffraction. Soon, however, the interest shifted to the structure of inorganic materials, in particular metals and alloys. Bragg had an extraordinary gift of interpreting X-ray patterns in terms of structures by a direct visual approach and trained his pupils in these methods. The X-ray patterns were obtained with cameras using the rotating powder rod and Bragg did not waste time with elaborating this technique as long as it was yielding patterns good enough for interpretation.

Sir William Bragg occasionally came to Manchester to give a lecture, in which the most complex physical facts were presented in the simplest of terms. Like his father, W. L. Bragg possessed to an extraordinary degree this ability of finding forceful analogies. He was very much in demand for giving outside lectures.

While measurements from powders present several advantages, the method of the powder rod, exclusively used at the time, allowed to utilize intensity data for the interpretation of diffraction patterns to a limited extent only, since the relative intensities of lines at different angles depend on the shape, density distribution, and absorption coefficient of the individual powder rod, which factors are difficult to allow for. We became thus interested in the problem of developing a powder method for obtaining intensity data. We could establish intensity expressions for a powder layer conforming to the para-focusing condition, such that relative line intensities become independent of absorption in the powder. A goniometer was constructed for photographic recording which satisfies the requirements of these expressions; it makes use of a slit in a screen rotating in front of the film, which for counting becomes the counter slit. This instrument gave very sharp lines while using much wider beams than permitted with the powder rod technique. It could be adapted to the cases of having the X-ray entrance and counter slits equidistant or not from the powder.

For photographic recording—then generally used—there remained the handicap that photographic opacity and X-ray exposure are not proportional, the evaluation of each line requiring thus a conversion by an elaborate process of point to point scanning. We found that for small densities a linear relation exists between X-ray exposure and the number of silver grains produced. Instead of counting the grains we evaluated them by measuring the light scattered from a scanning beam. The goniometer and the method of scatter densitometry gave us the means for making quantitative intensity measurements from powders. Several investigations have been made of extinction, differential absorption and atomic scattering power using these methods. A. Baxter investigated the anomalous scattering near the L-edges, which for the K-edge had been investigated theoretically by Hönl.

My official duties in Manchester consisted in giving a course in thermodynamics and kinetic theory and running an advanced laboratory for honours students. Such a laboratory with set experiments becomes very dull to students and teacher alike, so it was gradually converted into short research, or rather pseudo-research, projects; more often the latter, because in most cases similar results had been obtained previously by other methods. The students presented their work in a colloquium which Professor Hartree, the mathematician, attended. He gave valuable comments on mathematical treatment and theory. This activity maintained contact with a broad front of physics.

The students impressed their interviewers for industrial jobs, but I wonder how much this was due to their better training and how much to their greater ability of expression.

When Bragg was called to succeed Rutherford in Cambridge, P. M. S. Blackett became head of the department. He had a winning personality and great enthusiasm for his work, which he transmitted to staff members and students. Apart from a passion for cosmic rays, he had widespread outside interests extending to art and politics.

In this period—spring 1938—also falls a visit to Vienna, where I was to establish contact with Mark and Schrödinger with a view of helping them to leave Nazi territory. Hitler had just invaded Austria and everybody was wearing distinctive buttons. Having none I found myself at a handicap, the more so since my hotel happened to be the headquarters of the German General Army Staff. Grubbing in my pockets I discovered an enamel button B.I.F.—British Industries Fair —which I had just visited on my way. This produced miracles: with the button pinned to my coat the soldiers presented arms, officers saluted and many heels clicked whenever I passed. I found that

Mark was taken care of; I was able to see relatives of Schrödinger and to make some arrangements.

Soon the war was to engulf the whole of Europe. I had just been given a sabbatical leave and was on my way to the U.S., where I intended to make arrangements for the manuscripts of my father, the late philosopher Franz Brentano. With the outbreak of war I returned to Manchester to find actually that there was no special need for me. So after one year I went to the U.S. for the summer months, taking the manuscripts of my father with me. I was offered a position at Northwestern University, where a large expansion of the physics department was planned and an X-ray physicist needed. This opened up the tempting opportunity of planning my own laboratory. I accepted to stay and found myself once more in new surroundings. In England the students are polite on the borderline of being submissive, the sharpest criticism is expressed in the words: 'how interesting', and the superiority of the teacher is so much respected that a student standing next to me watched my sleeve catching fire in reaching over a burner without so much as making a sign. To my reproach he answered: 'I did not want to be rude, Sir.' In the U.S. the students have much less restraint, instead of saying: 'how interesting', they would say: 'I don't see that', and would back up their doubts by a lengthly exposition in my office, giving me the opportunity to clarify the point and to satisfy them. This made for refreshing relations. At the beginning of my stay it also happened that a student, to whom I had given a low grade, commented: 'I am never given a low grade, you must change it.' On asking a colleague about this strange behavior and giving the name of the student, he said: 'this is the name of a well-known Chicago gangster, you have to be very careful.' I did not change the grade and I am still here to report it. Probably the father gangster condoned the newcomer for his lack of experience. It did not happen again, so I had no opportunity of testing whether I had learned how to behave. The vitality and keenness of the students, and the availability of technical facilities, were encouraging. The situation would arise when I could say to the head of the department that I had all I wanted and did not require anything more. When did this ever happen in Europe! For the first time I did not have to make my own X-ray tubes and in this connection I was pleased to be able, on a very minor scale, to make a contribution in the sense of Röntgen: a manufacturer of diffraction tubes thought he had to avoid a more efficient design in order not to infringe on patents of another tube maker. I could point out that I had published the description of this design much earlier and that there was no valid patent on it. A students' laboratory similar to the one in Manchester was established, out of which some research items resulted. I may dwell on one because it is of some significance in X-ray technique. Many attempts had been made to introduce scintillation counting to X-rays, which has obvious advantages, particularly for short wave-lengths. They failed because the X-ray count was drowned in the general background of thermal and other random emission from the crystal and the photomultiplier commonly used. The difficulty was overcome by considering that each X-ray photon produces the emission of several photoelectrons (about 13 with sodium iodide crystals and Mo K radiation according to West, Meyerhof and Hofstadter), while thermal electrons are emitted individually at random. By chopping the electron emission into sufficiently short intervals it can be arranged, that for each such interval the probable number emitted by random processes is small compared with the number emitted by an X-ray photon. Using a discriminator it is easy to count only the pulses due to X-rays. This was done by Ladany and reported at the 1953 Rochester Meeting of the American Physical Society.

We also tried to improve photographic X-ray techniques. By projecting the line pattern on a film, which is being rocked through a small angular range (a fraction of the line width), narrow high peaks could be broadened out and the quantitative evaluations improved. On the other hand, when correct line contours in terms of intensities were required, a microdensitometric method was developed for automatically converting opacities to X-ray exposures. These methods were used in investigations on small-angle X-ray scattering and in particular, jointly with Spencer, in the study of any change of the AgBr lattice, when the latent image is formed. This required all the refinements of powder techniques; it led to distinguishing between two theories. The formation of the latent photographic image, of the electrical conductivity of pressed powders, and of electrical properties of highly disturbed lattices, were investigated. This is reported in Ph. D. theses by the students who undertook this work, but this leads away from the field of X-ray diffraction.

During my stay at Northwestern von Laue came for a visit to the U.S. (1948). It was wonderful to see him stepping out of the train, in Chicago, somewhat aged but with his radiant expression of idealism unbroken, in spite of all he had gone through. He had exposed himself by protecting many who had been persecuted under the Nazi régime, he had also opposed submitting to its external ritual: entering the university he used to carry one heavy book in either hand, so that he could not give the Nazi salute.

He came shortly after the war from the country where the war devastation of institutions and laboratories was greatest to that where they had suffered least. The seeming opulence of America hurt him. When we took him to a Woolworth shop he noticed toy balloons being blown up with helium gas, to be handed over to children. Laue said with a sigh: 'if we had only two of these helium cylinders in our laboratories, what could we not do!' But this was not his lasting impression. He was given full opportunity to state the plight of German physics, which he was so well qualified to present. It was given to him to see the resurgence of scientific activity in his country.

On reaching retiring age the university gave me the use of a laboratory. I then reverted to X-rays and the spinning powder layer which I had introduced a long time ago as a means of increasing the number of crystalline domains taking part in diffraction for reducing the background of short-wave radiation.

A teaching position tends in a way to develop a split personality, the extrovert teacher, versus the introvert researcher. To be relieved from this duality may seem a certain simplification, but a university position is an ivory tower and when this is abandoned soon other tasks appear; in my case this is the care of the manuscripts of my father, so that there are two sides to the change.

References

To para-focusing and powder goniometers: J. C. M. Brentano, *Arch. Sc. Phys. et Nat.* (4) *44* 66 (1917); (5) *1* 550 (1919); *Proc. Phys. Soc. 37* 184 (1925); *49* 61 (1937); *Journ. Appl. Phys. 17* 420 (1946).

To extinction, differential absorption and anomalous scattering: Baxter and Brentano *Z. f. Phys. 89* 720 (1934); *Phil. Mag.* (7) *24* 473 (1937); J. C. M. Brentano Proc. *Phys. Soc. 47* 932 (1935), *50* 247 (1938); Brentano, Honeyburne and Berry, *Proc. Phys. Soc. 51* 668 (1939).

To quantitative photographic methods: Brentano, Baxter, Cotton, *Phil. Mag.* (7) *17* 370 (1934); J. C. M. Brentano, *Journ. Opt. Soc.* A. *35* 382 (1945); *Journ. Phot. Sc. 2* 150 (1954).

To scintillation counting: Ladany and Brentano, *Phys. Rev. 92* 850 (1953); *Rev. Sc. Instr. 25* 1028 (1954).

Personal Reminiscences

M. J. Buerger

My interest in science was started by a volume entitled *The Complete Chemistry, a Text Book for High Schools and Academies* by Elroy M. Avery (Sheldon and Company, New York and Chicago, 1883). This was my father's high school chemistry text in 1893, and I discovered it in our family library about 1910, when I was seven years old. This explained a good deal about matter, physical and chemical changes, and of course, chemistry. This seemed to be just what my mind needed, for it explained the complicated things about me in terms of simpler units, so that it became obvious that the world was a matter of chemistry. I was soon well acquainted with the contents of the book, and astonished my parents and their friends with this curious knowledge.

In due time I studied chemistry in high school and found it, along with geometry, my most interesting study. My father, noting this interest, felt that M.I.T. was the place for me, and when I finished with high school, I continued my studies there.

Because of the uninspiring teaching of chemistry, I looked around for another related field. I found that mining engineering made use of chemistry through mineralogy, so I entered the study of mining. This was a fortunate choice, because I immediately came in contact with Mr. Walter H. Newhouse, at that time a graduate student studying for his doctor's degree under Professor Waldemar Lindgren. Every student needs a wise and inspiring teacher, and Newhouse became this to me. What I had missed in chemistry, Newhouse made good in mineralogy, for that was what he taught. But more than that, Newhouse had a feeling for relevance, and always stripped a matter of its extraneous wrappings and went directly to the core. I recognized this characteristic and tried to emulate him. Although I graduated with a degree of B.S. in mining engineering, I took my bachelor's thesis under him. At just about that time he was given funds for a research assistant and he offered this post to me. My job was to relieve him of teaching elementary mineralogy while he devoted the corresponding time to research.

This was a turning point in my career. During the last two years, I had had tution scholarships, but my mother had supported me. It had never occurred to me to go on with graduate study, since there was no money available for it, so this assistantship changed my career from that of a practising mining engineer to that of a teacher. In two years I took my master's degree in geology, and in two more my doctor's degree in mineralogy, all under Newhouse, who had been advanced to an assistant professor. Newhouse and I became close personal friends, and shared an office. I absorbed much from him in the philosophy of science. We both basked in the research atmosphere created by Professor Waldemar Lindgren, dean of geologists and head of our Department of Geology.

In 1927, before I obtained my doctor's degree, Bragg spent a term at M.I.T., and I attended his lectures. Bertram E. Warren, a fellow graduate student in physics, became a student of Bragg's at that time, and with him worked out the structure of diopside. After receiving his doctor's degree, Warren continued in crystal-structure analysis. I had no laboratory facilities, but also wished to try out crystal-structure analysis, so, with Warren's kind permission to use his equipment, I took a set of rotating-crystal and oscillating-crystal photographs of marcasite. I solved this simple two-parameter structure alone in 1930–31 by applying the techniques I read about in the papers appearing in the *Zeitschrift für Kristallographie*.

This marked the second, and perhaps the most important, turning point in my career. I had been teaching mineralogy, optical crystallography, and petrology, so that up to this point I had been a mineralogist. But the structure of marcasite represented my first excursion into pure crystallography. I was delighted with the certainty of the conclusion reached by crystal-structure analysis, as compared with the arguable results published by the geologists and mineralogists of that period. Just following my delightful experience with marcasite, the new president of M.I.T., Karl Taylor Compton, obtained a large grant for research from the Rockefeller Foundation, and eventually I was asked by Professor Henry Shimer, acting head of our department at that time, what I needed for research. I immediately dreamed up $10,000 worth of X-ray equipment, had it approved, and I was shortly in the business of investigating the crystal structures of minerals.

In 1937 I was a member of Donald MacMillan's expedition to

Baffin Island in the Arctic. This trip up and back gave me the leisure to start my book *X-ray Crystallography*. The book had two objectives. The crystallographic preliminaries of crystal-structure analysis had never been treated in any detail, but had always been compressed into a small portion of a book covering all of crystal-structure analysis and the results the art had achieved to that date. *X-ray Crystallography* was intended to pay attention to the important preliminaries of unit-cell and space-group determination. At the same time it was written to the mineralogical crystallographers and as a protest against their current practices. At that time, the mineralogists made optical goniometric studies of the face development of crystals, and from these they attempted to deduce, on the basis of various theoretical ideas, mostly ill-founded, what the lattice of the crystal was. They were so confused that they tried to distinguish between a 'morphological lattice' and a 'structural lattice'. *X-ray Crystallography* was intended to present them with a fool-proof way of finding *the* lattice. I wrote a preface in which I stated my opinion of their groping methods. The preface was too frank to publish, so I had to write another moderate one, which now appears with the book, but I had the pleasure of writing what I thought of the mineralogical crystallography of the day. Writing two prefaces (the second publishable) was a practice I continued in later books.

I have found teaching a worthwhile career. In the first place, it has been my experience that students react to me as I reacted to my teachers. They need the teacher not only to guide them in technical matters, but to transmit a philosophy to them, partly by precept, partly by providing an appropriate atmosphere. Students are susceptible, and the teacher has a great responsibility. Many times I have had the experience of having a former student unconsciously quoting back to me my own philosophy. On the other hand, I have also been taught much by my students. The close rapport between student and teacher makes it possible for the teacher to absorb from his students knowledge which has developed since the teacher was involved in formal study, or which the student, with his youthful viewpoint, has seen fit to cultivate. More generally, I have found that a group of students and their teacher are members of a small select society, and that they teach and inspire one another, especially if the group has reached the critical size of, say, five. Within this group it becomes fashionable to advance in knowledge and to publish newly acquired knowledge, so that the members of the group emulate and stimulate one another.

Paul Niggli was a man I admired very much. I had the good fortune to spend a week with him when we both took part in the mineralogical excursion to Lapland following the 1951 meeting of the I.U.Cr. in Stockholm. I have had occasion to study his writings many times, and have been impressed by his characteristic of being ahead of his time. I encountered this rather recently when I was concerned about reduced cells. Niggli had worked out the whole matter years before it was used, and apparently just for fun. As editor of the *Zeitschrift für Kristallographie* he also demonstrated his high caliber. He either accepted a paper or rejected it, but did not require the author to make it fit his ideas or style. When I was asked to become a co-editor of the *Zeitschrift für Kristallographie*, I accepted with pleasure, remembering that Niggli had long occupied that post. Needless to say, Niggli's actions as editor have had a strong influence on my own. Because of my admiration for him, I also felt it a high honor when I was elected a Foreign Member of the Academy of Sciences of Torino to fill the place left vacant by Niggli's death.

I believe most scientists, including myself, do not spend enough time in an attempt to gain perspective. Usually we are too busy finishing the many projects we have started. I had this view impressed upon me as a result of a four-month stay in Brazil, while I was visiting professor in the Faculty of Philosophy in Rio. During that period my office routine could not intrude upon me, and I had a chance to think about the phase problem. Starting with Dorothy Wrinch's simplification of a Patterson map to a set of points, I was able to generalize her primitive result of solving the vector set based upon a triangle, and, before I left in January 1949, lectured to my class on how to solve a general vector set. The paper, typed on board ship while returning home, was sent immediately to *Acta Crystallographica*, but, unaccountably, did not appear until March 1950. By this time others were finding solutions of Patterson maps by other ingenious methods. My own viewpoint eventually lead to the broad class of image-seeking functions. The first of this class was the product function. The idea of a minimum function occurred to me as I was laboriously multiplying pairs of Patterson values to form the product function. The first minimum functions were therefore made by comparing pairs of values of the Patterson function as written down as if ready for contouring a Patterson projection of the crystal. I compared the two values at the ends of a line image as I allowed the image to range over the cell, selecting, at each location of the image, the minimum, and reading it aloud to my daughter Marla who wrote down this value in its proper place on another map, the M_2 map. This tedious procedure, after being practiced a few times, gave way to the graphical method. All this was stimulated by a visit to Rio, and I could mention other ideas developed by appropriate loafing in Florida, etc.

The precession camera was developed as I was writing the part of Chapter 10 of *X-ray Crystallography* concerned with the limited symmetry information obtainable from oscillation photographs. It was evident that the symmetry of the oscillation motion limited the symmetry of the record. Why not increase the symmetry of the motion? Precession ideally provided a radially symmetrical motion. The oscillation and precession motions are compared in a figure on p. 207 of *X-ray Crystallography*. It was easy to make the first camera, shown on p. 208, and the first photographs, shown on p. 209, justified the whole idea. All this was carried out in 1937. In 1938 de Jong and Bouman showed how to avoid radial distortion as well, and this improvement was quickly added to the primitive precession camera to give the first precession camera as we now know it. This just missed appearing in *X-ray Crystallography*, the manuscript of which was finished in 1940. It was described in *Monograph No. 1*, 1949, of the American Society for X-ray and Electron Diffraction. The availability of manuscripts of this sort brought forth the *Monographs of the ASXRED*, and these were later taken over and continued by the American Crystallographic Association.

Before World War II, most crystallographers published their serious papers in the *Zeitschrift für Kristallographie*. Editors of journals in the better recognized sciences did not always encourage crystallographic contributions. About 1939 I began to agitate among my colleagues for the establishment in the United States of a Journal of Crystallography. There was some support for the idea, but most were afraid to go ahead for fear of financial troubles, and certain societies, notably the Mineralogical Society of America, began to be alarmed at possible defection of some of its membership. Maurice L. Huggins and I corresponded about the possibility of a journal, and eventually I found myself a chairman of a subcommittee of the National Research Council Division of Chemistry charged with investigating the possibility of establishing a journal. In the end, the Americans did not want to proceed alone, but the episode drew together a group of X-ray crystallographers and this lead to their forming a society, to be called the American Society for X-ray and

Electron Diffraction (ASXRED), in July, 1941, at Gibson Island. I believe that Huggins and I pushed the rest into this.

Meanwhile there was already an active small group in Cambridge, Mass., who called themselves the Crystallographic Society, and later, the Crystallographic Society of America. My correspondence indicates that this group met as early as 1939–40. It had such speakers as Fankuchen and Bridgeman, and eventually held successful national meetings in Northampton, Mass. (21–23 March 1946), Annapolis (19–21 March 1947), New Haven (1–3 April 1948) and Ann Arbor (7–9 April 1949). This society paralleled the ASXRED, but did not tie itself down to the tool of X-ray diffraction. It held a joint meeting with the ASXRED at Yale University, 1–3 April 1948, and the two societies merged in 1950.

*The Genesis and Beginnings of X-ray Crystallography at Caltech**

C. L. BURDICK

That Caltech was one of the first two centers in America where X-ray crystallography was started and continued was due entirely to the imagination, vision, and conviction of Dr. Arthur A. Noyes. In 1916, which was very early in Caltech's beginnings, Dr. Noyes was nominally visiting professor of chemistry, but actually he, with Drs. Millikan and Hale, was planning and shaping the future of this whole great institution. This particular specialty was only one of the many in which he saw great promise and forthcoming utility.

Prior to 1915 Dr. Noyes was occupied wholly as director of the Research Laboratory of Physical Chemistry at MIT in Boston, and it was there in 1913–14 that I came to know Dr. Noyes and to be a graduate student under him at the laboratory in MIT's old 'Engineering C' building. This was near the end of the halcyon days of classical physical chemistry when on the staff could be noted among the younger men such names as G. N. Lewis, C. A. Kraus, and F. G. Keyes.

Dr. Noyes encouraged me to go abroad for doctorate work and postdoctoral study, and it turned out that I sailed from New York in July 1914 on the last regular German liner to get through to Hamburg. Because of war conditions my doctorate period was divided between Professors Fichter and Rupe in Basel and Professor Willstätter at the Kaiser Wilhelm Institute for Chemistry in Berlin-Dahlem. It might be noted that there were then two young, comparatively unknowns by the names of Hahn and Meitner in this same Institute. Following this, the writer had a brief period of postdoctoral work at the Kaiser Wilhelm Institute of Physical Chemistry under Professor Fritz Haber and then a later opportunity to work on 'Reststrahlen' with Rubens at the University of Berlin and to sit in on the lecture courses of Nernst,

* Reprinted, with kind permission, from *Physics Today*, Oct. 1958.

Planck, and Einstein. What untold recognition and appreciation these names call forth today.

Early in 1916 it was only with some circumlocution, occasioned by the separate British and German censorship of all mail from America, that advice came to me from Dr. Noyes to try to find a way to spend my last months abroad working in the X-ray laboratory of Professor William H. Bragg at University College in London. In his letter Dr. Noyes expressed his strong belief in the importance of X-ray atomic structure analysis for the future of theoretical chemistry, and his wish to get something of the kind started at MIT.

It was not so simple for an impecunious, young PhD of American neutrality without connections to get from Berlin to London during the critical period of the Zeppelin raids and unrestricted submarine warfare. But due to the volunteered help and influence of Professor Kamerlingh Onnes of the Cryogenic Laboratory in Leiden, it was accomplished.

The six months in the laboratory of Professor Bragg were of inestimable value, even though its leader was largely occupied with war work. With what time Dr. Bragg had to spend with us in guidance, Dr. E. A. Owen and the writer were able to work out a tentatively satisfactory X-ray structure for carborundum, which was later published.

An interesting presentation could be made of the primitiveness of the equipment then available, the old-fashioned induction coil with Leyden-jar condensers and a mercury interruptor, the gas-filled X-ray tubes of unpredictable and uncertain output and 'hardness', and gold-leaf electroscopes with the strangest static aberrations when it came to measuring ionization intensities.

On returning to MIT, which was in process of moving from Boston to Cambridge, my assignment from Dr. Noyes was to build a Bragg X-ray spectrometer with any improvements which the state of the art would permit. This meant, first, having at hand a really good X-ray transformer with rotary-disc rectifier and secondly, and of most importance, an only just then developed, new Coolidge X-ray tube fitted with a palladium target. Dr. James A. Beatty, then a student, was principal collaborator in building, putting this machine into initial operation, and testing.

In the latter part of 1916 when Dr. Noyes left for his annual tour of duty as visiting professor at Caltech he asked me to go with him and build another spectrometer embodying the refinements which our tests had shown could be made. The Caltech team on this was James H. Ellis, Fred Hensen (instrument maker), and myself. The things which made the original Caltech spectrometer probably the best of its day were its high-power input and relative constancy of measured electrical energy to the tube. This gave possibilities for narrower spectrometer slits, precise angle measurements, sharp reflection peaks, and better measurement of relative reflection intensities of the spectral orders than had probably ever been made before.

The measurements carried out in these months permitted the determination of the crystal structure of chalcopyrite and resulted in two papers by Ellis and Burdick. These were submitted to the *Journal of the American Chemical Society* and the *Proceedings of the National Academy of Sciences* early in 1917, and appeared as publication Number 3 of the Gates Laboratory of Chemistry.

Then the United States entered the war. Dr. Noyes' time was spent in large measure in Washington, Burdick was called from the US Ordnance Reserve into the nitrogen fixation program, and for a time the X-ray crystal work at Caltech lagged. After the war, with the arrival of Dr. Roscoe G. Dickinson, the wheels hummed again. High-voltage sparks flew and research results by many collaborators came out apace. Today the successorship of Dickinson is in the notable hands of Pauling. Dr. Noyes' vision of the early days of 1916 has paid off and will continue to do so in generous measure.

Moseley's Determination of Atomic Numbers

C. G. DARWIN

One of the earliest important consequences of Max von Laue's great discovery of X-ray diffraction was its use in verifying the theory of the atomic numbers of the elements. This was done by H. G. J. Moseley in Manchester towards the end of 1913, when he measured the frequencies of the K-spectra of the elements from calcium to zinc. It may be interesting to tell how this came about, and in describing it I propose not to adhere to the rule that no discovery can be claimed until it has been published. It seems to me that a juster picture will be given by a record of the opinions prevailing in the laboratory, and the gradual growth of the conviction with which they had been held often months before any publication.

It must be hard for anyone now to realize the troubles facing physicists in those days on account of the conflict between the classical mechanics and the quantum theory. They seemed not to complement each other as we now recognize, but rather to present impossible contradictions. Thus in the case of the X-rays J. J. Thomson had assumed that they were straightforward electro-magnetic waves. He had calculated how they would be scattered by matter, and his predictions were very satisfactorily verified by Barkla. On the other hand the ionization produced by X-rays gave electrons of velocity so high that it was clearly contradictory to the wave theory as then accepted. Indeed for a time W. H. Bragg inclined to the view that the X-rays must be some kind of corpuscle. It was in this atmosphere of mutual contradictions that we were working in those days.

Up to the end of 1912 Rutherford's laboratory in Manchester had been almost entirely devoted to the study of radioactivity. The work covered many fields, but much the most important one undoubtedly was the discovery of the nucleus of the atom. Through counting the scintillations of scattered α-particles, work mainly carried out by Geiger and Marsden, Rutherford's brilliant calculation of their statistical distribution was verified, and it was also possible to determine roughly the nuclear charge, equal to about 100 electrons for gold and about 10 for aluminium. It was round this time that Niels Bohr first came to Manchester, and greatly influenced our thinking by his deep appreciation of the limitations of the classical theory. He did indeed then begin his theory of the hydrogen spectrum, but at first, though it looked most promising, we could not judge how far it would extend.

When we first heard of Laue's discovery, it appeared obviously of the very greatest importance, but there were still the conflicting views about the nature of the X-rays, and we had little idea of what would ensue, or that there would be any connection with the nature of the nucleus. Moseley had come to Manchester in 1910, and he had been working on radioactive subjects like the rest of the laboratory, but he now proposed to take up the study of X-ray diffraction as being the most exciting new field in physics, and I agreed to join him in this work; we did not in the least know what might come out of it. When we first approached Rutherford, he was distinctly discouraging on the ground that nobody in Manchester knew much about the technique of X-ray experiment, so that we would be severely handicapped as compared with other laboratories where they were familiar objects of study. However we decided to go ahead, and in particular to go beyond the photographic method by which the diffraction had been first discovered by Friedrich and Knipping. It seemed to us that ionization methods would give the advantage of being able to get quantitative measures, and that these would be necessary in order really to understand the subject.

Working with Moseley was a most impressive experience. He was without exception the hardest worker I have ever met. He had two principles in his work. The first was that when one starts to set up an experiment one must not stop for anything until it is set up. The second was that when one starts the experiment itself one must not stop till it is finished. There were of course no regular meals, and work often went on for most of the night. Indeed one of Moseley's expertises was the knowledge of where one could get a meal in Manchester at 3 o'clock in the morning. In spite of these strenuous rules, it was most agreeable to work with him, and there was a constant stimulus all the time in our discussions about the fundamentals of the subject.

Our interest was the general character of X-rays, and in particular we were inclined to accept the idea of white X-rays, which had been proposed by W. L. Bragg. We also adopted his method of studying the oblique reflection from the flat face of a large crystal. The two chief ones we worked with were of rocksalt and potassium ferrocyanide, both about 5 cm square. After various changes in the arrangements the work was done mostly with the crystal mounted on the axis of a turn-table. It was exposed to a narrow beam of X-rays coming from a platinum anticathode through a very narrow slit, and the detection was done in an ionization chamber set up on the turn-table at double the glancing angle of the crystal. With a view to getting strong effects with our narrow slits we used a narrow ionization chamber charged with helium which was raised to such a potential that it was on the verge of sparking. We had at first a good deal to learn about the technique of X-rays, but by a long variety of experiments we convinced ourselves that the rays we were studying were entirely similar to white light.

Through having adopted very fine slits we were unlucky in one respect, for by sheer chance at first we never happened to set our instrument in the position to reflect any of the characteristic rays of platinum. However it was not long before we were told about them by W. H. Bragg when he was visiting Manchester, and we could then easily verify his work and could identify the different orders of spectra of course much more readily than we had been able to disentangle the white X-rays. A small further point may be mentioned, in that W. L. Bragg had not yet worked out the crystalline structure of rocksalt, so that there was uncertainty about the absolute wave-lengths to a factor that might be even as much as $\frac{1}{2}$. However not long afterwards he solved this problem.

While we were engaged on this work Bohr had been developing his great theory of spectra, and as time went on there arose the possibility that the various X-ray spectra might be analogous to light-spectra but contributed by the internal electrons of the atom. Moseley therefore set out to determine these wave-lengths, but I had been spending some time calculating the optical effects to be expected and I decided to go on with that, since Moseley was now quite confident that he could do this work by himself.

As to my own work, we had made one set of experiments which gave specially interesting results; this was the determination of the ratio of the intensity of the reflected beam to the primary. The general calculations of the effects of arrays of atoms were not simple, including as they did the disentangling of the various orders of reflection of the white X-rays, but they hardly involved more than the mathematics of optical diffraction theory. It emerged that the pencil of reflected rays could not be more than 6″ broad. Now even if one assumed that the reflection was perfect in this range, the total amount reflected was at the most generous estimate less than a tenth of what we had measured again and again. The calculation was made on the assumption that each atom scattered the rays independently of its neighbours, and it was therefore necessary to examine the effect of such cooperation, but this proved that the result would be to weaken the beam still further. Moreover in this second calculation the scattering proved to be proportional to f, the scattering coefficient of the single atom, whereas with the earlier calculation it had been proportional to f^2, and there seemed to be evidence that this last was usually the correct answer. It was these results that led a good many years later to the study of imperfect crystals, with their primary and secondary extinctions.

Reverting to the work of Moseley, I followed it with great attention, visiting his room frequently, and noticing that nearly every time I did so the experiment looked completely different; this did not surprise me because I knew that he was always ready to take everything to pieces again if he thought of any possible improvement, no matter how trivial. At first he spent a lot of time using ionization methods in view of our previous experience with them. But in the end he turned to photography, and he found that for his purpose it was much easier. He had a large glass tube containing a railway with blocks of the elements from calcium to zinc mounted in a row along the carriage. At its ends there were stopcocks each carrying a reel at its inner end, so that he could move the carriage to and fro without losing his vacuum. Each element could thus be used as anti-cathode in turn by bringing ti opposite the cathode of the X-rays, and also opposite the slit which let the rays emerge. These rays then fell on his crystal so that they were reflected at the appropriate angles on to the photographic plate. The result was the photograph that has become famous in books on the subject. Before this time Bohr had explained the relation of the hydrogen spectrum to Rydberg's constant, and he had conjectured that the K-spectra might be essentially the same thing as the simple hydrogen spectrum though this could not be verified until their wavelengths were known. However, applying these ideas to his photographs Moseley was immediately able to say that the nuclear charges of his elements ranged by integers with quite high accuracy from 20 for calcium to 30 for zinc.

After this work at Manchester he moved back to his home in Oxford early in 1914 and, working in the Oxford University Physics Laboratory (Clarendon Laboratory), he continued the experiments there by studying the L-spectra of heavier elements. He thus completed this great task of establishing the principle of atomic number from hydrogen to uranium. He finished this work in time to join the party of the British Association that was to visit Australia in that year. But the war started and he returned home as fast as he could. He took a commission in the Royal Engineers, and was despatched to Gallipoli. There he was killed during one of the landings, a personal tragedy and a great tragedy for the progress of science.

For auld lang syne

J. D. H. Donnay

I came to Crystallography through Mining Engineering, Geology, and Mineralogy. By training I thus belong to the Old School. At the University of Liège, my first encounter was with a crystallographer, G. Cesàro, born in 1849 but still in charge of the entrance examination. When I entered my third year and time came to take the required courses in crystallography and mineralogy, Prof. Buttgenbach had just been appointed. My first reminiscence is a sobering one—flunking the first quiz! This failure may well have been the turning point of my career: in order to make up for my ignominious showing, I had to take a good second look at symmetry, and the derivation of crystal forms, by the traditional method of truncations, gave me a thrill.

My next professor of Crystallography was A. F. Rogers, at Stanford University. Although an old-time mineralogist, he recognized the importance of X-ray diffraction and was conducting a joint course with M. L. Huggins, then in the Chemistry department. He also emphasized the theory of groups and taught the derivation of forms by Gadolin's stereographic method. His enthusiasm for Geometrical Crystallography was boundless. At a meeting of the Mineralogical Society of America, a (then) young and up-and-coming 'X-ray crystallographer', M. J. Buerger, had remarked that he did not understand how people could still maintain any interest in the old geometrical crystallography, that once the crystal structure was unraveled everything worth knowing about the crystal was known. Rogers got up, expressed his disagreement, and concluded in a vibrating voice (I still hear him), 'Geometrical Crystallography has had a glorious past, and it will have a glorious future!'

My chief debt to Rogers is perhaps that he introduced me to Georges Friedel's admirable *Leçons de Cristallographie*, the second edition of which had just appeared (1926), for no other book has had on me as strong an influence. After 35 years I still find that it makes profitable and challenging reading.

The man who really showed me the shear pleasure of research was H. W. Morse. After six months of prospecting for oil in Morocco, I had come back to U.S.A. and, as 'Research Associate in Geology and Teaching Fellow in Mineralogy', I was spending another year at Stanford. Morse had prepared artificial three-dimensional spherulites of hundreds of compounds and had observed their interference effects between crossed nicols, in parallel light. I was told to go and see the beautiful phenomenon that simulated the conoscopic uniaxial figure. It was my good luck to derive the equation of the retardation curve of the spherulitic figure — though not until J. V. Uspensky (the Stanford mathematician) and H. A. Kramers (the visiting physicist from Utrecht) had shown me how to solve an elliptic integral. How elated I felt! My Ph.D. thesis, *The Genesis of the Engels Copper Deposit*, paled into insignificance—it had been a *job*, but *this* was pure joy! How well I remember the many happy nights that followed, during which I was working with my old friend Dr. Morse in the little house that was his laboratory, under the eucalyptus trees at the far end of his garden... (And how often the milkman's arrival in the early morning reminded me of bedtime!)

In the summer of 1931, I visited G. Friedel in Strasbourg. It was a short visit, for he was not well, but it left a deep impression on me. I had shown him our work on spherulites and other aggregates, and he had said, 'You have gathered a large number of very interesting facts.' I had been disappointed, 'Perhaps, yes, but—the interpretation?' With a faint smile and a gesture of powerlessness, he had answered, 'Ah, cela....' It was not until much later that I came to realize that his disappointing comment had *not* been disparaging and that, to his way of thinking, uncovering facts that could not have been predicted was worthwhile, was indeed the very foundation of scientific research.

In September I joined the Johns Hopkins University as 'Associate in Mineralogy and Petrography'. From the start I found friendly advice and help in the Chemistry Department, where Emil Ott, then M. L. Huggins, and in 1936 David Harker taught X-ray diffraction; at the U.S. Geological Survey, where W. T. Schaller showed me how to use a 2-circle goniometer; at the Geophysical Laboratory of the Carnegie Institution of Washington, where G. Tunell (who had just succeeded R. W. G. Wyckoff) and Tom F. W. Barth were taking great interest in structural crystallography. My friendship with M. A. Peacock, who was then with Charles Palache at Harvard, also began at that time. Somehow I got deeply immersed in the problem of crystal habit.

Henri Ungemach, of Strasbourg, had been using *multiple indices* to designate the forms of a trigonal crystal with a rhombohedral lattice: he would write $(30\bar{3}0)$ for the prism $(10\bar{1}0)$, reserving the latter symbol for crystals with a hexagonal lattice. As $(10\bar{1}1)$ would have been valid in either lattice, he even proposed to drop the minus sign over the third index in the rhombohedral case. He therefore expressed the rhombohedral criterion as '$(hkil)$ with $(h + i + l)$ a multiple of 3'. I had asked Huggins, at lunch time, whether Ungemach's criterion was the same as that used in X-ray diffraction. Huggins had been slightly baffled by the unfamiliar formulation, but he had answered that, off-hand, yes, he thought it *was* the same. This had led me to use multiple indices to express lattice criteria in other systems: (200) for the cube in I and F lattices, (222) for the octahedron in the I lattice, and so on. I had also been intrigued by Baumhauer's *zonal series*, of which Ungemach had found many striking examples, but I thought (erroneously, as it turned out) that they were strict consequences of the law of Bravais.

In June 1936 I went to see Ungemach, who was very ill at the time. He had insisted that I come immediately after landing, 'otherwise you might never meet me', he had written. As I entered his room, I found him studying my paper on calaverite. For three days he talked to me of the morphology of minerals, and I marveled at his knowledge. I do not think it is much of an exaggeration to say that he knew all the forms of all the minerals! He reeled off zonal series after zonal series. And he also said, 'Remember the base: why is it absent in so many species?'... Friedel's 'unpredictable facts'—he had them all! As I was about to leave, that Wednesday afternoon, I suddenly felt deeply moved, and I told him that, as long as I would live, I would remember the Strasbourg crystallographers who had taught us so much: Friedel and Ungemach. And he replied, modestly, 'Ah, ça, c'est beau! Car Friedel, voyez-vous, il était grand... comme cela (and he lifted his emaciated hand as high as he could), tandis qu'Ungemach, il est grand (and he let his hand drop to a few inches of the bed cover)... comme ceci.' The next day I was to give a paper on the form birefringence of chalcedony to the French Society of Mineralogy, in Paris. As he called the meeting to order, the chairman broke the news: Ungemach had passed away during the night. He had left me all his measured crystals and all his notebooks.

On my return to Hopkins in the fall, I went to salute David Harker who, heralded by the newly discovered Harker Section, had just arrived from Cal Tech to replace Huggins. It was friendship at first sight. He sat in my course, and during the second semester I would attend his. Little by little it was finally dawning on me that the Baumhauer-Ungemach series might well, in some cases, contain *more* than the law of Bravais. Knowing very little about space groups, I asked Dave whether there did not exist space-group criteria similar to the lattice criteria. So he told me about the existing tables. Then I showed him the morphology of orthorhombic sulfur in Friedel's *Leçons*: the law of Bravais unmistakably pointing to an F lattice, and the many anomalies in the list of forms arranged according to decreasing frequency of occurrence (the pinacoids and most forms whose symbols contained a zero appeared too high in the list). Would he look up the space group of sulfur and find its systematic extinctions for me?

Days passed, then weeks. I was beginning to wonder whether Dave was taking me seriously, but one late afternoon he appeared with a broad grin, 'Every symbol with a zero in it must have the sum of the indices divisible by 4.' Quickly I grab my Friedel, fling it open—Eagerly we pour over it—All the anomalous symbols violated the rule! We had it! And feverishly we started looking for more, and more, examples—The abstract on the Baumhauer-Ungemach series that I had sent to the Mineralogical Society of America for its coming Christmas meeting was already printed, but the paper was never given: instead, I was able to announce our generalization of the law of Bravais, which appeared in the *Comptes Rendus* in February 1937.

It was at that same Christmas 1936 meeting that Peacock presented his 'Harmonic-arithmetic rule', the powerful tool that enables one to recognize the dominant face in any *simple zone* (one governed by a primitive reciprocal-lattice net). He had concentrated his efforts on the triclinic system, which he thought would be the most general, but in which all zones are simple. We spent the summer of 1937 together at Harvard, working on the 'new Dana' between memorable discussions on the relationships between crystal morphology and crystal structure. I wrote my own paper on the development of crystal zones in the summer of 1938, which, like many other happy summers, was spent in the company of J. Mélon at the Institute of Crystallography and Mineralogy at the University of Liège. By taking the zonal extinction criteria into account, Peacock's Harmonic-arithmetic rule could be generalized to give a perfect explanation of the series of Baumhauer and Ungemach.

In 1939 I was called to Laval University in Quebec, where I got my first X-ray unit (a Baird gas tube and a Buerger Weissenberg camera). The three years I spent in Canada were highlighted by the visits to Peacock in Toronto. I would take my students to see him, in the hope that some of his perfectionism would rub off on them. He would show us how to plot a gnomonic projection, pricking the face poles half-way through the thickness of the drawing paper with a fine needle mounted on a chuck. Duly impressed we watched in reverend awe. Then he would open the drawer and produce a honing stone—to sharpen the needles ('they *do* get blunt')!—What exquisite figures he could draw! What simple and beautiful English he could write to record his highly conscientious observations!

After the war—in the wake of thousands of powder patterns taken at the Hercules Powder Company— I blissfully returned to Academe and to Hopkins. One of the first jobs of 'reconversion' was to create an International Union and a Journal. This happened in London in July 1946. I attended this first postwar meeting on my way to Belgium (where I had just been appointed a professor at Liège). The British participants wished to have a journal devoted to 'X-ray analysis', but they finally compromised and settled for the name 'Journal of Structural Crystallography'. A minority objected to the word 'structural' as unduly limiting the scope of the journal. The spokesman for the majority argued that the word 'Crystallography', alone, would be too restrictive! Hoping to save the unity of Crystallography, I countered that, since it was the function of an adjective to restrict the meaning of the noun, 'Crystallography' without any adjective would be more general. But the motion to delete 'structural' was defeated, and Dave Harker, who was presiding, concluded, 'Donnay has fired his last cartridge!'—What is in a word! It was obvious that, to most people present, the term Crystallography did not evoke the study of crystals in its broadest meaning, but connoted only 'hemihedry, holohedry, and all that sort of things' (as W. T. Astbury astutely put it). Two days later the Russian delegation arrived, who had been delayed en route. The discussion was not re-opened, but the name of the journal was changed, behind the scenes, to *Acta Crystallographica*. Single handed Academician Shubnikov had turned our rout to victory by remarking that, in Moscow, the Institute of Crystallography of the Academy of Sciences of the USSR comprised many sections besides that of Structural Crystallography; he had also recommended Latin as a good language for titles of international journals.

In 1912 Laue presented us with a magnificent tool. Thanks to him, Crystallography has reached undreamed-of heights. But the study of crystals did *not* begin in 1912: it started with Kepler—as Laue himself told us. The results of the past remain the foundation on which we build. Throughout my career I have striven for the rapprochement of classical crystallographers and diffractionists—teaching crystal structure to geologists and crystal morphology to chemists, pushing the amalgamation of CSA (Crystallographic Society of America) with ASXRED (American Society for X-Ray and Electron Diffraction) to get ACA (American Crystallographic Association). Today, in admiring the majestic edifice of crystallography, I would like to think that, if others have brought in the freestones, I have contributed some of the mortar.

Epilogue

In 1949 J. D. H. D. married a crystallographer. Together Donnay and Donnay worked happily ever after. Their latest papers (1961) deal with a Second Generalization of the Law of Bravais.

Personal Reminiscences

R. Glocker

Early in 1912 I began research in the Physics Institute of the University of Munich under Geheimrat Dr. W. C. Röntgen. This gave me the opportunity of following the fundamental experiments of M. von Laue, W. Friedrich and P. Knipping from close by in the most literal sense. For the scene of these experiments was the adjoining Institute for Theoretical Physics whose Head was Professor Dr. A. Sommerfeld. The latter had, shortly before, and in a far-seeing way, appointed the experimentalist W. Friedrich, a pupil of Röntgen, to the newly created position of second assistant. M. von Laue was connected with the chair of theoretical physics as Privatdozent.

When the disturbing hissing of arc lamps in projectors in some of the main lecture rooms of the University was traced to an interrupter in Sommerfeld's Institute, this was the first indication that something mysterious was in the making there. It soon became known that the experiments had at long last revealed the wave nature of X-rays by causing them to be diffracted in a crystal. In July 1912 M. von Laue gave the details in a talk to the Physics Colloquium in Sommerfeld's small lecture theatre. Some forty physicists and crystallographers attended, among them Professor Ernst Wagner, Röntgen's senior assistant. He was stimulated by the talk to propose the following crucial experiment for testing the correctness of Laue's theory: if one of the many secondary rays obtained by diffraction from a first crystal is picked out by a diaphragm behind the crystal and allowed to fall on a second crystal parallel to the first, then only very few secondary rays should be produced behind the second crystal, and their wave-lengths should stand in simple ratios, like those of the harmonics to the fundamental wave-length in acoustics. The performance of this experiment was assigned to me as my thesis subject by Röntgen.

Because the secondary rays are so much less intense than the primary rays, exposure times of many hours were to be foreseen. Only he who has himself worked with gas-containing X-ray tubes can appreciate what was implied in such exposure times. Every five or ten minutes a little palladium tube, whose open end was sealed through the glass wall of the X-ray tube, had to be heated to dull red glow by holding a small spirits flame under it, so that minute quantities of hydrogen entered the tube by diffusion through the palladium walls. Besides, the current breakers, and the induction coils were not built for lasting loads and soon broke down. This was considerably improved when, at about that time, rotating high-voltage rectifiers became available. In view of the long exposure times all attempts were made to increase the sensitivity of the detection of X-rays. Photographic plates were prepared with two sensitive layers, one on top of the other. In order to achieve a uniform development the top layer was detached after the normal time of development, and the development of the bottom layer continued. For viewing, the dried layers were again superimposed.—A considerable gain in intensity was achieved by letting the beam impinge on the cube face of the rocksalt crystal under a glancing angle of only a few degrees.

At the end the desired effect was obtained, in spite of the initially gloomy prospects. The photographic plate behind the second crystal showed a diffraction pattern of which all spots could be explained by the monochromatic wave-lengths λ, $\lambda/2$, and $\lambda/3$.

E. Wagner was a physicist who liked to work with the utmost precision. I always remember one of his sayings: 'I like to work so precisely as if I were to live forever.' Since Röntgen too was very critical when examining whether a paper was mature for publication, I have often been wondering, during my four decades of scientific production, what parts of an investigation are of real importance for the intended aim, and which are not? To deal with everything in uniform detail would lead to a low efficiency of publication. The younger generation is prone to follow the opposite extreme.

The two-fold importance of Laue's discovery for Physics was quickly recognized. While the determination of the wave-lengths of the characteristic X-ray emission gave a great lift to atomic theory, the exploration of the internal structure of crystalline, and later of amorphous bodies soon assumed an extent of applications in science and engineering which surpassed all early expectations. A third, more recent branch of engineering applications of the same principles of diffraction is, however, less well known. This is the determination of elastic stresses in engineering work pieces. Under stress, the distances of the atomic netplanes of a crystalline material change, and this minute change can be measured by back-reflection X-ray methods. By means of the known elastic constants the stresses in the material can be found from these changes. In other words, X-rays detect changes of distance, with the atoms serving as scale marks. This method differs from the usual mechanical ones of finding tension in that only the elastic strain is being measured, and not the sum of elastic and plastic deformations. It further gives the absolute value of the stress and is, for this reason, particularly well suited for the determination of residual and other internal stresses. Also the method leaves the work piece intact. A drawback of the method is that the small penetrating power of the necessarily very soft X-radiation producing back reflections limits the stress determination to a thin surface layer.

If a metal is subjected to alternating loads, the incipient 'fatigue', which precedes fracture, can be recognized from the fact that the stress indicated by the X-ray method diminishes while the alternating load remains unchanged. This shows that the strains produced by the load are no longer elastic, but partly plastic. The region in which fatigue occurs is extremely limited; in a steel rod with a pressure-tension load along the axis fatigue develops on the surface in an area of a few square millimeters or less.

Since the discovery of X-rays was of such eminent importance for medicine, one might well ask what further importance X-ray *diffraction* added. The influence is less obvious but undeniable. One of the pillars of X-ray therapeutics is dosimetry, the measurement of the radiation energy penetrating to a given point in the human body and becoming biologically efficient there. The interaction between X-rays and matter could be clarified only after one had succeeded to measure X-ray wave-lengths and to pick out or produce definite wave-lengths. All physical, chemical, and biological action of X-rays is an indirect one, carried out by way of photoelectrons, Compton electrons, or, for extremely hard radiation, electron pairs. Given matter of known chemical composition, the dosis can be calculated for any X-ray wave-length by applying known physical laws. Alternatively, it is possible to prepare test materials, and in particular fluorescent ones, which have the same composition as muscle or bone. If the intensity of the fluorescent light is measured by means of a photomultiplier, the dosis can be obtained which will be received at the same spot in muscle or bone tissue. The long discussed problem of the existence of selectively efficient kinds of X-rays, similar to the selective action of ultraviolet light on skin erythema, has been answered by showing that the absorption spectrum of X-rays in biological material is a continuous and not a line spectrum.

References

E. Wagner, 1913. *Physikal. Zeitschrift 14,* 1232; R. Glocker, 1914. *Physikal. Zs. 15,* 401 and *Annalen d. Physik* (Lpz.) 1915, *47,* 377.

Personal Reminiscences

A. Guinier

When I began to think of scientific research in 1935, it was Ch. Mauguin whose advice I asked for the choice of a subject. Honestly, I would find it hard to tell today what first took me to Mauguin. I had never taken courses in mineralogy, but it was common knowledge that Mauguin, who had himself come from chemistry, was a good patron also for a physicist. Anyway, looking back today over a span of 25 years, I can appreciate how very much to the point and profound were the first directives he gave me, and if I find my name figuring in this collection of reminiscences it is so because Ch. Mauguin had a feeling for the importance which the physics of solids would acquire in future beside crystal structure analysis, and of the rôle X-ray diffraction was bound to play in this new branch of science.

For the last two or three years a young physicist, J. Laval, had been making measurements in Mauguin's laboratory which at the time were very original. Their aim was to find the intensity of X-ray scattering between the directions of Bragg reflections. The measurements were made with an ionization chamber and the scattered intensity was so weak that Laval had to go to the very limit of sensitivity of his electrometer. He had, by that time, found very remarkable results which one was far from understanding.

While Laval continued his experiments Mauguin wanted the same problem attacked by a different method. For that reason he asked me to fix up a device for registering photographically the phenomena Laval had discovered.

In the then usual Debye-Scherrer diagrams the background intensity between the powder lines was caused by unwanted side effects, such as the non-monochromatism of the primary radiation, the scattering by the air in the camera, etc., and these effects also masked the simple scattering by the sample. One had therefore to get rid of these background producing causes, and at the same time it became essential to work with a primary beam of the highest intensity, since the aim was to study scattering a thousand times weaker than the normal scattering in powder lines.

The answer to this came with the use of curved crystal monochromators for the purification of the X-rays given out by the tube. The techniques of curved crystals were well known in Paris through the spectrographic work of Y. Cauchois, and furthermore, Laval already used them. In theory the combination of a curved crystal monochromator and a diffraction camera of the focussing type, like that of Seemann-Bohlin, should give powder diagrams of great intensity combined with high resolution. The attempt was, in fact, satisfactory, and the decisive success came by placing the target focus of the X-ray tube in the position of the point source on the circumference of the camera, and by making the focus as small as possible. Since we were using demountable X-ray tubes, it was possible to modify the cathode until a much finer target focus was achieved than in the then available sealed-off tubes which were built for high load rather than small focus.

This is the origin of the diffraction camera now called Guinier camera. Its use has spread mainly in Europe where several variants have been constructed, as by Hägg in Sweden, De Wolff in the Netherlands, and Jagodzinski in Germany. In U.S.A. it became known at a time when the counter diffractometer had already produced a considerable improvement of the powder method, so that it is not much used there nowadays.

But what I was aiming at was not so much the improvement of powder diagrams than the study of the diffuse scattering by crystals. Soon Laval and I continued research along different lines: while he made a fundamental study, experimental and theoretical, of the diffuse scattering of perfect crystals in thermal vibration, I studied diffuse scattering due to imperfections of the crystal structure. The first examples of this I obtained on substances which I had occasion to examine in my camera in 1937/38.

The camera with monochromator permitted the study of scattering at very low angles, down to some ten minutes of arc—which had been quite impossible in the usual cameras. No wonder therefore that even in the very first trials on various samples I could observe details which had not previously been detected. Some samples gave in the immediate neighbourhood of the direct beam a diffuse spot, often of high intensity and limited to an angular range of the order of a degree. At that time B. E. Warren had observed the same for carbon black and had attributed it to the state of dispersion of the scattering matter. Simple experiments convinced me that small-angle scattering is indeed characteristic for the division of matter into submicroscopic particles (< 1000 Å) and does not depend on the atomic structure. Thus it is observed for very small crystals in colloidal metals as well as for amorphous particles (silica gel) and colloidal solutions (albumen).

A simple theory, valid for a sufficiently dilute system of particles, relates the decrease of intensity with scattering angle to the value of a parameter which is characteristic of the size of the scattering particles. This so-called Law of Guinier is successfully applied for instance in the study of protein solutions.

Nowadays small-angle scattering by any kind of particle is well understood in reciprocal space; it is a particular case of the broadening of diffraction spots for the case of very minute crystals. P. Debye and G. Porod have extended the simple theory to heterogeneous systems without the restriction to very dilute ones. Low-angle scattering is used extensively in the study of macromolecules and fine grain catalysts.

Small-angle scattering has also revealed a 'large scale structure' in many solid high-polymers. This structure is intermediate between a very irregular heterogeneousness (which would produce continuous scattering) and strict periodicity (which would give sharp diffraction spots). At present there exist plenty of well-agreeing observations, but we do not as yet understand the origin and conditions of stability of such quasi-periodicities whose scale is out of proportion to that of the basic structural cell. One can only guess that there lies hidden here an important element for the full understanding of the structures and properties of bodies built from molecular chains.

As I was studying low-angle scattering in 1938, two metallurgists, Calvet and Jacquet, asked me to make an X-ray examination of aluminium-copper alloys of which they were studying the changes during age-hardening. With his new method of electrolytic polishing Jacquet achieved to detect precipitates at an earlier stage of age-hardening than his predecessors. Calvet and Jacquet gave me samples, age-hardened at room temperature, in which they could not yet find any precipitation. These samples gave rise to a problem: what structural change could explain a considerable change in the mechanical properties, while even the most meticulous microscopical study revealed no change of the alloy before and after the ageing?

By good luck the samples of Calvet and Jacquet were of such coarse grain that in my camera only three or four grains were hit. If they were kept fixed during the exposure to a monochromatic X-ray, the patterns contained, besides the usual diffraction spots of the Al-structure, very faint spots or radial spikes, which could not be explained by the ordinary diffraction theory whatever the crystal lattice might be. On rotating the sample, some of these spots wandered like spots on a Laue diagram even though the rays used were strictly monochromatic. I was very happy when I observed these unexpected spikes, for I felt that I had hit upon a diffraction phenomenon characteristic for the hardened crystal for which the other experimental methods had failed.

The essential features of the observations became understandable if one assumed that the AlCu crystal contains a series of small bits of (100) planes which produce specular reflection of the X-rays. Knowing that the copper atoms have the tendency to separate out, it was natural to surmise that the little reflecting flakes are formed by the segregation of copper. The detailed study of the reciprocal space of a single crystal soon showed that this model was too simple and that the copper-rich flake produces distortions of the adjoining aluminium planes.

The same observations on AlCu and a very similar interpretation were published simultaneously and independently by G. D. Preston, then of the National Physical Laboratory. Preston, however, did not continue this line of research after the interruption by the war.

This was the first example of a structure which is found in many oversaturated solid solutions in the course of their returning to stable equilibrium. The atoms of the excess solute do not segregate at once to form a well crystallized second phase, but they gather in small regions, of the order of some ten Ångström across and imprisoned in the crystal. These regions where the solid solution becomes heterogeneous and which are finally accompanied by smaller or greater distortions of the adjoining matrix crystal, are called the 'Guinier-Preston Zones.'

A structure of this kind corresponds to a state which is called pre-precipitation, and many examples are known. The interpretation of the initial X-ray results has now been confirmed by direct pictures obtained with the electron microscope. Furthermore, one has observed similar structures in crystals subjected to neutron irradiation; in this case what segregates out are the point defects caused by the irradiation, be they empty sites or interstitial atoms.

In a more general way, 'G. P. Zones' are a particular example of destroyed periodicity, and it is well known that such imperfections play an important rôle in the physics of solids. As another example let us mention the imperfect ordering, or antiphase domains, of alloys that are capable of being perfectly ordered. X-rays are sensitive to such lack of periodicity, provided a sufficient number of atoms is affected. The G. P. Zones of AlCu were a particularly favourable example because the flake form, by enhancing diffuse scattering along a line in reciprocal space, led to a relatively large intensity. Many other alloys, studied under the conditions we had originally, would have not let us observe anything.

Dislocations are, as defects, too rare and too small for producing abnormal scattering. Like many others, we looked for them in vain. However when dislocations pile up along subboundaries they effectively produce a disorientation between the adjoining sub-grains. In 1948 I showed that a modification of the simple Laue method, known as the Guinier-Tennevin method, permits the study of disorientations of the order of half a minute of arc. This method, together with microscopic observation, allowed to make visible the polygonized state of a metal which occurs during the annealing of the deformed crystals as long as true recrystallization does not occur.

Personal Reminiscences

G. Hägg

My first intimate contact with X-ray crystallography took place in 1927. Fifteen years had then passed since von Laue's discovery of X-ray diffraction, but the new discipline had not developed much at this stage, largely because of the First World War. Although the methods were beginning to permit the analysis of structures of medium complexity, the number of fairly simple compounds not yet investigated was still very large. Great fields could, therefore, be opened for study as was also pointed out to me by Arne Westgren. He had just obtained the chair in General and Inorganic Chemistry at the University of Stockholm, and asked me if I was willing to work for the 'filosofie doktor' degree in X-ray crystallography. I had earlier done research in other branches of chemistry, and I have a vivid memory of Westgren describing the thrills of X-ray work, where 'one nearly has the feeling of touching the atoms.'

Westgren had by then collaborated for some years with Gösta Phragmén, at this time connected with the Institute of Metal Research in Stockholm. Westgren's great imaginative power, skill in numerical calculations and the speed and easiness with which he worked was complemented very neatly by Phragmén's critical attitude, thorough knowledge of thermodynamics and great experience in workshop problems and instrument construction. Their cooperation at this time was very important for the early development of X-ray crystallography in Sweden. Although Phragmén was later on mostly occupied with other problems, especially as head of the Institute of Metal Research, his death in 1944 at the age of 46 meant a very great loss to X-ray crystallography.

In about 1927 Westgren and Phragmén were investigating alloy systems exclusively. The success with which they solved fundamental structural problems in this field was to a large extent the product of their great experience in physical metallurgy, their sound methods of preparation, and in particular the excellent powder cameras constructed by Phragmén. These cameras were of the Seemann-Bohlin focussing type, but following an analysis of the focusing conditions Phragmén had succeeded in obtaining a resolution which for many years could not be attained by any other camera. This put the Stockholm laboratory in a very favourable position at a time when most equilibrium diagrams had to be revised by means of phase analysis with powder photographs and when a great many structures were determined by powder methods. In single-crystal work, Laue photographs still played an important role for crystal adjustment and symmetry determination. For the subsequent structure analysis, however, they had already been superseded by oscillation and rotation photographs.

In these surroundings I started to investigate the binary systems of iron with the elements of the fifth group. Here I came at once in contact with metallic phases formed between transition elements and non-metals, a group of compounds to which I was later to devote much time. My thesis was ready in 1929, and when it was to be discussed in public the Faculty chose Professor Gregori Aminoff as its opponent. In this way I got to know Aminoff more closely and had many connections with him in the years that followed. Aminoff was a very modest man of retiring disposition. This, together with the non-educational character of his institution, tended to isolate him from the wider circles of Swedish science. Moreover he had very few pupils. In crystallography he was much aided by his deep sense for spatial relationships which was probably connected with his artistic gifts. Before 1913 he practised several years as an artist and had studied arts in France and Italy.

After having obtained my degree in 1929 I was appointed lecturer in General and Inorganic Chemistry at the University of Stockholm. From a scientific point of view the following seven years in Westgren's institute were the best of my life. They were filled with work on many interesting problems, the intercourse with Westgren, Phragmén and our fellow-workers was stimulating, the equipment was fairly adequate and administrative work, which was later to be a heavy burden, was practically non-existent.

During these years I continued the studies of metallic phases between transition metals and non-metals, started work on solid interstitial solutions with the structures of iron sulphide and selenide, tungsten bronzes and spinels, and took up the investigation of molybdenum and tungsten oxides. Satisfactory diffraction data for the latter oxides could only be obtained following the construction of a Weissenberg camera, the first of its kind in Scandinavia. Since during this period I was also connected with the Institute of Metal Research, it was my duty to do some research in physical metallurgy. In this field, X-ray diffraction studies bearing on the formation and decomposition of martensite as well as on nitride hardening were carried out. To this was added work of a more incidental character, e.g. structure determinations of dithionates.

In order to obtain experience in the preparation of metallic hydrides I spent a couple of months in 1930 in Professor A. Sievert's laboratory in Jena. I also saw various German institutes where X-ray diffraction was practised. My visit to V. M. Goldschmidt in Göttingen had the consequence that some months later Goldschmidt invited me to join his staff. In view of what happened in Germany a few years later I am glad that I had to refuse his kind invitation.

At the end of 1936 I was appointed professor in General and Inorganic Chemistry at the University of Uppsala. Research in inorganic chemistry had not been carried out there for more than twenty years, so it had to be organized from zero level including the training of a scientific staff. This made work very slow at the beginning, and effective research could not be started before 1940. Since then, however, the volume of research has increased at a fairly rapid rate. This has been possible above all through the cooperation of a number of enthusiastic and clever collaborators.

Autobiography

ALBERT W. HULL

In the spring of 1912, I had been teaching physics at the Worcester Polytechnic Institute for four years, and was beginning to worry about the future. My midnight photoelectric research had yielded one or two scientific papers, but I was aiming at such perfection that I hesitated to report more. The meeting of the American Physical Society was coming in June, and my wife persuaded me that it was time to give a paper. So I went to the meeting in New Haven, Connecticut, and there I met Dr. Irving Langmuir and Dr. William D. Coolidge, of the General Electric Research Laboratory. There followed soon an invitation to speak at the weekly Laboratory colloquium. I spent the evening with Langmuir, who told me about his space-charge experiments.

Then came an invitation to spend the summer at the Laboratory, and following that a letter from Dr. Willis R. Whitney, Director of the Laboratory, inviting me to join the staff. I hesitated, saying that I didn't think I was capable of doing anything practical. His reply was wonderful: 'I like you all the better for your hesitation. Don't worry about the practical part, that is my job.'

When I came to the Laboratory in 1914, Langmuir had discovered the law of electron space-charge, and Coolidge, following closely Langmuir's discoveries, had utilized the unique electron emission of tungsten to invent his hot-cathode 'Coolidge' X-ray tube. It was appropriate that Coolidge should make this invention, for he was one of the first in this country, while at M.I.T., to experiment with the original Crookes' X-ray tube, and he still bears the scars of the burns from those pioneer experiments.

Langmuir continued his electronic research, with his many well known brilliant contributions, the reprints of which, currently being published by the Pergamon Press, fill twelve volumes.

Coolidge continued for many years the development of his X-ray tube; first with the copper-backed tungsten target tube, for field use during World War I; next the oil-immersed dental tube which is standard today in all dentists' offices; then the line of highpower oil-immersed tubes, culminating in the multi-section, million volt, gas-insulated, transformer-enclosed tube which is standard high power radiographic equipment today; and finally the 100 million volt betatron.

I began working under Dr. Langmuir's direction, and soon discovered the negative resistance 'dynatron.' [1] As with Coolidge, this was a 'natural' because of my experiments at Worcester with secondary electron emission.

At this point something fortuitous happened. Sir William Bragg visited our laboratory and spoke at our colloquium, telling us about the X-ray crystal analysis work which he and his son were doing. In the discussion I asked if he had found the crystal structure of iron, which I though might be a clue to its magnetism. He might have answered, 'no, but I think we shall have it soon', and that would have ended it. But he replied, 'no, we have tried but haven't succeeded.' That was a challenge, and I decided to find the crystal structure of iron.

It was a rash decision, for I was totally unfamiliar with both X-rays and crystallography. But I had the Coolidge X-ray tube, and the new Kenotron rectifiers, which Dr. Saul Dushman of our laboratory had just developed. With these rectifiers I constructed a 100,000 volt d.c. power equipment, filtering the rectified current by a pair of condensers with an inductance between them. One of our young patent attorneys, Mr. W. G. Gartner, noticed this filter and patented it for me. Ten years later I was surprised to learn that all the manufacturers of radio receivers were licensed under my patent.

From the start I had planned to use powder for my X-ray crystal analysis, since it was common knowledge that single crystals of iron had not been produced. I visualized that all the Bragg reflections would be recorded simultaneously, and might be unscrambled.

With iron filings, which were rotated continuously in order to produce randomness, I soon had some good powder patterns. These I gave to an assistant, a very able young lady, to compare with Bragg values for the three cubic systems. She reported that none of them fitted.

The reason that I entrusted these calculations to an assistant, rather than making them myself, was that I was still 'holding on' to the dynatron project, studying applications—a lesson on what not to do, from which I was able to profit later. One of the outstanding qualities of Dr. Langmuir's research which was partly responsible for his tremendous accomplishment, was his habit of stopping when he had made a discovery or invention, and going on to the next job, leaving applications to others.

I spent the next few months studying the X-ray spectrum of tungsten, and the law of absorption of X-rays at high frequencies.

At this point I was fortunate in having a two-week's visit from Dr. Fred E. Wright, well-known mineralogist of the Geophysical Laboratory in Washington, D.C. With his help I mounted a single crystal of 3.5 percent silicon iron, and determined its structure by Bragg reflections. It turned out to be body-centered cubic. Immediately I became suspicious about the interpretation of my iron diffraction patterns, and while riding home on my bicycle at noon I made the calculations and found that the patterns agreed perfectly with a body-centered lattice.

I proceeded then to work out the theory of powder crystal analysis—at home, nights and Sundays, for I never did any writing on 'laboratory time'—and published my paper on 'A New Method of X-Ray Crystal Analysis,' [2] in 1917; and in 1919 'A New Method of Chemical Analysis.' [3]

The First World War interrupted this work, the Research Laboratory working as a team on submarine detection.

We got very few foreign journals during the war, and when it was over I was surprised to learn that Debye and Scherrer, in Switzerland, had independently discovered the X-ray powder method of crystal analysis, and had published it nearly a year ahead of me. Hence it is very properly known as the Debye-Scherrer method, although Sir William Bragg, with true Anglo-Saxon loyalty, continued for some years to call it the Hull method.

At the end of the war I went back to X-ray crystal analysis, and soon had analysed nearly all the common metals. [4,5] Dr. Wyckoff very properly characterized my analyses as non-rigorous, for I am no crystallographer. But I believe that all my results are correct.

In the meantime, Dr. Wheeler P. Davey of our laboratory had suggested the use of logarithmic plots for the solution of cubic, hexagonal and tetragonal structures, and had superintended the construction of these plots. They were widely used. With them it was not necessary to know the axial ratio of the crystal; one simply marked off on a strip of paper the positions of all the lines in the experimental diffraction pattern, to the same scale as the plots, and moved the strip over the plot until an exact correspondence was found for *all* the lines. (Some experimental lines might be lacking but there could be no extra ones, unless the sample contained an impurity.

At the time these plots were finished I had just obtained diffraction patterns of zinc and cadmium, but had not analysed them. With the plots the correct axial ratio, quite different from the published metallurgical value, was found quickly and easily and was routine. The use of these two examples, as illustrations of the use of the charts, was responsible for my writing our joint paper as senior author, which I later regretted out of consideration for Dr. Davey.

I might have gone on analysing more structures indefinitely, but I sensed from Dr. Whitney's attitude (he never 'directed' me) that I had gone far enough; and Langmuir advised me that I could continue for a lifetime, since there was no lack of materials to be analysed; but that there were more interesting problems in our laboratory. Therefore, I went back to electronics, and proceeded to invent the magnetron,[6] the screen-grid tube,[7] and the thyratron.[8]

I call them inventions for want of a better term; but they are so simple that they seem scarcely to deserve that austere designation.

The magnetron was simply an extension of J. J. Thomson's crossed-field calculations for electron orbits. J. J. had shown that electrons, moving from a plane cathode toward a parallel plane anode, would be bent into cycloid paths by a magnetic field parallel to the planes, and would fail to reach the anode. But of course this analysis ceased to apply when they reached the edge of the plates, and they all went to the anode.

I was curious about what would happen if the plates were bent into cylinders, so that there was no edge. The electrons might be permanently prevented from reaching the anode. I tried it soon after coming to the Laboratory, but failed. Presumably I was not sufficiently careful about vacuum and symmetry.

After deciding to terminate crystal analysis work, I came back to the 'magnetron' problem, and made the calculation of the paths of the electrons. It proved to be quite simple. It showed that they really could not reach the anode below a certain critical voltage. After this calculation the experimental proof was easy.

The screen-grid tube was a by-product. Our radio department had appealed to the Laboratory for an explanation of the 'noise' in their super-heterodyne receivers. Langmuir and I discussed it, and he suggested that it might be the 'shot-effect' of the individual impacts of electrons on the anode, which had been predicted by Schottky. I decided to measure the shot-effect.

The obvious method was straight radio frequency amplification. A voltage amplification of 100,000 fold was required at a frequency of one megacycle. I knew that the maximum total amplification that had ever been obtained by series triodes was about 200 fold; but I also know that the reason for the limitation was 'feedback' due to the capacity-coupling of plate and grid.

Faced with the problem, it seemed obvious that this coupling could be completely eliminated by thorough screening of the grid from the plate, both internally and externally. We had excellent construction facilities in the laboratory, and such 'screen-grid' tubes were quickly made. Of course they were completely successful.

I still can't understand how such a simple solution of the feed-back problem had eluded engineers for ten years—including myself, for I had worked on the problem and had constructed special tubes in an effort to *reduce* the capacity.

The principle that made possible the 'hot-cathode thyratron' was a chance observation of a very simple kind. I was studying the characteristics of a discharge from a 'thoriated' tungsten filament to a concentric anode in low pressure argon, and my assistant, Mr. W. F. Winter, showed me an unusual volt-ampere characteristic. As the anode voltage was increased the current increased quickly to a maximum at 20 volts, and then decreased rapidly to almost zero at 100 volts. I recognized at once that the bombardment of the filament by argon ions was knocking off the thorium atoms, which gave the filament its large electron emission, faster than new ones could diffuse to the surface. That was to be expected. But I also noticed the important fact that the thorium was *not* knocked off below 20 volts.

Such a simple observation was the solution of a problem that had bothered engineers, including myself, for more than 10 years; namely that when low pressure gases were introduced into hot-cathode rectifiers, the filament coatings were completely stripped off in from 10 to 100 hours, and even pure tungsten filaments were reduced in diameter sufficiently to materially change their resistance. This problem was completely solved by the specification of a 'disintegration voltage' of 20 volts, above which the anode voltage should not be allowed to rise. This simple precaution made possible the development of hot-cathode gas rectifiers and thyratrons that followed.

My only other major research problem was an analysis of the stresses in glass-metal seals. Large seals could be made only with very thin copper, which yielded to the stresses before they broke the glass.

With the help of Mr. E. E. Burger, and analytical assistance from

Dr. Hillel Poritsky, I made a thorough study of these stresses, including methods of measuring them; and developed a special alloy called Fernico which made seals to a special glass, developed by Dr. Louis Navias of our Laboratory, that were completely stress-free. The fact that a somewhat similar alloy had been discovered by Westinghouse scientists some two years earlier, so that we now buy our Fernico from Westinghouse, is relatively unimportant. Our analysis, and the development of Fernico, made possible the large power rectifiers and thyratrons of today, including Coolidge's multi-section X-ray tube. This 2 million volt X-ray tube has 20 metal sections, separated by glass sections; and the metal is Fernico.

References

1. The Dynatron, A Vacuum Tube Possessing Negative Resistance *Proc. I.R.E. 6*, 5–35 (1918).
2. A New Method of X-Ray Crystal Analysis *Phys. Rev. 10*, 661–96 (1917).
3. A New Method of Chemical Analysis *Jour. Am. Chem. Soc. 41*, 1168–75 (1919).
4. The X-Ray Crystal Analysis of Thirteen Common Metals *Phys. Rev. 17*, 571–88 (1921).
5. The Crystal Structures of the Common Elements *Jour. of the Franklin Inst., 193*, 189 (1922).
6. The Effect of a Uniform Magnetic Field on the Motion of Electrons Between Concentric Cylinders *Phys. Rev. 18*, 31 (1921).
7. Characteristics of Shielded Grid Pliotrons *Phys. Rev. 27*, 432 (1926).
8. Hot-Cathode Thyratrons *G. E. Rev., 32*, 231 (1929); *32*, 390 (1929).

Personal Reminiscences

R. W. James

I was born in London on 9 January 1891. My father, who had had no formal scientific education, had nevertheless a deep and lively interest in scientific matters, particularly in natural history. I was brought up to believe that one took an interest in such things as a matter of course, and I certainly owe a very great deal to this early training. I went first to the local elementary school, then to the Polytechnic School in Regent Street, and finally to the City of London School, where I had the good fortune to be taught chemistry by Henry Durham, one of the finest teachers I have known, who was able to foster in his senior boys an enthusiastic attitude of inquiry.

From the City of London School I won a scholarship at St John's College, Cambridge, where I read for the Natural Sciences Tripos, specializing in physics. In my year there were only about a dozen taking the second part, one of them being W. L. Bragg. Our practical work was supervised by C. T. R. Wilson, who expected from us the standards of accuracy and patience he had set for himself, and if we fell far short of this ideal, the attempt to reach it was good training.

I took my degree in the summer of 1912, and started work in the Cavendish Laboratory under Professor Sir J. J. Thomson, very shortly after the publication of the paper announcing the discovery at Munich of the diffraction of X-rays by crystals by Friedrich, Knipping and v. Laue. My own attention was first drawn to this by an account given by J. J. Thomson to the research students' colloquium, and I think I am correct in saying that W. L. Bragg was also present on this occasion. Not very long afterwards his first paper, embodying the reflection idea, appeared in the *Transactions of the Cambridge Philosophical Society*. I was very interested in diffraction problems, and had been stimulated by C. T. R. Wilson's lectures on the subject for the second part of the Tripos, which, if they left something to be desired in their delivery, yet formed an excellent basis for a serious student to build on; but I had already started experiments in a different field and did no work on diffraction at Cambridge.

In the summer of 1914, just before the outbreak of war, I had joined Sir Ernest Shackleton's expedition to the Antarctic as physicist, and sailed in the *Endurance*. The ship was ultimately crushed in the ice and sank, and the party were fortunate to reach land in small boats, so that much less scientific work was possible than had been hoped for. My own work consisted largely of magnetic observations, and of work on the movement and physics of the pack-ice done in conjunction with J. M. (now Sir James) Wordie.

On my return to England at the end of 1916 I joined the army, and early in 1917 went to France to do sound-ranging; and there I renewed contact with W. L. Bragg, who was in charge of the technical side of its development. There too I met for the first time C. G. Darwin and C. H. Bosanquet, both of whom were sound-rangers. When in 1919 Bragg succeeded Rutherford he offered me a post in his department, and it was then that I started to work on the diffraction of X-rays.

While still in the army, but knowing that I was going to Manchester, I had bought a copy of the first book by W. H. and W. L. Bragg, *X-rays and Crystal Structure*, and I have still a vivid recollection o- starting to read it one afternoon, and becoming so completely imf mersed in it that I appeared at Mess that evening an hour late. I think my Colonel found my excuse a little difficult to understand; but the first reading of that book was one of my really exciting scientific experiences.

Although my first two papers were on the structure of antimony and bismuth, both very inadequate by modern standards, although they gave the correct structure, my main interest in Manchester was not in structure determination, but in the physical and optical basis of the subject. I worked for a time with Bragg and Bosanquet on the absolute intensities of the spectra from rocksalt, on secondary extinction, and on the measurement of the atomic scattering factors for sodium and chlorine, work that is discussed elsewhere in this volume. I was particularly impressed with the importance of the scattering factor as an aid in structure determination, and used it in an analysis of the structure of barytes. Then, in 1925, I started to make more accurate observations of the effect of temperature on the intensities of X-ray spectra, taking measurements from the temperature of liquid air up to about 600°C with rocksalt, and later extending the low-temperature measurements to aluminium and potassium chloride. In this work I was associated with Miss E. M. Firth, R. G. Wood, and G. W.

Brindley. In collaboration with D. R. Hartree and I. Waller we were able to confirm experimentally the existence of zero-point energy in the thermal vibrations of the rocksalt lattice, and to estimate their amplitude.

The work on rocksalt was communicated to the Royal Society in 1927, and I was asked to read the papers. I have never forgotten the way in which Lord Rutherford, then President, put me at my ease before the meeting. Taking me by the arm, he said 'You haven't read a paper here before, have you? Do you mind if I give you some advice?' I said I should be most grateful if he would do so. 'Well' he said, pointing at the same time to the presidential throne, 'For Heaven's sake don't be too difficult. If you knew what I have to put up with, sitting in that Chair!' Very typical of Rutherford, entirely human, and very sensible.

In 1931 and 1932 I spent about ten months in Debye's laboratory in Leipzig, working on the influence of temperature on the scattering of X-rays from silicon-tetrachloride vapour. Debye encouraged me to tackle the problem theoretically at the same time. The theory confirmed the absence of appreciable effect which the experiments showed. I have always been very grateful to Professor Debye for helping to turn my interests, which had been almost entirely experimental, towards the theoretical side of the subject, and for giving me confidence that I found very useful later.

At about this time the series of volumes to be called the *Crystalline State* that were to replace *X-rays and Crystal Structure* were being planned. The first volume, by W. L. Bragg, appeared in 1934, and it was proposed that I should be responsible for the optical part of the subject in a second volume. Rather less than half of this had been written when in 1937 I was appointed to the Chair of Physics in the University of Cape Town, in succession to B. F. J. Schonland, and my eighteen years in the Manchester laboratory came to an end. It had been a very happy period, and I should like to say here how much I owe to Professor Bragg for the inspiration of his friendship and leadership during that time. It is a debt I cannot sufficiently acknowledge.

One of my predecessors in the Cape Town Chair, Professor Alexander Ogg, who still had a room in the department, from which he directed the Union Magnetic Observatory, had carried out work on crystal structure, publishing papers on the structure of antimony and bismuth at about the same time that my own two papers had appeared; and he had also done pioneer work on the structure of potassium sulphate. Crystal work had however stopped at the time of my arrival, and had to be revived. I decided to continue work on the aromatic nitro-compounds in which I had become interested in my last years at Manchester. In 1943 the first structure that of 4 : 4′ dinitrodiphenyl was published by J. N. van Niekerk, later to become head of the X-ray Division of the National Physical Laboratory at Pretoria. Some account of the work at Cape Town is given elsewhere in this volume. My load of teaching and administration became too heavy for me to do much work myself, but I was fortunate in having from time to time a number of able assistants, among them Dr. D. H. Saunder, Miss E. M. Archer (Mrs. D. H. Saunder), and Dr. Aaron Klug, who did his first X-ray work in Cape Town.

In 1945 the second volume of the *Crystalline State* was completed at Cape Town, and it was published in 1948. I retired from the Chair in 1956, and during 1956 and 1957 was Vice-Chancellor and Acting Principal of the University. X-ray work continues under my successor, Professor Walter Schaffer.

Personal Reminiscences

H. Lipson

Fourier Strips

In 1932, C. A. Beevers and H. Lipson at Liverpool were attempting to work out the structure of $CuSO_4 \cdot 5H_2O$—a triclinic crystal with two formula units in the unit cell. They had established that the copper atoms were in special positions, but the problem seemed to be much too difficult for the trial-and-error methods that were then currently in use. Beevers then suggested trying the Fourier methods that W. L. Bragg (1929) was advocating for the determination of structures; they had been tried out on diopside but they had not yet been used for an unknown structure.

The problem however was how to carry out the summations. There were 95 intensities and it was decided that, in the light of the limit of resolution, over five hundred points would have to be considered in the asymmetric part of the unit cell. This was considered to be a formidable task.

Beevers made the first exploration. Taking one point only, he worked out the contribution of *all* the F's; it took him 40 minutes. On the basis of this, assuming a 20-hour week (they had other duties as well!) they reckoned that it would take about eight months to make the complete survey. This would not do! They therefore tried to think of improved methods, involving long strips of paper with sine-curve ordinates written on them; one read off numbers in prearranged sequences and the other multiplied them by the observed F's on a slide rule basis. With these artifices the first Fourier synthesis was completed in a month (with, it may be remarked, most pleasing success).

During the process new ideas were forthcoming and by means of these the synthesis was checked by performing it again in rows along the other axis, and the mission was accomplished in two weeks. The results showed suprising agreement.

The methods still seemed unsystematic and clumsy, mostly because each curve of $\cos 2\pi(hx + ky)$ parallel to the x axis had to start at a separate value of ky. Lipson then had the idea of expanding the function into $\cos 2\pi hx \cos 2\pi ky - \sin 2\pi hx \sin 2\pi ky$, so that the ky terms could be considered as amplitudes rather than phases. This simple idea—which at first sight seemed retrograde—made the whole computation much easier since all the amplitudes for one k could be added together once and for all; it could be carried out quite efficiently by one person who had the necessary arithmetical ability.

The procedure involved working out the various sine curves, entering them in a book, and adding the columns of figures (Beevers and Lipson, 1934). Lipson thought it rather wasteful to lose the figures each time; obviously a curve with a particular amplitude was likely to recur, and so was worth-while preserving for further use. Therefore he entered the numbers on strips, and filed them for future use. After a few summations he had an imposing collection, which he added to systematically in any spare moments.

Thus the Fourier strips originated. Beevers, who had then moved to Manchester, made a copy and then suggested that further printed copies be made. Professor W. L. Bragg and Mr. R. W. James were very sympathetic and after much consideration of the enormous cost involved (about £100!) seventy sets were produced and given or sold to other laboratories (Lipson and Beevers, 1936). As usual in research, the production of the strips involved much more work than was expected, but the final results seem to have well justified the trouble and cost of their preparation. Beevers has continued to produce sets of strips and by now (1961) three hundred sets have been supplied to laboratories all over the world.

Intensity Statistics

The idea of making use of a statistical survey of the intensities diffracted by a crystal arose in 1942. *Nature* received two letters from S. H. Yü (1942a, b) in China claiming that absolute intensities could be derived from relative ones. The letters were sent to be refereed at Cambridge, but they did not give enough detail for the methods to be properly assessed. A. J. C. Wilson thought that the claim was unreasonable, but H. Lipson pointed out that if one assumed any arbitrary positions for the atoms in a unit cell the general level of the F's should be the same as for the correct structure. He suggested the rather clumsy procedure of working out such a set of F's and scaling the observed ones to give the same total; Wilson however looked into the theory and found a simpler procedure based upon Patterson's ideas of interatomic vectors.

Wilson (1942) tried out this procedure on data for copper sulphate and the alums, and found that, although it gave correct orders of magnitude, it was not very accurate. The methods were worth using, but they had to be modified as more accurate atomic positions were obtained. The work was then dropped in favour of more immediate practical problems.

After the war, Wilson returned to the problem that had been worrying him: why did not the alums and copper sulphate give better results? He traced the discrepancy to the fact that some heavy atoms lay on special positions and this discovery induced him and Rogers to explore the influence of symmetry or intensity distributions. From this exploration arose the extensive list of papers published by them and their co-workers at Cardiff.

References

1. Lipson & Beevers, *Proc. Roy. Soc.* A 1934, *146*, 570.
2. Lipson & Beevers, *Proc. Phys. Soc.* [London] 1936, *48*, 772.
3. Wilson, A. J. C. (1942). *Nature*, 150, 152.
4. Yü, S. H. (1942a). *Nature*, 149, 638.
5. Yü, S. H. (1942b). *Nature*, 150, 151.

Reminiscences

K. Lonsdale

I took my first degree (B.Sc., Physics) from a womens' College when I was 19. W. H. Bragg, Quain Professor of Physics at University College, London, had been one of the examiners, for the papers were common to all Colleges of London University. Nowadays I would have been booked, either for an industrial or for a University post, six months at least before sitting the final examination. Then, there was a real shortage of scientific jobs (there was practically no industrial research in Physics and one of my colleagues applied for 150 posts before he got one, even though he had a good higher degree). Although I was told that I had headed the University list with the highest marks in ten years, I had no work in prospect except schoolteaching, which I did not fancy at all, when Sir William Bragg, then 60 years of age, sent for me to come and see him at University College. He first asked me how I had managed to do so well in the B.Sc. examination? I can't remember my answer to this unanswerable question, but he went on to offer me a place in his research team with a D.S.I.R. (that is, Government-provided) grant of £180 a year, and the opportunity of working for my Master's degree. It was luxury and I jumped at it. To give some idea of money values: at that time a two-course lunch at University College cost 1/6, tea extra. No-one drank coffee then, and morning coffee was unheard-of.

Twenty years later, when W.H.B. died, I was still with him at the Royal Institution as a Research Fellow on a year-to-year grant of £400 a year. Those who left his group for University teaching posts added substantially to their salaries, but they were always most reluctant to go, and often W.H.B. had practically to push them out. He could pick his men, and did; and the inspiration of working with such a team and under such a leader was wonderful.

Yet in fact he did not appear to lead at all, and we were not a team in the modern sense of the word. He did not suggest a subject for my research. He told me that it would take perhaps 3 months to collect together the apparatus that I would need (ionization spectrometer to be constructed in his workshop by the formidable Mr. Jenkinson, a huge lead-covered box to hold and shield the Coolidge tube, gold-leaf electroscope, induction coil with Hg interrupter, condenser, high-tension battery, spark-gap and milliammeter) and then I would put it all together myself. In the mean time he advised me to study Hilton's *Mathematical Crystallography* and think about some problem I might tackle. Professor H. Hilton had been my mathematics teacher at Bedford College, but I had had no idea that geometrical crystallography had been his special subject. I was fascinated by his book, which gave all the elements of symmetry but never mentioned general equivalent positions. I drew half-arrows in all the diagrams to represent asymmetric units; and still have the book, 40 years later.

It was understood that my problem would be the structural analysis of some organic compound (the inorganic compounds belonged to W. L. Bragg and his school in Manchester). But I knew *no* organic chemistry and very little of any other kind. So I asked W. T. (Bill) Astbury and he suggested succinic acid. He was investigating tartaric acid himself and succinic acid seemed likely to be a bit easier! The crystals had to be about 2 mm linear dimensions and I had beginner's luck and got a wonderful crop at the first attempt, by dissolving the whole bottle in hot water in an immense flask and cooling it overnight. I can still remember Astbury's astonishment when I showed him about one hundred perfect specimens, all of the right size, the very next day. He soon convinced me that this *was* pure luck.

A little while later W.H.B. asked me to measure the density of anthracene for him, by the flotation method. I gave him the result in some trepidation, for it was 10 per cent higher than that given in the literature. But he was pleased, for it corresponded with his calculated value on the basis of two molecules in the unit cell, whereas the published value would have given $n = 1.8$.

By the beginning of 1923 I had the bits and pieces of my ionization spectrometer equipment and a large table. A fellow research student (now my husband) offered to do the soldering for me. I agreed that he should show me how to do it, but said that then I would like to finish it myself. Cutting the gold leaf for the electroscope and attaching it (with spit) was a tricky job, but if you got a good leaf it was very sensitive and reliable. There was no 'beam-trap' on the spectrometer. How we failed to collect substantial doses of radiation (if we did fail) I simply cannot imagine, for in spite of the lead box the beam emerging through the slits that formed a collimator was partially divergent and of formidable intensity and dimensions. We certainly did get first-layer-line reflections into our ionization chambers, although these were only intended to collect the zero layer line; and our determinations of space group were sometimes in error through this unsuspected fact. We sat for hours with one eye glued to a microscope, taking readings of the movements of the gold leaf; occasionally we argued about crystal symmetry and less scientific subjects.

Our structure analysis was largely guesswork aimed at satisfying spatial and symmetry requirements and the more intense reflections. When I asked my chemistry friends whether succinic acid was likely to be in the trans- or cis-configuration they naturally replied that the question was meaningless, since there could be free rotation about the central single bond. However, I pointed out that in the solid state free rotation seemed not to be possible, and it was then suggested that since maleic and succinic anhydrides were isomorphous, succinic acid might well be in the cis-form when crystallized. I eventually produced a structure which gained me my M.Sc. although it was subsequently shown to be wholly wrong. In a sense I feel that my whole career has its foundations on sand. Can one give back an undeserved degree?

Meanwhile my reading of Hilton's *Mathematical Crystallography* and our many laboratory discussions had given us the idea that it should be possible, by systematic investigation of zero-intensity reflections, to determine the crystal space group if not uniquely at least within narrow limits. Without realizing that this had been already done by P. Niggli and R. W. G. Wyckoff (for alas, our library work was no better than that of many of today's young 'up-and-comings'), Astbury and I proceeded, independently but in parallel, to work out the 230 space-group tables, and we persuaded W.H.B., not without difficulty, to send them for publication. The usefulness of the Astbury-Yardley Tables, which the Royal Society had to reprint—a very rare event for a *Phil. Trans.* paper—perhaps lay in the fact that they were intended for immediate practical use. W.H.B., although himself a mathematician of distinction, had little use for any abstract mathematical theory, unrelated to experiment.

In 1923 most of us had moved with Sir William Bragg, lead boxes and ionization spectrometers and all, to the Royal Institution; and I was installed in a ground-floor laboratory with two other ionization spectrometers and some early gas-tube equipment for the photographic technique. My most joyous recollection of this period is of J. D.

Bernal on his knees looking for his one and only crystal of (I think) γ-brass. My chief regret is that I have no photograph of that probably unique occasion. I also remember the one occasion on which I saw Sir William even mildly angry, although I cannot remember the date. We always had tea together in the Royal Institution library (free: so nearly everyone came regularly) and there were often visitors. One visitor, a very supercilious young man, was amused at the fact that an elderly scientist at University College, London, had spent his whole life measuring the boiling point of sulphur with greater and greater accuracy. W.H.B. commented sharply that such research workers were the salt of the earth and that it would do no harm if their critics were equally conscientious. Yet he was kindness itself when I had to admit that I had made (and published) a mistake.

One of the more colourful characters at the R.I. in those early days was the cleaner, Nellie Collins, who was invariably cheerful and talkative. She and I exchanged letters once or twice a year until her sudden death on 6 February 1961. She was a staunch member of the Salvation Army and lived at the incredible address of 1, Frugal Cottages.

I left the R.I. in 1927 on my marriage to Thomas Lonsdale, who had a job with the Silk Research Association, then housed in the Textile Department of the University of Leeds. It had been my intention to give up scientific research work and settle down to become a good wife and mother, but my husband would have none of it. He had not married to get a free housekeeper; and if he were a good husband and father I could quite well keep on my 'Arbeit', as we called it to distinguish it from the domestic chores. It was great fun, although we were hard up, too much so to be able to afford University lunches. I kept very exact accounts and I have them still. We shopped together weekly at the Leeds Market, prepared vegetables together in the morning before leaving home and I allowed myself 1½ hours for lunch (he only got 1 hour) so that I learned to specialize in meals that took only 30 minutes to prepare. With gooseberries at 10d. a stone (14 lbs!) we made jam and on Christmas Eve we hung around the Market until it closed at midnight and got 2 geese for 5/– in the Dutch auction, of which we promptly sold one for 2/6 and went home rejoicing, with the other. I was not a vegetarian in those days, but I was very happy! My husband set up some apparatus of his own design at home with which to measure the torsional strengths of metal wires, for his Ph.D. degree; and while he experimented I did crystallographic calculations.

Professor Whiddington had welcomed me in the Physics Department at Leeds University, where W. H. Bragg had formerly been the Professor and where he and W. L. Bragg had carried out their early and intensive studies of crystal structures. The advent of World War I had prevented the establishment of any research school there, and Whiddington was interested in a different field, so I had once more to build up the X-ray diffraction equipment with the bits and pieces I could find and with the help of an apparatus grant of £150 from the Royal Society. A radiatory-type Coolidge tube cost £46, a Bragg ionization spectrometer with gold-leaf electroscope, microscope and two goniometer heads was £67.4s.6d., patterns and castings for photographic apparatus came to 12s.6d., and I returned the balance of £46.3s. to the Royal Society in 1929. C. K. Ingold, then Professor of Chemistry in the University of Leeds, had offered me some large crystals of hexamethylbenzene and although these turned out to be triclinic, they formed the subject of what was, perhaps, my most fundamental and satisfying piece of research. There was only one molecule in the unit cell and it was possible, without any assumptions except that there were 12 carbon atoms present in the unit cell, to show that the molecule was effectively hexagonal and plane. I have a treasured letter from Sir William dated 30 October 1928, in which he says: 'I think your new result is perfectly delightful: many compliments upon it! I like to see the benzene ring 'emerging'. I thank you too for sending me a special account of your work: it is very welcome. I have been showing what you have written to Muller and Cox and Sir Robert Robertson and there was great interest. I will write again about it before long: this is just a note of acknowledgment.'

While I was away Sir William wrote me a number of letters, often by hand, although he had a devoted secretary, Miss Winifred Deighton who stayed with him throughout his time at U.C.L. and the R.I., over 20 years, and of whom I was rather afraid. She was tall and impressive-looking—and I was not!

In 1929 my first baby came and I found it rather difficult to do everything in the home and also find time for 'Arbeit'; so I wrote to W.H.B. and he persuaded the Managers of the Royal Institution to give me a grant of £50 for one year with which to hire a daily domestic helper. Her name was Mrs. Snowball (it really was!) and, with her to wash and clean, I managed to care for the baby, cook and continue the structure analysis of C_6Cl_6.

In 1930 my husband got a new job at the Road Research Laboratories near to London and we moved back. I was now expecting another baby and I spent most of my time finishing off (by hand, of course, using log tables) the Fourier analysis of hexachlorobenzene. In view of the oft-repeated claims of organic chemists that the crystallographers told them nothing that they did not already known, it may be interesting to reprint a more generous tribute from C. K. Ingold (dated 2 November 1931).

'Ever so many thanks for the reprint of your wonderful paper on Hexachlorobenzene. I have just been studying it in the Royal but am very glad to have it in a handy form. The calculations must have been dreadful but one paper like this brings more certainty into organic chemistry than generations of activity by us professionals.'

I have some correspondence to show that I was a little worried about the triangular contours of the carbon atoms (*Proc. Roy. Soc.* A *133*, 536 (1931)) and wondered whether these were an artifact due to termination-of-series errors, as they almost certainly were. I was also engaged in correspondence with Astbury, W.H.B. and others about a comprehensive book on Crystal Structure Analysis that we proposed to write (this was later abandoned in favour of the first *International Tables*), and on calculations of the 'structure factor formulae' of the 230 space groups, later published as *Structure Factor Tables, 1936* in photo-litho-printed firm from the handwritten MS: 181 quarto pages for 10s.6d.

In November 1931 Sir William wrote to me again 'A piece of good news: Sir Robert Mond is giving me £200 with which you are to get assistance at home to enable you to come and work here. Can you come and see me soon?' I did some hasty sums and produced an 'Estimated Annual Budget' which convinced me (and him) that £200 was not enough.

Estimated Annual Budget

	£.	s.	d.
Wages of Nurse-housekeeper	60.	0.	0.
Board and lodging, uniform and insurance for nurse-housekeeper	52.	0.	0.
Extra charring and laundry	26.	0.	0.
Extra cost of housekeeping (at least)	26.	0.	0.
(This includes more extravagant use of fuel, light, power, food, and the cessation of my own activities, such as gardening—fruit and vegetables—,bread and cake-making, jam-making, and dressmaking for the children)			

Personal expenses:			
Fares (48 weeks at 10/– weekly)	24.	0.	0.
Lunches (less cost of my lunch at home—48 weeks at 8/9 weekly)	21.	0.	0.
Extra cost of clothes (5/– weekly)	13.	0.	0.
Books, notebooks, equipment, subscription and all extras	26.	0.	0.
	248.	0.	0.
Plus income tax	29.	0.	0.
	277.	0.	0.

I think he got me £300, but by now my careful keeping of weekly accounts was going to the dogs and I have no record of my exact salary; I only know that I rejoined him in the summer of 1932 and left again (but only temporarily) in 1934 for the birth of my third and last child. On my return there was no X-ray apparatus to spare and it was suggested that I might be interested in a huge old electromagnet instead. I was. K. S. Krishnan had just published some fascinating papers on the diamagnetic anisotropy of aromatic crystals and I at once threw myself into this new field. Krishnan had shown that the method provided evidence of molecular orientation; I confirmed and extended this to aliphatic compounds and showed also that by using the simple Larmor-Langevin equation one could calculate the effective radii of the σ and π electronic orbits in aromatic compounds and show that the latter were of molecular dimensions (*Proc. Roy. Soc.* A *159*, 149, (1937)).

Early in 1940 Dr. I. E. Knaggs showed me some Laue photographs that she had taken of benzil $(C_6H_5 \cdot CO)_3$, using unfiltered radiation from a Cu target. On these there was a beautiful and intense trigonal pattern of diffuse scattering; and on looking at our collections of Laue photographs several of us found similar though less intense patterns from other substances. We wrote a letter to *Nature* about it (*Nature*, *145*, 820, (1940)) in which we questioned whether the extra reflections found on Laue photographs of diamond by Sir C. V. Raman and P. Nilakantan could really be given the explanation he had offered. This pulled me into two fields which still interest me greatly today: on the one hand that of thermal vibrations in crystals and on the other the behaviour of diamond under different conditions.

It happened that about this time Dr. A. Muller was absent from the R.I. for some months through illness and I was able to have the almost exclusive use of the 5 kW and 50 kW X-ray tubes that he had developed, as well as the invaluable assistance of H. Smith (Smithy), who was one of the few outstanding laboratory technicians who subsequently achieved an obituary in *Nature*. Another X-ray tube for which I was indebted to Dr. Muller was a very small tube, originally intended for X-ray microscopy, giving a wide-angle divergent beam. With this I was able, without the use of precision apparatus of any kind, to measure the C—C distance in individual diamonds from their divergent-beam patterns with an accuracy of 1 in 70 000 and to show that they could vary between themselves by as much as 1 in 7000, from a very pure gem specimen to a highly impure industrial diamond.

Even after 40 years as an X-ray crystallographer I am not yet as old as Sir William Bragg was when I joined his team in 1922, and I hope therefore that I still have many years of usefulness ahead of me, if I have managed to catch anything of his enthusiasm and ability for clear thought and concise exposition.

Recollections of Dahlem and Ludwigshafen

H. MARK*

In 1921 there were essentially three groups in Berlin-Dahlem which worked in the X-ray field: Becker and Jancke who were associated with Professor Herzog; Böhm and Zocher who cooperated with Professor Freundlich, and Brill and myself who worked with Polanyi and Weissenberg.

At that time the theory of X-ray diffraction was as good as completely developed in a series of important contributions beginning with the original articles of M. von Laue and W. H. Bragg and culminating in the dynamic theory of P. P. Ewald, which described the interaction of an X-ray beam with a crystal lattice in all details. From the experimental point of view, however, the field was still in a rather infant state. There were no X-ray tubes available which could be operated at high intensities over longer periods without permanent careful supervision, there did not exist precision cameras in which crystals or crystalline objects could be conveniently mounted for irradiation in all possible orientations and there were no instruments designed which would permit X-ray diagrams to be taken at extravagant conditions such as at very low or very high temperatures, in a vacuum or in an atmosphere of a chemically very reactive gas. As a consequence most structure determinations prior to that time either operated with well developed single crystals (mostly of inorganic nature) or with polycrystalline powders of relatively simple structure (elements or compounds having lattices of high symmetry). The groups in Dahlem, however, were interested in colloidal systems of inorganic nature, in organic substances, most of which form crystals of low symmetry, and in polymeric materials, which represented a class of organic compounds that was rather unexplored at that time. The main concern of all of us

* Since the author of this review has been previously connected with a large part of this work from 1921 to 1931, it is a special pleasure for him to follow Prof. Ewald's suggestion in giving a brief account of the activities during this period.

was, therefore, to improve our experimental procedures and the evaluation of the data in all possible ways.

A first important step in this direction was the development of the techniques of fiber diagrams and, in general, of diffraction patterns of very small crystals having some degree of orientation in space. Polanyi and Weissenberg opened the way for this approach by a careful and exhaustive analysis of the way in which spatial order inside of a sample reflects itself in spatial order of the diffraction points on the photographic plate and a series of papers on materials of all kinds utilized their formulas for the elucidation of the deformation of metals, of the growing of colloidal suspensions and of organic crystals, and of the basic structure of natural fibers, such as cellulose and silk. One interesting result was that the crystallographic elementary cell of cellulose and silk was found to be very small. This led Polanyi and Brill to the conclusion that these materials are either built up by small cyclic units or by long chains, which pass in a continuous manner through a number of consecutive unit cells.

These early results of the Dahlem X-ray group have later been often quoted in an incorrect manner by assuming that the X-ray diagram of cellulose and silk was an indication of a low molecular weight of these materials.

Another important area of improvement was the construction of X-ray tubes which would permit to obtain monochromatic radiation (mostly Mo, Cu, Fe, or Cr) of high intensity (10–100 mA) at moderate voltages (30 000–60 000 volt) by a more or less automatic operation so that these tubes could either be run at peak intensities for a short time or at moderate intensities over long periods. It took several years and the combined efforts of our best experimentalists (Brill, Boehm, Ehrenberg and von Susich) to arrive at a situation where strictly monochromatic diagrams of normal samples could be made in a few seconds and, correspondingly, very small specimens could be X-rayed within reasonable times of exposure. As a consequence, the structure of B_2H_4 was determined from powder diagrams of the crystallizate of less than half a cubic centimeter of the gaseous materials at $-180°C$ by Pohland and the atomic weight of hafnium was computed by Hassel from the dimensions of the unit cell and the density of a $(NH_6)_3HfF_7$ single crystal which weighed only 0.2 gammas. Relatively fast chemical reactions, such as the alkalization of cellulose and the hydration of certain silicates were cinematographed by Katz, Rosbaud and Susich and the rapid phase change during the crystallization of rubber was quantitatively studied by Hauser.

But the ease with which X-ray patterns could be obtained in short times even under somewhat extravagant conditions (temperature, vacuum, pressure, chemical environment) led not only to interesting results in many special cases, it also produced an important change in the general use of this technique. Until then, the preparation of a good and clear diagram was a relatively difficult experiment which required considerable skill, time and effort. Hence, it was customary to use one or two X-ray patterns as the basis for a scientific paper and to evaluate these diagrams in all possible directions with the aid of assumptions which were not always absolutely justified. Now the X-ray method became one of the standard analytical tools, which was used to check almost as with a microscope, the progress of a scientific investigation step by step, with the aid of orienting snapshots, which were most valuable to pin down the situation and to help decide the next step of the work. For example more than fifty individual diagrams were made during a study of the swelling of cellulose by Katz, but only the five or six best ones were actually included in the publication; almost one hundred orienting pictures were made by Polanyi and Schmid in their systematic studies of deformation mechanism of single metallic crystals. This diagnostic use of the X-ray method as an infallible guide in solid state research of all kind is taken for granted today but it needed considerable skill and effort to develop it during the years from 1923 to 1928.

One of the most valuable products of this era was the *X-ray Goniometer* which was conceived and constructed with all its details by Weissenberg and first put to practical use by Boehm. Even the most recent instruments of this type are not much better than the first original Weissenberg camera which is still in use in the laboratories of the Fritz Haber Institute in Berlin-Dahlem. Other important instrumental innovations which have greatly contributed to the usefulness and applicability of the X-ray method were cameras for exposures at very low and very high temperatures, in vacuo and under pressure, furthermore crystal monochromators with flat and curved crystal plates, automatically recording ionization chambers and arrangements to produce highly parallelized and very intensive beams of a diameter down to a few microns (Kratky's micro X-ray camera).

For several years (1921–1924) the interest in structure determinations prevailed, but then the key role of X-rays and gamma rays for basic problems concerning the structure of radiation and matter (Compton effect, Geiger counter, light quantum hypothesis, dispersion theory) focussed attention on general questions of the physics of X-rays and electrons. Fortunately the highly developed technique of producing and registering strictly monochromatic and parallel X-ray beams was exactly what was needed for successful experiments along these more fundamental lines. Thus under the influence and with the active cooperation of Einstein, Kallman, von Laue, Szilard, and Wigner studies on the width of X-ray lines, the anomalous dispersion of X-rays, the intensity, width and polarization of the Compton lines, and on polarization of characteristic X-ray radiation were carried out and published in a series of articles. Equally fundamental was a thorough test of Ewalds' dynamic theory with the aid of very careful and delicate precision measurements of the angular width of X-ray reflection, which was carried out by Ehrenberg, together with Ewald and Mark. In connection with this work another instrument was developed—the double crystal X-ray spectrometer—which kept on playing an important role in precision measurements of high energy spectroscopy until now and was recently improved and refined for the use with gamma rays by Jesse Dumond and Hans Mark.

In 1927 the author and part of the Dahlem group moved to Ludwigshafen on Rhine and attempted to introduce X-ray diffraction methods into the realm of industrial application. With the cooperation of a few new members several problems of practical importance were attacked but fundamental studies in the field of X-ray and electron physics were not neglected. Dohse, Dunkel, Hengstenberg, Wierl, and Wolf made interesting contributions to the nature of the adsorbed state of small organic molecules on solid surfaces, the mechanism of deformation of crystals, the scattering of electrons by simple molecules in the gaseous state (Wierl) and the adsorption of light by coloured organic substances. The techniques for the preparation and registration of intense monochromatic beams of X-rays and electrons were further improved, and K. H. Meyer initiated work on the establishment of a uniform picture concerning the molecular structure of high polymeric material and synthetic substances.

In the laboratories of the I. G. Farbenindustrie A. G. in Ludwigshafen the method and reasoning of solid state physics had to be applied to the characterization and testing of industrially important materials in competition with the already well known and firmly established classical procedures of inorganic and organic chemistry. These branches were in the hands and under the supervision of highly qualified industrial experts such as Alwin Mittasch and Otto Schmidt and had proven their usefulness and indispensibility in many cases.

The new group had to demonstrate its own value for the company in the face of many existing strongly entrenched working teams. This was a difficult position and not even the benevolent and continuous support of such inspired industrial leaders such as Bosch, Gaus and K. H. Meyer would have prevented a gradual starvation of the new approach. Fortunately, however, there were some early successes which convinced even the most skeptical observers that the new methods and the new atmosphere, which they carried with them into the somewhat stagnant air of the great plant were of undeniable value and almost necessary for a successful progress in the future fields of chemical technology. Brill demonstrated convincingly the value of X-ray analysis for the study of catalysts in the ammonia and methanol synthesis. Susich and Valko were able to characterize synthetic fibers and rubbers with X-ray and electron diffraction far easier and better than anyone else in the wide-spread industrial organization of the I. G. Farbenindustrie. Dohse and Dunkel made substantial contributions to the understanding of polymerization reactions and Wierl and Wolf surprised the dyestuff experts by showing them how much they can learn about the objects of their interests and efforts by the use of X-rays and electrons. Practical people are slow in acknowledging the value of new approaches but very fast in using them extensively once their value has been established. As a consequence it took only four years and the modern methods of spectroscopy, scattering and photographing with X-rays and electrons were a part of process and product control not only in Ludwigshafen but also in most other plants of the Company.

Personal Reminiscences

Isamu Nitta

When I was a student of chemistry in the University of Tokyo, I had a vague ambition to devote myself in future to the question of how far one could go into the phenomenon of life from the standpoint of chemistry. Thus I had no presentiment that I should be later engaged in research and education in X-ray crystallography and crystal chemistry. However, this actually came about when M. Katayama, one of my teachers, recommended me to S. Nishikawa to work under him in the field of X-ray crystal analysis. This was in 1923, the year of my graduation from the University. After that I worked for about ten years in his laboratory in the Institute of Physical and Chemical Research, Tokyo, occupying myself primarily with the analysis of organic crystals. From the beginning I thought it quite significant to make any contribution to the confirmation of stereochemistry by such a direct physical method as X-ray diffraction analysis. With such expectation it was inspiring to me to learn from Nishikawa the experimental techniques of crystal analysis using Laue photographs and an ionization spectrometer, though it was not very easy for me to throw myself suddenly from chemistry into physics. It was very fortunate that in my undergraduate course I had attended the lectures of Professor Nagaoka on the Maxwellian theory of electricity and magnetism, and also those of Professor Nakamura on the classical theories of crystal structure and space groups. This greatly facilitated the access to X-ray analysis for the chemistry student I was. The three earliest classical books on X-ray crystallography, namely *X-rays and Crystal Structure* by W. H. and W. L. Bragg, *Kristalle und Röntgenstrahlen* by P. P. Ewald, and *The Structure of Crystals* by R. W. G. Wyckoff, rendered great help to me as they did to all others of that time.

My first paper, which appeared in 1926, was on the crystal structure of iodoform,[1] a substance I chose in an attempt to prove the tetrahedral nature of the carbon atom. The atomic parameters, x and y, for the iodine atoms in this hexagonal crystal ($P6_3$) could be determined exclusively from the qualitative comparison of intensity inequalities for pairs of reflections such as hkl and khl. Although I could of course not find the positions of hydrogen and carbon atoms owing to the overwhelmingly strong scattering power of iodine, the probability of the tetrahedral nature of the carbon atom could be seen in the structure. Only the hexagonal unipolar space group did not appear to conform with the observed bipolar external form of the crystal. I became gradually aware that this might be due to the orientational disorder of the molecular axes along the hexagonal axis. In my early period I was convinced that one should begin with the simplest compounds of organic chemistry; i.e., with methane derivatives and those containing only one carbon atom. Such a thought was due to youthful overvaluation of logic and lack of experimental experience, and it made it difficult for me to find, in that period of development, suitable crystals for analysis.

My second paper, which appeared in the same year, was on the crystal structure of pentaerythritol,[2] and contained some comments on the previous investigations of the same crystal by other authors. One of the unfortunate conclusions of these authors was due to the inappropriate description of the unipolar, tetragonal symmetry of the crystal given in Groth's *Chemische Krystallographie*. By observing carefully the crystal growth of pentaerythritol from aqueous solution, I realized that crystals frequently grow with their tetragonal axes hanging perpendicular from the surface of the solution, so that the crystals appeared as if they had unipolar, tetragonal axes. Bearing this in mind I chose the space group $I\bar{4}$, which enabled the central carbon atom of the molecule to conform with the tetrahedral distribution of the valence bonds.

From 1928 to 1931 I was sent abroad for further study by the Institute of Physical and Chemical Research. I first visited Professor Coster at the Physical Laboratory, University of Groningen, and stayed there more than one year in the hope that I might learn something about X-ray spectroscopy, which would be of chemical significance in the problems of the structure of atoms and of the nature of chemical bonds. When I first met Coster, he proposed that I carry out an experiment proving the non-validity of Friedel's law using zincblende, a surprising coincidence, because such an experiment had just been finished by my teacher, Professor Nishikawa, and my colleague, Matsukawa, by the time I left Japan. Without repeating such an experiment, I tried to see if the effect reported by Bergen Davis, namely the Raman effect in the X-ray region, was a real one, and came to the conclusion experimentally that such an effect did not exist, although a negative proof is always very difficult.

After studying further the problem of multiplicity in the M series of rare-earth elements, I then visited professor Ewald at the Institut für theoretische Physik, Technische Hochschule in Stuttgart, with a hope of learning a bit of crystal lattice optics, which Ewald had developed in its full elegance in 1916. My visit was motivated by the deep impression I had when I read Ewald's book mentioned above and the classical paper by W. L. Bragg in 1924 on the refractive indices of calcite and aragonite. I thought that the general theory of Ewald might be of use in the course of X-ray crystal analysis of molecular crystals, such as organic. Indeed, Ewald's theory was very difficult for me to master, but the personal contact with him during my stay there was a great addition to my later life. In this period of my stay in foreign countries I saw many eminent men, of science in general and of the special field of X-ray crystallography, during personal visits, at university colloquia and at public lectures. These contacts acted as an incessant source of scientific encouragement for me.

In 1933 I was called to a professorship of physical chemistry in the Department of Chemistry, Osaka University, which was newly established. I was quite fortunate to have a collaborator in X-ray crystallography as good as Tokunosuké Watanabé, with whom I have worked more than a quarter of a century. From my experience up to that time, I thought it very significant for me, as a chemist, to try as far as possible to mediate between physics and chemistry in university education. As for X-ray crystallography, I felt it to be my duty to popularize among organic chemists the X-ray method for determining molecular structure, not as a method of physics, but as one of organic chemistry, although this method was not so far developed at that time. Our first contribution from the laboratory was the crystal analysis of cubic barium dicalcium propionate published in 1935.[3] In this crystal we had to make the propionate group possess digonal symmetry. Thus we were led to assume a rotational or orientational disorder so as to furnish the group with a statistical digonal character. I then came back again to pentaerythritol and in 1937 with Watanabé carried out two-dimensional Fourier syntheses of the electron density in this crystal,[4] quite independently of the similar work by Llewellyn, Cox and Goodwin. This was the first two-dimensional Fourier synthesis in Japan, and this was also the first hydrogen-bonded structure found in Japan. The anisotropy of thermal expansion of pentaerythritol in this low temperature tetragonal phase was then determined by means of X-ray diffraction, and the result showed the characteristic features of the crystal structure of hydrogen-bonded layers. Further X-ray investigation revealed that the transformation of pentaerythritol at about 190°C from tetragonal to cubic was of a most complicated type, composed from the onset of intramolecular rotations and of orientational disordering of the molecule as a whole.[5] These various findings stimulated us to further studies: investigations of thermodynamical, electrical and optical properties of pentaerythritol in both phases,[6] X-ray and crystal-chemical studies of orientational and rotational disorder in molecular crystals,[7] X-ray analysis of organic as well as inorganic hydrogen-bonded crystals, etc. At all times I have been very fortunate to have had many good students who have worked with me on these and various other projects and to whom I am indebted for their devoted collaboration.

References

1. I. Nitta, *Sci Papers, Inst. Phys. Chem. Research*, Tokyo, *4*, 49 (1926).
2. I. Nitta, *Bull. Chem. Soc. Japan*, *1*, 62 (1926).
3. I. Nitta and T. Watanabé, *Sci. Papers Inst. Phys. Chem. Research*, Tokyo, *26*, 164 (1935).
4. I. Nitta and T. Watanabé, *Nature*, *140*, 365 (1937); *Sci. Papers Inst. Phys. Chem. Research, Tokyo*, *34*, 1669 (1938).
5. I. Nitta and T. Watanabé, *Bull. Chem. Soc. Japan*, *13*, 28 (1938).
6. I. Nitta, S. Seki and M. Momotani, *Proc. Japan Acad.*, *26* (9) 25 (1950); I. Nitta, S. Seki, M. Momotani, K. Suzuki and N. Nakgawa, *ibid.*, (10) 11 (1950); I. Nitta, T. Watanabé, S. Seki and M. Momotani, *ibid.*, (10) 19 (1950); I. Nitta, S. Seki and K. Suzuki, *Bull. Chem. Soc. Japan*, *24*, 63 (1951); R. Kiriyama, S. Yabumoto and I. Nitta, *ibid.*, *27*, 115 (1954); I. Nitta, *Sci. Papers Inst. Phys. Chem. Research, Tokyo*, *37*, 114 (1940).
7. I. Nitta, *Z. Krist.*, *112*, 234 (1959).

Experiences in Crystallography – 1924 to Date

A. L. PATTERSON

It is an almost impossible task to try to analyse the reasons why one becomes involved in any field of scientific endeavour. One can only describe the circumstances, significant or otherwise, and the actions taken, logically or otherwise, under these circumstances. In 1924 when I wrote to W. H. Bragg asking him to accept me as a Research Worker in the Davy-Faraday Laboratory, I was writing an M.Sc. thesis at McGill on the production of hard X-rays by the β-rays from radium active deposit. I had had a very sound general education at Tonbridge School in England. This covered all the languages and literature and the chemistry required of a specialist in mathematics and physics at McGill, and gave me a good start on the mathematics and physics as well.

A number of circumstances influenced things at this time. I had taken a Second Class Honours B.Sc. degree in 1923 and only one who has attended a British or Colonial University can realize the depth of ignominy attached to such a thing. This disgrace was correctly ascribed by my professors to too many friends in Montreal and an addiction to skiing, bridge, and dancing and other related activities. Some quite fantastic suggestions were made as to how I might *live this down* but knowing myself these suggestions were a little difficult to take.

The second point was that the department was haunted by Rutherford who had left McGill in 1907. Our courses were interlarded with references to work 'done in this laboratory'. This was all very well but the stomach of an undergraduate is limited in its capacity in this direction. Present and future activity were much more important. But the fact was that all the faculty had worked either with Rutherford at McGill or at Manchester or at the Cavendish. I had met Rutherford when I spent the summer of 1922 in Cambridge counting α-particles for a Canadian friend and taking Thirkill's famous laboratory course in optics. I was much more impressed by E. R. in person than I was by him as a haunt but I remember being very vehement at the time that I was more interested in waves than I was in particles. Little did I know that they were to get together so soon. Apart from all this the undergraduate training in physics at McGill was quite good. Perhaps stronger in classical theoretical physics than in other fields. There were some excellent lecturers in the department and they emphasized very heavily the training of students as lecturers. Physics students were required to carry a heavy load in mathematics. This was excellent and I learned a great deal particularly in the British tradition, but a great deal of 20th century continental mathematics had not penetrated.

And again I had become very interested in my thesis problem. There were two things involved: what happened to the β-rays as they collided successively with the atoms of the solid, and what was the nature of the radiation they produced. Good questions! I figured out two things I might do about it. I might learn how to analyse the spectrum of the X-radiation. This was preferable to measuring the total radiation in the classical tradition of my thesis experiments. The other thing was to learn something of the theory both of the statistics of the β-rays going through the matter and of the details of their encounters with the matter.

When in the spring of 1924, McGill announced the endowment of two Moyse Travelling fellowships, one in literature and another in science it was with this background that I made my application. In my first draft, I listed four possibilities in order: 1) W. H. Bragg at the Royal Institution; 2) M. Siegbahn in Stockholm; 3) N. Bohr in Copenhagen; 4) E. Rutherford in Cambridge. The order of this listing aroused dire consternation. Rutherford simply could not be last, and also for some reason I was allowed only three choices. I stuck to my guns on the first and finally wrote Bragg, Rutherford, and Bohr. Since there was no air mail and the timing for the application was close there was no question of being accepted anywhere before making the application. I was a little surprised and very pleased to be awarded the fellowship as I had been told that the Second Class would weigh heavily against me. It seemed that a very good friend on the mathematics staff had fought hard for me. I decided to go to England where my parents were anyway, and wrote a two-page letter to Bragg telling him why I wanted to work in his laboratory and asking him if I could see him when I got to London. Because I was sailing within a week or so I asked Bragg to write me care of my ship on arrival in Southampton. His reply was prompter than I could have expected and his letter had met the ship on its preceding trip. When I went aboard in Montreal, I found the following letter in my cabin on the paper of the Davy-Faraday Laboratory:

3rd July 1924

Dear Mr. Patterson,

I will expect you in London between the 23rd and the 26th, when I shall be glad to see you.

Yours sincerely,
W. H. Bragg

Fortunately before I left McGill, Dr. Eve's secretary had showed me his letter from Bragg which I quote only from memory:

Dear Eve,

Shall be very glad to have Patterson.

Yours sincerely,
W. H. Bragg

This with my own letter gave me a very happy Atlantic crossing. I arrived in time to see Bragg on the 23rd. He told me that Jenkinson was making two new cameras, one of which was assigned to me. It would have a single circle goniometer head! He suggested that I work on the phenyl aliphatic acids as a tie in between his work on naphthalene and anthracene, and the work Muller and Shearer were doing on long chain compounds. He then told me to go take a good long holiday, possibly to read the two volumes of Tutton, and follow them by Hilton so that I would be ready to read the new paper by Astbury and Yardley on space-groups when I came back in September. I did read Tutton, but Hilton and the paper took the best part of the next year.

When I returned in September I set out to build my own X-ray equipment. The tube was a gas tube designed by Shearer and was built in the shop. The pumps were of local design and I was told that when I had made one in soft glass, I would be given some of a very expensive new glass called Pyrex to make the two that I needed. The pumps were made with metal water jackets closed by split corks at the ends. They always leaked and my camping experience suggested wet asbestos cord as a means of conducting the drips down the drain. The high voltage source was an induction coil. The core was a heavy cardboard tube packed with soft iron wire. The primary was cotton covered heavy copper wire wound on the outside of the cardboard tube and taped in place. This primary and core was mounted vertically in the center of a square base-board on which three or four secondary coils were mounted, separated by small ebonite insulators. Fortunately for us, the secondary coils could be bought by the dozen and we did not have to wind them ourselves. The interrupter for the DC current was also a local product. An aluminium hot-water bottle was half-filled with the electrolyte and a platinum wire shielded by a pyrex tube was inserted through a vented cork. When operating, these interrupters wailed like banshees so that at least ten or twelve of them were set out in rows on the lab roof. Occasionally one would develop an excess of gas and blow up. The screams of the interrupters could be heard on the arc of the lantern in the lecture room and often a worried Davy-Faradayite would hastily leave a Tuesday or Thursday afternoon lecture because he had heard his own well-recognized note stop abruptly. I was very proud of myself when I first had X-rays some time in November.

There was a great deal of activity at the Royal Institution at this time. Astbury and Yardley had written their space-group paper and all of us were learning to apply it to practical examples. Bernal's paper on rotating crystal methods was already in manuscript. Astbury applied these methods to the acetyl acetones and I to phenyl propionic acid. He and I spent hours in working through the details of the paper sometimes getting different answers from one another and from Bernal. Bragg and Gibbs worked out the structure of α and β quartz, Müller and Shearer were making some sense out of long-chain compounds. But most of us were determining space groups and obtaining very little structural information. It was only when a molecule had some symmetry that some definite information could be given. This was the case with the work of Knaggs and of George. We were all hoping for some 'clue' such as a change in one axis between two related compounds as had given Bragg an idea for anthracene, or the 'enhancement principle' which suggested interleaving of molecules. Many hours of work and perhaps sometimes more of discussion were spent in trying to devise ways of analysing organic structures. Of course many of these hours did pay off in the work of the Royal Institution in the early thirties, when Lonsdale solved hexa-methyl- and hexachloro-benzene and Robertson, who arrived after I had left, got durene, naphthalene and anthracene, and then developed the heavy atom and isomorphous replacement methods. In addition to the people I have mentioned already, others of the very close group were Burgers, Jackson, Mathieu, Orelkin, Plummer and Ponte. In addition we all had wonderful chemical backing from Saville, Smith and Lawrence.

A very great impetus to our discussions was added with the appearance of Duane's paper in 1925 on the crystal as a three-dimensional Fourier series, reviving W. H. Bragg's suggestion from 1915. None of us seems to have known of the latter, and although George Shearer was doing beautiful work in using one-dimensional Fourier series in explaining intensity distributions in long-chain paraffins, acids, and ketones, no one had the notion of using two- or three-dimensional series.

It had been my plan to return to McGill after two years in Bragg's laboratory to complete the work for Ph.D. and I had applied for a Canadian National Research Council Fellowship to support me at McGill. In the meantime the Pulp and Paper Industry Research Laboratory at McGill had become interested in the work of Herzog and Jancke on the X-ray diffraction from cellulose, and suggested that I go to Dahlem to join the group headed by Hermann Mark which had been built up in Herzog's Institute to study such problems. Mark's group at the Kaiser-Wilhelm-Institut für Faserstoffchemie involved Bredig, Ehrenberg, Gottfried, Herrlinger, Kratky, Naray-Szabo, von Susich, Weissenberg, with Szilard and Kallmann in sort of orbital attachment. Another person in Herzog's Institute who was a great inspiration to me was Gerda Laski who headed an infra-red group. Down the street was Haber's Institut für Physikalische Chemie where I made many personal friends as I did in Eitel's Institut für Silikatforschung, next door. There I gave my first seminar in 'German' as I spoke it. The subject was W. L. Bragg's paper 'The Structure of Certain Silicates.' I can only quote Herzog's comment after my talk: 'Vorlesen können Sie vielleicht, aber Deutsch leider nicht.'

Of course the real excitement of each week was the Physical Colloquium with von Laue as chairman and Einstein, Planck, and Nernst sitting in the front row. Others that were around were Bothe, Hahn, Meitner, Pringsheim, Wigner and so many others that one cannot begin to list them all. Schrödinger had not yet come to Berlin, but his influence and Heisenberg's dominated the scene. The word was Quantenmechanik. But although I was excited about all this new physics I was still trying to understand von Laue's paper on the diffraction of X-rays by small particles. This had been suggested to me by Herzog and Mark as my project with the hope that I could make an accurate determination of the 'particle size' of cellulose. I was fascinated by the notion which von Laue introduced of the space surrounding a reciprocal lattice point and I could not see how to do anything about it except by the methods of approximation which he had used. Under the influence of my quantum mechanical friends I bought a copy of Courant-Hilbert and discovered that something called the Fourier Transform existed and at the same time learnt what a potent mathematical entity an orthogonal set of functions really was. These two concepts had not been developed at all in the courses I had taken. Each new set of functions had appeared independently of the others and the fact that they all had common properties was not mentioned. I had been taught about a Fourier Integral Theorem but always in its double integral form. The fact that it could be split into two, one the spectrum of the other was something very startling and illuminating. In a few weeks after reading the first few chapters of Courant-Hilbert I wrote a brief note for the *Zs. f. Physik* which was the basis for my later work on particle shape functions. After reading von Laue's paper I became very interested in trying to extend the theory. The work that I did made me very pessimistic as to the meaning of a particle size determination. As a result I did not have the courage to publish any interpretation of the experimental work which I had done on line-broadening in Dahlem. A theoretical paper did, however, come out of it all.

I never really got to know von Laue until Easter time in 1927. The copy of *Nature* which I received on Good Friday contained Davisson and Germer's paper on the diffraction of electrons by nickel. I rushed around to tell all my friends about it but found that everyone was away for the holiday. The only thing I could do was to read and re-read the paper, check all the calculations and do some of my own including an incorrect interpretation of why the diffraction angle did not check with the lattice spacing. I interpreted this a as change in lattice spacing near the surface of the crystal whereas Bethe correctly explained it a few weeks later in terms of a refractive index effect. When von Laue suspended the normal programme of the next colloquium and reported the paper, I disagreed with some of the points in his presentation and said as much. Most of the Professors in Germany valued their dignity very highly and to have disagreed with them in public would have been suicidal. Von Laue was not this way at all and said that I apparently knew more about the paper than he and asked me to review it. This I did, so scared that the first line I drew on the blackboard came out dotted. After the session, von Laue invited me to come out to his home a day or so later, when we had about three hours of discussion, first about electron diffraction, and then about the work on particle size and on the Fourier interpretation of the reciprocal lattice which I was trying to do in Berlin. Thereafter he was very friendly to me and I was able to see him frequently during the rest of my stay in Berlin. Discussions with him often began with a vaguely formulated problem and ended with something clear cut, even though a solution was perhaps not obvious.

At the same time as I arrived in Germany I had acquired my first and only McGill Ph.D. candidate. Thomas N. White, Jr. had taken his undergraduate degree in 1926 and it had been arranged for him to work with me on X-ray diffraction for M.Sc. and possibly Ph.D. When my plans were changed and my return to Montreal was delayed from 1926 to 1927 it was decided that he would set up some equipment and start X-ray work on his own. This he did to his great credit and obtained his M.Sc. in 1927 just before I returned from Germany to McGill.

During our collaboration at McGill, White and I became interested in the cyclohexane hexols, a remarkable group of compounds which occurs throughout nature. About seven or eight isomers are known and one and only one of these isomers or a related methoxy compound occurs in a given species. Closely related species may have different isomers, and widely separated species the same isomers. It was perhaps the background reading for these studies which made me so sure that X-ray diffraction had an important future in the support of biochemistry. However, all we could do then was to determine a lot of space groups.

In the spring of 1929 I visited G. H. Cameron at DuPont in Wilmington and on my way back stopped off in New York to see Ralph Wyckoff at the Rockefeller Institute. As a result of this trip two things happened. I started collaboration with Harvey Cameron which resulted in the monograph we published together in 1937 and in two theoretical papers which I wrote in 1939. Second, I was asked to join the staff of the Rockefeller Institute and Tom White was asked to come with me to hold a National Research Council post-doctoral fellowship at the Institute. While at the Rockefeller, White and I continued the work on the cyclohexane compounds and looked at a number of other substances.

But my obsession with the notion that something was to be learned about structural analysis from Fourier theory continued. Early in 1930, I looked through the tables of contents of all the mathematical journals in the New York Public Library. Whenever the title was promising I looked through the paper and whenever the paper was promising I read it carefully. A lot of the papers which I tried to read I didn't really understand and so missed the point of many papers (e.g., Toeplitz) which were later to be of importance for the 'phase problem'. I did pick up a number of ideas in the process and wrote two papers for *Zs. für Krist.* largely concerned with location of maxima of Fourier series and also on the 'enhancement principle'.

In the fall of 1931, I accepted a job at the Johnson Foundation for Medical Physics in Philadelphia where I hoped to do something with X-ray diffraction on biological materials. I did take some powder pictures of horse hemoglobin using a camera cooled by Prestone circulating between it and a barrel filled with ice and salt. Most of the work of the Foundation was concerned with nerve physiology and I was very interested in Bronk's work in this field and that of Hartline on the eye of limulus. I had a good time collaborating with Ray Zirkle in some experiments on the effect of X-rays on the growth of fern spores, but my other excursions into biology were not too happy and I decided in 1933 after two years in Philadelphia that somehow I had to get back into crystallography and Fourier series. The appearance of the early Fourier papers from the Royal Institution and West's beautiful paper on KH_2PO_4 only served to whet my appetite.

Fortunately I had saved some money during the years in New York and Philadelphia and thought I had enough to keep me for one year on my own. I didn't really understand about depressions and did not contemplate three years out of a job.

In this 'year' I hoped to do some structure work and to take another shot at the Fourier series. I knew that Bert Warren had an active and running X-ray lab at M.I.T. and that Norbert Wiener probably knew as much about Fourier integrals as anyone in the world. So I asked Bert to take me on as guest. No pay but no fees for use of the lab. The Physics Department of M.I.T. was in a big burst of activity under K. T. Compton as President of the Institute and John C. Slater as head of the Department. I cannot go into any detail about the rest of the Department but in Warren's group were Gingrich, Hultgren, Serduke, and G. G. Harvey. There were many bright seniors doing undergraduate theses and several very good Ph.D. candidates. When I was first there the main interest of the lab was tending toward liquids. I started again on particle-size work because of a seminar I had to prepare and had many opportunities to talk with Wiener. This latter was a laborious process but a very intriguing one. There was then and is now no subject which can be brought up on which Wiener does not have something interesting to say. And with him the subject is always changing. I estimate I got in about one question on Fourier theory per two or three hours conversation, but the answers were usually pay dirt.

I very soon learnt from him the fact that I had to work with the Faltung, but it took me more than a year to catch on to what it was all about. I spent most of my time looking at Faltungen of step-functions such as those which had been used by Shearer in the study of the aliphatic acids and ketones. And this was largely because I could make the Faltungen of step-functions by simple geometry and did not need to compute the series by the tiresome methods then available.

The understanding of the Faltung came, of course, from the work on liquids and their radial distributions. Warren with Gingrich and others had perfected the techniques used by Debye and Menke in the study of the X-ray scattering from liquids. These were of course based on the original suggestions of Zernike and Prins. Warren and Gingrich had already had the idea that these methods, applied to powders, would give the radial distribution in a crystal. While trying to learn about their work I noticed that the mathematical form of the theory given by Debye and Menke would be identical with that of the Faltung if the integrations over random orientation were left out and the randomness of choice of origin was left in. What was immediately apparent was that the crystal contained atoms and that the Faltung of a set of atoms was very special in that it would consist of a set of atom-like peaks whose centers were specified by the distances between the atoms in the crystal. It was fortunate that this was clear from the beginning and it was in this form that the interpretation of $|F|^2$-series was proposed. It is unfortunate that the notion arose later that the maxima of the Faltung were determined by the distances between the maxima of the Fourier series. This is clearly untrue in general and was never suggested by me. However, to go back to the story, all this happened on a Tuesday, and Friday was the deadline date for the Washington spring meeting (1934) of the American Physical Society. An abstract had to be prepared in a hurry to go in with that of Warren on the radial distribution in carbon black and that of Gingrich and Warren on the radial distribution in powders which was basic to my work. The only $|F(h)|^2$-series which I was able to compute in the month between the deadline and the meeting was the (hk0) of KH_2PO_4 and a one dimensional series for a simple layer structure. All three papers were very well received and had very full discussion with A. H. Compton in the chair and W. L. Bragg in the audience to ask the right questions.

Soon after the Washington meeting W. L. Bragg visited M.I.T. and told us about the copper sulfate structure of Lipson and Beevers and that they had a method for computing series by using 'strips'. We soon received the data for copper sulfate from them and computed the F^2-series and the sharpened F^2-series from their data. We had also computed the F^2-series of hexachlorobenzene from K. Lonsdale's paper.

All these series were computed by a method suggested to me by George Kimball which involved multiplication of every term in the series by its appropriate sine or cosine. We recognized the tremendous repetition involved. Since we had not yet heard the details of Beevers and Lipson's strips the method later developed in more detail by Tunell and myself was set up.

In retrospect it is a source of satisfaction to me that I did so much work on the F^2-series method before publishing the second paper. This made it possible for me to draw attention to many of the difficulties which were likely to and which did arise. I must say that I was very annoyed at myself for missing the beautiful extension of the method made by Harker. I guess that I really could not get out of the plane.

While I was looking for jobs during my second and third years at M.I.T., I was offered fellowships in physics at two quite well known universities on the condition that I change my field of research. It was therefore very gratifying in 1936 that I was offered an assistant professorship in physics at Bryn Mawr College, with the express purpose of developing X-ray analysis in parallel with the wider interests of Walter Michels in the solid state.

Bryn Mawr College throughout its history has been remarkable for the fact that with the small faculty usually associated with a liberal arts undergraduate college it runs a full program of graduate work for the master's and doctor's degrees. It is a most stimulating place with the tradition that the faculty run the academic end of the college without interference from deans or presidents. However, it is difficult for the faculty members (particularly in science) to get any research done themselves. Their only way of keeping research going is through the work of the graduate students. Of these there are not many, but they are of high quality and the X-ray work was kept going largely through the collaboration of two very good Ph.D. candidates and a number of M.A.'s.

The kind of work which a faculty member could do himself was typified by two particle size papers, started at M.I.T. but finished at Bryn Mawr. The work on homometric structures was ideal, since I was able to have the benefit of wonderful collaboration from mathematical colleagues. I also did some work on generalized transforms, and finally Walter Michels and I ended my stay at Bryn Mawr in the full glory of a book on elementary physics.

When in 1949 I was given the opportunity by Dr. Stanley Reimann to start an X-ray diffraction group at the Institute for Cancer Research, I realized that the urge to apply these methods to biological problems could now be satisfied. At McGill, when the interest first developed, the prospect of X-ray analysis, even for a crystal of a very small organic molecule, ended with a space-group determination. Now it is possible to say to a biochemist that almost any molecule containing a score or two of atoms that can be prepared in crystalline form can be analysed in a finite amount of time. And now too an attack on the largest molecules is succeeding with the development of special techniques.

*Early Work on X-ray Diffraction in the California Institute of Technology**

Linus Pauling

The first crystal-structure determinations to be made in the United States were those of Hull, the determination of the structure of some metals by the powder method, and of Burdick and Ellis. The work of Burdick and Ellis was done in Pasadena, in the Gates Laboratory of Chemistry (constructed in 1916) of the California Institute of Technology, which at that time was called the Throop College of Technology. C. Lalor Burdick, after obtaining his M.S. degree in the Massachusetts Institute of Technology in 1914 and his Ph.D. degree in Basel in 1915, had spent a few months with W. H. Bragg in University College, London, learning the technique of using the X-ray ionization spectrometer. In 1916 he built an ionization spectrometer in the Massachusetts Institute of Technology, and then moved to Pasadena, where he built a second one. He and James H. Ellis, who was Research Professor of Physical Chemistry in the California Institute of Technology, then carried out an X-ray investigation of a single crystal of chalcopyrite, 8 mm in diameter. Their paper, published in *J. Am. Chem. Soc. 39*, 2518 (1917), was entitled 'The Crystal Structure of Chalcopyrite Determined by X-rays'. The integrated intensities of 13 forms were measured. The structure was reported to be a superstructure of the sphalerite structure (chalcopyrite is $CuFeS_2$). The reported structure was almost but not quite right; fifteen years later it was found that the distribution of Cu and Fe among the Zn positions of the sphalerite structure is somewhat different from that proposed by Burdick and Ellis. A paper on the structure of silicon carbide (cubic form) by Burdick and E. A. Owen (Professor of Physics in University College of North Wales, 1926–54, now Emeritus) was published in *J. Am. Chem. Soc. 40*, 1749 (1918). It describes work done by them in London and continued by Burdick in the Massachusetts Institute of Technology.

Roscoe G. Dickinson came to Pasadena and began his X-ray work

* Contribution No. 2755 of the Gates and Crellin Laboratories of Chemistry.

with the Burdick spectrometer in 1918. In 1920 he received his Ph.D. degree, the first one awarded by the California Institute of Technology; his thesis has the title *I. The Crystal Structures of Wulfenite and Scheelite. II. The Crystal Structure of Sodium Chlorate and Sodium Bromate.* The ionization spectrometer was also used by Richard M. Badger for his undergraduate thesis (B.S. 1921), which has the title *The Effect of Surface Conditions on the Instensity of X-ray Reflections from Crystal Planes.*

Dickinson then began making use of photographic methods. He prepared 'spectral photographs' (rotation photographs) and Laue photographs, and interpreted them by the 'method of Nishikawa (1915) as developed by Wyckoff (1920)'. R. W. G. Wyckoff spent one year, 1921–22, in the California Institute of Technology.

Richard M. Bozorth was the second man to receive the Ph.D. degree from the California Institute of Technology (1922). His thesis contained determinations of the structures of $PbMoO_4$ (wulfenite), KCN, $(NH_4)_2SiF_6$, and CdI_2, all made with use of spectral and Laue photographs.

X-ray diffraction was for some years the principal field of research in the Division of Chemistry and Chemical Engineering of the California Institute of Technology. By the end of 1922 twenty papers had been published by members of the Division, of which fifteen were on the determination of the structure of crystals. In 1922 structure determinations were published as follows: potassium and ammonium chlorostannate, potassium zinc tetracyanide and the corresponding crystals containing cadmium and mercury, phosphonium iodide, potassium chloroplatinite and potassium and ammonium chloropalladite, by Dickinson; potassium cyanide, ammonium fluosilicate, and cadmium iodide, by Bozorth; and nickel hexammine chloride, bromide, and iodide, nickel hexammine nitrate, zinc bromate hexahydrate, sodium hydrogen acetate, and silver molybdate, by Wyckoff.

Other fields of research began to be prosecuted, and in a few years the papers on the determination of the structure of crystals by X-ray diffraction dropped to about twenty percent of the annual total from the Division of Chemistry and Chemical Engineering, and then to about ten percent, where they remain. Over the period of 44 years since 1917 about 350 papers on X-ray diffraction have been published from the Gates and Crellin Laboratories of Chemistry, representing the determination of the structure of about 350 crystals. Many American X-ray crystallographers received their training in the California Institute of Technology.

I had developed an interest in the structure of crystals before I came to Pasadena. During the year 1919–20 I had a full-time appointment as Assistant in Quantitative Analysis in Oregon State College. My writing desk was in the chemistry library, and I was thus encouraged to read the journals. When the papers by Irving Langmuir on the electronic theory of valence were published in the *Journal of the American Chemical Society* in 1919 I read them with much interest, and also the paper that Gilbert Newton Lewis had published on this subject in 1916. My interest in the structure of molecules and crystals was awakened to such an extent that in the fall of 1919 I gave a report on molecular structure and the nature of the chemical bond in the research conference of the Chemistry Department of Oregon State College.

In 1920 I re-entered college, as a junior, and in 1922 I received the degree of Bachelor of Science in Chemical Engineering. I was then given an appointment as a graduate assistant in the California Institute of Technology. Professor Arthur Amos Noyes at once wrote to me to advise me about my activities during the summer. I was employed that summer as a paving plant inspector, helping to pave a highway along the Pacific coast near Astoria, Oregon. Professor Noyes pointed out that my training in physical chemistry was not good, and suggested that I work the problems in the textbook *Chemical Principles* that he and Sherrill were publishing that summer. Using the proof sheets of the book, I worked the problems in my spare time. I also read the book *X-rays and Crystal Structure* that W. H. Bragg and W. L. Bragg had written in 1915. The decision was made before I reached Pasadena that I would carry on research in the field of the determination of the structure of crystals by the X-ray diffraction method and would have Dickinson as my supervisor.

When I arrived in Pasadena near the end of September 1922 Noyes suggested that I make a structure determination of the lithium hydride crystal in order to obtain evidence as to whether or not this crystal contained the hydrogen anion. He thought that evidence on this point would be provided by the scattering power of hydrogen for X-rays. I began work on the synthesis of lithium hydride and began the construction of an apparatus to take powder photographs. I gave up the lithium hydride problem shortly, on learning that the structure determination had just been completed by Bijvoet and Karssen.

In the meantime, Dickinson suggested that I also prepare some other crystals for X-ray study. The method that Dickinson and other workers in the laboratory were using had as its first step the preparation of a spectral photography by rotating a rather large single crystal in the X-ray beam from a tube with rhodium or molybdenum anticathode in such a way as to permit the Bragg reflections from a large developed or ground face to be recorded on a photographic plate (plates, rather than films, were used during the first year or two of my work). The reflections from one face of a cubic crystal gave the edge of the unit cube of the smallest possible unit, the other possible units being integral multiples of this value. A specimen for the preparation of Laue photographs was then made by grinding a thin section of the crystal, usually about 0.3 mm thick, with which several Laue photographs were made. These photographs were indexed by making gnomonic projections. The values of $n\lambda$ for the spots were then calculated for the smallest unit permitted by the spectral photographs and for multiples of this unit. In the preparation of the Laue photographs an X-ray tube with tungsten target was used, with peak voltage 54 kV, corresponding to a short wavelength limit of 0.24 Å. The smallest possible unit for the structure that would account for the reflections observed on the Laue photographs (by giving no $n\lambda$ values less than 0.24 Å) was assumed to be the correct unit. The Laue symmetry of the crystal was found by taking Laue photographs with the crystal oriented in such a way that the X-ray beam was parallel to a principal symmetry axis of the crystal, and the theory of space groups was applied to find all possible arrangements of atoms compatible with the Laue symmetry, the composition and density of the crystal, and the observed X-ray reflections. The effort was made to eliminate all but one of these possible atomic arrangements and to evaluate the parameters determining the atomic positions. Because of uncertainty about the atomic form factors, use was in general made only of qualitative comparisons: when it was observed that one plane gave a stronger X-ray reflection than a second plane with larger interplanar distance it was concluded that the structure factor (with the form factors for the atoms assumed constant) must be greater for the first plane than for the second plane.

This method permitted one or two parameters to be determined rather easily, and more with considerable difficulty. Wyckoff and Dickinson had found that interesting crystals with high symmetry often had structures determined by only one or two parameters, whereas those with low symmetry might involve many parameters. Dickinson accordingly suggested to me that I search the literature for interesting substances that were described as forming cubic crystals and make preparations of them for X-ray investigation. I selected a number of substances, synthesized some of them, grew crystals suitable for X-ray investigation (in general, I tried to grow crystals two or three millimeters in diameter, in order that I might grind them to obtain a good-sized thin plate for Laue photography), and attempted to determine their structure.

My luck was not good. I made an electric furnace in which anhydrous sulfates could be melted and slowly cooled, and in this way got some nice octahedra of $K_2Ni_2(SO_4)_3$; and on 28 October 1922, one month after my arrival in Pasadena, I prepared a photograph of the spectrum of X-rays from a tube with rhodium target reflected from an octahedral face of the crystal and at the same time, to calibrate the apparatus, reflected from the cleavage face of a calcite rhombohedron mounted just above the other crystal. The spacing for (111) of the sulfate was found to be 5.663 Å, corresponding to 9.808 Å for the side of the unit cube. The density of the crystal, as determined by the pycnometer method, was 3.33 g/cm^3, corresponding to four formulas per unit cube. A Laue photograph made on 31 October 1922 showed the symmetry expected for point group T or T_h, and also showed no reflections requiring the unit to be larger than 9.8 Å on edge. The space group is T^4. This space group has positions for four equivalent atoms, with one parameter, or twelve equivalent atoms, with three parameters. Hence there is one parameter for every four atoms in the unit cube. There are 76 atoms in the unit cube, and the structure is accordingly determined by 19 parameters. I at once decided to work on another crystal. (By study of an isomorphous crystal, langbeinite, $K_2Mg_2(SO_4)_3$, this structure has recently been determined.)

I then decided to attack an intermetallic compound. No structure determination for an intermetallic compound had yet been made, and as an undergraduate I had been interested in metallography and had at one time considered the possibility of specializing in the chemistry of metals and alloys. A search of the literature revealed that big single crystals of a cubic compound of sodium and cadmium, $NaCd_2$, could be prepared by dissolving cadmium in an excess of molten sodium and allowing the melt to cool slowly; on treatment with absolute ethanol the excess sodium could be converted to sodium ethanolate, leaving large octahedra of the intermetallic compound. In this way I obtained fine crystals about three millimeters in diameter. I made X-ray photographs of one of these crystals in November 1922, and found that the unit cube is about 31 Å on edge and contains about 400 sodium atoms and 800 cadmium atoms. I did not know how many parameters would be required to determine its structure, but it seemed evident that I should attack some other crystal. I have returned to the study of $NaCd_2$ from time to time, and other X-ray crystallographers have also investigated it, but its structure is still unknown.

My third effort was with cubic crystals of $CaHgBr_4$, which are easily grown from aqueous solution. The cubic unit was found to have edge 19.14 Å and to contain 32 atoms of calcium, 32 of mercury, and 128 of bromine. This crystal, too, seemed to be too complicated for successful attack at that time. The structure has not yet been determined by anybody.

Dickinson then proposed that a mineral be investigated, and he selected molybdenite. He cleaved a section about 0.3 mm thick from a chunk of the mineral, available in bulk in the chemistry stockroom. Excellent X-ray diffraction patterns were given by this cleaved section and the structure determination was carried out in a month. This structure involves one parameter.

I then attacked several other crystals with success. In particular, another intermetallic compound was prepared and was found to have a simple structure. This compound, with formula Mg_2Sn, was made by melting together the metals in the proper ratio and allowing the melt to cool slowly. The compound is brittle, and by striking the solidified mass with a hammer it was broken into fragments, which were found to be single crystals bounded by octahedral cleavage faces. The structure is that of fluorite.

During the following decades more and more powerful methods of crystal analysis have been developed. Each year brings many interesting reports about new structures, with novel features. I think, however, that none of them causes in me such a feeling of pleasure as was produced by the successful determination of the structure of molybdenite and the discovery that the six atoms of sulfur are arranged about the molybdenum atom at the corners of a trigonal prism, rather than of the octahedron characteristic of sodium chloride and some of the other simple structures known in 1922. (For some reason that I do not remember, but that may have been my feeling of astonishment about the sizes of atoms, the coordination polyhedron is described in the paper that Dickinson and I published on the structure of molybdenite as a 'small' triangular prism.)

Forty years ago very few structures had been determined. Any crystal that could be successfully attacked had a good chance of giving a surprising result. Now many structures have been determined, and we know a great deal about structural principles; but we do not know everything, and the X-ray crystallographer, as he carries out the study of another crystal, still has a good chance of being surprised by the result of his work.

My Time with X-rays and Crystals

Michael Polanyi

A great German poet once said, 'Where kings are building, carters find work.' Most of history is written about kings, and that is as it should be, but the work of carters has its own history too, and that too is important. For great discoverers would achieve little unless followed by enterprising settlers. I shall recollect here a brief phase of the early colonisation of the immense domain of knowledge on which Max von Laue first set foot 50 years ago, followed by W. H. and W. L. Bragg, whose discoveries first revealed its major treasures.

The example of great scientists is the light which guides all workers in science, but we must guard against being blinded by it. There has been too much talk about the flash of discovery and this has tended to obscure the fact that discoveries, however great, can only give effect to some intrinsic potentiality of the intellectual situation in which scientists find themselves. It is easier to see this for the kind of work that I have done than it is for major discoveries, and this may justify my telling this story.

At the time when I started this work, a chain of research institutes existed in Germany supported by the Kaiser Wilhelm Gesellschaft. One of these was the newly founded Institute of Fibre Chemistry in Berlin-Dahlem, to which I was appointed in the autumn of 1920. My first contacts revealed the peculiar character of the company I had joined. Following German custom, I called on the Directors of the other institutes in Dahlem, and first of all on the great Fritz Haber, Director of the Institute of Physical Chemistry. Haber, who had seen my speculative papers on reaction kinetics, referred to them with a stern admonishment: 'Reaction velocity' said he 'is a world problem. You should cook a piece of meat.' He meant that first of all, I should prove my capacity as a craftsman; the rest would follow. Rather different was the impression conveyed by Carl Neuberg, Director of the Institute of Biochemistry. 'Don't stay here, dear Colleague,' he said. 'Accept the first offer of a Chair at a university. If you don't make a discovery for a couple of years here, you are just an old ass; at a university you have always your academic laurels to rest upon.'

Discovery requires in fact something beyond craftsmanship, namely the gift of recognizing a problem that is ripe for solution by your own powers, large enough to engage your powers to the full, and worth the expenditure of this effort. Haber had admonished me, because he thought I had taken on a problem that was not yet ripe, and in any case too large for me. Perhaps he was right; but at any rate, my new job with the Institute of Fibre Chemistry led me in a different direction. On my arrival at the institute its Director, Reginald Oliver Herzog—whom I remember warmly for his kindness and wide intellectual perceptiveness—immediately gave me what Haber said I needed, a piece of meat to cook. Following his discovery of the crystalline nature of cellulose (paralleled by Scherrer) Herzog—aided by his assistant Jancke—had just found that a bundle of ramie fibres irradiated by an X-ray beam at right angles produced a diffraction pattern composed of sets of four equivalent dots symmetrical to two mirror-planes, one passing through the primary beam and the axe of the fibres, and the other normal to the former. There was excitement about this diagram and I was asked to solve the mystery. In the next few days I made my first acquaintance with the theory of X-ray diffraction, of which, owing to wars and revolutions and my exclusive interest in thermodynamics and kinetics, I had heard little before. However, I soon returned the result that the four-point diagram was caused by a group of parallel crystals arranged at random around one axis, and this interpretation was included in a joint publication with the work of Herzog and Jancke. So I had cooked a piece of meat—and this transformed my position. Herzog, with kindly enthusiasm, showered me with every facility for experimental work, most precious of which were funds for employing assistants and financing research students. In this I was incredibly lucky. I was joined by Herrmann Mark, Erich Schmid, Karl Weissenberg, all three from Vienna, by Erwin von Gomperz and some others; the place was soon humming. It was the time of runaway inflation and poor Herzog found it difficult to pay all these people. Protest meetings were held, resolutions passed, Weissenberg in the lead; the Institute earned the name of an 'Assistenten-Republik'. We had a glorious time.

I shall now try to give an account from memory—without access even to my own papers—of the way some of my further contributions were born in those days. I found that all the dots forming the fibre diagram lay on a series of hyperbolae, each hyperbola comprising dots reflected by planes having identical indices with respect to the crystal axis parallel to the fibre. I established the formula determining the series of these hyperbolae, as a function of the identity period parallel to the axis of the fibre. Thus equipped, I evaluated the elementary cell of cellulose and drew the conclusion that the structure of cellulose was either one straight giant molecule composed of a single file of linked hexoses, or else an aggregate of hexobiose-anhydrids; both structures were compatible with the symmetry and size of the elementary cell—but unfortunately I lacked the chemical sense for eliminating the second alternative.

This foolishness had also an amusing consequence. When I first stated my conclusions in the Colloquium presided over by Haber, there was a storm of protest from all sides. The assertion that the elementary cell of cellulose contained only four hexoses appeared scandalous, the more so, since I said that it was compatible both with an infinitely large molecular weight or an absurdly small one. I was gleefully witnessing the chemists at cross-purposes with a conceptual reform when I should have been better occupied in definitely establishing the chain structure as the only one compatible with the known chemical and physical properties of cellulose. I failed to see the importance of the problem.

A failure of the same kind was my treatment of sero-fibroin. Herzog had discovered its fibre diagram and handed it to me for evaluation. I determined its elementary cell and observed that there was only room for glycine and alanine in it. But I could not make up my mind what to think of the other observed decomposition products. I did not recognize the immense importance of the question, and passed it on half-baked to Brill, for the doctoral thesis he was doing under my supervision.

Such failures are worth recording, in order to correct the current theories of the scientific method, based altogether on success stories. It is interesting to recall in this connection, that the weakness of my initiative was due in part to the fact that my confidence in this line of inference was impaired by a slight deviation of the position of two dots on the equator of the cellulose diagram from the theoretical values predicted by my analysis. These loose ends were debilitating; but I am glad that at least I did not obey the current Sunday school precepts of the scientific method, which would command you to reject a theory if a single piece of evidence contradicts it. (I still don't know what caused the discrepancy in those dots.)

Following my discovery of the hyperbolae in the fibre diagram of cellulose and my evaluation of its elementary cell on this basis, the principles thus established were transposed into the rotating crystal method in collaboration with Weissenberg and Mark. The former far surpassed me in mathematics, while the latter lent me his manipulative skill bordering on genius. To the best of my recollection, the project, including the suggestion of using an elongated Debye-camera for the purpose of including the higher layer lines, came from me. Weissenberg generalized the layer line relationship, which I had only established for the directions of the crystal axes and perpendicular incidence of the beam, to include identity periods in any direction and all angles of incidence; Mark carried out the first experiments with the new method. The first use of the rotating crystal apparatus for the determination of an unknown crystal structure was made, I think, by Mark and myself in 1923 on white tin. It was used in the first place by Mark, Schmid and myself for elucidating the plastic flow of zinc crystals.

The strength of solids had now become my principal interest. The technological purpose of the Institute had thrown this great problem into my lap. I saw that the characteristic feature of the solid state, namely its solidity, was yet unexplained, and indeed, hardly explored by physicists. I found that in the light of the recently discovered structure of rocksalt, such a crystal should be thousands of times stronger in resisting rigid rupture or plastic deformation, than it actually was. In facing this paradox I appealed to two features of modern physics (1) that a crystal of rock salt was one giant molecule and (2) that inside a molecule energy could be transferred by quantum jumps not controlled by the laws of classical mechanics. From a calculation based on the actual strength of rocksalt I concluded that the energy required for producing the new surface formed by breaking the crystal would have to be supplied from the stress stored up on either side of the future break, in an area extending two or three millimeters in both directions of it. So I set out to show experimentally that crystals shorter than a few millimeters were stronger than those of greater length. The result was inconclusive and the whole idea may have been wrong, but in pursuing it I stumbled on an important aspect of the strength of materials which seemed to reflect my original paradox and to encourage the way I was trying to solve it; I came to know about the hardening of materials by cold working.

I was deeply struck by the fact that every process that destroyed the ideal structure of crystals (and thus reduced the areas which could be regarded as single molecules) increased the strength of crystalline materials. This seemed to confirm the principle by which I explained the low resistance of crystals to stress and to refute the rival theory—inspired by the work of Griffith on quartz threads—that the weakness of crystals was due to cracks or other imperfections of structure. The cold working of rocksalt crystals by vigorously filing their sides and of tungsten crystals (obtained from filaments for incandescent lamps) by drawing them through a die, confirmed this. The results, presented in September 1921 under the title 'The Hardening of Crystals by Cold Working' to the meeting of the Bunsen Gesellschaft, were received with uneasy surprise. Gustav Tammann, speaking as an elder statesman, expressed this in the discussion. Yet, as later work was to show, my observations were fundamentally sound.

Meanwhile, I took up antecedent questions. Some metallurgists, interested in my work on the hardening of single crystals, told me of a method invented by Czochralski for producing metal crystals in the form of wires. It consisted in pulling out a thread from a pool of molten metal, so that the thread continued to solidify at the rate at which you were pulling it out. Erwin von Gomperz, who was doing his thesis with me, was put to growing single crystals of tin and zinc in this way. Unfortunately, the metal tended to come out in lumps, and the project was saved only by the intervention of Hermann Mark who covered the liquid metal by a sheet of mica with a hole in the middle, through which the thread came out as a smooth cylindrical wire. But for this ingenious intervention, our subsequent investigations of the plastic flow of metals might not have come about.

The next stage of our work elucidated the crystallographic laws of plastic flow in zinc and tin. These are now well known. Of this work I should like to say only that it is a rare instance of something supposed to be a common occurence, namely of the participation of as many as three scientists as equals in a fairly important piece of work. We were lucky in hitting on a problem ripe for solution, big enough to engage our combined faculties, and the solution of which was worth this effort. The ripeness of the problem was confirmed, when a few months after our paper on zinc had came out, a similar investigation was published by G. I. Taylor and Miss Elam in England, that solved the same problem for an entirely different system, namely aluminium. Though these two parallel papers applied very different methods, they both evaluated identical possibilities concealed in a common intellectual situation.*

* The wooden model of slip in crystals which I often see reproduced in current litterature,

The following episode might illustrate this principle on a smaller scale and show incidentally also, how crude was the knowledge of crystal structures on which we were relying at the time. Shortly after Mark and I had published the structure of white tin, we received the visit of the Dutch scientist Van Arkel, who told us that our result was wrong, for he had established an entirely different structure of the metal. Only after hours of discussion did it become apparent that his structure was actually the same as ours, but looked different because he represented it with axes turned by 45 degrees relative to ours.

Most of this work was completed in a little over two years from the day when I first 'cooked a piece of meat' in Dahlem—all of it being supported as a rather odd kind of fibre chemistry by our noble-hearted director, R. O. Herzog. Having established the geometrical mechanism of deformation in crystalline solids, we could now take up effectively the physical problems of deformation that had first drawn me into this field. But Mark had become engrossed in structure analysis and Weissenberg had also taken up problems of his own; it fell, therefore, mainly to Schmid and myself to embark on the physics of solid strength, equipped with the crystallographic results to which our whole group had contributed before. Schmid established his law of shearing stress at the yield point. Together we observed that minimal deformation of crystals can lead to noticeable hardening and discovered the fact of 'recovery' which cancelled hardening without recrystallization. Schmid established hardening for greater deformations, by applying his law of shearing stress. Jointly we proved by strains applied under hydrostatic pressure, that stress vertical to a slip plane leaves its resistance to shearing unaffected. Experiments with W. Meissner and E. Schmid at the temperature of 1°K demonstrated the athermal quality of slip in crystals, and this brought out the fundamental contrast of crystal plasticity to the deformation of amorphous solids, which become perfectly rigid at absolute zero. Observations with G. Masing on fine grained zinc ruptured at liquid air temperature, confirmed the fact that the internal fragmentation of a crystalline material increased its resistance to brittle rupture. In every instance so far the strength of crystals was found rising from its paradoxically low value, towards the much higher theoretical strength which Griffith has actually observed in amorphous quartz threads—to

was made on my instructions in the small workshop of a joiner in Dahlem. At one time the mathematician R. v. Mises remarked acidly on its widespread use by members of our group: 'All the problems of plasticity are apparently to be solved by pushing this model to and fro.'

the extent that the disturbance of the crystalline order shifted the condition of the crystal towards the amorphous state. This was also the explanation I found for the curious Joffé-effect: I showed by experiments with W. Ewald, that the water dissolving the surface of a rocksalt prism reduces its resistance against plastic flow; and that it is the onset of this flow, acting as cold working, that increases the crystal's resistance to brittle rupture, as observed by Joffé.*

I shall pass over our enquiries starting from the discovery of fibre structure in cold worked polycrystalline metals, by merely mentioning that it was nice to be able to account for this phenomenon by the crystallography of plastic deformation as observed in single crystals. More interesting perhaps was the fact (found with P. Beck) that while the annealing of a bent aluminium crystal caused it to recrystallize, no recrystallization took place if it was straightened out before annealing. These were sidelines, for they threw no light on the nature of solid strength, which remained shrouded in the paradox that all effects which would tend to restore the ideal crystal structure appeared to weaken the material far below its ideal strength, whereas every disturbance of this structure tended to raise its strength towards its ideal value.

My fascination with this fact had borne fruit—but it had proved excessive. From a paper written by Erich Schmid on the occasion of my 70th birthday, I gather that the picture I had formed of the hardening and weakening of crystals made me overlook an important clue for modifying it. My experiments with W. Ewald (1924) show that bending a rocksalt crystal in one direction hardens it only for further bending in the same direction and actually weakens it for bending it back. Schmid says that such mechanical recovery has subsequently been effected in various crystals, including those of metals. Had I noticed the fact that deformation may actually weaken a crystal, my mind would have been more receptive to the idea that the extremely low resistance of crystals against plastic deformation might be due to the kind of irregularity in the structure of the crystals that is now known as *dislocation*. However the idea of dislocations causing a high degree of plasticity did gradually take shape in my thoughts. I gave a full account of this theory in April 1932 in a lecture addressed to the members of Joffé's institute in Leningrad, who received it well. On returning to Berlin I talked about my theory to Orowan who told me

* Joffé himself contested my explanation, especially since the lowering of the yield-point of rocksalt by dissolving its surface with water had not been observed before. But I gather that this effect is now well established.

that he had developed a similar idea in his thesis about to be submitted for a degree. He urged me to publish my paper without considering his rival claim, but I preferred to delay this until he too was free to publish. (This explains why my communication appeared in print in Germany and in German a year after I had left the country and had already published many papers in England.)

Meanwhile the principle that scientists only reveal hidden knowledge which has become accessible by the intellectual situation of the moment, reasserted itself. Once more the same intimations had matured to the same solutions in another, totally different, mind, that of G. I. Taylor in England. What I had published as 'Versetzung' in German, he published simultaneously as 'dislocation' in English.

By the autumn of 1923 I left the Institute for Fibre Chemistry through promotion to independent Membership in the Institute of Physical Chemistry. Haber received me now with full confidence in my ability to work as a scientist and I immediately plunged back into reaction kinetics. My time with X-rays and crystals had lasted three years.

Personal Reminiscences

J. Monteath Robertson

During the 1914–18 war my father lost his eyesight so I had to leave school and manage the farm at home. I liked this work but was always deeply interested in science, having read about J. J. Thomson's work on the atom and the electron in Arthur Mee's *Children's Encyclopedia* and other popular works, which my brother and I used to discuss very fully and often far into the night. We both felt a tremendous excitement about these great discoveries. Only afterwards, when my brother returned from the war, was it possible for me to do some private study and gain University entrance. Without a proper school training I found the science part easy but languages difficult.

At the University of Glasgow I read rather widely in Mathematics, Physics, Chemistry and Geology, but when it came to research I went into Chemistry, mainly I think because it seemed closer to the atoms and molecules in which I was deeply interested. My Professor, G. G. Henderson, persuaded me to start working on the structure of the sesquiterpenes and so I took my Ph.D. in organic chemistry. This was a difficult field and little progress was made, although L. Ruzicka about this time had been able to discover the carbon framework in some of these structures by dehydrogenation experiments. I could not help feeling that there should be some purely physical method by which one might be able to find the spatial positions of the atoms in such complex molecules. I knew about the X-ray method. W. H. Bragg had published his early measurements on naphthalene and anthracene, while Muller and Shearer had published some striking work on the long chain compounds. This seemed to be the method I was looking for, so with difficulty I obtained some good crystals of my complicated sesquiterpene derivatives and sent them to W. H. Bragg. Nothing much happened, so in 1926 when I was able to obtain a Fellowship of sufficient value (£250 per annum for two years) I followed my crystals to the Royal Institution, fondly hoping that in the two years available

I might work out these structures. This was in fact finally accomplished, but the task took over thirty years instead of two! An interesting commentary on this situation was made at a Chemical Society meeting in London a few months ago where I was reading a paper on the structure of some terpenoids. In thanking me, Professor D. H. R. Barton, who knew something of this early history, congratulated me on having at last completed my Ph.D.!

My decision to go to the Royal Institution was not an easy one to take. At this time I was also greatly attracted by work on atomic structure and I thought of seeking a place in the Cavendish Laboratory where Aston was working on isotopes. But in the end it was the Royal Institution that drew me because I wanted to finish the job I had begun in Glasgow. The transition from organic chemistry to physics was also difficult, especially as the equipment was primitive. I had to make most of my own X-ray tube with glass and sealing wax, and run it from an induction coil. The pumps were also primitive and home-made, so that one had quickly to become expert in high vacuum technique and leak detection. But everyone at the Royal Institution was helpful. Sir William Bragg asked me to start on a photographic survey of the anthracene intensities, while he himself worked in the next room with his ionization spectrometer. My more immediate tutors were W. T. Astbury, J. D. Bernal, Kathleen Lonsdale, A. Muller and G. Shearer; and I owe them all a great debt. It was an atmosphere in which one had to learn quickly, because these people did not suffer fools gladly.

By 1928 I had not achieved much, apart from publishing two incorrect structures, but had learned a good deal. I was then elected to a Harkness Commonwealth Fellowship and went off to the United States for two very happy years, spent mainly in the Physics Department at Ann Arbor, Michigan. I learned some theoretical physics with Uhlenbeck, Goudsmit, Dennison and Laporte, and tried to keep up some chemistry with Gomberg, in whose laboratory I had a bench which I seldom occupied. I read A. H. Compton's *X-rays and Electrons*, and also became acutely aware of the phase problem. This I tried to solve by a scheme that would take too long to describe, but computational facilities were totally inadequate for the approach I was trying. In 1930 the most important event in my life occurred when I married Stella Nairn in Toronto, in contravention of one of the conditions of my Fellowship, which was tenable only by unmarried persons.

Sir William Bragg then offered me a post in the Davy-Faraday

Laboratory and we returned to the Royal Institution. The next nine years were perhaps the busiest in my life, and I was able to solve many organic structures accurately and completely. The work was hard and the computational difficulties severe. We devised an elegant system of moving strips of figures on a large board, and many tedious hours were spent in summing Fourier series and making structure factor calculations after the children were in bed. This work was important and exciting, because we were measuring bond lengths and putting the chemical formulas on a true absolute scale for the first time. However, there was still an element of dissatisfaction, because the chemist could justly claim that he was already very familiar with these structures and that we were in a sense only confirming what he already knew. Our ultimate goal was to be able to solve structures that were chemically unknown.

Here the big event occurred a few years later when one day Sir Jocelyn Thrope and R. P. Linstead brought some excellent crystals of the newly discovered phthalocyanines to the Royal Institution. I was immediately summoned by Sir William Bragg and asked to examine this problem. To begin with it seemed an almost impossible task, because the unit cells were large and the reflections exceedingly numerous. But there was a striking isomorphism between some although not all the derivatives, and I felt that some help could surely be obtained from this fact. To cut a long story short, the isomorphous substitution and heavy atom methods of phase determination were evolved in this work and the structure completely determined without any reference to chemical theory. Great new vistas of possible progress in the direction of solving chemically difficult or unknown structures were thus opened up.

The organization at the Royal Institution during this time certainly helped a great deal in the progress I was able to make. There was an excellent workshop under Jenkinson. H. Smith who operated Muller and Clay's 5 kW rotating anode X-ray tube was most helpful, and B. W. Robinson and R. H. V. M. Dawton devised an integrating photometer which I used a great deal. I had no research students, although towards the end of my time at the Royal Institution I enjoyed active collaboration on certain problems with A. R. Ubbelohde, Ida Woodward, J. J. de Lange, and L. O. Brockway. (Why do young and immature members of University staffs always demand help from large numbers of graduate students from the moment they are appointed? If they would first work for ten years with their own hands they might then be better able to make a distinctive contribution.)

I left the Royal Institution early in 1939, not through choice but because with a rapidly growing family it was now urgently necessary to earn more money. Before leaving I had got as far as writing a paper to suggest that if mercury could be substituted for zinc in insulin, which seemed chemically feasible, then even a structure of this complexity might ultimately be solved by the heavy atom method. But now I went to Sheffield University, as senior lecturer in Physical Chemistry, and in a sense had to begin life again, in a laboratory that had none of the equipment I needed. However, the still greater interruption of the Second World War now began. I joined the Air Force and stopped all X-ray work for a number of years.

I returned, not to Sheffield, but to Glasgow as head of a large Chemistry Department. For my research this again meant starting from nothing, in a laboratory entirely devoid of X-ray equipment and with considerable responsibilities for teaching chemistry. However, I was now in a position to get what I wanted, and the University generously supported me. I soon began solving structures again and was pleased to find that I had not forgotten the way. I got out the notes on earlier work that had been hurriedly compiled and put away in a safe place as the first air raids began. Also, it was now not only desirable but a duty to organize the training of research students in many fields. I am glad to think that these efforts have been successful. Among those who began X-ray work with me in Glasgow after the war are A. McL. Mathieson and J. D. Morrison (now mass spectrometry) in Melbourne, Maria Przybylska in Ottawa, J. G. White in Princeton, Jack Dunitz in Zurich, Sidney Abrahams with Bell Telephones in New York, Walter Macintyre in Boulder, James Trotter in Vancouver, M. G. Rossmann in Cambridge, H. M. M. Shearer in Durham, and J. S. Broadley and D. M. Donaldson at Dounreay, Thurso (now atomic energy). Ian Dawson (now electron microscopy), J. C. Speakman, George Sim, Tom Hamor and Andrew Porte (now nuclear magnetic resonance) are all back in Glasgow again, after adventures in other places, and are actively participating in our latest and most exciting work.

We are, I think, now achieving a spectacular revolution in the X-ray analysis of organic molecular structures. The methods first developed in the 1930's are now coming to full fruition. This is largely because, with the invention of fast electronic digital computers, it is now possible to work in three dimensions with the same or even greater facility than in two dimensions ten years ago, and so for the first time we can really use the X-ray method at its full power. This is, of course, an over-simplification. Much new and important development work has had to be undertaken during these recent years and many new problems solved. However, if we adopt the chemical approach which I have always advocated, and prepare a series of suitable phase determining derivates (and in general we know enough chemistry to be able to do this), then we can with confidence obtain complete solutions of organic molecular structures containing up to 100 atoms or more. However, this work is not yet, nor is it likely to become in the immediate future, a routine tool which anybody can apply. A good deal of judgement, and often intuition, are required in the first stages of deciphering the dimly resolved electron density distributions. It is always unsafe to predict the future, but I would say that crystal analysis is likely to remain both an art and a science for some time to come.

Personal Reminiscences

Paul Scherrer

When I arrived in Göttingen in 1913 as a student of Physics, Göttingen was still a small quiet University town, full of charm and without factories; life was centered in the University, and under the still surface lay hidden an intellectual life of unsurpassed intensity. The circle of mathematicians surrounding Felix Klein,—David Hilbert, Carathéodory, Landau, Runge, Toeplitz, Herglotz, Hecke, Weil, Courant and Noether—as well as the particularly richly staffed Physics with Riecke, Voigt, Wiechert, Debye, Born, von Traubenberg, Madelung, Simon, Prandtl, von Kármán, and Tammann, rendered the Faculty very attractive to mathematicians and physicists. This was the time when at last the young Quantum Theory of Light began to be taken seriously although it could by no effort be straightened out with wave theory. One tried hard to become convinced of the reality of Bohr's electron orbits in the atoms in spite of all the hesitations the physicist felt in accepting the hypothesis that the electron on its stationary orbit about the atomic nucleus does not radiate,—a flagrant contradiction to Maxwell's theory. The next job to be done was therefore to find a check on Bohr's hypothesis which worked so simply and directly in the case of spectral emission, by looking for a direct evidence of the reality of the electronic orbits. Debye had come to the conclusion that specific diffraction effects should be produced with X-rays by the regular spacing of electrons on circular orbits: Using monochromatic X-rays, one should find the curve of diffracted intensity vs. angle of diffraction to show broad maxima and minima, provided the elctrons kept at fixed distances from each other in the course of their motion. Debye proposed to me that we try together such diffraction experiments. We used at first a gas-filled medical X-ray tube with platinum target which happened to be available in the collection of the institute. For power source we used an enormous induction coil with mercury interrupter and a gas-filled rectifying valve. The whole set-up appears nowadays like a show piece taken from a museum. The first diffraction photographs, with paper and charcoal as the scattering substances, showed no diffraction effects. The reason for this may have been that the thick glass wall of the tube absorbed the Pt L-radiation and transmitted only the continuous background. The film was relatively insensitive for the K-radiation, which, besides, was not strongly excited, so that possible maxima were covered up by the continuous background.

This prompted me to construct a metal X-ray tube, water-cooled and with copper target. The tube remained connected to the rotating Gaede mercury pump. An aluminium window, 1/20 mm thick, permitted the rays to emerge. I also constructed a cylindrical diffraction camera, of 57 mm diameter, with a centering head for the sample, of the type which is being used still nowadays.

For the sample I used the finest grain powder of lithium fluoride; Debye and I were most surprised to find on the very first photographs the sharp lines of a powder diagram, and it took us not long to interpret them correctly as crystalline diffraction on the randomly oriented microcrystals of the powder. The diffraction lines were much too sharp than that they could have been due to the few scattering electrons in each single atom. That in lithium fluoride we picked a cubic crystal powder with exceptionally favourable scattering properties was a piece of good luck.

Modifying the application of Debye's original idea we also obtained diffraction photographs with fine jets of benzene and cyclohexane. The perfectly clear broad interference rings produced by these were interpreted as the diffraction diagrams of the carbon hexagons in these molecules. In later years, and with utmost success, Debye has refined and followed up this first X-ray study of the structure of liquids.

The possibility of determining structures without the need for macroscopic crystals was obviously of great value, since for many substances single crystals were not available, or one wanted to investigate agglomerations of crystals.

There was a forbidding multitude of substances that were of interest. While the determination of cubic structures was easy (MgO, Fe, etc.), I did not succeed to find, for instance, the lattice of boron. Professor Stock of Berlin had supplied me with pure boron powder, and I devoted much effort on this substance. Even the method of indexing which Professor Runge proposed proved unsuccessful in application to the boron diagrams although using sufficiently accurate data one should arrive by it nearly automatically at the correct lattice. On the other hand it was not difficult to determine the structure of a complex-salt like K_2PtCl_6. The arrangement of the atoms followed exactly the scheme which Professor Werner in Zürich had surmised from chemical reasoning: the small Pt ion at the centre of a octahedron formed by the six Cl ions.

In conjunction with Debye the structure of graphite was investigated, since this substance was not available in large crystals. Only later did we learn that graphite had also been studied by the Braggs. We prepared the samples in a variety of ways and found in all cases the same lattice independently of the crystal habit. The famous six-sided graphite flakes of Groth's mineralogical collection in Munich were kindly put at our disposal and, to our immense amazement, they turned out to be molybdenum trioxide crystals!

The powder method was also invented independently about a year after the beginning of our work by A. W. Hull in Schenectady, and was used by him mainly for the structure determination of metals; during the war the communication with U.S.A. was interrupted.

In spite of the many crystallographic applications which the method suggested, it seemed more suitable to Debye and myself to take up problems of physical interest in preference to those which fell naturally to crystallographers.

It was the atomic scattering factor that interested us most. Evidently the diffraction diagram of a crystal could at the best yield the scattering amplitude of an atom only for the discrete angular values of a crystal diagram, and not the continuous scattering curve down to quite small angles. But even so, by extrapolating the experimental intensities to the scattering angle zero, it should be possible to find the actual number of the scattering electrons in the atom. The evaluation of the LiF diagrams demonstrated with gratifying certainty that both kinds of atoms were present in the crystal as ions: the extrapolation to $\theta = 0$ gave scattering amplitudes which stood exactly in the ration 2 : 10. Lithium had only 2, fluorine however 10 electrons, corresponding to ionized atoms. At the time this was an interesting and by no means obvious result, and I remember well Professor Voigt's enthusiasm when he heard of it.

Another problem of great interest since the time when Planck first quantized the oscillator was that of the existence of the zero point energy $\frac{1}{2}h\nu_0$ of an oscillator. The corresponding motion of the atoms in a crystal, which should persist even at the absolute zero of temperature, ought to show up on an X-ray diagram. Debye, and Waller, had calculated the effect of atomic motion on the diffracted intensity.

Diamond with its high Debye temperature seemed a particularly well suited crystal for demonstrating zero-point motion. Accurate intensity measurements were made and gave an unqualified confirmation of the existence of zero-point energy of the oscillating lattice atoms.

My thesis for becoming a lecturer (Habilitationsschrift) dealt with the determination of the structure and size of colloidal particles by means of X-rays. Up to then colloids could be investigated only with the ultramicroscope, and of course no information could be gained on the structure of the particles. I started with colloidal gold and silver because of these substances samples with a definite particle size could easily be prepared by Zsigmondy's nucleation method. I derived a formula for the broadening of powder diffraction lines with decreasing particle size and succeeded not only in measuring the size of minute particles of only 20 Ångström linear dimension, but also proved that particles of such a small number of atoms already crystallize in the normal structure.—Organic colloids too gave interesting diagrams. Using the fibres of ramie I observed for the first time the interesting fact of fibrous structure, namely the regular arrangement of cellulose crystallites along the direction of the fibre axis. Unfortunately, this paper became very little known owing to its delayed publication in Zsigmondy's book on Colloids.

Later on, in Zürich, I continued working on X-ray diffraction for a while before getting involved in other lines of work, mainly on ferroelectrics and on nuclear physics. This work dealt in part with the determination of the atomic scattering factor. The interest in such measurements was revived once that wave mechanics and the Fermi statistical model of the atom led to more precise predictions regarding the scattering curve of the atom. The electron density distributions were thus determined for metallic lithium, copper and gold. Of even greater interest, however, were the scattering measurements on mono-atomic gases, for instance a mercury gas jet, by which the monotonic character of the scattering curve, i.e. the absence of maxima and minima, was confirmed in agreement with wave mechanics.

Other work, with Wollan and Kappeler, was aimed at establishing the details of the intensity distribution within the Compton line in the scattering by the inert gases. Because the electrons are not at rest within the atom, the Compton line is broadened by a kind of Doppler effect, and it becomes possible to correlate the momentum distribution of the electrons in the atom to the details of the shape of the Compton line. Professor Sommerfeld kindly had the theory for the discussion of our experiments worked out.

Together with Dr. Dubs I carried out comparative scattering experiments on water at rest, in laminar flow and in turbulent motion, the latter at high Reynolds numbers, looking for small differences in the scattering diagrams.

As a matter of curiosity I would like to mention that, while still in Göttingen, I gained the impression that perhaps electrons in passing through a crystal might be diffracted and show interference like waves. I borrowed from Professor Lise Meitner a strong beta ray source and used it for transmitting a beta ray through rocksalt—without, however, finding an unambiguous result. This experiment was prompted by one of those silly ideas which, according to Professor Runge, one should always follow up, without, however, feeling disappointed by lack of success.

Autobiographical Data and Personal Reminiscences

A. V. Shubnikov

I was born in Moscow in 1887 into the family of a bookkeeper and obtained my secondary education in the Commercial School, graduating in 1906. My favourite subjects in school were geometry, physics, and chemistry. My interest in crystallography was aroused early during my years at high school after I attended a popular course in crystallography at the Polytechnical Museum, given by G. V. Wulf for a large audience. After having passed an additional examination in Latin, I entered the Moscow University, Department of Natural Sciences of the Physico-Mathematical Faculty, determined to specialize in crystallography. This particular subject attracted me, and still does so because it combines my three favourite subjects in great harmony and because of its deep underlying philosophy. Wulf started his research work in 1909. He already knew me as one of his rather outspoken students. My career was started with these two phrases: 'Iurii Viktorovich, I should like to work for you...' '...very well, set up the thermostat...,' and I started to work. I still remember the great satisfaction with which I turned up my sleeves in order to carry out my first task. My first paper 'On the Symmetry of Potassium Bichromate Crystals' was published in 1911 in German. In the same year, Wulf and many other professors left the State University and transferred, as a whole group, to the Free Peoples' University organized by private means. In the fall of 1912 the news of Laue's discovery reached Moscow. At that time I was enlisted into military service in an infantry regiment stationed in Moscow. Every now and then I could visit Wulf's laboratory in the evenings or do book-work at home. At that time I was particularly interested in the works of Fedorov and Schoenflies concerning space groups; these works had already had a twenty years history. In those days not one of the crystallographers doubted the lattice structure of crystals, that is, the validity of the law of rationality of parameters which was verified experimentally. Even

Laue himself could not doubt this (at that time he worked in Munich together with the leading German crystallographer P. von Groth). Therefore, Wulf and his associates saw in Laue's discovery primarily the triumph of theoretical crystallography and further the possibility of actual determination of crystal structures. Soon afterwards Wulf acquainted his associates with the formula $\frac{\lambda}{2} = \frac{\Delta\varepsilon}{m}$ derived by him, which in essence was identical with Bragg's well known formula. The unfortunate circumstance that Wulf was formally denied the priority for this discovery left me with a very deep impression. The following year, 1913, Wulf spent organizing experimental work of X-ray diffraction of crystals in his poorly equipped laboratory and I spent it in military service. From my evening visits to the laboratory I clearly remember the figure of the equipment agent shuffling his feet, I still remember his morning coat, his turned up moustache and his sleek blonde head which seemed to have been forced through the opening of the tight, high collar. With this agent Wulf led endless discussions about equipping the laboratory with the most necessary X-ray diffraction instruments from Germany. K. V. Vasiliev helped Wulf to install these instruments. My military service was over by the end of 1913 and for half a year I assisted Wulf during his lectures. It was my duty to prepare all possible demonstrations and experiments which usually accompanied these lectures. Sometimes I took the liberty of interrupting Wulf's speech and personally explaining the phenomena I demonstrated. Wulf jokingly called this 'lectures in duet'. In these lectures we used experiments and models for explaining the phenomena of the diffraction of X-rays when they pass through a crystal. Wulf designed the following model for explaining the distribution law of the spots on the Laue pattern. Several glass slides were glued together along the edges in such a manner that a system of planes belonging to a zone was obtained. The model could be rotated around the axis of this zone. When a narrow light beam was directed on to this model at a certain angle to the axis, the light spots on the screen described an ellipse when the model was rotated.—In the summer of 1914 I was again called into military service and sent to the front near Warsaw in August. After having been seriously wounded I was assigned to noncombat duty at a war plant. There during my free time I was able to work a little on geometrical crystallography. One of my papers from this period, published in the *Izvestiia Akademii Nauk* in 1916, was brought to the attention of Fedorov. Unfortunately, my correspondence with Fedorov concerning this subject, together with extensive unpublished material, was lost. Part of this work was later published (in 1922) in the paper 'The Basic Law of Crystal Chemistry' which discussed the relationship which I discovered between the chemical formula of a substance and its crystal structure. In 1918, after having been demobilized, I returned to my work as Wulf's assistant at the Peoples' University in Moscow, but during those difficult days we could not begin any serious research work. In 1920 I went to Sverdlovsk to work in the Mining Institute which was being established there at that time. My assignment at that higher School was to lecture on crystallography. There, in Sverdlovsk, the center of the stone cutting industry of the Urals, I was personally in charge of the preparation of the thin plates of natural crystals for the courses in crystal optics. I developed an interest in the problems of cutting, grinding, and polishing of crystals and stone. This served as the starting point for many of my papers on quartz, later published in my book *Quartz and its Applications*. In order to teach crystallography in the spirit of my teacher Wulf, I had to create, with my own hands, new models and demonstration devices, repair polarization microscopes, prepare quartz wedges, and sometimes even simple glass lenses to replace those that were used as burning glasses (there was a shortage of matches in the smaller towns). A turning lathe was urgently needed for work of this kind. I was able to find a person who owned a small lathe and was willing to sell it for a million rubles. While I was attempting to obtain this money at the Mining Institute the price of the lathe was increased to two million rubles. I needed the lathe very badly in the laboratory and paid the second million out of my own pocket. My affection for the lathe was not looked upon with approval by my 'geological' superiors who considered crystallography as a science which could be taught with a piece of chalk and wooden models and nothing else. As I had to prepare the crystal optical specimens myself, I worked out the sorting of emery powders according to grain size and became interested in the moiré patterns which appeared when two sieves were superimposed on each other. Comparing the sieves with crystal lattices I determined some rules pertaining to symmetry and to the interference of waves. The outcome of this work were two papers concerning new phenomena of moiré optics. I remained in Sverdlovsk for five years. During this time I succeeded in introducing the teaching of crystallography and wrote a textbook of crystallography, published in a limited edition, for the students. Because it was impossible to order X-ray equipment from abroad, organized research of crystal structures did not materialize in Sverdlovsk. At the invitation of Academician A. E. Fersman I went to Leningrad in May 1925 and accepted a position in the Mineralogical Museum of the Academy of Sciences (USSR). From that time on my work has been in the Academy of Sciences (USSR). At the Museum in Leningrad I had a room of which the floor was destroyed by the flood of 1924. After the floor was repaired and I started to organize crystallographic research, I again had to live through the trials and tribulations of acquiring a lathe, but in a slightly different version. The finances of our government had been put in order, millions of rubles had been converted into ordinary rubles, but to obtain them for the purchase of a lathe was impossible for the same old reason. The generally accepted opinion was that any developments in the realm of crystallography required only a brain but no hands or physical devices, and to use a lathe or milling machine was unheard of. Fedorov was cited as an example and argument; the majority of his crystallographic work was produced by the 'armchair method' alone and without any 'experimental stage settings'. Unexpectedly, the problem of the lathe found a very simple solution. Without much ado Fersman gave me some money from his personal funds and so helped me to purchase a second-hand lathe which must have dated back to the time of Frederick the Great at least. I still remember my great satisfaction and pleasure when, after having undertaken the repairs myself, the lathe was ready for work. The example set by Fersman found followers within the walls of the Mineralogical Museum. In addition to the immediately necessary items, many rather valuable objects were donated by various associates. Later, we gave the Administrative Department of the Academy of Sciences (USSR) many a headache when the objects acquired in this manner had to be listed in the inventory. I am frequently reminded of those wonderful bygone days by my own bookcase in my office at the Institute of Crystallography which still carries the inventory number affixed to it 35 years ago.

In December 1925 Wulf passed away. With his passing, the chair of crystallography at the Moscow State University became vacant; at the same time the chair became vacant also at the Leningrad University. I was elected to both chairs but accepted neither for various reasons, the main one being that I did not wish to relinquish or reduce my work at the Academy of Sciences (USSR). As a result, the chair of crystallography was discontinued in Moscow, and the one in Leningrad was headed by Prof. O. M. Ansheles.

In 1927 I was sent to Germany and Norway for the purpose of

becoming acquainted with the X-ray diffraction analysis work in these countries. This trip coincided with the 'Week of Russian Science' in Germany which made it possible for me to meet Einstein, Planck, Laue, and other leading figures of German Science. At that time I asked Laue to show me his laboratory where he investigated crystal structures; to my great amazement Laue told me that he did not have a laboratory at all. In Munich I had the opportunity to meet Groth (several months before his death), to work for about four weeks on X-ray diffraction of crystals with Rinne in Leipzig, and to contact Schiebold. In Norway I also had the opportunity to acquaint myself with Goldschmidt's X-ray diffraction apparatus and to meet the then little known Zachariasen. On my return trip from Oslo to Hamburg I spent a day in Copenhagen where I visited Bohr's laboratory where a certain amount of optical research went on in the basement.

On my return I started an active campaign for X-ray structural analysis in my own laboratory. As a result I was again sent to Germany in 1929 for the specific purpose of ordering an X-ray diffraction apparatus for studying crystals. After about a year, the huge apparatus consisting of a 10 kilowatt transformer, six enormous condensers, instrument panel, large X-ray tubes, rectifiers, and so on, finally arrived. The apparatus was set up by my fellow laboratory workers under the supervision of a specialist sent from Germany. Soon a young physicist, a recent graduate of the Leningrad University, B. K. Brunovskii, was found who was willing to work with this device and independently specialize in the field of X-ray structural analysis. Thus, after a 15 year period caused by the war and the death of Wulf, research of crystal structures again came to life. But only to be cast aside again because of World War II ten years later. In 1934 the government decided to transfer the Academy of Sciences (USSR) from Leningrad to Moscow. This undoubtedly wise decision had, however, a rather undesirable effect on the progress of the research on crystal structures. The X-ray device had to be dismantled and reassembled in Moscow. Brunovskii's paper 'X-ray Analysis of Catapleiite', the first paper of its kind in the Soviet Union, was published in 1935. In this paper Brunovskii acknowledged the assistance of J. Bernal who, when visiting the Soviet Union, always made a point of visiting our laboratory which at that time, as a smaller research group, was a part of the Lomonosov Institute of Geochemistry, Mineralogy, and Crystallography founded by Fersman. Towards the end of 1937 the laboratory was separated from the Institute and became the independent 'Laboratory of Crystallography of the Academy of Sciences

USSR.' This created the most advantageous conditions for the future development of the research on crystal structures. The work started by Brunovskii was unfortunately terminated by his untimely death. I then asked N. V. Belov to take over and direct the X-ray structural analysis of crystals with the existing device. Belov was then already known in our country as a great expert of investigated structures and, in particular, as the translator into Russian of the book *Kristallchemie*, 1934, by O. Hassel; during its translation he added a great number of his own remarks and 27 figures. Belov did not consent right away, his refusal being motivated by the fact that he had never been concerned with or really interested in the experimental side of X-ray structural analysis. After a certain time our negotiations were resumed; the outcome was that Belov consented to start the work but on the condition that for two years I should not intervene while he was 'learning' X-ray structural analysis. Some time later Z. G. Pinsker joined the Laboratory of Crystallography. Pinsker, having been associated with another institution of the Academy, had already designed his electron diffraction device. Thus, our laboratory had two very promising divisions of experimental research on crystal structures. The war which started in 1941 did not enhance the continuation of this work. When Hitler's armies approached Moscow there were only 15 people in the laboratory. The majority had been evacutaed to the Urals where they were engaged in defense work. Although in charge of the latter, I still found some time to work on some abstract ideas (antisymmetry, piezoelectrical textures). Some of the X-ray structural analysis work, started under Belov's supervision, managed to keep alive. And at the height of the war L. M. Beliaev defended his candidate's dissertation on the structure of the mineral Ramsayite from Chybiny.* In 1943 the staff of our Crystallography Laboratory, inspired by the victories of our army, returned to Moscow confident that the war would end with the complete defeat of fascist Germany.

In 1944 the laboratory had become so well staffed that the question of reorganizing it into the 'Institute of Crystallography of the Academy of Sciences USSR' with all the rights and responsibilities resulting from such a move was contemplated. It was decided to organize this institute within the Department of Physico-Mathematical Sciences of the Academy of Sciences and not in the Department of Geological-Mineralogical Sciences under which the laboratory had functioned during the war. The Institute of Crystallography began its work in

* Chybiny—Mountainous region near Murmansk (Kola Peninsula, USSR) where Apatite and rare minerals are found (translator).

1945. After the end of the war, research on the structure of crystals started to develop rapidly in our and other Soviet Institutes and has continued to do so right up to the present time. In addition to the problems of theoretical and experimental crystallography, I was also interested in the technical application of single crystals: quartz, rubies, Rochelle salt, and others. The result of my interest in crystallization was the publication of my book *How Crystals Grow* in 1935. My books *Quartz and its Applications* and *Symmetry* were published in 1940, *Piezoelectrical Textures* in 1946. In the following year, 1947, the pamphlet *Crystal Growth* appeared. *Optical Crystallography* was published in 1950 and revised in 1958 under the title *Fundamentals of Optical Crystallography*. The book *Symmetry and Antisymmetry of Finite Figures* was published in 1951. *The Investigation of Piezoelectrical Textures* (written in collaboration with my associates) appeared in 1955. Altogether, including papers which appeared in journals, I have published over 250 items concerning different branches of crystallography. The Institute of Crystallography which I helped found now boasts of a large number of people working under its roof. A third large building for the Institute has recently been completed. The Institute has shops equipped with all kinds of lathes. The journal *Kristallografiia* has been published since 1956 under my editorship. This journal is translated into English and is also available in the United States.

Personal Reminiscences

M. E. Straumanis

On 10 January 1925 I graduated from the University of Latvia (in Riga) with the degree of chemical engineer. The University was the successor to the Polytechnic Institute of Riga, and its chemistry department had a very good reputation in Europe. To mention some of its former members: Wilhelm Ostwald (a native of the country) started his splendid career (after studies and assistant years in Tartu, Estonia) at the department and went then to Leipzig, Germany; his successor was the organic chemist, C. A. Bischoff; for a number of years Professor P. Walden, a student of Ostwald, taught various courses in chemistry and became famous though his 'Walden inversion' and the Ostwald-Walden rule; in chemical technology C. Blacher and Glasenapp were well known names.

My major professor and advisor was Dr. M. Centnerszwer (of Poland), a student of Ostwald and Walden. He was well regarded because of his work on phosphorous, hydrocyanic acid, critical phenomena and kinetics of dissolution of metals in acids. I worked in Professor Centnerszwer's laboratory for one year on a metal dissolution problem, for which I received an award from the University and for half a year on my thesis on hydrogen overpotential on various metals. Out of this work three publications in the *Z. f. physik. Chemie* resulted. After graduation I decided to continue the same work and prepare a thesis for a Dr's degree in chemistry. In Latvia, as in Russia and some other European countries, it is not at all necessary to enroll as a student in order to work for a Dr's degree. Thus, I could continue my experimental studies at the chemistry department under Professor Centnerszwer, receive a small fellowship and work the whole day as no courses were required. The laboratory was well equipped (we even had a glass blower and a good mechanical shop at the physics department), and this made working a pleasure and I was happy. At that time I developed an approach to the study of metal corrosion problems by exploring the behavior of simple single cells, e.g. Zn-acid-Pt. This method which earned us the name of the Baltic School (Centnerszwer-Straumanis) in Germany and Russia, was further developed by G. W. Akimow in Moscow and is widely used even now for solving various corrosion problems. After the acceptance of the thesis, comprehensive examinations followed before the department, which consisted of 40 professors and teachers, and then I had to defend the thesis before the professors and the public.

During my study I came to the conclusion that the rates of corrosion and dissolution of metals in various corrosive agents depend largely upon the concentration of impurities and imperfections in the metal and its structure in the case of alloys. Therefore I wanted to acquire more laboratory experience regarding the structure of metals. The Rockefeller Foundation gave me this opportunity by granting a generous stipend for physical metallurgy studies in Göttingen (Germany) in the institute of the famous physical chemist and metallurgist, G. Tammann. Before going to Germany I married Eva Reinhards, who was and still is a great help to me and who firmly stood at my side through all the hard years that followed. We went to Göttingen together in the spring and into a fine country which had just recovered from the grave consequences of World War I.

The research work in the institute of Professor Tammann was mostly qualitative at that time (1927/28). However, this work was of great importance for the German metal industry, as it showed clearly how to solve industrial problems on the basis of scientific research. In this respect I learned much from professor Tammann. The other aspects were less satisfactory, since he liked old methods of research and there was no X-ray equipment in the institute, although professor Tammann in discussions liked to refer to results obtained by X-ray methods. There was a large group in the institute: about 30 from many countries, including U.S.A. and USSR, working for their Dr's degree. There I first met professor Curtis L. Wilson, the present dean of the School of Mines and Metallurgy in Rolla, Missouri. I made many friends there. Upon my return to Latvia, I continued my research, was elected assistant professor, did some lecturing (complex compounds) and teaching in the preparative laboratory. For my corrosion studies I needed single metal crystals and, therefore, started to grow them from the melt and also from their vapour. Several publications on metal crystal growth resulted, which were very favourably accepted. The meshlike structure of impure single zinc crystals fascinated me[1] and so I decided to go once more to Germany to study such structures by X-ray methods.

This time I went to the K.W.I. für Metallforschung, then (1931) located in Berlin, for 5 months with the support of the U. of Latvia and the Cultural Fund. There I encountered a fine scientific atmosphere: Professor G. Sachs had just left the Institute, but there were such famous scientists as E. Schmid (now Vienna), G. Wassermann (Clausthal), W. Boas (Melbourne), J. Weerts (died), and P. Beck (Urbana). Dr. M. Valouch (Praha) was also working there. I started to make my first precision lattice constant determinations (back-reflection method Sachs-Weerts), and realized the significance of such measurements and the disadvantages of the present methods.

Upon my return to Latvia I immediately got busy establishing an X-ray laboratory at the department of chemistry. The necessary funds were obtained from the Rockefeller Foundation and the University and an X-ray machine and equipment was ordered from Dr. H. Seemann (Freiburg i. Br., Germany). In 1932 the new X-ray machine was at work. The next year I did some studies in Professor E. Zintl's and in Dr. Seemann's laboratories (Freiburg i. Br.) and there I also had occasion to meet Professor M. von Laue (Berlin). Back in Riga the question of lattice parameters was put on the research program. I started the work with Dr. O. Mellis (now Professor of Mineralogy at the Univ. of Stockholm), by measuring the lattice constants of silver, copper and of sodium chloride. To eliminate the line shift due to the absorption of X-rays by the samples, they were made as thin as possible by sticking the powder to a thin fiber of Lindemann glass, much in the same manner as we are doing now. I further proposed, for exclusion of the uncertainty in line measurement due to film shrinkage after development, to make two powder patterns of each sample: one regular Debye pattern and one back reflection according to van Arkel. From a combination of the films the effective circumference of the film cylinder could be calculated [2] and from it the number of degrees per mm. This method worked well but not as well as I expected: one reason for the fluctuations of the calculated constants was the inaccurate knowledge of the sample temperature during the exposure and the second was the unequal shrinkage of the separate films. Although I considered the possibility of replacing both films by one but in an intermediate position, I did not grasp the full advantage of this possibility at that time.

Meanwhile, on the death of professor W. Fischer (1934) I was nominated his successor in analytical chemistry, and director of the analytical and X-ray laboratories. In this position I had more time for research than ever before. Students came to work for their engineer's thesis and there were some candidates for Dr's degrees. Among the latter was my former classmate at the University, Alfreds Ieviņš, a very able man (later Professor of Inorganic Chemistry at the U. of Latvia and now at the reestablished Polytechnic Inst. of Riga). We decided to attack the precision determination of lattice parameters again, with the intention of using the results for Ieviņš doctoral thesis. As the work proceeded, the errors endangering precision determinations were gradually removed, new precise cameras were designed (mostly by myself), thermostats for the cameras were built and comparators for film measurement were constructed. Ieviņš proposed again to use one single film in an intermediate position instead of two and this time I realized at once all the advantages connected with this film placement. Since the patterns, powder as well as single crystal, taken on such films had an asymmetric appearance, the method was called 'asymmetric' in the book of Halla and Mark.[3] We decided to keep this name as an appropriate one for our method of precision determination, which, however, consisted not only in the use of films in the asymmetric position, but also in using very thin samples, nearly free of absorption in the high back reflection region, high precision cylindrical cameras, kept at constant temperatures before and during the exposure, and high precision comparators for film measurement. It turned out that large cameras have disadvantages concerning operation and no advantages concerning precision, as compared with smaller ones, 57.4 to 64 mm in diameter. The experience gained in precision determination of lattice parameters was later described in a booklet, published by Springer in Berlin in 1940. Now more than 25 years have elapsed since the first publications of the asymmetric method [4] and I must say that a much better agreement in lattice-constants determination between various laboratories would have resulted (see report of Dr. Parrish[5]) if my colleagues and the camera manufacturers had followed the directions outlined in this booklet.

However, the lattice parameters were not the goal, but only a step toward it. As previously mentioned I was, and still am interested in the basic reactions of attack of metals by corrosive agents. How is the rate of attack influenced by impurities and their distribution in metals, and what is thereby the role of imperfections and point defects in metallic crystals? (At that time very little was known about dislocations.) In this respect a simple equation attracted me very much:

$$M_x = v\, d\, \mathcal{N}/n,$$

where v is the volume of the unit cell, d is the density of the material, n is the number of molecules per cell, N is Avogadro's number, and where M_x may be regarded as the X-ray molecular weight of a compound or element. If there is a difference between M_x and M_y, the molecular weight as determined from chemical analysis, it could be attributed to the imperfect structure of the respective crystalline material. However, in order to use the equation, a precise knowledge not only of v, which could be determined to a high degree of precision by the asymmetric method, but also of N and d was necessary (n is supposed to be an integer, the magnitude of which is easily found). So, I started to determine the most probable value of N and began to think about precision density determinations.[6] However, it took more than 10 years to advance to this point. The reason was that I had to leave Latvia. This was very hard for me, because my research laboratories were very well equipped, there was money for research and I had a staff of well trained and reliable chemists.

The Russian Communists occupied Latvia already in June 1940. All was changed in line with the Moscow pattern. Although the best conditions of life were those of the University staff, the changes were, nevertheless, very extensive: many professors were dismissed, the Divinity department was closed, the administration of the University completely altered. Personally, I could not complain. Due to my scientific activity I was chosen to be professor (head) of inorganic chemistry and I had a staff of 14 assistants who were to be trained as future chemists in this field. I was sent to visit Moscow by the rector of the University and for a second time (in January 1941) at the invitation of Professor A. Frumkin of the Academy of Sciences. I delivered two lectures (in Russian) on lattice constant determination and on corrosion of superpure aluminium, both with results new to Russian scientists. Nevertheless at home, because of all the numerous changes and the general intimidation of the population, the research activity was at low ebb.

In July 1941 the country was taken over by the German armies. Under the German regime the old order was restored only to a limited extent, much to the great disappointment of the population. However, the University could continue its teaching and research activity nearly in the old manner with only a few German supervisors. As time progressed, new difficulties arose (e.g. food shortage) and it became more and more clear that the Russians would return. So, one day in 1944 we had to decide whether or not to stay in Latvia. Why did we leave our native country? There were three main reasons:

1) The German administration ordered the professors and teachers of the University to leave for Danzig (Germany) and gather at the Institute of Technology.

2) The population of Latvia was so intimidated in only one year by the ruthlessness of the Communist regime that it started in masses to leave Latvia.

3) We had four daughters and one son.

Thus, we left all our possessions, including two houses, and began our troublesome exit to Danzig in the fall of 1944. I said good-bye to my laboratories in which in 10 years 100 publications were produced, not only on lattice constant determination but also on dissolution and corrosion of metals, on description of new complex compounds (with Dr. A. Cirulis), on structure of eutectics (with Dr. N. Brakšs) and on various questions of colloid chemistry (Dr. B. Jirgensons, now at the University of Texas, Houston), and analytical chemistry (Dr. E. Eegrive).

Professor W. Klemm of the Danzig Institute of Technology was kind enough to take care of us and in a short time, I received several offers for employment. I chose a position as far as possible to the West, the Institute for Metal Chemistry at the University of Marburg, directed by the 'Geheimrat' Professor R. Schenck. On his suggestion I started research on the sodium tungsten bronzes. The Institute was well equipped, also for X-ray work, but no high-precision measurements could be made there. For my family and myself I received a couple of small rooms in a village at a distance of one hour by train from Marburg. The train was nearly never on time and more and more difficulties piled up toward the close of the war; finally the Institute was damaged by bombs. After the German surrender the Institute was closed by the Allies, but after some months I received permission from the local American Government to continue my research. It was very difficult to work, although Dr. A. Dravnieks (now at Armour Research Foundation, Chicago) helped me. In addition, the political development became very unfavourable for refugees, as it turned out clearer each day that there was not the dimmest hope for the restitution of an independent Latvia. Thus, like hundreds of thousands of other refugees we could not return to our native country, to the great surprise of many American and English authorities. There was only one way out: emigration to the U.S.A. I began studying English, of which I had very little knowledge, and wrote letters to professors of American Universities. My wife was a great help to me in my English studies. Among the offers received, the best one came from Dean

Curtis L. Wilson of the Missouri School of Mines and Metallurgy at Rolla. He still remembered me from the studies with Professor Tammann in Göttingen and offered me a position as Research Professor of Metallurgy with very little class work.

In December 1947 we arrived in Rolla and, with the invaluable help of the Dean and other members of the Faculty, we soon settled in and began slowly to enjoy the advantages of life here: we acquired a house and a few years later a second hand Steinway grand piano, a very fine instrument. Although I have a bad musical memory, I, nevertheless, like to play the piano. The reason may be the deep content and the brilliance of many musical pieces, the beautiful tone of good pianos and a sense of satisfaction felt after a piece has been mastered. Also, while practicing, I think about my research and laboratory problems. My children had no difficulty in high school and they all graduated with good grades and attended college. Of course, there were some things which were less to my liking: many of the graduate students were not too well prepared for research, some articles for publication had to be tailored according to the wishes of the referees and I was astonished by the amount of teaching required from most professors. However, I am pleased to notice that this is now gradually improving.

Finally precision lattice parameter determinations could be made. I compared some of the values obtained with those determined previously in Riga (e.g. Al, Ag) and they agreed within the limits of error. Hence, I could go farther and begin with precision density determinations after an interruption of more than ten years. I also realized that it is more advantageous to determine *n*, the actual number of molecules per unit cell (see equation mentioned above), instead of *d* or M_x for the determination of the perfection of the lattice of a crystalline substance.[7] Good results were obtained with zone refined Al, with dislocation free Si (which showed a perfect lattice) and with oxide phases TiO, Ti_2O_3 and TiO_2 which showed the presence of constitutional vacancies and interstitials. In this work I enjoyed the effective support of Dr. W. J. James, Professor of Chemistry at the School. However, difficulties arose when, against all expectations, the lattice constant and density method indicated the presence of interstitial atoms for deformed Al (hard wire). These results have been reported by me in Berlin, Vienna, Bern, Neuhausen (Switzerland) and Rome, while a Fulbright Professor in Austria at the invitation of Professor H. Hohn. Aluminium is still under examination here, but I hope now that in other places, where dislocations are being investigated, more attention will be paid to the very important experimental precision density determinations. They are difficult and this is the main reason why only few precision determinations are to be found in the literature.

References

1. M. Straumanis, *Z. anorg. Chemie 180,* 1 (1929).
2. M. Straumanis and O. Mellis, *Z. Physik 94,* 184 (1935).
3. F. Halla and H. Mark, *Röntgenographische Untersuchung von Kristallen,* J. A. Barth, Leipzig, 1937, p. 177.
4. M. Straumanis and A. Ievinš, *Naturwiss. 23,* 833 (1935); *Z. Physik 98,* 461 (1936).
5. W. Parrish, *Acta Cryst. 13,* 838 (1960).
6. M. Straumanis, *Z. Physik 126,* 65 (1949).
7. M. E. Straumanis, *Chimia 12,* 136 (1958).

Some Personal Reminiscences

Jean Jacques Trillat

There are moments in one's life when one should be able to stop and look back, just as an alpinist scaling a mountain looks back on the accomplished part of his way and gathers breath. The fiftieth anniversary of X-ray diffraction presents me with such a moment, and it is with a feeling of profound admiration that I look upon the immense work performed from the very first moments onwards by such pioneers as Max von Laue, W. H. and W. L. Bragg, and Maurice de Broglie. Without them, without their basic work, nothing much could have been achieved in the fields of Crystallography and of Applied X-rays.

The idea of having participated, even if to a very minor extent, in the building up of these fields gives one a feeling of elation. For is there a better reward for a researcher than to be aware of his having known the Great Masters, worked under their direction, and, sometimes, of having extended their work along new lines? To me, who belongs to this generation, this is an occasion for re-visiting the past and, in doing so, for giving expression to all my gratitude to those without whom my development undoubtedly would have been quite different.

It happened by pure chance that I was led to take an interest in X-rays. My father who was a departmental manager at the Institut Pasteur inculcated in me the love of research and of scientific rigour, and my entire development was determined by this starting point. This was the first piece of good luck, and, after my involvement in the First World War had come to an end, it made me enter, as a matter of course, the École de Physique et Chimie in Paris for a training in chemical engineering. My father, however, was of the opinion that an engineering diploma should be followed up with a doctoral thesis, and this is why in 1923, at the age of 24 years, I entered the Institut Pasteur with the object of starting research on the synthesis of ephedrine, under the direction of Jacques Trefouel who is the present director of this famous Institute.

I took a lively interest in organic chemistry, and I certainly obtained at the Institut Pasteur an excellent training as a chemist which stood me in good stead throughout my career. The difficulty of the chosen subject filled me, however, with some anxiety. It was at this point that the second piece of good luck occurred, namely my meeting Maurice de Broglie with whom, one day, I had a long conversation which was to decide my scientific orientation.

I became a member of his famous laboratory in 1924 and had the good fortune of remaining there until 1933. I had just got married and was confronted with many financial problems; my wife, to whom I owe so much, encouraged me to accept the assignment, in spite of her being worried by my having to work with these mysterious rays.

Maurice de Broglie, realizing my training was in chemistry, proposed from the start that I should study certain problems related to the structure of long-chain organic compounds (fatty acids, paraffins, soaps). After a few months with him, I had understood the basic notions regarding X-rays and was applying them to the study of fatty acids and their soaps, using the rotating crystal method. At that time very few papers had been published on this subject, mainly those of Alexander Müller and of Shearer, and these served as my starting point.

I worked at that time in a small maids-room in the attics of the private mansion of the Duc de Broglie in rue Châteaubriand. I did everything myself, even cleaning up the room, and if I needed a slightly more elaborate piece of apparatus, I asked Alexis, formerly the Duc's footman now turned mechanician. And all at once results accumulated, in quantity and quality so good that after two years, in 1926, I handed in my thèse de doctorat on the subject: 'X-rays and long-chain organic compounds; studies on their structures and orientations'.

What a wonderful group was at that time working in de Broglie's laboratory: A. Dauvillier, J. Thibaud, R. Lucas, L. Leprince-Ringuet, and later Magnan, Cartan and many more. Most illustrious of all was the Duc's very much younger brother, Prince Louis de Broglie. Every week a colloquium brought us together in the office of Maurice de Broglie where each of us presented his results and we discussed recently published papers. It is thus that I had the privilege of witnessing the emergence of wave mechanics, and this made me take great interest, later on, in electron diffraction and its applications, as well as in electron microscopy.

Never would I be able to describe appropriately the agreeable atmosphere of trust in this laboratory, whose chief knew not only how to assist his pupils by his experience, but also to impart to them, what is perhaps even more important, the right attitude towards Science.

After having finished my degree work I continued working in de Broglie's laboratory. Some simple ideas made me invent new methods for special purposes, such as the method 'de la goutte tangente' where X-rays fall on a hanging liquid drop under varying glancing angles. I tried my hand at numerous subjects which, however, had it in common to stem from chemical problems. To me, X-rays were a means, not an end, a means for studying particularly chemical transformations and surface structures; I had a hunch that from these a great number of applications might result.

The presence of Louis de Broglie in this laboratory gave me the opportunity of learning electron diffraction which had just been shown experimentally to exist by G. P. Thomson in a form very closely resembling X-ray powder diffraction along the lines of Debye and Scherrer. Starting from the apparatus which Maurice Ponte had constructed, I built several more and more refined versions and used them, naturally, in the first instance for obtaining new insights into the long-chain organic compounds. I had the priviledge of seeing von Laue take interest in some of this research, and had several very friendly discussions with him.

This entire first part of my career stood under the sign of Maurice and Louis de Broglie; it was essential for the time that followed. On the other hand, my training as a chemist and engineer gave me from the very start a bearing towards practical and even industrial applications of the new methods, and in that connection I have to say how grateful I am to certain industrial concerns (Kodak-Pathé, Péchiney, Michelin and many others) for having proposed to me, from the very beginning, research on industrial problems, which frequently became to me the starting point of new studies. I recall that in 1928 the Director of the Compagnie Péchinet, Raoul de Vitry, asked me to install X-ray equipment in one of his works (the 'Duralumin') 'for structural control and research', he said, without fully anticipating the kind of result. After a few years the results were such that today there exists in France hardly a single industrial plant which does not use X-ray or electron diffraction and electron microscopy in its laboratory.

The third piece of good luck in my life was a lecture I gave in 1932 to the Society of Industrial Chemistry of Nancy. There I heard that at the Faculty of Science in Besançon a physics chair was vacant and that voting on the candidates was to take place in a few days. I had never felt the urge to instruct; this, however, was a chance, and on the advice of my wife I presented myself as a candidate, and was chosen.

In 1933 I therefore left the de Broglie Laboratory, sad, it is true, but taking with me some of the instruments which the Duc de Broglie was good enough to give me for equipment. In Besançon, I found spacious rooms with nothing in them, and not even supplied with a.c. current.

In a year's time the laboratory began to function, and I spent happy years there up until the outbreak of the Second World War.

This Besançon laboratory of which I was now the director trained a large number of students, some of whom have become masters in their fields: Merigoux, Saulnier, Oketani, Fritz and others. It was there that, together with René Fritz, I built in 1935 the first French electron microscope. There again, I directed my research to problems of mineral and organic chemistry by applying jointly the diffraction of X-rays and electrons. Sometimes these methods proved yet insufficient, and in that case I had to devise novel means, as for instance the registering interfacial tensiometer which yielded many interesting results in the study of lubricants, or microradiography by photo-electrons.

After the Second World War I was appointed at the Sorbonne, first to a chair of P.C.B. (Physics-Chemistry-Biology), then to one of Physical Chemistry and finally to one of Electron Microscopy and Diffraction. I needed a laboratory, and the Sorbonne had none to give me. But thanks to the Centre National de la Recherche Scientifique (CNRS) I was enabled to install myself in Bellevue and to organize there the beginnings of what is today one of the best-equipped laboratories of this Institution.

It was here that I began a new and exciting period of my life; everything had to be built up from the ground in the difficult post-war period. From a researcher I was obliged to become, as it always happens beyond a certain age, a Director of Research, with all the loathsome but necessary duties such a job implies, such as administration, report writing, financing, social problems, etc. I have, however, always managed to save the maximum possible time for looking after the research that is going on in my laboratory, and to take a personal part in it. But very often I feel sad for having no longer the time for experimenting myself, for adjusting instruments, for waiting with anxious expectation in the red light of the dark-room for a diagram to come out in the developer. I consider it now as my primary task to help young people in gaining their full measure, to provide them with the necessary apparatus for their own research, and, first and foremost, to give them subjects to study, to discuss their experiments with them, and to teach them in my turn what I myself learnt from Maurice de Broglie: the proper attitude towards Science.

In looking back over the path travelled, the large number of years spent in the laboratory, the friendship with which famous Scientists have honoured me, I say to myself that there is no finer life than that of a research man, for it implies the most beautiful social function and the greatest rewards one can hope for in this world.

Personal Reminiscences

B. E. Warren

It would be very interesting to know the extent to which chance occurrence leads people into the field which they follow for their life work. While I was a student at the Massachusetts Institute of Technology, Professor Max Born made a short visit, and gave a set of lectures on Crystal Lattice Dynamics. I probably did not understand one quarter of what he said, but nevertheless the set of lectures started an interest in crystal physics.

Having been awarded a small fellowship for a semester in Europe, I chose Stuttgart so as to combine the X-ray diffraction course of Professor Glocker with theoretical work under Professor Ewald. On arrival at Stuttgart, I was quickly sized up as a completely inexperienced individual, and assigned a little problem calculating the indices of refraction of a salt whose structure was at that time in doubt. The calculation was to be done by the method developed in Born's *Atomtheorie des Festen Zustandes*, and I was left to dig it all out by myself. This might sound like fiendish treatment of an innocent young student (and it was), yet I have always felt a deep debt of gratitude to Paul Ewald because he gave me exactly what I needed most at that time. The calculation was of no importance, but what I took away from Stuttgart at the end of the semester was a feeling of confidence, having dug out this small problem completely on my own, I would not hesitate to tackle any subsequent problems. It is generally true that the best teaching is the kind that forces a student to teach himself, and thereby learn to think for himself.

Back at the Massachusetts Institute of Technology as a graduate student, there came another and extremely important chance occurrence. Sir Lawrence Bragg spent four months as a visiting professor giving a set of lectures on X-ray diffraction and crystal structure determination. For his lectures Professor Bragg needed crystal structure models, and since these were not available it was necessary to make them in a hurry. I was lucky enough to win the opportunity to be Professor Bragg's chore boy in charge of building models in a hurry. Since this involved a great deal of discussion and planning, it was an opportunity to get acquainted with Professor Bragg.

As is well known the really great men of the world are friendly, considerate, and completely devoid of pretension. One day Professor Bragg dropped the remark that being entertained and invited out to dinner nearly every night in the week was rather strenuous and tiring. So with the naiveness of youth, I promptly asked if he would like a change such as going to the hockey game that night between the Canadian Mapleleafs and the Boston Bruins. He eagerly accepted and we went, and it was an evening which I shall never forget. First he had to know which was the home team, and from then on he cheered as loudly and enthusiastically for the Boston Bruins as any Bostonian in the audience. Another time were we taking a Sunday hike in the mountains of New Hampshire. After a long hike in the snow we built a camp fire to cook the hot dogs which we had planned for lunch. On opening the packsack, I found that I had brought the rolls but had forgotten the dogs (frankfurters). Professor Bragg still laughs about this when I meet him, he had never supposed that any person could have a face as long and sad as mine when the awful discovery was made.

In addition to building models during these four months, I had the opportunity to work with Professor Bragg on the structure of diopside. An excellent set of quantitative intensity measurements had been made by West, and Professor Bragg had brought the data with him. This was my first contact with structure determination and I probably learned more in those four months than in any other period. It was an exciting experience, diopside was the first of the chain silicates, and the chain structure was completely unexpected.

A year or two later I had the opportunity to spend a few months at the University of Manchester. This was in one of those golden age periods, the membership of the laboratory including names such as Professor Bragg, James, West, W. H. Taylor, Bradley, Ito, Wood and Zachariasen. There was an opportunity to work for a while under James. I imagine that most people who have worked under this great man have had the same experience, it is not until some time later that one begins to realize how much he learned from working with James. At this time complex silicate structures were being done by quantitative intensity measurements using a spectrometer with ionization chamber and electrometer. The crystal was turned in steps to the beat of a metronome, while the electrometer deflection was held at zero by varying a balancing voltage, an operation which required a certain amount of manual dexterity. Present day students who are brought up to think that all experimentation must be done by completely automatic apparatus, would probably be horrified to think of working in this way, and they would probably have trouble in believing that really excellent measurements were made like this in the early days.

After Manchester I settled down at M.I.T. to work on complex silicate structures. This was the period when structures were done with a set of rotation and oscillation patterns. From the cell axes, the space group, and the laws of silicate chemistry, it was usually possible to guess the few likely atomic arrangements. The correct values of the coordinates were then obtained by adjusting the values until there was good agreement in relative intensities for a large number of spots on the oscillation patterns.

Although not yet realizing it, my real interest was in the physical optics of X-ray diffraction and not in structure determination. A chain of events soon brought me into a new field of research. Zachariasen's famous paper suggesting the random network structure of inorganic glasses was of great interest because of my familiarity with silicate structures. Debye's treatment of diffraction by gas molecules, and the Fourier inversion of liquid patterns outlined by Zernike and Prins and applied by Debye and Menke, suggested the possibility of X-ray studies of a non-crystalline material such as glass. This started a programme on the X-ray study of the structure of glass which lasted several years. Crystal monochromated primary beams were just starting to be used, and intensities were measured by film recording and microphotometering. Problems of technique such as the quantitative measurement of diffuse intensities, and the Fourier inversion of diffuse intensities, which were encountered here for the first time, have run through a great deal of the later work in other fields.

An interest in graphite and carbon black started at this time in a purely accidental way. The Fourier inversion method was being applied to the diffraction patterns of glass, and it seemed desirable to try the method out on other forms of amorphous matter. Knowing nothing about carbon black except that it was supposed to be an amorphous form of carbon, an X-ray recording and Fourier inversion was carried out on a sample of carbon black which happened to be in the laboratory. It turned out of course that carbon black is an extremely interesting material with its transition from the amorphous to the crystalline state through the two-dimensional random layer structures. The name 'turbostratic' which has come to be applied to this type of random layer structure was coined by my friend the chemist Nick Milas. Since he is a very competent Greek scholar, I asked him to coin a word for me from the Greek, but after some consideration Nick reported that this job could be handled better from the Latin.

I had never had any interest in metals until the idea occurred that short-range order in binary alloys should produce a modulated diffuse intensity from which the short-range order parameters could be obtained by a Fourier analysis. It was an obvious extension from the diffuse intensity of an amorphous material to the diffuse intensity which results from disorder in a crystalline material. Order-disorder in binary alloys has turned out to involve many interesting problems in technique and in the physical optics of diffraction by imperfect structures. The subject has suggested a large number of measurements and experiments, but curiously enough each investigation seems to open up new problems and indicate new complexities. It sometimes seems as if the study of order-disorder in binary alloys is like climbing a mountain where for each meter we rise in altitude, the peak goes up by two meters, so that although we are making great progress the end of the climb gets ever farther away.

Throughout the fifty years which we are reviewing, X-ray diffraction has never run out of problems, because each problem attacked introduced new questions which needed to be answered. Rather early in the study of short-range order in alloys it turned out that the correction for temperature diffuse scattering was the principal limitation to the accuracy of the short-range order parameters which could be obtained. Hence a study of temperature diffuse scattering was undertaken with the sole idea of learning how to correct for what was a nuisance in the measurement of short-range order parameters. But Laval and James had already shown that measurements of the temperature scattering of X-rays could be used to obtain the spectrum of the elastic waves which constitute the thermal vibrations in a crystal. And so it turned out that the temperature scattering was more than a nuisance, it was of great interest in itself as a method for determining the elastic spectrum of crystals. This seems to be part of the general principle that most experimental nuisances are potentially important tools for the study of some other effect. The nuisance results from some physical effect and it is therefore a means for studying this effect, furthermore the effect is big enough to measure otherwise it could not be a nuisance.

In recent years the work of my laboratory has been largely devoted to the study of imperfections in structure, such as the imperfections resulting from cold work in metals. Diffraction in a cold worked metal presents some very interesting problems in the physical optics of diffraction, and this was surely the motivating reason for going into this field. This last change is probably in line with a general trend in the application of X-ray diffraction, and in our point of view concerning the crystalline state. In the early days we tended to think of crystals as having perfectly repeating structures, and the interest was wholly in the determination of this ideal structure. Today we realize that there are many kinds of imperfections in crystals. For many physical properties it is the imperfections that are of primary importance, and hence a great deal of present day interest is centered around the X-ray studies of the imperfections in crystals.

Personal Recollections

A. Westgren

More than forty years ago, when I had finished my academic studies and had obtained my doctorate with a thesis on the Brownian movement I began to look for a new field of research. One of my best friends, a grandson of the famous spectroscopist Jonas Ångström, after whom the Å. U. has been named, advised me to take up X-ray investigations of crystals as a new line of research. 'This modern discipline,' he said, 'must be particularly profitable to a chemist. There is a tremendous lot to be done in chemistry with these new methods.'

At that time I was already married, and we had a son and I felt strongly the necessity of earning a livelihood. I was fortunate to get employment as metallographer at the recently founded, and already successful, ball-bearing company SKF in Gothenburg. In January 1918 on the long railway trip from Uppsala to Gothenburg I read a textbook on metallography but I also found time to study W. H. and W. L. Braggs' recently published book on *X-rays and Crystal Structure*, which I found to be extremely fascinating and stimulating. What I read there occupied my thoughts a great deal during the following year while I was learning the practice of steel microscopy.

I saw very clearly that the new X-ray methods could be used in metallography with great advantage and that the powder method of Debye and Scherrer was especially well adapted for this purpose. I had heard of this method first hand in 1916 in a lecture by Debye at Göttingen during the time when I was studying the Brownian movement and the coagulation of colloids at Zsigmondy's laboratory.

In the fall of 1919 I wrote a letter to Manne Siegbahn who at that time was still a young lecturer but who had taken over Professor J. Rydberg's duties as head of the physics department at the University of Lund during his illness. I asked Siegbahn if he would allow me to use his recently constructed metal X-ray tube for an investigation of steel and steel carbides. The permission was kindly granted.

An assistant of Siegbahn, a young student with the name of Axel Lindh, later Siegbahn's successor as professor of physics at Uppsala, introduced me to the use of the X-ray tube and together we produced X-ray powder patterns of iron, of steels heat-treated in different ways, and of cementite. The apparatus was primitive and the outcome of the investigations rather meagre. Unfortunately, we used a copper anticathode which, of course, gave highly blackened films the lines of which were hard to discern. Later, we changed to an iron anticathode and succeeded in obtaining better patterns. We made a camera with which it was possible to obtain photographs of an iron wire electrically heated to high temperatures. The structure of iron was determined at about 800°C (β-iron) and at about 1000°C (γ-iron). The former was found to be the same as that of α-iron; γ-iron was found face-centred cubic as was the iron of hardened austenitic steels.

At the beginning of 1921 I was appointed metallographer at the Metallographic Research Institute in Stockholm which had been organized by the steel and metal industries of Sweden. I met Gösta Phragmén there who was certainly a 'connaissance à faire'. He was a young student who had passed his first academic examination and wanted to deal with problems of technical interest during his further studies. Son of a prominent mathematician he had inherited much of his father's theoretical ability. Furthermore, he was an exceptionally able experimentalist, an ingenious constructor, a well-trained glass-blower, well versed in electrotechnics and handy in all kinds of mechanical work. Above all, he was a splendid character, being very modest and always ready to help. He was permanently surrounded by young research adepts whom he guided in their work with an inexhaustible benevolence. He was interested in thermodynamics and played an important role as adviser to metallurgists who wanted to apply its principles to the problems of steel production.

The mineralogical institute of the University, situated quite close to the Metallographic Institute, had a high-tension apparatus and equipment for taking Laue photographs which were used by G. Aminoff, at that time lecturer in mineralogy at the University. Phragmén had worked with him for some months and had found X-ray crystallography to be a most fascinating research field. We decided that together we should try to use its methods on metallurgical problems.

I was fortunate in obtaining grants from the university and some foundations, enabling me to buy another high-tension apparatus and instruments for taking powder and rotation photographs. The X-ray tube and the cameras were built according to our designs (mainly Phragmén's) by an instrument maker in Stockholm and at the workshop of the Metallographic Institute. The X-ray source was a metal tube of the Siegbahn-Hadding type which was evacuated by means of mercury pumps. The gas pressure of the tube was kept constant by means of an excellent capillary tube leakage constructed by Phragmén. This tube worked very reliably. The apparatus was used, at times night and day, during twenty-five years and thousands of X-ray patterns were produced with it.

I gave a lecture on my investigations in Lund at a meeting of Jernkontoret (The Iron Masters' Association) in Stockholm in 1920 and afterwards I had the pleasure to receive from some members of my audience a selection of beautiful carbide and silicide crystals grown in blow-holes of ferrous alloys. They were of great value in the following investigations. To enable us to make well-defined alloys we wanted to have a good vacuum furnace. We experimented some time with a cathode ray furnace and obtained very pure alloys with it but the sudden evolutions of gas from the heated specimens caused such violent fluctuations of the current as to hazard the existence of the high-tension apparatus. So Phragmén constructed a vacuum furnace based on electrical carbon tube heating which functioned very well. We used it for purifying magnesium, manganese and other metals by distillation and produced alloy melts of some hundred grams. I suggested that Phragmén should publish a description of this furnace so that his construction could be of use to scientists in other laboratories but I got a reply that is very characteristic: 'Anybody wanting to make a vacuum furnace must, of course, understand that it should be made somewhat in this way. A description of it is not worth the trouble of writing and the printer's ink.'

A collaborator during the first investigations with the new apparatus was Eric Jette, a jovial student from U.S.A., at least 6′3″ tall (Jette is an old Swedish soldier name and means 'giant'). Together we attacked the structure problems of the copper-aluminium alloys which was hardly a happy first choice as this system is rather complicated. The structure of many of its phases is, in fact, still unknown. During this investigation we found, however, a highly symmetrical phase of a kind that we later came across in many other alloys and with an atomic arrangement that has its analogy in γ-brass.

In the autumn of 1926 A. J. Bradley came from W. L. Bragg's institute in Manchester to take part in our work. He was interested in our research on α-manganese and during his stay here he solved its structure problem. When he returned to England at the end of the year he took with him a number of Laue, rotation and powder photographs of γ-brass and analogous phases and succeeded later in determining how the atoms are arranged in them. This was a real break-through in X-ray metallography and was a great stimulus for us in Stockholm to try to attack structure problems. Evidently, it was not impossible to solve them, even if there were many atoms present in the unit cell.

In 1927 Tr. Negresco of Bucarest who had lately studied in Paris visited our institute and took part in an investigation of the iron-chromium-carbon system. He spoke French which improved our knowledge of that beautiful language somewhat, but, alas, not sufficiently. He is now professor of metallurgy in his native city.

During the years 1921–25 the X-ray apparatus had found a place in the mineralogical institute of the University. Unfortunately, its dark-room could be put at our disposal only for a few years. Later, the loading of the cameras, the development of films and other photographic work, had to be performed in a very primitive, dusty and dirty dark room in a building belonging to the Metallographic Institute far away on the other side of the street. It was, however, always extremely exciting to develop the films and see what they had to tell us, so we willingly put up with this inconvenience. In 1926 all the equipment was moved into the Metallographic Institute. In 1927 I was, however, appointed professor of general and inorganic chemistry at the University and so we returned into its building but this time into its chemical department. A more rationally furnished X-ray laboratory was by and by fitted up there.

In other respects my department was very poorly equipped. If we had not had the benefit of collaborating with the Metallographic Institute the research possibilities would have been bad but, fortunately, the resources of that institute were kindly put at our disposal and my pupils thus had access to furnaces, microscopes and other facilities which they could use for their work. The research was therefore mainly orientated on metallographic problems, especially during the first years.

The number of students wanting to try their research abilities on X-ray crystallographic problems grew very rapidly and it soon became impossible for me alone to guide them in their attempts. I had, however, great help not only from Phragmén but also from several other collaborators who successively mastered the methods. One of them was Harry Arnfelt. Another was Gunnar Hägg who in 1929 got his doctor's degree and was appointed lecturer in general and inorganic chemistry. He was a firm rock in the turmoil of the young people. He went about his task of teaching so thoroughly that he even married one of the most able (and charming) lady-students whom I had confided to his care. Great assistance was also rendered by Lars Gunnar Sillén and Cyrill Brosset. Although retarded by much military service the former was ready with his doctoral thesis in 1940 at an age of twenty-three years. The latter got his degree in 1942.

From 1940 on I had the pleasure of working together with Anders Byström, a very talented student with a great ability for crystal structure research. Already at that time he suffered from consumption but we all believed and hoped that he would conquer his illness. He had time to write a fine dissertation on manganese and lead oxides but in 1952 he was overcome by his disease and died. Another young doctor who died prematurely was Olof Nial who shortly after having written a dissertation on alloys of tin with transition metals in 1945 was killed in a motor car accident. Both these scientists had been appointed lecturers at the university. Their death is greatly to be deplored.

Already before the decease of these collaborators Phragmén died suddenly during an operation in 1944. That was a severe blow and a great sorrow to all of us. He had been nominated head of the Metallographic Institute that was going to be reorganized and modernized and for which a new building was going to be erected not far from the Swedish Academy of Sciences. I had rejoiced in the prospect that Phragmén and his new well-equipped institute would be near the office where from 1943 onwards I had to perform my duties as secretary of the Academy. It might have been possible for me to carry on at least some research work there even if my time was occupied with administrative work. The death of Phragmén was a great loss to metal research in Sweden and to international science. He would have been an ideal head of the new institute where certainly great scientific conquests would have been made. The steel and metal makers in Sweden greatly deplored their loss.

During my first years at the Academy I tried in leisure moments to solve some X-ray problems which I had been engaged on earlier, but my efforts were in vain. My time was too much split up by administrative work. I retired, however, from the secretaryship in 1959 and since then I have got fairly well into the recent development of X-ray crystallography and I have taken up those problems again and, as I am very reluctant to give in, I hope I will succeed in solving them.

My Part in X-ray Statistics

A. J. C. Wilson

It is difficult to set down memories of how an idea was born. After twenty years they become befogged and coloured by the knowledge of later events, and the strict discipline of scientific writing is hard to shake off. Nevertheless, on the occasion of the commemoration of the discovery of X-ray diffraction and at the request of the President of the International Union of Crystallography, the attempt must be made.

In 1942 Yü submitted to *Nature* a paper on the determination of absolute from relative X-ray intensities, and the Editors of *Nature* sent the paper to the Cavendish Laboratory for an opinion on its merit. The method proposed was complex and depended on the use of a set of tables not then available in Britain, but Lipson and I did recommend publication (Yü, 1942). The proposal set us arguing over a practicable method of achieving the same purpose, and a hazy idea emerged that the general level of the intensities of the various reflections from a crystal must depend on the content of the unit cell and not on the details of the atomic arrangement. Lipson (unpublished, so far as I know) suggested calculating the F's for an arbitrary arrangement of the atoms in the unit cell and comparing $\Sigma|F_{calc}|$ with $\Sigma|F_{obs}|$ for suitable groups of reflections, but I wanted a tidier approach. Statistical calculations were in my mind in connection with diffraction by disordered structures like Co and $AuCu_3$ (Wilson 1942a, 1943), and it was soon evident that the appropriate statistical variables to use were the X-ray intensities, not the structure amplitudes. A very short calculation (Wilson 1942b) showed that the mean value of the intensity expressed in units of $(electrons)^2$ is equal to the sum of the squares of the scattering factors of all the atoms in the unit cell. Once obtained, this relation is practically obvious from conservation of energy, and is the first example of the blindness to the implications of what I knew, that has mingled a good deal of self-dissatisfaction with my pleasure in developing statistical methods.

Knowing the mean value of the intensities immediately suggests the problem of determining the probability distribution of the intensities about the mean. I derived what I thought was the general formula for this by an application of the method of induction, and found that it gave approximate agreement for copper sulphate (Beevers and Lipson, 1934). I drafted a paper on the subject, which I remember discussing with Ewald as we travelled to London together for some function or other. When revising the draft, however, I noticed that my argument made an implicit assumption of non-centrosymmetry in the atomic arrangement, and that a centrosymmetric arrangement would give a different result. This was an important finding, but I did not see its importance. Instead I put the whole matter on one side for four or five years, feeling that distribution functions that depended on symmetry were too complicated to bother with. I ought, of course, to have looked at the matter the other way, and have seen that the distribution function provided a valuable way of detecting those symmetry elements that do not cause systematic absences.

Enlightenment came some years later, when I was in Cardiff and responsible for a research student who found difficulty in distinguishing between a centrosymmetric and a non-centrosymmetric space group having the same systematic absences. There could have been many ways out of his difficulty, but while discussing the problem with Rogers I saw my work on distribution functions from the obverse, and fruitful, point of view (Wilson, 1949). X-ray determination of the absence of a centre of symmetry was received with a little scepticism at first—did not all the textbooks say that it was impossible?—and I well remember carrying a couple of slides in my pocket to a conference of the Institute of Physics, without being able to obtain an opportunity of projecting them. The friends to whom I showed them during the intervals hid their disbelief with varying degrees of success.

Statistical methods of determining the absence of mirror planes and rotation axes provide a third instance of blindness to the obvious. In my letter in *Nature* (1942b) I wrote:

'If two atoms are close together in the projection, they ought to be counted as a single atom with atomic factor equal to the sum of their respective atomic factors. · · · certain coincidences can be predicted from the space group only, and allowed for.'

I was then considering the matter in the direction: space group known; can one avoid statistical complications? It was not until many years later, in conversation with Rogers about ridges of high density in Patterson projections, that the reverse question occurred to me:

statistical anomalies detectable; what is the space group? Once the question had been posed it was easy enough for me to write down the factor multiplying the average intensity for the groups of reflections affected by various symmetry elements (Wilson, 1950), and with rather more labour Rogers (1950) was able to prepare the statistical equivalent of vol. I of the *International Tables*.

If there is any moral it is this: systematic work will usually discover the answer to a properly posed question, but discovery of the right questions to ask is a pretty erratic random variable.

References

1. C. A. Beevers and H. Lipson, 1934. *Proc. Roy. Soc.*, A*146*, 570.
2. D. Rogers, 1950. *Acta Crystallogr.*, *3*, 455.
3. A. J. C. Wilson, 1942a. *Proc. Roy. Soc.*, A*180*, 277.
4. A. J. C. Wilson, 1942b. *Nature*, *150*, 152.
5. A. J. C. Wilson, 1943. *Proc. Roy. Soc.*, A*181*, 360.
6. A. J. C. Wilson, 1949. *Acta Crystallogr.*, *2*, 318.
7. A. J. C. Wilson, 1950. *Acta Crystallogr.*, *3*, 258.
8. S. H. Yü, 1942. *Nature*, *150*, 151.

Personal Experiences of a Crystallographer

W. A. Wooster

My first contact with Crystallography took place in 1921 in a small room in Peterhouse, Cambridge, when I was a young freshman. I was being interviewed by a physicist with a view to deciding what subjects I ought to study during the first two years at the University. Having consulted my record he said, 'Well, of course, you will take Physics and Chemistry and then as a third subject I should suggest Mineralogy. It is a nice small subject.' As I had no knowledge of any science subjects apart from Physics and Chemistry, I gladly took my supervisor's advice. From the first lecture I found great interest in the subject. Hutchinson had the view that a lecture ought to be entertaining as well as being informative. We were taught classical crystallography with many practical exercises involving the drawing of accurate stereograms and the calculations of angles between faces and axes in all crystal systems. The optical goniometer in use by the students was of the same design as that described by Wollaston in 1813. On these instruments, which were quite adequate for the purpose, we measured a number of crystals of natural minerals. Hutchinson also gave a course on Mineralogy in which he went through the whole of the types of the mineral species. This tended to be rather like a recitation from a text-book but the special points were well brought out. Some students found descriptive mineralogy very tedious but I was quite fascinated by the development shown by natural crystals. Lewis was Head of the Department and he was remarkable in that he refused to render accounts in the form required by the University. A consequence of this was that funds for running the Department were dependent on students fees. Lewis also maintained an Appointments Registry for placing graduates in schools. This was located in the Department itself though it had no official connection with the University.

In 1927 I finished my Ph.D. thesis, under Rutherford's supervision, on the natural radioactivity of Radium B, C and E. Hutchinson asked me if I would become a University demonstrator in the Department of Mineralogy and regard as my special province the development of courses of lectures and practical demonstrations on Crystal Physics. This appealed to me very much and I gladly accepted. The salary was £125 per annum, though by supervising students it was possible to earn rather more. Hutchinson was a great raconteur and each day he gave tea to all working in the Department. He provided the madeira cake and made the tea and coffee himself and was the life and soul of the party. We all enjoyed his stories of life in Cambridge during his younger days. During one of the demonstrations in 1927 I was going out of the darkened room and one of the class, Nora Martin, was at the same time coming in. We collided and from that moment began the courtship which led to our marriage in 1928. My wife took up research in structural crystallography under Bernal who had just been appointed by Hutchinson to develop X-ray Crystallography in Cambridge. The necessary apparatus was just being developed and the X-ray tubes were continuously evacuated demountable gas-tubes of the design introduced by Müller at the Royal Institution where Bernal had worked before coming to Cambridge. My first piece of research in the Department of Mineralogy was occasioned by a visit of Mr. Alpheus Williams. He had an enormous enthusiasm for the study of diamonds and he normally carried in his pockets little sacks of stones which would be spread out on the table whenever he wished to illustrate a particular feature of a diamond. He had one large flat stone and I remarked that for the study of a possible piezoelectric effect in diamond this would be a good stone. He at once gave me the crystal and I did the measurement. Within the accuracy of my observation no piezoelectric effect could be observed. The courses in Crystal Physics and in X-ray Crystallography which were being built up occupied most of my time. There was not much technical assistance, either in making or running apparatus nor much money to buy materials. One small illustration serves to show how restricted funds were in those days. I wanted a reel of ordinary double-cotton-covered copper wire and I asked Hutchinson if I might get this. His reply was 'There are some odd pieces of wire in the laboratory, would these not do?'

At the beginning of the academic year 1930/1931 Hutchinson retired and Tilley was appointed Head of the Department. There had been an immense discussion going on since 1927 about the future of the Department of Mineralogy. A commission had been appointed by the University and this had recommended that two departments should be formed, namely, one of Crystallography and the other of Mineralogy and Petrology. Largely on grounds connected with the political views of certain people the former was not created while the latter was. The scheme was originally intended to leave research in X-ray Crystallography in its original place and to transfer the teaching in Crystallography to the new Department of Mineralogy and Petrology which was completed in 1931. As time went on the arrangement broke down and a considerable amount of research in X-ray Crystallography had to be done in the Department of Mineralogy and Petrology. Eventually the old department of Mineralogy was demolished to make room for the new Cavendish Laboratory.

One of my principal activities during the period 1931–35 had been the development of an automatic recording X-ray diffractometer. In the design of the instrument I collaborated with my brother-in-law, A. J. P. Martin. This instrument worked with punched celluloid film and recorded photographically. The data for the determination of the structure of gypsum, $CaSO_4 \cdot 2H_2O$, were obtained on this diffractometer. From January to September 1936 I had sabbatical leave in Norway, Sweden and Finland. My wife and ten-month-old son, Tony, came with me. In Oslo, Nora and I worked in Professor O. Hassel's laboratory of Physical Chemistry in the newly-built block at Blindern. We had our own portable X-ray tube and X-ray goniometer and it proved possible to do the work necessary to finish off the determination of the structure of gypsum. Here also I did much of the writing of *A Textbook of Crystal Physics* though this was continued in Uppsala and Helsinki. My first experience of skiing was gained round Oslo and this I very much enjoyed. Siegbahn was head of the Physics Department in Uppsala and much of the work was concerned with the long wave-length X-ray spectra. In Helsinki Wasastjerna was in charge of Physics and actively working on the theory of the physical properties of alkali halides.

After returning to Cambridge I concentrated on finishing *Crystal Phvsics*, and it was published in 1938. Political developments in Germany at last convinced Paul Ewald that he ought to come to Cambridge and he arrived, with none too much time to spare, in our X-ray Crystallographic Department, housed at that time in the Old Anatomy School during the rebuilding of the Cavendish. Ewald founded the discussion group which for many years went by the name of 'The Space Group'.

There were a large number of students in Cambridge throughout World War II and our courses continued in much the same way as before. My wife and I felt that any scientific work we did ought to bear some relation to the war effort. But professors were not encouraging and we sought out industrial crystallographic problems for ourselves. These were connected with diamonds, sapphire, coal and quartz. The diamond problems were all arising from the cessation of Dutch supplies of diamond dies for wire-drawing. Many problems arose concerning the texture of the crystals and we developed the topograph method of studying them. Sapphires had previously been obtained from Switzerland but now they had to be grown locally and problems of fracture and orientation of the crystals arose. Coal was being studied by the B.C.U.R.A.* research laboratories and on the practical and theoretical aspects of the measurement of the refractive and absorption indices of opaque substances they were glad of our help. The supply of quartz for piezoelectric oscillators had largely come from Brazil. During the war many cargoes of crystals never arrived and the supply position was at one time very bad. We were therefore asked by the General Electric Co. Ltd. to undertake the artificial growth of quartz crystals. We can claim to be the first to have produced quartz crystals by hydrothermal means in Great Britain. During the war our work was in the early stages but after the war a useful process was established. With the object of finding out how to make use of electrically twinned quartz we also undertook a research on the control of electrical twinning. This too was only completed after the end of the war. All this war-time work was done in the laboratory at our own home which we assembled in the early months of the war.

The institution known as 'Summer School' has been established a long time in many countries but I think the first Crystallographic Summer School was arranged in the Department of Mineralogy and Petrology in 1943. Together with Henry, Lipson and other members of the staff, we organized a course of lectures and practical exercises which in varying forms and in different places has continued until the present time. A consequence of this effort was the conviction that the material ought to appear in book form. In 1951 the *Interpretation of X-ray Diffraction Photographs*, which was based on the material used in the early crystallographic Summer Schools, saw the light of day.

International relations among scientists have always been a special concern of mine and I gladly supported the Society for Visiting Scientists when it was formed just after the war. One of its first actions was to invite a group of French scientists to come to this country and to give talks on the work they had been doing. At this meeting I heard Laval speak, for the first time, on the work involving diffuse reflection of X-rays from crystals. This subject appealed to me very much and I

* British Coal Utilization Research Association.

forthwith decided to make it a major part of my scientific work in the coming years. In this I was at a great advantage because of earlier experience with X-ray diffractometers. A succession of assistants and research students including, Macdonald, Ramachandran, Hargreaves Lang, Prince, Hoerni, Prasad and Sandor greatly assisted this work.

Two scientific instruments were indispensable for this work; one the X-ray diffractometer and the other the automatic recording microdensitometer. The 1936 design of diffractometer was not sufficiently well adapted for the purposes of diffuse reflection work and many changes had to be made. The production of these two instruments is connected with the development of the war-time work which Nora and I did together. While studying the control of electrical twinning in quartz we found it necessary to make ball-and-spoke models of the atomic arrangements. We made efforts to get other people to take up the manufacture of these models and the Company, called Crystal Structure Ltd., was formed to facilitate this transfer. In fact we found that it requires expert crystallographers to ensure that the models are properly assembled and we eventually took over the manufacture ourselves. Meanwhile I had had to construct a microdensitometer for my own work and we thought this would probably also be wanted by other crystallographers. The diffractometer has now been developed as an automatic setting and recording instrument in response to the invasion of crystallography by automation.

A separate chapter in my scientific work relates to the period 1956–58 when Dr. N. Joel was a research student of mine. It was again Laval who caused me to take an interest in the work on the breakdown of the classical theory of crystal elasticity. Joel and I studied the elastograms of ammonium di-hydrogen phosphate and found, as Zwicker had previously asserted, that the figures could only be interpreted on the assumption that velocities of transverse waves in the crystal, which must be equal on the classical theory, are in fact unequal.

I had originally been appointed to establish a course in Crystal Physics. The 1938 text book on the theoretical aspects was a consequence of this and so was *Experimental Crystal Physics* which appeared in 1957. This was subsequently translated into German, Russian, Polish and Chinese. After this I began to write *Diffuse X-ray Reflections from Crystals* based on the research work done since 1948. This book should appear early in 1962. I retired from my University post in 1960 because I wished to be able to devote a greater portion of my time to scientific work. This was only possible because of the laboratory Nora and I had built up during the war and because of the small factory making a limited number of crystallographic scientific instruments.

Personal Reminiscences

JEAN WYART

How did I come to crystallography, in 1925, when I had still to learn everything about this science? Certainly not by special inclination. Chance, casual circumstances which lead human destiny, made me a scientist and a crystallographer.

Living in a coal mining country, in the north of France, where only one primary school existed, I would have been at the end of my studies on reaching my twelfth birthday, and would have become a workman, like my father, had not war and the fire which burnt down our house, driven us out of our village, in October 1914. Finally we came to settle at Abbeville where there was a secondary school. I obtained a railway scholarship and thus studied for becoming a railway engineer. Consequently I prepared the competitive examinations of admission to the highest schools which, in France, play a most important part besides the Universities. Having passed in 1923 both the entrance examinations of the 'Ecole Polytechnique' and the 'Ecole Normale Supérieure', I chose the second one which leads to educational careers, and thus put a definitive end to the railways. The 'Ecole Normale Supérieure' is in the Latin quarter of Paris and the students, free of all worries for the necessities of life, attend the courses of the Sorbonne and some additional lectures at the school; most of all, they enjoy the advantage of fine laboratories and of the daily contact with their professors. I spent there four marvellous years.

The third year of the studies at the Ecole Normale is used for prepararing some laboratory research work, the conclusion of which must be a memoir called 'Diplôme d'Etudes Supérieures', which has to be maintained in an oral examination at the Faculty of Sciences. At the end of my second year, my professors of physics, Henri Abraham and Eugène Bloch, had offered me to work at the laboratory of the School. I had accepted when Eugène Bloch asked me to go and see a professor of mineralogy of the Sorbonne, Charles Mauguin, who wished somebody to help him out with some calculations. This is how, with one of my friends, a mathematician, I entered for the first time, on a spring day of 1925, a laboratory where I was to spend my whole life. Charles Mauguin wanted to draw up numerical tables in order to interpret his rotating-crystal patterns. He used a cylindrical camera and applied formulas of spherical trigonometry to infer the indices of the reflecting reticular planes from the position of the X-ray spots. The calculations required by Charles Mauguin were tedious, long and laborious, and a month later I went back to his laboratory to tell him that we had neither means nor time to finish his numerical tables. He then gave me the news that these tables had become obsolete. It had occurred to him to bring in the reciprocal lattice and this made the graphical interpretation of X-ray patterns immediate. He has described in a detailed memoir, published early in 1926, the methods of interpretation of X-ray patterns which are still in use nowadays. I think that this idea had occurred to him without having brought in Ewald's work. He used for a long time, as most of the crystallographers do, the polar lattice introduced by Bravais in 1848, in order to solve the crystallographic computations based on goniometer measurements. It was at the time of writing his memoir that he noticed that the construction well known now by every student under the name of 'Ewald construction' appeared in fact in an article by Ewald, published in 1921 in *Zeitschrift für Kristallographie*, as well as in his book, *Kristalle und Röntgenstrahlen* published in Berlin in 1923. Like all other students of physics and chemistry at that time, I was absolutely ignorant even of the elements of crystallography. That day, Charles Mauguin showed me his X-ray equipment and his X-ray rotating-crystal patterns, but most of all I was amazed by the Laue diagrams. Mauguin possessed an incomparable charm and had a wonderful talent to make difficult relations perfectly clear. He explained to me the connections between the crystal lattice and the reciprocal lattice, its application to the interpretation of X-ray patterns, and the rose-like arrangement of spots on the Laue diagrams. On that day I spent three hours with him and came back to the school enthusiastic for crystallography. It was rather difficult to explain to Henri Abraham and Eugène Bloch that I preferred to spend my year of research work under the direction of Mauguin. When arriving in October 1925 at the Sorbonne mineralogy laboratory, I first was somewhat bewildered. Compared to the physical and chemical laboratories of the Ecole Normale, this one was extraordinarily clean and silent. Only Frédéric Wallerant and Charles Mauguin worked there, and two aged laboratory servants who spent all their time at cleaning. Mauguin had asked me to make the synthesis of a basic zinc acetate which had just been discovered, in order to study it with X-rays. I had to distil acetic acid anhydride under vacuum; I soiled a very clean room and the servants, upset by this unusual trouble, were not at all pleased. I hastened back to the Ecole Normale to prepare there my fine almost perfect octahedra of basic zinc acetate. Above all, I found there my friends, the young teachers who were training us and renewed our brisk discussions. I had read the works of Laue and of the Braggs; I was enraged to see that none of my friends believed in the reality of the arrangement of atoms such as the Braggs proposed it. The Bragg's structure, they thought, was only a clever hypothesis to explain the X-ray diffraction, just as if two sorts of atoms existed: the atom of the chemists and the atom of the Braggs. The majority of the chemists don't think much of such research and one of them asked me if I would really enjoy playing at cup-and-ball with atoms.

By stressing this point, I wish to underline the state of mind which prevailed in our laboratories in 1925. Our country, terribly ravaged by the war (almost all the young scientists enrolled at the Ecole Normale during the years 1911 to 1915 were killed) was in this field very far behind the others. Mauguin was almost the only one in France who wanted to use X-rays for chemical aims. Meanwhile, amongst the students of the Ecole Normale, Ponte had just come back from the Royal Institution in London where he had spent a year with Sir William Bragg; my friend Marcel Mathieu had followed him and was still there in that year.

When I came back to the Sorbonne in January 1926, I took my first photograph of an oscillating crystal. I had chosen a too wide oscillation angle for this complex chemical compound and I still remember the thousand diffraction spots on the film.

Wallerant and Mauguin had asked me to attend the lectures on mineralogy. I was not at all prepared for natural sciences, but I thoroughly enjoyed listening to these two remarkable teachers who mainly stressed the crystallographic aspect of mineralogy. We were only about thirty students, most of them geologists or chemists; and yet it was the only teaching of crystallography that existed in Paris. It is not suprising that almost all the students in France were completely unaware of the importance of the discoveries of Laue and the Braggs.

These few months spent in the mineralogy laboratory have determined my scientific career. I was the only research student and Charles Mauguin gave me a great part of his time. Rapidly I learnt all the elements of crystallography and the X-ray techniques; Mauguin, with his wide-spread learning, explained to me also the current event of that time, atomic physics, which occupied his mind. When I left him in July 1926, in order to prepare the competitive examination of 'agrégation' of physical sciences for the next year, and after that to do my military service, I had firmly made up my mind to come back to work with him. This I did as early as October 1928. I came back to this laboratory of the Sorbonne which I never left since. The following year Stanislas Goldsztaub and Jean Laval also came to work under the direction of Mauguin. It was Frédéric Wallerant who proposed to me, as subject for my Ph.D. thesis, the study of zeolites. He was a short, skinny, limping elderly man; with a severe and stern countenance, yet he was great-hearted and I was very fond of him. Just as he arrived, at nine in the morning, I used to hear the noise of his walking-stick. Monsieur Wallerant was coming in for a few minutes talk with me. This would often concern some matter near to his heart. He told me how impressed he had been by the discovery of Laue, how he had unsuccessfully tried to repeat it with his friend, the well known physicist Villard. He showed me the insufficiently powerful tube and chiefly the diaphragm with a too small aperture for an X-ray beam of sufficient intensity. He used to say rather bitterly that he had spent much work on crystallography, but had come or too late or too soon, for the main discoveries had already been made by Haüy, Bravais, Sohncke, Schoenflies, Fedorov. He was deeply convinced that X-rays were going completely to transform this science; but he did not believe in atomic structures. He saw in X-rays only a particularly powerful goniometer, which disclosed the lattice by a kind of statistical effect. He urged me not to persist on childish attempts of determining the atomic structure of chabazite but rather to use X-rays as a convenient goniometer for studying the modifications of the lattice brought about by the replacement of calcium atoms by other cations and by the diffusion of zeolitic water by temperature action.

When my thesis was finished, I sent a copy to Georges Friedel, professor at Strasbourg who some years earlier had done excellent work on zeolites. He wrote me a long letter with some compliments and some criticism. He congratulated me for a point which I had considered as secondary and of lesser importance. I had shown that the silica skeleton left by an acid attack on heulandite, though it retained some optical anisotropy, is amorphous and is not at all a silicic acid characteristic for heulandite as some chemists and some mineralogists asserted. Regarding the arrangement of atoms in chabazite which had cost me a tremendous work, he let me understand the uselessness of such an effort. Georges Friedel, like E. Mallard, regarded the crystal as more complex, and the crystal symmetry as having only a statistical significance and as resulting from twins of very small individuals, most of triclinic symmetry.

While I was working on my thesis, I was led to collaborate for the first time in an international enterprise of crystallography. Charles Mauguin was a member of a Committee of crystallographers who, under the chairmanship of von Laue and Sir William Bragg, had been commissioned to draw up the *International Tables for the Determination of Crystal Structures* published in 1935. Mauguin entrusted me with the drawing of the symmetry groups other than the quadratic groups which had been taken over by Astbury in Leeds. It was an excellent exercise for me; in the evenings, after supper, I used to work with the help of a designer, in a quiet and silent Sorbonne. Everything went very well up to the cubic groups and the same drawings, hardly modified, are still to be found in the new edition of the *International Tables*. The representation of the cubic groups was more difficult. Mauguin had long discussions with me and at last I gave up this work which had given me the opportunity of meeting at the laboratory Ewald, Bernal and C. Hermann.

Frédéric Wallerant retired in 1933 and I became lecturer attached to the chair held by Mauguin. Being fond of teaching, I was very sorry to see that our lectures were so little attended; therefore, at the beginning of each academic year, we began sending a letter to all the teachers in physics and chemistry calling their attention to the physico-chemical importance of crystallography and asking them to advise their students to attend our lectures. This propaganda was soon effective and the number of our students started to increase steadily.

I wish to say some words about crystallography at the International Exhibition of Paris of 1937. The idea had occurred to Jean Perrin to create a 'Palais de la Découverte' in order to display the great strides of science to the general public. There was a special room for crystallography; we had promised some fine minerals which would be displayed next to big models representing their atomic structures and the corresponding X-ray photographs; we had also offered some fine experiments on liquid crystals. We were late, particularly for the large models. I remember having spent many a night in supervising the work of solderers from the inland water transports in order to prepare, from thousands of brass balls, the models of sodium chloride, of left and right quartz, mica, felspars, etc.... As the 'Palais de la Découverte' still exists, one can still look at these models which gave me so much trouble.

The war years, from 1940 to 1945, were very difficult at the Sorbonne. The laboratories were not heated and the only electric heating was for the mercury-arc rectifiers which we had on our high voltages. The only way for us to carry on our work was to organize a mechanical workshop where we built ourselves most of our devices, X-ray tubes, autoclaves for hydrothermal synthesis, etc.

In Paris we were actually cut off from the rest of the world and one must have experienced such an isolation to realize how necessary in research work is a minimum of information. The South of France was more favoured and some scientific periodicals arrived there. We organized a secret passing on of these periodicals and I undertook the responsibility, at the National Centre of Scientific Research (CNRS), to create and run a Documentation Center which published an Abstract Bulletin of the received periodicals and supplied microfilms of the articles to the scientists who asked for them; since then the Documentation Centre has spread out to a considerable extent and it has cost me a great part of my time.

The greatest event as soon as the war was over, was the meeting of the X-ray Analysis Group of the Institute of Physics which was held at the Royal Institution, London, in July 1946. Crystallographers form a large family and we were happy to meet again. Sir Lawrence Bragg was the chairman, Laue was present. On this occasion the foundation of the International Union of Crystallography was laid, and I was a member of the small group who met little later at the Cavendish Laboratory in order to work out the first statutes. The Union has much expanded since and the number of the participants to the General Assemblies is ever increasing. At the laboratory of mineralogy of the Sorbonne of which I became the director on the retirement of Mauguin in 1948, the number of research workers has become too large. We are far beyond the stage when we were obliged to ask our colleagues to send students to our lectures. More than six hundred of them regularly attended our teaching and passed the examinations in 1960. Laue and the Braggs have made of crystallography a flourishing science.

Reminiscences

Ralph W. G. Wyckoff

It is not easy to write of what happened fifty years ago because the things remembered are so rarely what we now imagine as most important. In my case crystals were objects of early fascination but in memory this interest was aesthetic and not at all alloyed with what might underlie their beauty of form. Though such childhood concerns undoubtedly colour what we do later in life, my choice of crystal structure as a doctoral thesis was purely accidental. I had been proposing to see how J. J. Thomson's investigations of positive rays could be turned to chemical advantage. But it happened that S. Nishikawa, one of the first to follow the Braggs in determining atomic positions, was spending a couple of war years in the physics department at Cornell; and the chairman of my graduate committee thought I should take advantage of his presence to learn something about the then-new methods of X-ray diffraction and their chemical potentialities. Nishikawa and I became friends and he agreed to guide me in working through a couple of structures—of sodium nitrate (I have forgotten why) and of cesium dichloroiodide (chosen because of current interest in the chemistry of cesium and because this compound had a mystery which it has since lost). It was from Nishikawa that I learned how to prepare X-ray spectra and Laue photographs and to use their data to select between possible atomic arrangements. He also brought a knowledge of the theory of space groups, acquired from a Japanese professor who had been one of the few to work in Germany during the 1890's when the theory was being created. In looking back one expects to remember some details of the inevitable discussions through which these things became clear; but in fact what remain vivid are a laboratory consisting of a decrepit medical X-ray machine and homemade instruments in lead-covered wooden boxes, and trips together over the countryside in a still more decrepit Ford—or lasting insights into the very different ways an easterner and a westerner looks at life's basic problems.

On getting my degree in 1919 I went to the Geophysical Laboratory to begin there an application of X-ray diffraction to minerals. Much of the first two years was spent in getting together the Analytical Expression of the Theory of Space Groups and in trying to have it published. It has always been a matter of gratitude to remember how, in the end, it was Dr. Woodward, not as President of the Carnegie Institution but as a mathematician, who intervened to authorize its publication. About then Ellis, who with Burdick had published the first structure to be obtained in the United States, stopped on his way to California. He arranged for Dickinson, then transferring to the California Institute of Technology, to spend a short time in Washington familiarizing himself with the use of Laue photographs and space-group results; and out of Dickinson's visit came the year I spent in Pasadena in 1922. During that time Bozorth, working on his thesis, Dickinson and I turned out a number of structures. In those days one could find excitement in success with the simplest of crystals but more vivid than these excitements is the memory of the still-unspoiled beauty of California's mountains.

With the completion of my *Structure of Crystals* in 1924 came the freedom to spend summers in Europe. At this time recovery from the war was fostering contacts across the ocean and the costs of travel were low enough to permit two or three months abroad each year, spent partly in visiting laboratories and partly on vacation. I had been corresponding with Professor Schoenflies since shortly after the war ended and these trips gave the chance for visits to him in Frankfurt and in Austria. It was also during one of these summers that Ewald organized what must have been X-ray diffraction's first international conference in his mother's studio on the Ammersee. Meetings were not then the almost daily occurrences that they have now become and this one had all the advantages given by its small size and informal character. A snapshot taken at it shows von Laue, Darwin and W. L. Bragg, Mark and H. Ott, Waller, Fokker, Debye and one or two others I no longer recognize.

There must be few of us who do not find their interest shifting as their work develops. I found that, as the power of X-ray methods grew, the possibility of their ultimate application to biological systems became increasingly alluring. When, after eight years at the Geophysical Laboratory, the opportunity unexpectedly arose to go to the Rockefeller Institute to start X-ray work on such substances, it therefore seemed the obvious thing to do. Studies there began with aliphatic-substituted ammonium salts and gradually extended to amino compounds. During most of the ten years in New York Corey worked with me and when the work was halted in 1937 he took with him to Pasadena a partly finished study of glycine; the structures of this and other amino acids he has since published are an essential part of the understanding of proteins that is now emerging. During the last years at the Rockefeller Institute we were developing air-driven ultracentrifuges to purify and crystallize the proteins and other biologically significant substances from which we were beginning to get X-ray patterns; and I found that with these ultracentrifuges animal and plant viruses too unstable for chemical treatment could be isolated. From one of the viruses thus purified Beard and I prepared an effective vaccine against the sleeping sickness of horses which was at that time epidemic in the United States. When I had to stop work at the Rockefeller Institute, the job that presented itself was in industry and involved the attempt to prepare large amounts of such a vaccine. It turned out that a satisfactory vaccine could be made without costly ultracentrifugation and the next two years were spent in devising large scale methods and making it in million-dose quantities. The work was exciting because of the dramatic way in which widespread use of our vaccine effectively ended a disease which the year before had killed more than 170 000 horses and was beginning to take a considerable toll of human victims. It was scientifically important in providing the first successful 'killed' vaccine against a virus disease and the first vaccine of this sort to be manufactured in chicken embryos. By the time this job was finished the war was imminent and we turned our experience of embryos to the making of a rickettsial vaccine against epidemic typhus fever. After preparing several million doses for the U. S. Army, my laboratory undertook the large scale freeze-drying of human blood plasma that the Red Cross was beginning to collect for the armed forces. A new apparatus was developed for doing this and two plants incorporating it and each processing individually more than a thousand bacterially sterile bleedings per day were built and operated. When this was over I returned to academic life and during a brief stay at the University of Michigan began using the electron microscope to visualize the virus and other protein molecules earlier investigated by ultracentrifugation. It was here that Robley Williams and I discovered the advantages of metal shadowing for this purpose. From that time till I retired two years ago (with the exception of two years spent in the Foreign Service as Science Attaché in London) my work at the National Institutes of Health, with frequent visits to the virus laboratory of P. Lépine at the Pasteur Institute in Paris, has been largely devoted to the electron microscopy of virus and other macromolecular particles and to observing how some of these are produced in living matter.

Throughout the 25 years since leaving the Rockefeller Institute I had no facilities for continuing the determinations of crystal structure which were a primary concern of early years. It was in order not to lose complete touch with the subject that I started, in odd moments during the war, the compilation of data that has since been appearing as *Crystal Structures*.

As our electron microscopic methods have grown more powerful it has become increasingly possible to employ them as a new way to examine the molecular and atomic order in crystals. With this in view I showed a number of years ago how the individual molecules could be seen in crystals of virus and other proteins. The molecular marshaling proved to be substantially what one would expect, but there was a deep satisfaction in actually seeing what X-rays predict and in effecting a meeting at the molecular level between direct observation and the elaborate deductions that relate these data to the order in nature producing them. It is unlikely that photographs such as these will greatly aid in our determinations of crystal structure but they can give a new insight into the anatomy of crystal faces and a direct picture of crystalline perfection. With microscopes that now attain resolutions of almost atomic dimensions such studies are being made of crystals with small molecules and of metallic crystals where effects due to individual atoms are sometimes seen. At this level direct visualization and diffraction commingle to furnish a technique whose evidence about individual atoms and molecules can complement the statistical information of X-ray diffraction.

This high resolution electron microscopy is, however, only for those whose eyes are still young and with my retirement from the National Institutes of Health personal research in this field has, like the determination of crystal structures, become an affair of the past. There are, however, applications of X-rays still to be made which do not require either the instrumental and computational elaborateness of modern structure determinations or the sensory acuteness of electron microscopy. Soft X-rays seem to offer such possibilities and I am now occupied in their reinvestigation using the various modern experimental procedures that have so greatly furthered what can be done in the ordinary X-ray region.

PART VIII

The Consolidation of the New Crystallography

The Consolidation of the New Crystallography

Before the First World War the number of scientists who were engaged in X-ray diffraction and crystal structure analysis was small: perhaps ten in Germany, ten in England, five in France, four in Japan, five in USA, two in Russia and five in the Netherlands and the Scandinavian countries. In the years of the war, these numbers did not materially increase, although the field itself expanded by the invention of the powder method, improved apparatus design by closer attention to the X-ray geometrical optics, and by the dynamical theory.

After the war, from about 1920 onwards, the number of research workers increased, at first mainly in England, but soon also in USA and the Scandinavian countries. The determining feature for this increase was the presence of scientific leadership, such as that offered by W. H. and W. L. Bragg in England, Siegbahn and Westgren, later Phragmén, in Sweden, Wyckoff, Pauling and A. H. Compton in USA. In all these instances 'schools' developed quite naturally, and the advance, both with respect to subject matter and to numbers of workers, was rapid once a nucleus was formed. The interest in the particular crystal, which was most pronounced in W. L. Bragg's large series of structure determinations, soon brought the subject close to the border between physics and chemistry, and wherever this was the case, it flourished. It is interesting to remark how few structure determinations came from Germany, where X-ray diffraction belonged to physics, and the great majority of chemists, proud of the achievements gained by their traditional methods, for a long time did not tolerate the intrusion of a physical method in their research or their teaching. This lack of structural results contrasts sharply with the important extension of physical methods originating in Germany: powder diagrams, layer-line rotation and oscillation diagrams, the Weissenberg and Schiebold-Sauter goniometer method, wide-angle diagrams and micro cameras; Niggli's adaptation of the theory of space groups from the form in which Schoenflies had left it to a more suitable shape for practical work also belongs here. These methods, now widely used everywhere, found a very slow acceptance in England, where the merits of the Bragg spectrometer were exploited to the full; conversely, except for early work of Mark, the spectrometer was little used during the twenties in Germany.

Much of the personal contact between the workers in the new field was gained by visits to other laboratories for shorter or longer periods. The two British schools at the Royal Institution and in Manchester were the focal points to which especially American, Canadian, French and Dutch adepts converged. In Germany, Mark and Polanyi's laboratories in Dahlem, Rinne-Schiebold's institute in Leipzig and also Glocker's and Ewald's schools in Stuttgart attracted foreign students, and the same holds for Mauguin's laboratory in Paris – but all this on a very much smaller scale than what the British laboratories had to offer. Later on, the seedlings of the two main schools, such as J. M. Robertson's school in Glasgow, G. E. Cox' in Leeds, Bernal's and Lonsdale's in London, joined with the older ones in the dissemination of the experimental and theoretical methods of crystal analysis. An important part in making possible the international exchange of young scientists was played by several Foundations, in particular in the 1920's by the International Education Board (Rockefeller Foundation).

It was a fortunate circumstance that many of the workers in the new field were of a rather homogeneous age group, and so found it easy to get to know one another and to discuss their problems. Sir William Bragg (*1862), Charles Mauguin (*1878) and Max v. Laue (*1879) were the elders; there followed an intermediate group – born 1888–90 – with Andrade, W. L. Bragg, Darwin, Brentano, Glocker, Ewald and Westgren among them, and this led over, via James and Bijvoet, to a large group born between 1895 and 1905 whose main training took place after the end of the war. This last group contains the principal pupils of Sir William at the Royal Institution and of W. L. Bragg in Manchester, besides the veterans in USA like Wyckoff, Pauling, Patterson, Warren, Buerger, Donnay and others, and leading early figures in other countries, such as Mark in Germany, Mathieu and Wyart in France, Hägg in Sweden, Nitta in Japan.

Whether the reason be the humane and friendly example set by the elders, the similarity of age and interests, or the immense field of exploration open to everyone – the fact is that the crystallographers soon became a very friendly crowd, from laboratory to laboratory and from country to country.

A first step towards the international consolidation within the new crystallography was taken by Ewald in 1927 when he proposed to the other editors and the publisher of the *Zeitschrift für Kristallographie* at a meeting held in Romanshorn on Lake Constance to 'internationalize' the journal by allowing papers in English and French besides in German. The ready acceptance of this proposal by the editor-in-chief, Niggli, the co-editors Laue and Fajans, and the publisher, K. Jacoby, was followed by a similar reaction on the part of the authors whose papers had up to then been translated for publication. The unique role which the *Zs.f. Krist.* had gained in the pre-X-ray days under Groth as *the* specialized journal of crystallography was preserved by this step. In later years, when many renowned German scientific journals had to bow to the racist decrees of the Nazi 'Chamber of Culture' the *Zs.f. Krist.* could refuse to recognize their applicability to a journal of international character and having a Swiss editor-in-chief.

The further history of the *Zs.f.Krist.* need be mentioned only briefly. Like all other scientific journals in Germany the *Zeitschrift* had to close down towards the end of the war; the last issue was number 1 of Vol. 106, dated February 1945. Thus it came that the second number of Vol. 106 appeared in October 1954, after a break of nearly ten years, just in time for Laue's 75th birthday. From then on the *Zeitschrift*, under the editorship of G. Menzer, M. Buerger, F. Laves and I. N. Stranski, has developed very well, continuing the tradition of an international authorship. The *Strukturbericht*, which in the pre-war days of the *Zeitschrift* had been a supplement to its volumes, as a service to the consolidation of the new crystallography, had meanwhile become an independent undertaking under the title *Structure Reports* which was not connected with any journal.

As a second step in the international consolidation, the *International Tables* (*Internationale Tabellen zur Bestimmung von Kristallstrukturen*) should be mentioned. The following situation led to their planning. By the late 1920's a number of books had appeared on X-ray diffraction. There was the Bragg classic *X-rays and Crystal Structure* (1st ed. 1915, 4th ed. 1924), which was in the main a summary of the work of the Braggs in classic simplicity but without any attempt at being comprehensive of the whole field of work. The second book to appear was Ewald's *Kristalle und Röntgenstrahlen* (Springer 1923), followed but little later by Ch. Mauguin's *La Structure des Cristaux* (Paris 1924). Wyckoff published *The Structure of Crystals* in 1924; Mark in 1926 *Die Verwendung der Röntgenstrahlen in Chemie und Technik*; G. L. Clark in 1927 *Applied X-rays*, and in the same year R. Glocker's *Materialprüfung mit Röntgenstrahlen* appeared. A. Scheede and E. Schneider were preparing two volumes *Röntgenspektroskopie und Kristallstrukturanalyse* which came out in 1929. Besides, there were comprehensive Handbook articles, by Ewald (*Hndb. d. Physik*, Springer 1927, Bd. 24) and by H. Ott (*Hndb. d. Experimentalphysik*, Lpz. 1928, Bd. 7). A number of these books aimed at including all the material that would be needed for practical work, and this led not only to the repetition of lengthy wave-length and space-group tabulations, but, worse, to confusion caused by different numerical values and designations. This was worst in the case of space groups. Niggli, in his *Geometrische Kristallographie des Diskontinuums* (Bornträger 1919) had taken over the Schoenflies nomenclature. There existed, among the mineralogists, a wide variety of designations of the 32 crystal classes, and this meant that there were tendencies in diverging directions for the closer attachment of space-group symbolism and terminology to those of the point groups (e.g. Rinne and Schiebold). Soon after Niggli's mode of representing the space groups there appeared Wyckoff's much used tables *The Analytical Expression of the Results of the Theory of Space Groups* (Carnegie Institute of Washington, 1922; 2nd ed. 1930), and in 1924 Astbury-Yardley (Lonsdale)'s *Tabulated Data for the Examination of the 230 Space Groups by Homogeneous X-rays* (*Phil. Trans. Royal Soc.* A 224) which used a pictorial representation of space groups very different from Niggli's. One of the perplexing features for the user was that neither the axial directions nor the position of the origins were always the same in these descriptions.

The prospect that textbooks, in order to be useful in the laboratory, would have each to present the *complete* tabulations, each one very likely in its own form and with its own faults and misprints, weighed so heavily on Ewald that after a meeting of the Faraday Society in London (March 1929) which had brought many crystallographers together, he laid the matter before a representative group of crystallographers whom Sir William Bragg had invited to the Royal Institution. It was decided to form three committees: one for the nomenclature of space groups, one for the abstracting of crystallographic papers, and a third one for the simplified and standardized publication of structural work. These commissions were to report to Sir William Bragg. In the course of their work the first and last of these commissions merged into a *Tables* Committee with Astbury, Bernal, Hermann, Mauguin, Niggli and Wyckoff as members. The commission on abstracting, consisting of Bernal and Ewald, worked out rules for abstracting crystallographic papers on one-page set forms, and delivered a report, but it turned out that the tabular form of abstracting was not adaptable enough for general use, and nothing came of this work.

Ewald and Bernal, however, prepared a detailed agenda for a conference on standard crystallographic tables, and when Wyckoff announced that he would be coming to Europe in the summer, Bernal, as chairman of the *Tables* Committee, called a working conference to Zürich in July 1930 on the invitation of Niggli to hold the conference in his institute.

In three or four days of intensive work the members of the Committee, reinforced by Kolkmeijer, James and Mrs. Lonsdale, Mark, Schneider, Schiebold and Brandenberger, discussed the material prepared by Bernal and Ewald under the latter's chairmanship and agreed on point after point. A tight-lipped Niggli watched the proceedings without taking much part, until on the third day he exclaimed: 'Gentlemen, you are stealing my book. I will not agree to any such publication unless it is given to my publisher, Gebr. Bornträger in Berlin, in lieu of a new edition of my book'. Neither this surprising accusation, nor the curious condition was allowed to stand in the way of materializing the *Tables*. Ewald, feeling that his further participation in the work on the *Tables* was likely to jeopardize the good understanding between him and the chief editor of the *Zeitschrift*, withdrew officially from the *Tables* and took part in their growth only indirectly, through their editor, C. Hermann. Sir William Bragg and M. von Laue accepted the Honorary Editorship and through their efforts, as well as those of Mauguin, Kolkmeijer and Wyckoff, substantial subsidies for the preparation and publication of the *Tables* were obtained from academies and other learned societies. The Rockefeller Foundation enabled Hermann to go to USA for a month in connection with the sections prepared by Wyckoff and by Pauling. The two volumes appeared in 1935, and their consolidating influence on the field of crystal analysis is undeniable. It includes the general acceptance of the more meaningful designation of the space groups which was worked out between Hermann and Mauguin.

A third step towards the international consolidation of the new crystallography occurred at the end of the war. By 1943 the British crystallographers had formed an 'X-ray Analysis Group' (XRAG) within the Institute of Physics, and two similar groups existed in USA, the 'American Society for X-ray and Electron Diffraction' (ASXRED), founded in 1941, which was biased towards physics, and a mineralogically inclined 'Crystallographic Society of America' (CSA), founded in 1945. In 1944 the yearly meeting of XRAG was held in Oxford, and Ewald, who then taught in Belfast, was invited to give the evening lecture. In it he gave a historical survey of some stages of X-ray crystallography and ended with a strong plea for the formation of an international society or union which would represent the new crystallography. This idea was followed up by the British crystallographers, and in particular by Sir Lawrence Bragg, the chairman of XRAG. In June 1946, within a year of the termination of the fighting, he arranged for an international meeting of crystallographers in London which was attended by some 120 crystallographers from most of the allied countries. In spite of the general travel restrictions still imposed on Germans at the time, it was possible to have Laue there. The Russian delegation arrived at the end of the meetings and could take part in various committee meetings following the plenary sessions. It was a wonderful re-union of old friends who had been separated during the years of the war, and a first meeting of many colleagues known to one another only as authors of important papers. The result was the decision to form an *International Union*, if possible an independent Union of Crystallography, and until this be consolidated, to form an *interim* representative committee; – to prepare national committees for crystallography in the participating countries; – to charge R. C. Evans and Ewald to explore and prepare the necessary steps with the International Council of Scientific Unions (ICSU) for the recognition of this Union by drafting statutes and by preparing the constitutive First General Assembly as soon as possible which, on the invitation of the USA delegation, was to be held at Harvard University. Furthermore it was decided to prepare the publication, by the Union, of

(i) a special journal for the new crystallography, belonging to the Union and under the editorship of Ewald and R. C. Evans, I. Fankuchen, J. Wyart and A. V. Shubnikov as national co-editors;

(ii) *Structure Reports*, under the Editorship of A. J. C. Wilson, as a continuation of *Strukturbericht*; and

(iii) *International Tables for X-ray Crystallography* under the editorship of Mrs. Lonsdale. (The German *Internationale Tabellen* had been reprinted during the war in USA but were no longer to be had; besides, a revision of the first *Tables* seemed desirable.)

Of these projects the quickest to materialize was the journal. It received its name, *Acta Crystallographica*, on the proposal of Shubnikov, at a meeting of the journal commission in Cambridge a few days after the London meeting where all attempts at finding a suitable name had led to nothing. Shubnikov withdrew as co-editor after the first few issues had appeared as there was no probability of his transmitting papers in Russian in the near future. The publication of a new journal of unknown appeal was a somewhat risky affair in the early post-war period, with shortages of paper and labour and instability of prices. Sir Lawrence Bragg appealed to British industrialists for a subsidy over the first five-year period, and obtained it, and USA sources matched this, while smaller guarantees came from other countries; in addition substantial subventions to the journal and the other publications launched by the Union were received from UNESCO. Specifications for the production of *Acta Crystallographica* were prepared and sent to firms in seven different countries, and tenders received. Finally, the production was entrusted to the Cambridge University Press, with whom Evans, who undertook to act also as the technical editor, had the easiest contact. When in 1951 the Cambridge Press found it impossible to accomodate the growing volume of *Acta* with their other obligations, the publication was transferred to the Danish firm of Ejnar Munksgaard, in whose hands it still is. The first number of the new journal appeared in March 1948, and at the time of the Harvard meeting (28 July to 3 August 1948) the third issue had appeared.

Since then *Acta* has been growing steadily, the first twelve volumes under Ewald's editorship, with Evans (later Asmussen), Fankuchen, Wyart and later also Hughes, Lipson and Nitta as co-editors; and from 1960 onwards under A. J. C. Wilson's editorship with the same co-editors and Jagodzinski. Vol. 1 of 1948 had 348 pages, the latest complete volume, 14 (1961) has 1318 pages. The number of subscribers has been growing steadily year by year reflecting the still increasing interest in crystal structure, and so has the demand for back volumes which contain many essential papers.

The formal establishment of the *International Union of Crystallography* (IUCr) was greatly facilitated by the very helpful interest the then Secretary General of ICSU took in it, the Cambridge astronomer F. J. M. Stratton, an old personal friend of both Ewald and Evans. During the war, most of the international unions lay dormant, and they slowly emerged again to activity in the period 1945–50. Stratton, who had devoted many years to these Unions, was much in favour of small unions which he considered closer-knit and quicker of action than those which covered enormous and in some respects heterogeneous parts of science. The application of the provisional representation of crystallography for the recognition of an International Union of Crystallography was granted in 1947, pending acceptance of the provisional statutes by the constitutive assembly. This First General Assembly was called for Cambridge (Mass.) where the University of Harvard had offered to be host to the meeting (28 July to 3 August 1948).

Here again, the joy of finding old friends and new, in even greater numbers than two years previously in London, the pleasure of discussing with expert colleagues after a long pause, and the progress reported on crystal structures and methods – all this made the meeting to those participating a truly memorable one. Added to this was the satisfaction that the business meetings decided on statutes and thereby changed the *interim* representation of crystallographers into the first formal Executive Committee of the International Union of Crystallography, with R. C. Evans as General Secretary, elected Sir Lawrence Bragg as its first President, and M. von Laue as its first and only Honorary President. The establishment of the Union meant the scientific, and to a certain extent also professional organization of the large number of workers in the ever expanding field that had been opened up 36 years earlier by the genius of M. von Laue, W. H. and W. L. Bragg.

From that day on the Union has steadily developed. The following Table giving details of the five General Assemblies and International Congresses shows this:

IUCr General Assemblies and International Congresses

	Year	Place	Country	President	Papers	Participants
1	1948	Harvard	U.S.A.		86	310
2	1951	Stockholm	Sweden	Sir L. Bragg	225	340
3	1954	Paris	France	M. J. Bijvoet	420	630
4	1957	Montreal	Canada	R. W. G. Wyckoff	325	600
5	1960	Cambridge	U.K.	J. Wyart	600	1250
6	1963	Roma	Italy	P. P. Ewald	?	?

The number of countries which adhere to the Union through their national academies or other learned societies has grown to 26. R. C. Evans, to whom the Union owes so much for its establishment, served as secretary to the Union from 1948 to 1954; he was succeeded by D. W. Smits, the present secretary.

In the years between the triennial congresses a number of other meetings were arranged. Thus there was a symposium on 'Structures on a scale between the atomic and microscopic dimensions' in Madrid (2–6 April 1956) and a conference on 'Precision determination of lattice parameters' in Stockholm in 1959. IUCr further participated in the Fedorov Commemoration meetings in Leningrad in 1959, and also co-sponsored an international symposium on 'Electron diffraction' in Kyoto in September 1961. Generous financial contributions from UNESCO facilitated the attendance of many participants in the congresses and other meetings.

These meetings mean a great deal to all crystallographers, by acquainting them with their colleagues abroad and their work. This is particularly true for those who still are in a pioneer position in their countries and ordinarily have little chance of regional contact with other crystallographers.

The most important work of the Union lies, however, with its various Commissions. The editing of the journal by the *Acta Crystallographica Commission* (chairman A. J. C. Wilson) has been mentioned above. Before taking over *Acta Cryst.*, Wilson was chairman of the *Structure Reports Commission*, and under his editorship and with the splendid cooperation of a large number of section editors and abstractors, ten volumes of *Structure Reports* were prepared, covering the years 1940–1954. Now Wilson has been succeeded by W. B. Pearson in Ottawa who, with the help of an increased board of section editors and abstractors, is attempting to halve the present six-year gap between the publication of a paper and of its review in *Structure Reports*. With the ever increasing rate at which crystal structures are being determined, their uniform and complete registration in *Structure Reports* becomes the longer the more indispensable.

The *International Tables Commission*, still under the chairmanship of Dame Kathleen Lonsdale, has revised and enlarged the former *Internationale Tabellen* to a three-volume, not yet completed work which assembles all the geometrical, analytical and physical data required for crystal structure analysis by X-ray diffraction. Again, the existence of this standard work is a real service to the community of crystallographers in all countries.

Besides these publishing commissions of IUCr there are others for which publication is only incidental. Only a few examples of their work will be mentioned:

In the *Commission on Crystallographic Apparatus* (present chairman W. Parrish) questions of standardization have been discussed – matters that were investigated also by special national commissions, for instance of XRAG, but should be agreed upon internationally. The commission also conducted an international testing of the accuracy of high-precision lattice parameter determination and held a conference in Stockholm (1959) on this topic, and on counter methods for crystal structure analysis. Its latest activity is a second experimental comparison of the properties of commercially available X-ray films; a first comparison had been carried out under the auspices of the commission some five years earlier. Besides an *Index of Crystallographic Supplies* has been compiled by the commission, of which the third edition is being prepared for the Rome Congress.

One of the most effective contributions to the international consolidation of crystallographers comes from the initiative of the chairman of this commission: the compilation of a *World Directory of Crystallographers*, the first edition of which appeared in 1957. In the second edition, which was edited by the General Secretary of IUCr, D. W. Smits, more than 3500 scientists from 54 countries are listed who employ crystallographic methods. A new edition will be prepared soon after the Rome Congress in 1963.

The increasing use of electronic computers led in 1960 to the formation of a *Commission on Crystallographic Computing*, with G. A. Jeffrey as chairman. The best utilization of computers, the standardization and exchange of programs, and problems of publication were among the topics discussed at a meeting of this commission near Frankfurt, in preparation of a full-scale symposium in connection with the Rome Congress in 1963.

The *Commissions on Crystallographic Data*, *on Crystallographic Nomenclature*, *on Electron Diffraction*, and *on Crystallographic Teaching* are further commissions of IUCr which show the type of important international work to be done. IUCr is also represented on some Commissions of the International Unions of Pure and Applied Physics (IUPAP) and of Pure and Applied Chemistry (IUPAC), and on the ICSU Abstracting Board.

* * *

Bearing in mind the picture of world-wide activity and cooperation, of which this Commemoration Volume presents another example, it is in a mood of pride and gratitude for the past and of serene confidence in the future of their science that crystallographers convene in Munich, the birth place of their particular branch of science, for celebrating the fiftieth anniversary of M. von Laue's, W. H. Bragg's and W. L. Bragg's pioneer work which started the development of

THE NEW CRYSTALLOGRAPHY.

APPENDIX

Biographical Notes on Authors

Biographical Notes on Authors

Andrade, Edward Neville daCosta

*27 Dec. 1887

ed.: Univ. Coll. London 1905–10; Heidelberg 1910–11; Cambridge 1911–12; Manchester 1913–14

B.Sc.; Ph.D. Heidelberg 1911: *Über Wesen und Geschwindigkeit metallischer Träger in Flammen* (Lenard); D.Sc.

Fellow U. Coll. London 1915; Prof. of Physics, Artillery Coll. Woolwich 1920–28; Quain Prof. of Physics London U. 1928–50; Dir. R. I. and Davy-Faraday Lab. 1950–52; emer. Prof., London U. 1950– .

Banerjee, Kedareswar

*17 Sept. 1900 in Sthal (Pabna), East Pakistan

ed.: Jubilee School, Dacca; Calcutta U.

B.Sc., M.Sc., D.Sc. Calcutta. Thesis: *Some Problems in Structures of Solids and Liquids* (C. V. Raman)

Reader in Physics, Dacca U. 1934–43; Prof. of Physics, Ind. Ass. Cult. Sci., Calcutta 1943–52; Prof. of Physics, Allahabad U. 1952–59; Dir. Ind. Ass. Cult. Sci., Calcutta 1959– .

Bastiansen, Otto C. A.

*5 Sept. 1918 in Balsfjord, Troms, Norway

ed.: U. of Oslo 1937–42

Cand. mag. 1940, cand. real 1942, dr. philos. 1949

Prof. of Theoretical Chemistry, U. of Oslo.

Bernal, John Desmond

*10 May 1901 in Nenagh (Eire)

ed.: Cambridge 1919–23; Nat. Sci. Tripos, M. A.

Research Asst., R. I. London 1923–27; Lecturer in struct. cryst., Cambridge; Asst. Dir. of Res. in struct. cryst., Cambridge 1936–38; Prof. of Physics, Birkbeck Coll. London 1938– .

Bijvoet, Johannes Martin

*23 Jan. 1892 in Amsterdam

ed.: Municipal U. of Amsterdam 1910–19 (1914–18 military service)

Ph.D. Amsterdam. Thesis: *X-ray investigation of the crystal structure of lithium and lithium hydrides* (A. Smits)

Asst. Gen. Chem., U. of Amsterdam 1922–28; also teacher of chem., Grammar School of Hilversum, 1923–28; Lecturer in cryst. and thermodyn., Municipal U. of Amsterdam; Prof. of Gen. and Inorg. Chem., State U. of Utrecht 1939–62; 1962 retired.

Bragg, Sir (William) Lawrence

*31 March 1890 in Adelaide, South Australia.

ed.: St. Peter's College, Adelaide 1900–05; Adelaide U. 1905–08; Cambridge U. 1908–11.

Lecturer in Physics, Trinity Coll. Cambridge, 1912; 1915 appointed Prof. of Physics, Manchester U.; 1915 Nobel Prize for Physics

Prof. of Physics, Manchester U. 1919–37; Dir. National Physical Lab. 1937–38; Cavendish Prof. of Exp. Physics, Cambridge 1938–53; Director, The Royal Institution and Davy-Faraday Lab. 1954– .

Brentano, John Christian Michael

*27 June 1888 in Vienna

ed.: Liceo Dante, Florence, and Engadina, Zuoz, 1904–07; Istituto di studi super., Florence 1908; U. of Munich 1909–14

Ph.D. Munich 1914. Thesis: *Über d. Einfluss allseitigen hydrost. Drucks auf d. elektr. Leitfähigkeit von Wismuthdrähten ausserh. u. innerh. des transvers. Magnetfeldes für Gleichstrom u. Wechselstrom* (W. C. Röntgen). DSc. U. of Manchester 1935

Asst. of M. v. Laue, Frankfurt/M 1915; Asst., Privatdoz. Eidgen. T. H. Zürich; Asst. and Senior Lecturer in Physics, U. of Manchester; Asst., Assoc., and full Prof. Northwestern U., Evanston, Ill.

Retired 1961; living in Blonay, Ct. Vaud, Switzerland.

Buerger, Martin J.

*3 April 1903 in Detroit, Mich.

ed.: High School Detroit 1916–20; Mass. Inst. of Technology 1920–22 and 1923–25; Graduate study at MIT (geolog.-mineral.) 1925–29

Ph.D. MIT 1929. Thesis: *Translation gliding in crystals* (W. H. Newhouse)

Asst. Geologist, US Geolog. Survey 1925; Assoc. Prof. 1929–37, Full Prof. of Mineral. and Crystallogr. at MIT 1937–44; Chairman of the Faculty 1944–56; Director, School of Advanced Studies MIT 1956– .

Bunn, Charles W.

*15 Jan. 1905 in London

ed.: Wilson's Grammar School, Camberwell, London 1916–23; Oxford U. 1923–27

B.A. 1926, B.Sc. 1927 in Oxford; D.Sc. 1955. Thesis: *The photochemical oxydation of alcohols by the dichromate ion* (E. J. Bowen)

Took up crystallography in 1927. Since 1928 with ICI (Imper. Chem. Industries, Ltd), first at Alkali Division in Northwich, Cheshire, now at Plastics Div., Welwyn Garden City, Herts.

Darwin, Sir Charles (Galton)

*19 Dec. 1887 in Cambridge

ed.: Marlboro' College 1901–06; Cambridge U. 1906–10

Jun. Lecturer Math. Phys., U. of Manchester (Rutherford) 1910–14; Army service in France 1914–18; Lect. in Math., Christ's Coll., Cambridge 1919–22; Vis. Prof. at Cal. Tech. 1922–23; Tait Prof. of Natural Philosophy, Edinburgh U. 1923–37; Master of Christ's Coll., Cambridge 1937–39; Dir., National Physical Lab. 1939–49.

Donnay, José D. H.

*6 June 1902 in Grandville, Belgium

ed.: Athénée Royal de Tongres 1913–14; Athénée Royal de Liège (Latin-Science section) 1914–20; U. de Liège 1920–25 (1922 candidat ingénieur, 1925 ingénieur civil des mines); Stanford U. 1925–26 and 1928–29

Ph.D. (Geology). Thesis: *The genesis of the Engels copper deposit* (A. F. Rogers and C. F. Tolman Jr.)

Geologist in Rabat (Morocco) 1929–30; Research Asst. Stanford U. 1930–31; Johns Hopkins U. Baltimore 1931– ; since 1946 Prof. of Cryst. and Miner.; three interruptions: Laval U., Quebec 1939–44, U. de Liège 1946–47, U. de Paris, Sorbonne, as Fulbright Lecturer 1958–59.

Ewald, Paul P.

*23 Jan. 1888 in Berlin

ed.: Viktoria Gymnasium, Potsdam 1900–05; Cambridge U. 1905–06; Göttingen U. 1906–07; Munich U. 1907–12

Ph.D. Munich 1912. Thesis: *Dispersion und Doppelbrechung von Elektronengittern* (A. Sommerfeld)

Assistant to Hilbert in Göttingen 1912–13; Asst. to Sommerfeld in Munich 1914–21; Privatdozent Munich 1918–21; Assoc. (1921) and full (1922) Prof. of Theor. Physics, T. H. Stuttgart 1921–37; Research Fellow, Cambridge U. 1937–39; Lecturer, later Prof. of Mathem. Physics, The Queen's U. Belfast, N. I. 1939–49; Prof. and Head of Physics Dept., Polytechnic Institute of Brooklyn (N.Y.) 1949–57; Part-time Prof. at Polyt. Inst. 1957–59; 1959 retired.

Glocker, Richard

*21 Sept. 1890 in Calw (Württemberg)

ed.: Humanist. Gymnasium 1899–08; U. Berlin 1909–10; U. Munich 1910–13; T. H. Stuttgart 1914

Ph.D. Munich 1914. Thesis: *Interferenz der Röntgenstrahlen und Kristallstruktur* (W. C. Röntgen)

Dir. of newly established X-ray lab. at T. H. Stuttgart 1920; Assoc. Prof. 1923, full Prof. of X-ray Technique 1925; retired 1960.

Guinier, André Jean

*8 Jan. 1911 in Nancy

ed.: École Normale Supérieure, Paris 1930–34

Dr-ès-sciences physiques 1934. Thesis: *Diffusion des rayons X aux très petits angles par les particules submicroscopiques* (Ch. Mauguin)

Chef de Service au Lab. d'Essais du Conservatoire des Arts et Métiers (Paris); Prof. au Conserv. d. A. & M.; Prof. à la Faculté des Sciences, U. of Paris.

Hägg, Gunnar

*14 Dec. 1903 in Stockholm

ed.: Nya Elementarskolan, Stockholm 1919–22; U. of Stockholm 1922–29; University Coll., London 1926 (work on surface chemistry with F. G. Donnan)

Ph.D. Stockholm 1929. Thesis: *X-ray studies on the binary systems of iron with nitrogen, phosphorus, arsenic, antimony and bismuth* (Arne Westgren)

Lecturer of Gen. and Inorg. Chem., U of Stockholm 1929–36; Prof. of Gen. and Inorg. Chem., U. of Uppsala 1936– .

Hull, Albert W.

*19 April 1880 in Southington, Conn.

ed.: High School Torrington, Conn.; Yale U. (A.B. 1905)

Ph.D. 1909. Thesis: *Initial velocities of electrons produced by ultraviolet light* (H. A. Bumstead)

Instructor 1909–11, Asst. Prof. in Physics, Worcester Polyt. Inst. 1911–13; Research Physicist, Gen. Electric Co., Schenectady, N.Y. 1914– ; Asst. Dir. Res. Lab. 1928–50.

James, Reginald William

*9 Jan. 1891 in London

ed.: Polytechnic, Regent Street, London 1903–07; City of London School 1907–09; Cambridge U. 1909–14

Physicist to Sir Ernest Shackleton's expedition to the Weddel Sea in the Antarctic 1914–16; with Royal Engineers in France 1916–19; Lecturer (1919), Senior Lect. (1921), Reader (1934) in Physics Dept., U. Manchester; Prof. of Physics, U. of Cape Town 1937–56; 1953 and 1956 Acting Principal, U. of Cape Town; 1957 retired.

Lipson, Henry

*11 March 1910 in Liverpool

ed.: Hawarden County School 1921–27; U. of Liverpool 1927–36

B.Sc. 1930, M.Sc. 1931 Liverpool

Jun. Sci. Off., National Phys. Lab. 1937; Asst. in Cryst., Cambridge 1938; Head of Dept., later Prof. of Physics, Manchester Coll. Sci. and Tech.

Lonsdale, Dame Kathleen (*née* **Yardley**)

*28 Jan. 1903 in Newbridge, Eire

ed.: Ilford County High School for Girls 1914–19; Bedford Coll. for Women (London U.) 1919–22; University Coll., London 1922–23

B.Sc. 1922 (Physics, Bedford Coll); M. Sc. Physics (Univ. Coll.) 1924. Thesis: *Structure of succinic acid* (Sir William Bragg); D.Sc. (Physics) 1929. Thesis: *Structure of Ethane derivatives* (Sir William Bragg)

D.S.I.R. Res. Asst. to Sir W. Bragg 1923–27; Amy Lady Tate Scholar, Leeds U. 1927–29; Res. Asst. to Sir W. Bragg at R.I. 1932–34; Leverholme Fellow 1935; Dewar Fellow 1945; Reader in Cryst., Univ. Coll. 1946–49; Prof. of Chem., U. Coll. London 1949– .

1945 elected F.R.S.; 1947 Special res. fellow, USA Federal Health Service.

Mark, Herman Francis

*3 May 1895 in Vienna

ed.: Vienna Volksschule and Gymnasium 1901–13; Service in the army 1914–17; U. of Vienna (Chemistry) 1917–21

Ph.D. Vienna 1921. Thesis: *Free Radicals in Organic Chemistry* (W. Schlenk)

Res. Assoc., K.W.I. f. Faserstoffchemie (R. O. Herzog) in Dahlem 1922–26; Res. Physicist, later Dir. of Res., I. G. Farben Labs., Ludwigshafen 1927–32; Prof. of Chem., U. Vienna 1932–38; Res. Mgr., Canadian Int'nl Paper Co, Hawkesberyy, Canada 1938–40; Prof. Org. Chem., Polyt. Inst. of Brooklyn 1940– ; Dir. Polymer Res. 1944–; Dean of the Faculty 1962– .

Nitta, Isamu

*19 Oct. 1899 in Tokyo

ed.: Daiichi Koto Gakko (First Prelim. School to U.) 1917–20; U. of Tokyo 1920–23

B. or M. Sci (Rigakushi) 1923; Dr. Sci. (Rigakuhakushi) 1930. Thesis-advisers Shoji Nishikawa and Masao Katayama

Jun. Res. Fellow at Inst. of Phys. and Chem. Research, Tokyo 1923

Prof. of Chem., Osaka U. 1933–60; Prof. of Chem. and Dean of Fac. of Sci., Kwansei Gakuin U., Nishinomiya 1960– .

Patterson, A. Lindo

*23 July 1902 in Nelson, New Zealand

ed.: High School, Montreal, Canada 1908–16; Tonbridge School, Tonbridge, Kent, Engl. 1916–20; McGill U. 1920–24; Postgraduate study at R.I. London 1924–26; at K.W.I. f. Faserstoffchemie in Dahlem 1926–27; McGill U. 1927–28

B.Sc. 1923, M.Sc. Phys. & Math. 1924; Ph.D. McGill U. 1928. Thesis: *The Application of X-rays to the Study of Organic Substances* (A. S. Eve)

Demonst. in Phys., McGill U. 1928–29; Res. Fellow, Rockefeller Inst. f. Med. Res., N.Y. 1929–31; Asst. Prof., U. of Penna., Philadelphia 1931–33; Assoc. Prof., MIT 1933–36; Prof. of Phys., Bryn Mawr Coll. 1936–49; Dir. of Res., Inst. f. Cancer Res. in Philadelphia 1949– .

Pauling, Linus Jr.

*28 Feb. 1901 in Portland, Oregon

ed.: Washington High School, Portland 1914–17; Oregon State Agric. Coll. 1917–19, 1920–22; Teach. Fellow Cal. Inst. Tech. 1922–25

B.Sc. in Chem. Eng.; Ph.D. Cal. Tech. 1925. Thesis: *The Determination with X-rays of the Structures of Crystals* (Roscoe G. Dickinson)

Postgraduate work at U. of Munich (Sommerfeld), U. of Copenhagen (Bohr) and Zürich, 1926–27; Res. Fellow, Cal. Inst. Tech. 1925–27; Asst. Prof. 1927–29, Assoc. Prof. 1929–31, full Prof. of Chemistry 1931– at Cal. Inst. Tech.

George Eastman Prof. in Oxford 1948; Nobel Prize in Chem. 1954.

Polanyi, Michael

*1891 in Hungary

Member K.W.I. f. Physikal. und Elektrochem., Dahlem 1922–33; Prof. of Phys. Chem., U. of Manchester 1933–48; Prof. of Social Studies, U. of Manchester 1938–58; 1958 retired, living in Oxford.

Robertson, John Monteath

*24 July 1900 in Perthshire, Scotland

ed.: Perth Academy 1914–17; U. of Glasgow (Chem., Geol., Math.) 1920–26

Ph.D. Glasgow 1926. Thesis: *Structural Relationships in the Sesquiterpene Series* (G. G. Henderson)

Commonwealth Fellow in Ann Arbor, Mich. 1928–30; Member of Staff of R.I. 1930–39; Senior Lect. Phys. Chem., Sheffield U. 1939–42; Gardiner Prof. of Chem., Glasgow U. 1942– .

George Fisher Baker Lect., Cornell U. 1951; Vis. Prof., Calif. U. 1958.

Shubnikov, Alexei Vasilevich

*29 March 1887 in Moscow

ed.: Moscow Commercial School 1898–1906; Moscow U. 1907–12; 1st. Grade Diploma 1913. Thesis: *Über die Symmetrie der Kristalle von Kaliumdichromat* (G. V. Wulf)

Asst. for Cryst. at the People's U. Moscow 1913; Prof. of Cryst. at Gornov Inst. (Ural); Sen. Res. worker, Miner. Museum of the Acad. of Sci. USSR, Leningrad; Head of the Inst. of Cryst. of the Acad. of Sci. USSR in Moscow.

Siegbahn, Karl Manne Georg

*1886

ed.: Lund U.

Prof. of Physics, Lund U. 1914–23; same Uppsala U. 1924–37; Dir. of Nobel Inst. f. Physics of the Royal Academy of Sci., Stockholm 1937–

Nobel Prize for Physics 1925.

Straumanis, Martin Edward

*23 Nov. 1898 in Krettingen, Lithuania

ed.: Gymnasium in Jelgava, Latvia; U. of Latvia in Riga 1920–25 (Chem. Eng.)

Ph.D. (Chemistry) Riga 1927. Thesis: *The dissolution of Zn, Cd, and Fe in acids, discussed from the point of view of local currents* (M. Centnerszwer)

Rockefeller Foundation Fellow in Göttingen 1927–28; Asst., Asst. Prof., Assoc. Prof. at U. of Riga; Full Prof. Chem., U. Riga 1942; Dir. Analyt. and X-ray lab. 1934–44; Res. Prof. of Metallurgy, School of Mines and Metallurgy, U. of Missouri, Rolla, Missouri 1947– .

Trillat, Jean Jacques

*8 July 1899 in Paris

ed.: Lycée Janson de Sailly in Paris, 1908–17; Sorbonne 1921–25; École supérieure de physique et chimie, Paris, 1921–23

Ph.D. Paris 1926. Thesis: *Rayons X et composés organiques à longue chaîne; recherches sur leurs structures et leurs orientations* (Jean Perrin)

Dir. Lab. de Rayons X du Centre National de la Recherche Scientifique, Bellevue, S. &O.

Warren, Bertram E.

*28 June 1902 in Waltham, Mass.

ed.: Waltham High School 1915–19; Mass. Inst. Techn. (Physics & Math.) 1919–25 and 1927–29; T. H. Stuttgart 1926; Manchester U. 1929

Sc.D. MIT 1928. Thesis: *X-ray Determination of the Structure of Metasilicates* (Sir Lawrence Bragg)

Asst., Assoc., full Prof. of Physics, MIT 1930– .

Westgren, Arne Fredrik

*11 July 1889 in Årjäng (Sweden)

ed.: High Schools in Stockholm and Uppsala 1900–07; U. of Uppsala 1907–15; U. of Göttingen 1915–16

Fil. Dr. Uppsala 1915. Thesis: *Untersuchungen über die Brownsche Bewegung, besonders als Mittel zur Bestimmung der Avogadroschen Konstanten* (The Svedberg)

Metallographer, SKF Ball Bearing Co, Gothenburg 1918–19; Metallographer, Metallographic Inst. Stockholm U. 1920–27; Prof. Gen. and Inorg. Chem., Stockholm U. 1927–43; Permanent Secretary, Royal Swed. Acad. Sci., Stockholm 1943–59; 1959 retired, doing X-ray cryst. res.

Wilson, Arthur James Cochran

*28 Nov. 1914 in Springhill, Nova Scotia, Canada

ed.: King's Collegiate School, Windsor, N.S. 1922–30; Dalhousie U. Halifax, Canada 1930–36, M.Sc. 1936; MIT 1936–38; Cambridge U. 1938–42

Ph.D. MIT 1938, Cambridge 1942. Thesis: *Heat Capacity of Ag, Ni, Zn, Cd, Pb*; *Heat Capacity of Rochelle salt*; *Thermal Expansion of Al and Pb* (Advisers H. L. Bronson, H. Mueller, B. E. Warren, W. L. Bragg)

Lecturer 1945, Sen. Lect. 1946, Prof. of Phys. and Dir. of the Viriamu Jones Lab. 1954, Univ. Coll, Cardiff, Wales.

Wood, Elizabeth Armstrong

*19 Oct. 1912 in New York, N.Y.

ed.: Horace Mann School, N.Y. 1918–30; Barnard Coll. 1930–33, B. A. 1933; Bryn Mawr 1933–39, M.A. 1934

Ph.D. (Geology, Chemistry) Bryn Mawr 1939. Thesis: *Mylonization of Hybrid Rocks near Philadelphia, Pa.* (E. H. Watson)

Teaching Geol. and Miner. at Bryn Mawr and Barnard 1934–43; Res. Member, Bell Telephone Labs, Murray Hill, N. J. 1944– .

Wooster, William Alfred (Peter)

*18 Aug. 1903 in London

ed.: Deacon's School, Peterborough 1912–21; Cambridge U. 1921–27

Ph.D. Cambridge 1928. Thesis: *Beta and gamma rays of Radium B, C and E* (C. D. Ellis and Sir E. Rutherford); Sc.D. Cambridge 1950

Demonstrator 1928–35; Lecturer in Dept. of Miner., U. of Cambridge 1935–60

Retired to own Lab. and firm, Crystal Structures Ltd., Bottisham, Cambridge, Engeland, in 1960

Wyart, Jean

*16 Oct. 1902 in Avion, France

ed.: Collège d'Abbeville 1916–23; and Lycée Saint Louis; École Normale Sup. Paris 1923–27

Agrégation des sciences physiques; Maître de conférences (Sorbonne) 1933. Thesis: *Recherches sur les Zéolithes* (Fr. Wallerant, Ch. Mauguin)

Asst. at the Sorbonne 1928; Prof. of Miner. and Cryst. at the Sorbonne 1948– .

Wyckoff, Ralph Walter Graystone

*9 Aug. 1897 in Geneva, N.Y.

B.S. Hobart Coll., Geneva N.Y. 1916; Ph.D. Cornell U. 1919

Instr. analyt. chem., Cornell U. 1917–19; Physical chemist, Geophys. Lab., Carnegie Inst. Washington 1919–27; Res. Assoc., Cal. Inst. Tech. 1921–22; Assoc. member, Rockefeller Inst. Med. Res. N.Y. 1927–37; Scientist at Lederle Lab. Inc. 1937–42, Assoc. Dir. of Virus Res. 1940–42; Techn. Dir., Reichel Labs. Inc. 1942–43; Lect. in Epidemiology, U. of Michigan 1943–45; Scientist Dir., US Publ. Health Serv., Bethesda, Md. 1946–52; Science Attaché, USA Embassy London 1952–54; Biophysicist, Publ. Health Serv. 1954–60; Prof. of Physics, U. of Arizona, Tucson, Ariz. 1960– .

Subject Index

List of Repeatedly Occurring Substances

Index of Localities and Schools

Name Index